# PLANTWIDE PROCESS CONTROL

**Wiley Series in Chemical Engineering**

---



**Bird, Stewart, and Lightfoot:** TRANSPORT PHENOMENA

**Brownell and Young:** PROCESS EQUIPMENT DESIGN: VESSEL DESIGN

**Felder and Rousseau:** ELEMENTARY PRINCIPLES OF CHEMICAL PROCESSES, 2nd Edition

**Franks:** MODELING AND SIMULATION IN CHEMICAL ENGINEERING

**Froment and Bischoff:** CHEMICAL REACTOR ANALYSIS AND DESIGN, 2nd Edition

**Gates:** CATALYTIC CHEMISTRY

**Henley and Seader:** EQUILIBRIUM-STAGE SEPARATION OPERATIONS IN CHEMICAL ENGINEERING

**Hill:** AN INTRODUCTION TO CHEMICAL ENGINEERING KINETICS AND REACTOR DESIGN

**Jawad and Farr:** STRUCTURAL ANALYSIS AND DESIGN OF PROCESS EQUIPMENT, 2nd Edition

**Levenspiel:** CHEMICAL REACTION ENGINEERING, 2nd Edition

**Malanowski and Anderko:** MODELLING PHASE EQUILIBRIA: THERMODYNAMIC BACKGROUND AND PRACTICAL TOOLS

**Masel:** PRINCIPLES OF ADSORPTION AND REACTION ON SOLID SURFACES

**Reklaitis:** INTRODUCTION TO MATERIAL AND ENERGY BALANCES

**Sandler:** CHEMICAL AND ENGINEERING THERMODYNAMICS, 2nd Edition

**Seborg, Edgar, and Mellichamp:** PROCESS DYNAMICS AND CONTROL

**Smith and Corripio:** PRINCIPLES AND PRACTICE OF AUTOMATIC PROCESS CONTROL

**Taylor and Krishna:** MULTICOMPONENT MASS TRANSFER

**Ulrich:** A GUIDE TO CHEMICAL ENGINEERING PROCESS DESIGN AND ECONOMICS

**Welty, Wicks and Wilson:** FUNDAMENTALS OF MOMENTUM, HEAT AND MASS TRANSFER, 3rd Edition

**Erickson and Hedrick:** PLANTWIDE PROCESS CONTROL

# PLANTWIDE PROCESS CONTROL

**Kelvin T. Erickson**
*University of Missouri—Rolla*

**John L. Hedrick**
*Automation and Control Technologies, Inc.*

JOHN WILEY & SONS, INC.
New York · Chichester · Weinheim · Brisbane · Singapore · Toronto

This book is printed on acid-free paper. ∞



Published simultaneously in Canada.



For ordering and customer service, call 1-800-CALL-WILEY.

**Library of Congress Cataloging-in-Publication Data**

Erickson, Kelvin T.
Plantwide process control/Kelvin T. Erickson, John L. Hedrick.
p. cm. – (Wiley series in chemical engineering)
Includes bibliographical references and index.
ISBN 0-471-17835-7 (cloth : alk. paper)
1. Chemical process control. I. Hedrick, John L. II. Series.
TP155.75.E75 1999
660′.2815–dc21 98-38903

10 9 8 7 6 5 4 3 2 1

*To Fran, Esther, and David (KTE)*

*To Eileen, Denise, Helene, David, and Nicole (JLH)*

# CONTENTS

**Preface** xi

**Chapter 1** Introduction to Process Plant Systems and Control Technology 1

1.1 Introduction 2
1.2 What Is a Plant? 3
1.3 Modern Process Control 11
1.4 Chapter Summaries 16
References 18

**Chapter 2** Control Engineering on Capital Projects 19

2.1 Introduction 20
2.2 Control Engineering 20
2.3 Plant Capital Project Management 21
2.4 Project Team Considerations 35
References 36

**Chapter 3** A Practitioner's Model for Automation and Control 37

3.1 Introduction 38
3.2 S88.01 Models 43
3.3 Information Management 72
3.4 User Interface 72
3.5 Control System Technology 76
3.6 Example Models 77
References 96

**Chapter 4** Process Modeling 97

4.1 Introduction 98
4.2 Dynamic Models Based on Fundamental Principles 100
4.3 Empirical Models 120
4.4 Discrete-Time Models 125
4.5 Simulation 133
4.6 Empirical Model Identification 137
References 165

**Chapter 5** Single-Loop Regulatory Control 166

5.1 Introduction 166
5.2 General Feedback Control 170
5.3 PID Controller 178
5.4 PID Controller Tuning 187
5.5 User Interface Considerations 211
References 212

**Chapter 6** Enhancements to Single-Loop Regulatory Control 213

6.1 Introduction 213
6.2 Cascade Control 214
6.3 Feedforward Control 227
6.4 Ratio Control 242
6.5 Split-Range Control 244
6.6 Override Control 248
6.7 Single-Loop Model Predictive Control 252
6.8 Batch Control Considerations 264
References 266

**Chapter 7** Multivariable Regulatory Control 268

7.1 Introduction 268
7.2 Multiloop Control 271
7.3 Decoupling Control 292
7.4 Singular-Value Decomposition 299
7.5 Multivariable Model Predictive Control 311
7.6 User Interface 322
References 327

**Chapter 8** Discrete Control 329

8.1 Introduction 330
8.2 Basic Discrete Control 330
8.3 Interlock Control 335
8.4 Discrete-Control Modules 339
8.5 Unit Supervision Discrete Control 350
References 370

**Chapter 9** Batch Control 371

9.1 Introduction 371
9.2 Batch Plant Operation Complexity 372
9.3 Batch Operation Model 373
9.4 Procedural Control in Batch Processing 375
9.5 Coordination Control 380
9.6 Recipe Control 380
9.7 Batch Campaign Management 383
9.8 Batch Data Management 385

9.9 Batch User Interface for Setup and Operation 385
9.10 Batch Control Example 386
References 401

**Chapter 10** Case Study: Pulp and Paper Mill 402

Pulp Mill Area 406
Paper Mill Area 499
References 510

**Appendix A.** Symbols Used in Piping and Instrumentation Diagrams 511

**Appendix B.** Ladder Logic 514

**Appendix C.** Sequential Function Charts 521

**Glossary** 526

**Index** 540

# PREFACE

This book presents the process of control system design for continuous and batch chemical plants starting from the initial concept to design, simulation, test, implementation, and operation. Recent industrial trends in the chemical, petrochemical, pharmaceutical, and food industries have forced design engineers to view control problems from a wider perspective than just a batch control unit or just a collection of single-loop controllers. Increased regulatory pressure on the chemical and petrochemical industry has meant that a larger number of variables must be controlled in order to satisfy certain environmental limits. In all industries, increased emphasis on product quality to maintain profitability has meant that certain product indicators must be regulated within shrinking limits. Both of these trends mean that control engineers are faced with designing more sophisticated control systems that must maintain multiple variables at desired values.

This book blends process control, batch control, and discrete control techniques into the overall control problem solution. The design process is presented for all plant control system levels: from the site level down to the device level. This book takes a practical approach to the design of plantwide control systems and presents many examples. Theory is used to back up the process control design techniques, but the emphasis is on the practical implementation of the design techniques. The book contains example problems throughout and culminates in a case study where the application of the techniques is illustrated.

The book starts by introducing the topic and explaining the overall process of plantwide control design. The first three chapters define the overall view of control projects and discuss the various issues in process control, including the role of standards. This section motivates the techniques to be presented in the remainder of the book. Chapters 4–7 present process control techniques, starting from the single-loop perspective and working up to unit wide and supervisory control techniques, including model predictive control. Design techniques for discrete and batch processes are presented in Chapters 8 and 9. This material is normally found in a manufacturing automation context and presents a structured, standard approach to discrete control problems and when processing material in batches. In Chapter 10, the entire process of plantwide process control design is brought together by the case study of a pulp and paper mill.

Every chapter begins with a story of a young control engineer as he learns about the world of plantwide process control by experience. These stories reflect the experience of the authors and their colleagues in the wonderful world of process control and we hope that you enjoy them.

### THE AUDIENCE

This book serves both professional and academic markets. For the new engineering college graduate, this book presents the aspects of control engineering design not normally taught in a university or college control course. However, experienced control engineers (primarily chemical and electrical) and managers also need to rapidly educate themselves in this area of control technology of which they are probably only somewhat familiar. This book is also directed to the senior undergraduate or first-year graduate student. The techniques taught at the introductory control course are put in their proper context of a typical control project that will be encountered in industry.

# ACKNOWLEDGMENTS

First, the SP88 committee members, especially Tom Fisher and Richard Mergen, are acknowledged for their dedication to the task of formulating a batch control model. The SP88 committee effort is the main inspiration behind this book.

The authors wish to acknowledge the beneficial suggestions and comments of many colleagues. Blu Englehorn, Esther Erickson, Dean Ford, Melissa Layton, Thomas Marlin, Khanh Ngo, and Daniel Rangel reviewed drafts of this book and provided many suggestions and corrections to improve the final product. Special thanks to Mike Cottrell and Daniel Rangel of International Paper, who allowed one of the authors to tour a pulp and paper mill, providing material for the case study of Chapter 10. We especially thank Fran Erickson for checking the entire manuscript for errors and for converting the figures to Adobe Illustrator™.

Portions of this material were taught in industrial short courses and university courses, and the students are acknowledged for their help in pointing out where the presentation was unclear.

The University of Missouri—Rolla and Magnum Technologies are acknowledged for the sabbatical leave for one of the authors while this book was being written.

Bob Otto and Ted Williams are acknowledged for their inspiration as pioneers in the field.

Finally, to our families goes our gratitude for their love and patience as we labored on this book. Above all we thank God for the talent and grace that enabled us to finish the project.

# 1 Introduction to Process Plant Systems and Control Technology

**Chapter Topics**

- What is a plant?
- What is process control?
- What is control technology?

*Scenario:* The key to a successful control system is to first understand the nature of the process to be controlled and then match the automation and control requirements with the right technology.

A number of years ago, a young and enthusiastic graduate engineer began his career in industry with great expectations. He was a determined individual and wanted to be involved in innovative and exciting technology. He chose process control as his career path—his family and friends are still trying to understand what he does.

As part of a new employee training program, he visited the company's plant sites to learn about automation and control technology. He was surprised to find that little was being used. With some apprehension, he reported his findings. A senior engineer asked him if he had uncovered why it was not being used. The novice engineer had not thought to investigate, so back out he went. The answer turned out to be simple—the installed control technology was not matched to the process it was trying to control.

The installed technology was designed for continuous processing but the process was not continuous. Overall, the process unit—a glass-melting furnace—would be classified as continuous, but periodically the energy supply was literally turned off, creating a major discontinuity in the energy supply. This occurred every 30 minutes when the energy flow—natural gas and air—was reversed to recover waste heat of combustion. The entire fuel–air path was changed and several minutes were required to complete the switchover. Meanwhile, the temperature and fuel–air flow automatic controllers became saturated due to reset windup because there was no flow to control during the switchover. The panel-mounted controllers spent most of the firing cycle recovering from reset windup and had little time to control the process before the next cycle began. The installed control technology was useless to operating personnel.

The young engineer learned early in his career that the fundamental nature of the process is a very important consideration in the design and implementation of successful process control systems. Furthermore, he learned early in his career that good control engineering starts by matching the control technology to the process requirements.

## 1.1 INTRODUCTION

The purpose of this book is to describe how to identify and specify successful control systems. The intention is to provide information and practical knowledge to meet the challenge of plantwide control in modern process manufacturing facilities.

**The Challenge.** *World-class process manufacturing plants have several expectations of process control systems. Besides enforcing control strategies of ever-increasing complexity, the systems are expected to provide real-time performance and status information and to warn of poor product quality or process upsets. Current systems are expected to provide secure operator workstations that are accurate windows to the process. Overall, they are expected to participate in plantwide automation that provides accurate and timely information to other enterprise systems. All these expectations must be very flexible and responsive to business and market changes. This is the challenge for today's control engineer.*

The purpose of this chapter is to define plantwide control. In practice, the term plantwide implies everything from a "lights-out" plant to a single production unit. The notion of a lights-out plant with no human attention is interesting but not very practical. In this book, the term plantwide control emphasizes supporting the operating personnel with the required amount of automation and control to achieve enterprise objectives. Plantwide control implies the effective integration of the human operator with the desired level of plantwide control technology. Process control technology by itself does not ensure this integration will happen. Expertise is required in order to effectively implement a process control system that achieves enterprise objectives.

Process control is the technology that provides mental leverage in plant production. For centuries, physical leverage has enabled humans to perform beyond their physical strength limitations. Mental leverage provides humans with the capability to perform beyond their mental limitations, such as a slow speed of response in mundane or complex control situations. Control technology will respond even when operating personnel are not present. It improves the ability of humans to understand the current condition of the process in order to make more timely and informed decisions. The enhanced ability is accomplished through decision support and direct control functions.

The need for plantwide process control is partly a result of the demand for increased mental leverage. Present-day plants require more sophisticated and complex processing coupled with increased concern for safety and environmental protection. In a modern manufacturing facility, a well-implemented control system must provide plantwide decision support and direct automatic control. It augments the capabilities of the operating personnel to better control and maintain a plant at its optimum potential to manufacture product.

The personnel often charged with the responsibility of designing, implementing, and maintaining a plantwide process control system are called control engineers. The field of control engineering is relatively new and crosses traditional engineering disciplines. A typical control engineer has a B.S., M.S., or Ph.D. degree in chemical egineering, electrical engineering, industrial engineering, or mechanical engineering. However, a control engineer may possess a degree from another discipline (e.g., chemistry or physics). Also, a person with a two-year engineering technology degree may obtain enough experience to be considered a control engineer. The typical control engineer is expected to work comfortably in the various disciplines, as needed by a particular project. For example, though a control engineer may have a B.S. degree in chemical engineering, he or she will probably also need to know some electrical engineering and some computer programming as part of the project implementation.

**The Scope of Plantwide Control.** *The scope of plantwide automation and control varies widely. To some, it is the control scheme of a particular plant unit. To others, it is the technology that provides a complete picture of plant operations within the plant site boundary. This disparity poses a challenge to the authors—how to address plantwide control for these diverse viewpoints.*

The plantwide viewpoint in engineering is represented in terms of thermodynamics, unit operations, streams, process equipment, structures, and control equipment. As described later, the chemical engineering community defines a plant as a collection of systems located on a site that belongs to an enterprise. The control system is just one of the many systems that must work together to fulfill the mission of the plant.

The plantwide viewpoint of business defines a plant in terms of financial information such as cost-of-goods-sold, order fulfillment, product inventory, and manufacturing capacity. The business that owns the plant expects it to deliver quality product on time to its customers. The survival of the enterprise and its plants depends on this goal. Plantwide control is only important if it contributes to this objective.

These different viewpoints suggest the operative definition for plantwide process control:

**Definition.** *Plantwide process control is defined as plant systems that enforce control strategies, provide a process information base, and enable operating personnel to fully monitor and command the plant process equipment to fulfill good manufacturing policies, support product marketing and customer sales activities, and protect the greater community and environment.*

Plantwide process control is the core technology to achieve enterprise production goals through automation of decision support and direct control. As such, it must encompass the appropriate groups of operating equipment. The emphasis in plantwide process control is on its relationship with operating personnel and its support of their efforts.

## 1.2 WHAT IS A PLANT?

Currently, there are probably several million plants in the world ranging from small single-product operations to large multiple-product complexes that cover many square miles. The size really does not matter in plantwide process control because even large complexes such as oil refineries, paper mills, and chemical plants are usually subdivided into separate operating sections that are small plants in themselves. The plant types range from those that manufacture discrete parts to those that produce bulk liquids and powders. Many plants are combinations of discrete-part manufacturing and fluid-processing operations.

Plants have different names in the various industries. For instance, a plant that produces beer is known as a brewery. Those that produce paper or steel are called mills. In the petrochemical industry, a plant complex is called a refinery. No matter what it is called, a plant is a facility that is located on a site somewhere close to raw materials and/or to distribution markets.

It is important to reemphasize that a plant is owned by an enterprise. The enterprise expects it to perform in support of the enterprise mission and goals. The enterprise may be government, commercial, or industrial. The purpose of a plant is to make products for an enterprise. A plant, therefore, is the physical and human assets of an enterprise that convert raw material into finished products by performing the appropriate manufacturing activities prescribed by the enterprise.

A plant acts on material in some physical way to produce the desired outcome—the product. The focus in this book is on profit-making enterprises that require thermodynamic, chemical, or biological unit operations to make products. This is the essence of the so-called process industries. The products are in the form of a gas, a liquid, and/or a solid.

A plant is physically located on a site. The site location is chosen to minimize raw-material transport costs or point-of-sale distribution costs. In some industries, a plant encompasses the whole site. In others, the site may contain more than one plant. Some sites include plants belonging to more than one enterprise. The combinations are endless in today's global economy.

A physical plant at a site is composed of equipment assembled in a manner that transforms raw material into product. The equipment is arranged and constructed to perform the desired plant activity to support a market. There are many ways to define a plant.

Chemical engineering defines a plant in terms of major process equipment items, chemical unit operations, and process material properties. Major equipment items include vessels, heat exchangers, pumps, and control valves. This equipment provides the environment to perform the desired unit operations, such as reaction, separation, absorption, and material transfer. The material properties include the desired thermodynamic state and physical properties of the product. The control content in the chemical engineering viewpoint is the stream properties such as flow rate and temperature coupled with the adjustments to the streams to achieve the desired operating conditions.

Plant construction engineers think of plants in terms of piping networks, building steel, and so on. This includes the utility systems, the mechanical systems, the electrical systems, and the waste disposal systems. This construction viewpoint includes the control instrumentation and final control elements and its installation in the physical plant.

Operating personnel think in terms of inventory, work-in-process, and recordkeeping. They are more concerned with the operating conditions and the information requirements of the enterprise systems than with how the plant is constructed. To them, a control system must help them make product and be easy to use. It must be robust and flexible to changing plant conditions.

This diverse thinking is a challenge to reconcile in executing capital projects for plantwide control. On projects, the chemical engineer, or his equivalent, makes the first design rendering of a process plant in the form of a process flow sheet and a process simulation. The simulation represents a plant as a set of unit operations connected by energy and material streams. The classical unit operations include separation, reaction, and so on. The other chemical engineering view of a process is the process flow diagram (PFD), which depicts the major plant equipment items and stream data. A sample of a PFD is shown in Figure 1.1. The PFD represents the major process equipment and the stream data at the various points of interest in a process design candidate. The stream data include the expected flow rates, temperatures, and composition of material and energy flow streams.

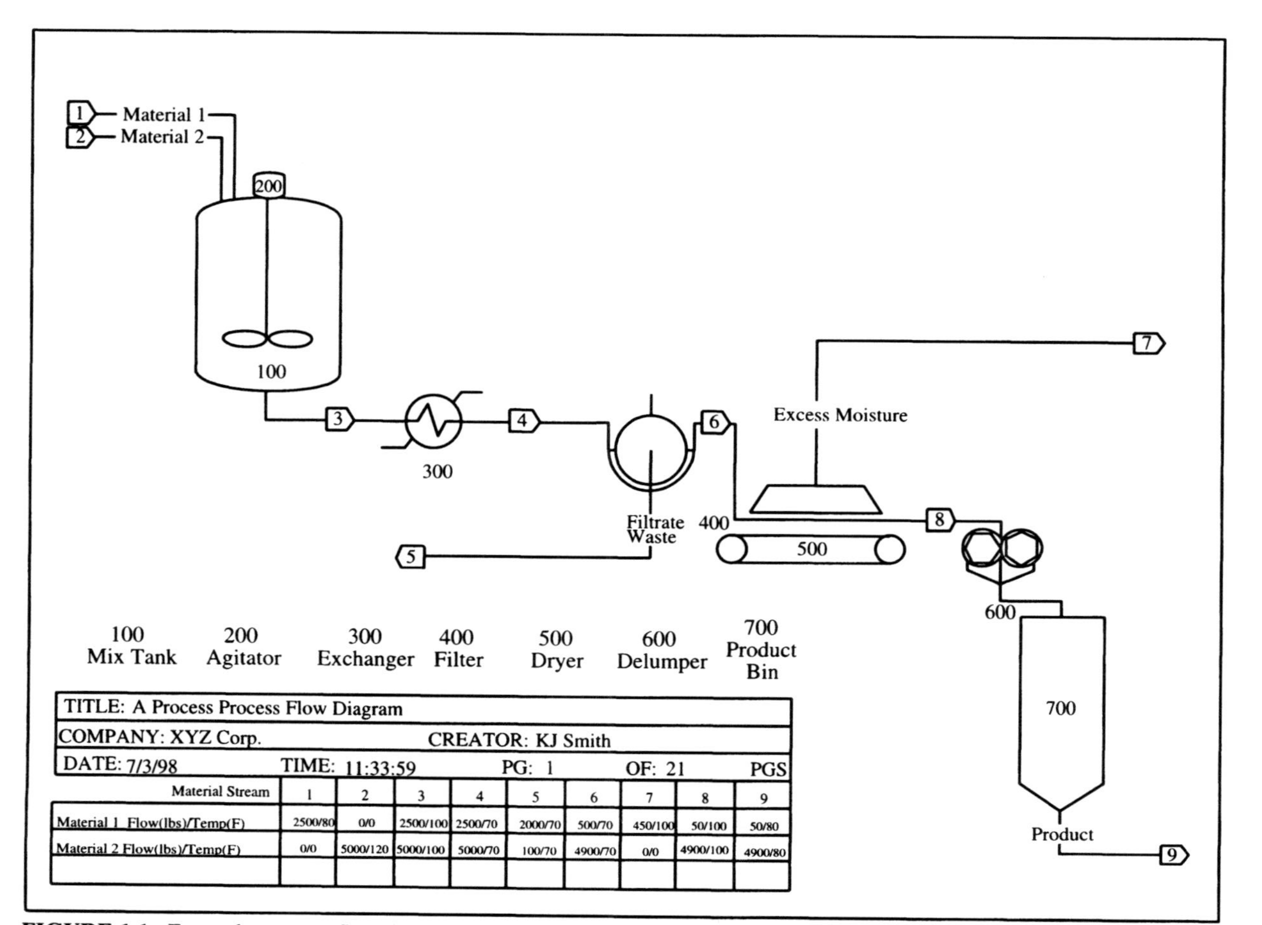


| Material Stream | 1 | 2 | 3 | 4 | 5 | 6 | 7 | 8 | 9 |
|---|---|---|---|---|---|---|---|---|---|
| Material 1 Flow(lbs)/Temp(F) | 2500/80 | 0/0 | 2500/100 | 2500/70 | 2000/70 | 500/70 | 450/100 | 50/100 | 50/80 |
| Material 2 Flow(lbs)/Temp(F) | 0/0 | 5000/120 | 5000/100 | 5000/70 | 100/70 | 4900/70 | 0/0 | 4900/100 | 4900/80 |
| | | | | | | | | | |

**FIGURE 1.1.** Example process flow diagram (PFD). (Reprinted by permission. Copyright © 1996, Automation and Control Technologies, Inc.)

The construction engineers use the chemical engineer's process design to develop a detailed plant design. The design is represented by a set of schematics and other drawings. The primary drawing used for construction is the piping and instrument diagram (P&ID). For examples of this design document see Figures 1.2 and 1.3. This representation depicts the detailed vessel configurations and the interconnected piping network. It also defines the control valves and motor controls for the vessels and piping network. However, it is not to scale and does not indicate how the equipment is installed. Sometimes the piping material is shown on the P&ID. Traditionally, the P&ID defines the instrumentation and controls for the process. However, only the basic single-loop controls are shown.

Until the emergence of the microprocessor-based control, the P&ID included representations of all the controls. Modern control schemes are too complicated to be properly represented on the P&ID, however. Today, more detailed design representations such as Scientific Apparatus Makers' Association (SAMA) diagrams, logic schematics, and sequence descriptions support the P&IDs.

Most operating personnel think of the process in terms of operating modules that convert raw material and produce product inventories. They tend to represent the process in terms of work flow, quality/assurance testing, and shipping/receiving activities. In plants that produce many products, the product routes or paths are important.

As control technology expands to a plantwide scope, it has taken over many traditional manual operations. As such, a more rigorous consideration of the operation viewpoint in the initial process design has become more important. As a result, a plant design representation is now required to define a set of operating modules that depict the product paths. This representation has proven valuable in organizing the design information throughout the lifecycle of the plant.

### 1.2.1 Operating Module Viewpoint

The operating module viewpoint looks at the plant as separate plant systems that operate as a coordinated group to make products. These plant systems have names such as plant sections, zones, cells, units, trains, and lines. The operating modules of interest in plantwide control applications are the process area, cell, and unit. These are defined more precisely in Chapter 3. The following descriptions provide a basis for the treatment of plantwide control in Chapters 1 and 2.

**Process Area.** *A process area refers to a logical grouping of process and control equipment that perform the desired unit operations to make a product. Process area boundaries are normally set by business or enterprise policies, such as operator jurisdiction, specific product, or other criteria. A plant may be composed of one or more process areas. Process areas are subdivided into process cells.*

**Process Cell.** *A process cell contains all of the equipment necessary to make a specific amount of product. In a batch process, for instance, a process cell can produce one or more batches. A process cell not only refers to a logical grouping of equipment but also defines a span of control within the process area. Production schedules and control strategies, for instance, are often developed on a process cell basis. Process cells are subdivided into process units.*

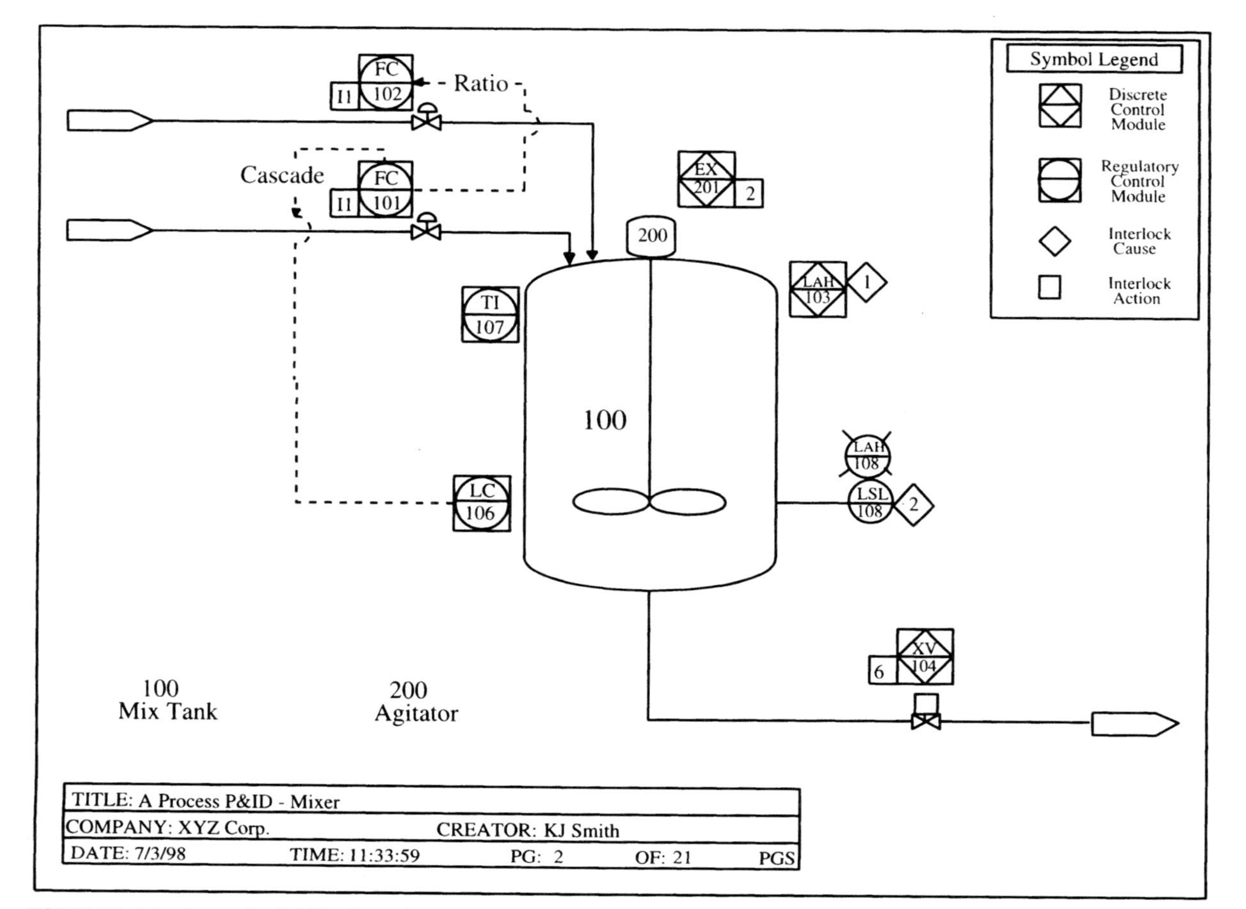


**FIGURE 1.2.** Example P&ID for mix tank. (Reprinted by permission. Copyright © 1996, Automation and Control Technologies, Inc.)

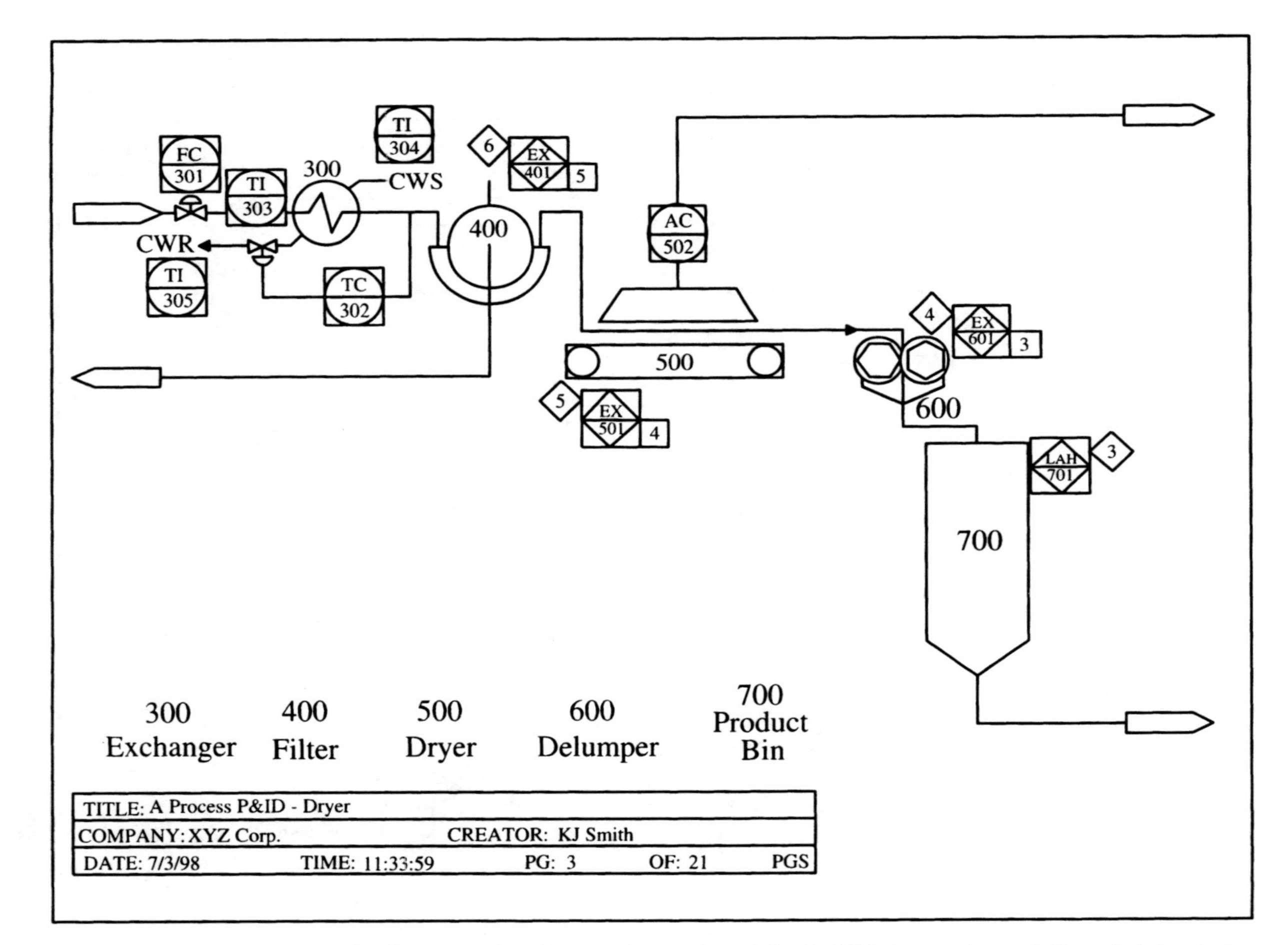


**FIGURE 1.3.** Example P&ID for dryer. (Reprinted by permission. Copyright © 1996, Automation and Control Technologies, Inc.)

**Process Unit.** *A process unit provides a further logical subdivision of physical equipment within a process area or cell. It contains all of the equipment needed to perform one or more major processing activities (react, crystallize, etc.). A unit is usually centered on a major piece of equipment, such as a tank or reactor. A unit is in only one operating state at a time (e.g.,* Idle, Startup, Production, Shutdown, Emergency Stop*). From an operating viewpoint the process unit is the lowest level operating module that is related to plantwide control.*

**Operating State** *(Definition).* The various operating modules possess operating states. The term *operating state* refers to a specific period of time during which characteristic operation and control activities are performed to operate the process equipment. These activities reflect the actions of a specific operating module. Many continuous processes, for instance, can be described simply by two operating states: Running and Idle. Additional refinements, however, are normally required to adequately describe the process operation. For example, the Running state may actually consist of a Startup state, a Production state, and a Shutdown state. Batch processes, on the other hand, nearly always exhibit several operating states.

To illustrate the operating module viewpoint, consider the material unloading process cell in Figure 1.4. The cell operating state is Idle when nothing is being unloaded. When an unloading activity is required, there may be up to three separate unloading systems in use, one from the truck, one from the bag splitter station, and one from the rail car. Each of

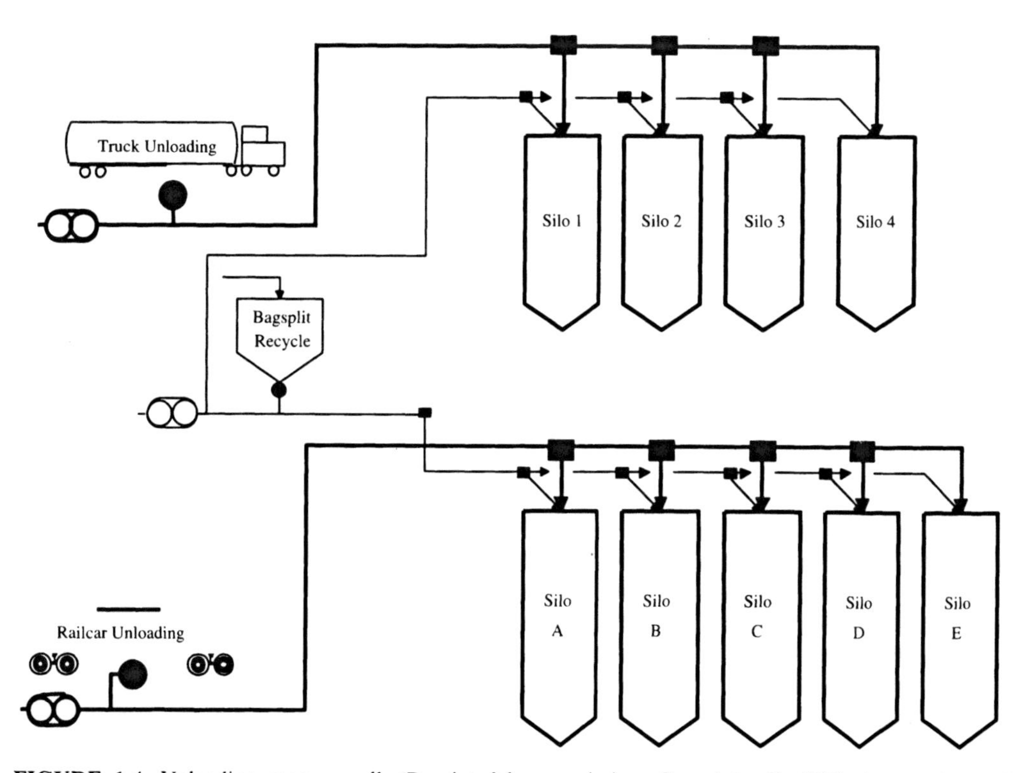


**FIGURE 1.4.** Unloading process cell. (Reprinted by permission. Copyright © 1996, Automation and Control Technologies, Inc.)

these can be defined with two states, Idle or Running. Each system may be operated separately as long as there are no silo route conflicts. This means there are three process unit operating modules in this process cell, each with a separate but coordinated operating strategy.

### 1.2.2 Physical Equipment Viewpoint

The most common view of a plant is in terms of the physical equipment. In the context of this book, there are two types: processing equipment and operating equipment. Processing equipment includes the vessels, pipes, pumps, valves, conveyors, and other machinery that is "wetted" by the material and energy streams. The operating equipment includes the control and business systems that support operating personnel in using the processing equipment to make product.

Control system equipment is connected to the associated processing equipment. This combination provides a specific production capability to make product. For instance, a batch reactor and the associated controls provide the capability to perform reactions to make a particular chemical. This process capability is the basis for describing the function of operating modules.

### 1.2.3 Industry Standard Plant Viewpoint

Recently, the process industry has endeavored to improve the exchange of engineering data in plant design activities. This includes activities from the initial concept design to as-operating modifications. This effort is large and spans many engineering disciplines. The effort is called the International Standards Organization (ISO) 10303 standard. It is a multipart standard to improve product data exchange in electronic systems. ISO 10303 is commonly referred to as STEP, the STandard for the Exchange of Product data. The scope of STEP (or ISO 10303) is large, spanning product data exchange in both commercial and military activities.

The effort in the process industries is spearheaded by the National Institute of Standards and Technology (NIST) as the STEP standards AP221, AP227, and AP231. AP221 (ISO, 1997a) is the standard for the exchange of process plant data and defines the data exchange for functional and schematic representations of a plant. This is the information associated with the P&ID. AP227 (ISO, 1994) is focused on data that defines the three-dimensional spatial representation of a plant. AP231 (ISO, 1997b) covers the data that define the conceptual process design in the front end of a project and usually is represented by the PFD and process simulation flow sheets. The reconciliation of these separate efforts has been combined, and this reconciliation has produced a more precise definition of a process plant in terms of design data. The authors have adopted this definition and will use the plant context model defined in AP231.

AP231 (ISO, 1997b) states that a plant is located on a site and is defined by a set of plant systems. These systems are composed of plant items. A plant item is the identification and description of a piece of physical equipment. The plant items may be single pieces of equipment or they may be complex assemblies of physical equipment that include both processing and operating equipment.

AP231 states that a plant system may be defined in a group/subgroup hierarchy. The hierarchy is not precisely defined, but the example in the standard is one whose top level is a process area. The process area plant system is composed of trains (cells) that are composed of process control units. This is consistent with the earlier definition of operating modules. These plant systems are defined to facilitate the plant design in an object-oriented manner to facilitate plant design data exchange. This representation recognizes that plant systems are operating modules that have an operating state.

In AP231, process descriptions are associated with the plant systems. The content of a process description is fully defined and includes the unit operation, component data, and basic controls. In fact, the standard recognizes the need to consider the operation and control of a plant in the process design activity. This means that plantwide control will be considered much earlier in a project when AP231 becomes the predominate methodology to conduct capital projects. This is a significant step forward in the pursuit of plantwide control.

The full AP231 data model suggests the following definition of a plant: *the physical and operating equipment located on a site with the capability to perform the desired operations to make product. It consists of one or more operating modules that process material to make product as part of the overall production activity in an enterprise.*

As discussed earlier, a site may have more than one plant belonging to more than one enterprise. Plantwide control must be considered within this context.

## 1.3 MODERN PROCESS CONTROL

The capabilities of modern control technology to support plant operation have progressed well beyond the panel-based control of the early days of process control. Panel-based systems were limited mainly to low-level control in plant operations. Activities such as startup, supervisory, recipe, and shutdown controls were left entirely in the hands of human operators. Control systems today have an expanded role, replacing manual manufacturing activities with full automation. Modern process control is the functional integration of real-time information management with closed-loop, interlock, and sequence control.

Despite the expanded role, a certain fundamental persistence exists in control, designated as control types. For instance, relay logic and the proportional-integral-derivative (PID) loop control types have been the primary functions of basic control since the earliest process control systems. Relay logic provides a well-understood functionality that is recognized as the best way to perform interlock control in both safety and nonsafety protection schemes. It is implemented today in both microprocessor-based systems and hardwired relay systems. Likewise, PID loop control has persisted despite the evolution of control technology from pneumatic to analog electronic and on to modern microprocessor-based technology.

It is important to understand that certain control types are technology neutral. They will persist regardless of the technology that implements them. Control technology, on the other hand, will change. With this change there is usually an improvement in the capability of the technology to provide more advanced control function. The PID function in basic control is an example. The microprocessor version provides more advanced features than the pneumatic technology form.

This section describes the fundamental control types of process control and defines the current technology that supplies the control types typically found in plantwide control.

### 1.3.1 Control Types

Process control is usually viewed as the plant systems that perform closed-loop automation. The primary role of a control system is to use automation in order to enforce the control strategies as prescribed by the enterprise. A second role has emerged with the proliferation of computer technology. Information is key in making appropriate decisions so the control system must also play a role in reporting plant performance information. Timely information is needed for tactical and strategic planning. The third role is to provide the proper span of vision for operating personnel to take corrective actions when required. Keeping this threefold role in mind helps ensure the control system more fully supports operations.

The roles of the control system in plantwide control are to:

1. Enforce plant control schemes.
2. Report plant performance.
3. Provide a user interface to the automation and controls.

To fulfill these roles, a plantwide control system cannot be one big monolithic algorithm. It is a carefully selected set of control functions that work in a coordinated fashion. The control function in a particular application is a function of the types of control that are implemented. The control types are fundamental to process control. They are selected to fulfill the primary role of the control system. The control engineer has the responsibility to select the appropriate combination of control types.

The control types in plantwide control are:

- Basic regulatory and discrete control
- Procedural control
- Coordination control
- Supervisory control
- Exception handling
- Alarm management
- Trend and event recording
- Production run and batch recording

Each of these control types also provides a user interface.

Basic regulatory and discrete control is the conventional loop and interlock control that has been the main workhorse for years. It is the foundation of all plantwide control systems. Controllability begins with this base. Besides loop and interlock control, this category also includes the indication and status measurements that are not directly controlled. This type of control is covered in Chapters 5, 6, and 8.

Procedural control is the conventional sequential control logic. This type of control enforces an ordered procedure of control actions when required in an application. Procedural control is common in batch applications and in any process that must be operated in a stepwise fashion. Procedural control is covered in Chapters 8 and 9.

Coordination control automates the arbitration of shared resources. The need for coordination control is a common situation in material transfer systems that feed more than one operating module. It is also required in batch applications. This type of control is covered in Chapters 8 and 9.

Supervisory control is commonly referred to as setpoint control using a model to determine the correct setpoint settings. This type of control is also referred to as multivariable model based control and is covered in Chapter 7.

Exception handling is the automation of the failure action and recovery when an abnormal process event or condition occurs. Exception handling is a consideration in all control types and includes both process and system failures. Chapter 6 covers exception handling for regulatory control and Chapter 8 covers exception handling for discrete control.

Alarm management is the automation that helps the operator identify abnormal conditions. As part of the user interface facility, alarm management annunciates and tracks process failures. Features such as alarm prioritization and alarm suppression avoid unnecessary and nuisance alarms.

Trend and event recording is the control function type that saves information about the process in a form that may be used for process event analysis. This information is especially useful in achieving continuous improvement.

Production run and batch recording is the control function type that saves production information in the form of time-based records. These records are used for validation and long-term histories.

The intent of this book is to cover these control types as they apply to plantwide control systems.

### 1.3.2 Control Technology

The control types are implemented in the control operating equipment of the plant. The capability of control equipment has changed rapidly in the past 20 years because of the computer. The pace of change continues, and any treatment of technology is quickly outdated. On the other hand, any discussion of plantwide process control must consider the control equipment implementation. This book intends to present control in a manner that is independent of the technology.

The use of computers in process control started in the late 1950s. These devices were business computers retrofitted for process control service. They were mainly used for supervisory control and data acquisition—the origin of the term SCADA. There were attempts to implement direct control of a process but the reliability and performance of computer technology was inadequate. The seeds of plantwide control, however, were planted and cultivated during these early years.

Early in the 1970s, the programmable logic controller (PLC) emerged to replace hardwired electromechanical relay panels. It was the first use of small-scale computer technology that specialized in the needs of control. In the mid-1970s, the microprocessor was introduced in direct control for PID loops. These two events began the age of computer-based direct control, which continues today as the control technology workhorse in direct process control.

In the late 1980s, the personal computer made its way into industrial applications. It was interfaced with the PLC and microprocessor technology as the user interface and supervisory control platform. Also in the 1980s the plant data communication system began interconnecting the control systems with other plant and enterprise systems. A plant communication web consisting of industrial computers and direct controller systems has become the predominate technology in the 1990s.

The 1990s will be remembered for the emergence of the control device network. To some, this network has enabled the move of the control functions back to their origin, in the physical equipment. As one may recall, early control was mainly basic regulatory and discrete control in the form of pneumatic technology located on the process equipment. Control device networks provide the same type of control but with expanded capability.

The technology of the user interface has undergone several changes as well. The early user interfaces were mechanical panels field mounted on the physical equipment. With the use of analog electronic controllers, the user interface moved into control rooms featuring the control panel with its array of meters, chart recorders, lights, and switches. Today, the predominate user interface is the shared display screen and keyboard/mouse driven by personal computers or mounted as dedicated operator panels. In either case, the predominate technology for the user interface is computer based.

The computer, therefore, has become the main technology in modern process control systems. Computer knowledge is critical in modern control systems. A brief look at the architecture of computer technology will set the stage for discussions of plantwide control.

***1.3.2.1 System Architecture.*** In general, computer technology is composed of hardware, embedded software, and application software. In process control applications, these components must work together in a very reliable and robust way. The hardware must endure vibration, heat, and other environmental conditions that are not present in the typical home and office locations. The embedded software must be fault tolerant and be able to execute in environments where many home personal computers would stop. The application software must provide the control types required in process control.

The typical process control system is a network of integrated computers. The communication system, likewise, must provide a high degree of throughput performance and fault tolerance. A typical system architecture for the all-digital control system is illustrated in Figure 1.5. The major hardware components depicted are:

- Control devices
- Controllers
- Operator stations
- Communication networks

Embedded software is the control system functions provided by the supplier. For direct controllers, this type of software is usually supplied in the form of read-only memory to avoid boot-up delays in time-critical control. For other computer devices, such as operator workstations, the embedded software is loaded from hard-disk drives on boot-up. Recently, some controllers are being loaded with embedded software modules from hard drives to provide flexibility in the type of control that is executed in a specific application. The key components of the embedded software are the operating system and the programs that execute the application software. The operating system software manages the hardware activities. In most control systems, the operating system of the various direct controllers is transparent to the user. It performs in the background, ensuring the system executes in a robust manner. For the application software, the embedded software provides the programming language generation and testing environment for the application software.

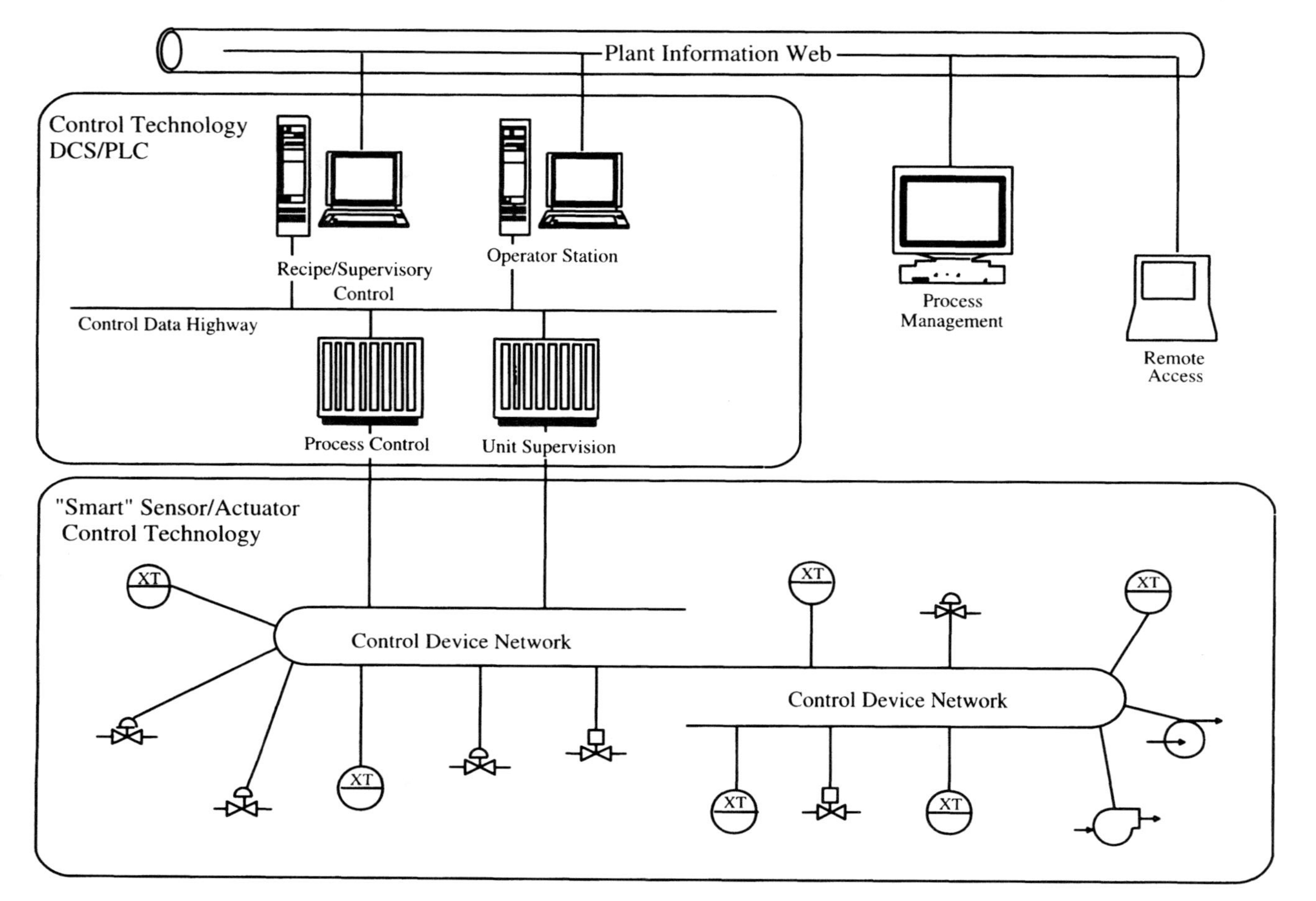


**FIGURE 1.5.** Typical modern system architecture. (Reprinted by permission. Copyright © 1997, Automation and Control Technologies, Inc.)

The application software is usually control-centric (centered around control). A good example is the PLC ladder logic language. Until recently, the programs that execute the application software were proprietary technology developed by each supplier. The new IEC 1131-3 standard (International Electrotechnical Commission, 1993) for PLC programming has changed this situation. This standard has enabled users to become supplier independent as they develop the application software in the IEC 1131-3 languages. Now control types are being implemented with standardized programming languages.

***1.3.2.2 DCS Versus PLC.*** Computer-based control evolved along two parallel paths. These paths followed the common practice of separating the basic control into discrete and regulatory control types. The discrete control type was provided by PLC technology. What became known as a distributed control system (DCS) provided regulatory control. This distinction persisted during the 1970s, 1980s, and early 1990s. While both control technology types claimed to do some of the functions of the other type, it was not until the IEC 1131 standards blended them that the distinction became obsolete. The blurring of the lines between them is continuing down to the control device network level. This merging of the DCS and PLC technology suggests a new term to describe the control technology for plantwide control: *programmable electronic system* (*PES*). This term is now used in many standards such as the ISA S84.01 (Instrument Society of America, 1996) for safety instrumented systems.

The PES provides the control-centric function types with a high degree of performance, reliability, and ease of maintenance. A PES often must respond in milliseconds to process events and at the same time handle situations with long time constants. It must do this with little or no downtime while running seven days a week, twenty-four hours a day. The demands on a current PES are significantly higher than on the panel-based systems in the early years of process control.

## 1.4 CHAPTER SUMMARIES

The book is divided into four main sections. The next two chapters define the overall view of control projects and discusses the various issues in process control, including the role of standards. This section provides a framework for the remainder of the book. The second section, consisting of Chapters 4–7, presents process control techniques, starting from the single-loop perspective and working up to unitwide and supervisory control techniques, including model predictive control. The third section, Chapters 8 and 9, covers design techniques for discrete and batch processes. The discrete design techniques in Chapter 8 lay the foundation for the batch process control techniques of Chapter 9. The entire process of plantwide control design is illustrated in the fourth section by the case study of a pulp and paper mill (Chapter 10).

Plantwide control is designed and implemented by control engineers through the execution of capital projects. Chapter 2 defines the activities on a capital project and the responsibilities of control engineering to design and implement a plantwide control system. Since plantwide control is an essential part of plant operations, the control engineering aspects of any capital project should be part of the preliminary design package and not delayed until later stages of the project. The control design information package consists of three major sections: (1) process operation description, (2) Control Concept, and (3) control strategy. The process operation description (POD) defines the operating strategy and control objectives of the plant systems. The Control Concept, a companion to the POD, defines the extent of the automation required to operate the plant system and achieve enterprise objectives. The control strategy is developed only after the control requirements, as expressed by the POD and Control Concept, are well understood. The control design information package becomes the guiding document during all activities of the capital project.

The control design information package is prepared in accordance with the reference model described in Chapter 3. This model, called an enterprise plant model, is used to assist in segmenting the automation system requirements into manageable pieces. The S88.01 batch control international standard is used as a starting point and expanded to include more traditional continuous process control and discrete control. This model divides the plant into hierarchical levels and incorporates both the operational and physical viewpoints of a process. Two examples illustrate the model. The concept of a state, possessed by each entity in the model, is then described. Following model development, aspects that affect all layers of the model are discussed: information management, user interface, and control system technology.

Modeling plays a key role in the successful application of process control and thus is discussed in Chapter 4, preceding the process control techniques. Without a model, the engineer can spend much time in a fruitless attempt to do the control in an ad hoc fashion. However, for many systems a simple, empirical model that describes the overall behavior of the system is more than sufficient to design the control system. In order to be complete, models based on fundamental principles are treated first, followed by simple empirical models. Discrete-time models, needed for the more advanced digital process control techniques, are treated next followed by methods to obtain the empirical model parameters. The material of Chapter 4 is intended mainly for background or review to support the process control techniques of Chapters 5–7 and can be skipped without loss of continuity.

Chapter 5 provides a foundation in the use and tuning of the regulatory controls used at the lower levels of the control hierarchy. General feedback control features are explained and then the basic PID controller is described. For most systems, a simple PID controller is sufficient to handle the regulatory control functions. The most popular PID tuning methods are also described.

While the PID single-loop controller works adequately in many process control situations, its performance can be improved if additional process information is known. These enhancements to the PID controller, and the situations in which they are applied, are discussed in Chapter 6. Cascade control, feedforward control, ratio control, split-range control, and signal selectors are described and placed with their appropriate application. This chapter concludes with model predictive control, an alternative to the PID controller algorithm that handles unusual process dynamics.

If a unit process is inherently multivariable, that is, it has more than one controlled variable and more than one manipulated variable, multivariable regulatory control techniques are often needed. These techniques are presented in Chapter 7. A multivariable controller is generally part of the cell or unit control strategy, taking measurements from indicators and manipulating single-loop PID controller setpoints. The two basic approaches to multivariable control—multiloop control and full multivariable control—are summarized and illustrated by examples. In multiloop control, single-loop controllers are connected to the process in such a way as to reduce the process interaction. Full multivariable control uses all measurements to simultaneously calculate the manipulated variables. Rather than being a comprehensive treatment, only the most common full multivariable control algorithm, model predictive control, is treated. The use of singular-value decomposition in choosing the sensors and manipulated variables is also illustrated.

The focus changes in Chapter 8 to discrete control and its role in plantwide control systems. Depending on the specific process, discrete control may be a large part of the control strategy. The sequential function chart is introduced as an important documentation and design tool for discrete control. Interlock control for both safety and nonsafety systems is also presented in the context of discrete control. Discrete-control examples are presented for discrete devices and unit-sequence control.

The discrete-control techniques in Chapter 8 become the foundation for the batch process control techniques described in Chapter 9. Batch plant automation requires a high level of discrete control. Also, some batch applications require complex regulatory control schemes such as temperature control in batch reactors. However, it is the recipe control requirement that distinguishes batch processes from continuous and discrete processes.

In Chapter 10, a pulp and paper mill is used to demonstrate the process of plantwide process control design. The control design information package for this plant consists of the three major sections described in Chapter 2: (1) the process operation description, (2) the Control Concept, and (3) the control strategy. Because of space limitations, the control strategy is presented only for selected plant units and cells. However, enough information is present to serve as a good example of a control design information package for a plant.

## REFERENCES

Instrument Society of America, *ISA-S84.01, Application of Safety Instrumented Systems for the Process Industries—1996*, Instrument Society of America, Research Triangle Park, NC, 1996.

International Electrotechnical Commission, *IEC 1131-3: Programmable Logic Controllers—Part 3: Programming Languages*, International Electrotechnical Commission, Geneva, Switzerland, 1993.

International Standards Organization, *ISO 10303: Industrial Automation Systems and Integration, Product Data Representation and Exchange—Part 227: Plant Spatial Data*, (also known as AP227), working project draft, December 9, Geneva, Switzerland, 1994.

International Standards Organization, *ISO 10303: Industrial Automation Systems and Integration, Product Data Representation and Exchange—Part 221: Functional Data and Their Schematic Representation for Process Plant* (also known as AP221), working project draft, February 23, Geneva, Switzerland, 1997a.

International Standards Organization, *ISO 10303: Industrial Automation Systems and Integration, Product Data Representation and Exchange—Part 231: Application Protocol: Process Engineering Data: Process Design and Process Specification of Major Equipment* (also known as AP231), committee draft, July 17, Geneva, Switzerland, 1997b.

# 2 Control Engineering on Capital Projects

### Chapter Topics

- Control engineering
- Plant capital project management
- Project team considerations

*Scenario:* The successful project begins with a thorough understanding of control requirements and with the ability to use the right frame of reference.

After a few years, the young engineer left the operating company where he started his career to join a "vendor" of process control equipment. His first few weeks at the new employer were exciting—finally, an enterprise devoted to process control. He expected to learn many things and contribute even more.

As he settled into his new position, he noticed a flurry of activity in a major project being conducted by his new employer. It involved a large control system application and, as sometimes happens, the customer was not pleased with the first demonstration session. It was not what they wanted. He heard his new colleagues protest that the customer had accepted a system specification and should not have been unhappy. After several heated discussions, the issue was resolved and a new specification emerged that more precisely defined the system responsibilities.

The mistake that caused the problem was fundamental. The original project specification did not clearly define what the customer wanted in terms the customer could understand. It was a classic case of "failure to communicate." An analysis of the original agreement revealed that his colleagues and the customer did not effectively reach agreement on what was to be delivered. The customer's terminology and structure of control was not the same as the vendor/contractor. Once the two parties agreed on what was to be delivered and the means of expressing it, the project was successfully executed.

The incident made a big impression on the young engineer. In his previous company, he had been both the customer and the contractor and not experienced the kind of miscommunication that caused the problem on this project. He learned the value of establishing a clear definition of customer requirements when performing services for any customer, inside or outside a company. This clarity is achieved through a formal design methodology and an appropriate frame of reference that all parties can understand. Since then his career has been focused on developing a common frame of reference for process control.

He encountered many examples of using the wrong frame of reference along the way. On one occasion he was faced with a typically late front-end design situation. He was confronted with a control design that specified the need for 400 highly interactive interlocks. His experience taught him that this would be very expensive to implement and start up. The process was a waste filtration system in a decommissioned plant. He had a hard time understanding why so many interlocks. After consulting with the customer he discovered the problem. The original designer's frame of reference was loop and interlock control. For all but a handful of interlocks, the control objective was to stop the inflow if the effluent flow is stopped. By taking a plantwide view, most of the interlocks were replaced with a supervisory scheme that stops the appropriate inflows when the outflows stop. The young engineer was able to significantly reduce costs by focusing on requirements rather than control details developed from the wrong frame of reference.

## 2.1 INTRODUCTION

Plantwide control is designed and implemented by control engineers through the execution of capital projects. This chapter defines the activities on a capital project and the responsibilities of control engineering to design and implement a plantwide control system.

Control engineering is a new discipline that has matured in parallel with the evolution of automatic control technology. As discussed in Chapter 1, the first control applications were limited to pneumatic control equipment and primary sensors. The technology eventually expanded with the emergence of analog electronic equipment. The level of control stayed about the same, but the capabilities of this later technology provided more options to deal with plant control challenges.

For many years, the capability of control technology remained at this level. It was the domain of the instrument engineer. The instrument engineer designed and installed the measurement and control loop technology on plant capital projects. As the enterprise demanded improved controllability, the use of control and measurement technology increased. This demand led to the emergence of the control engineering discipline.

As computer technology took over and became the main logic solver in control applications, the engineering skill set required to do projects increased dramatically. In the early days of process control, the control engineer worked in relative obscurity because his tools were limited to basic control. This situation is no longer the case. There is increasing demands for the control system to deliver enterprise information. The control system has become mission critical in the information side of the enterprise. The control engineer has a new, broader role to deliver plantwide control and information systems in cooperation with the business information systems organization.

## 2.2 CONTROL ENGINEERING

Control engineering is recognized as a professional engineering discipline. In the process industries, control engineering is the plant lifecycle activity that is responsible for the controllability of a process plant. This responsibility is shared with other plant disciplines in achieving overall operability of the plant to make product for the enterprise. In modern control systems, the control engineer interacts more and more with the information technology organization. The tools of the trade have changed from a screwdriver and multimeter to sophisticated computer software technology.

As a member of a capital project execution team, the control engineer is responsible for the design and selection of:

- Measurement and final control systems
- Safety instrumented systems
- Process control systems
- Unit supervision systems
- Process management systems
- Process information management systems
- Operating personnel workstations
- Production data management and data exchange with other plant systems

These systems are selected based on the extent of automation required to effectively operate a plant and fulfill the enterprise objectives.

Control engineering requires many skills. The modern control engineer must be well versed in disciplines like measurement science, fluid mechanics, electrical engineering, systems engineering, and software engineering. With the emergence of the fully integrated enterprise information system, the knowledge to design and implement information technology is also important. Today, the emphasis in control engineering is on delivering an integrated control and information system. The demands on the modern control engineer are significantly different than those in the early days of pneumatic controllers.

## 2.3 PLANT CAPITAL PROJECT MANAGEMENT

Project teams execute plant capital projects to build and improve manufacturing facilities. The capital project activities in plantwide control are those that are executed to improve the process controllability and information generation of an enterprise. The overall management of these projects is important.

As depicted in Figure 2.1, a plant is composed of three major areas as it relates to a capital project:

- Control and information technology
- Plant personnel
- Process technology

This simple model defines the overall flow on a capital project. As part of the requirements definition task, the extent of automation is determined. The extent of automation line is the boundary between the plant personnel and the other two major areas.

The process technology area contains the plant systems that process the materials to make product. In this area, automation provides the physical leverage to plant personnel. The extent of automation, therefore, includes the mechanical and pneumatic moving equipment, such as pumps, control valves, and conveyors.

The control and information technology area contains the plant systems that are designed and implemented by control engineering. The extent of automation here is the amount of operating activities that are performed by the control technology. The automation may be full automatic type, replacing manual activities, or it may be semiautomatic, augmenting the manual activities.

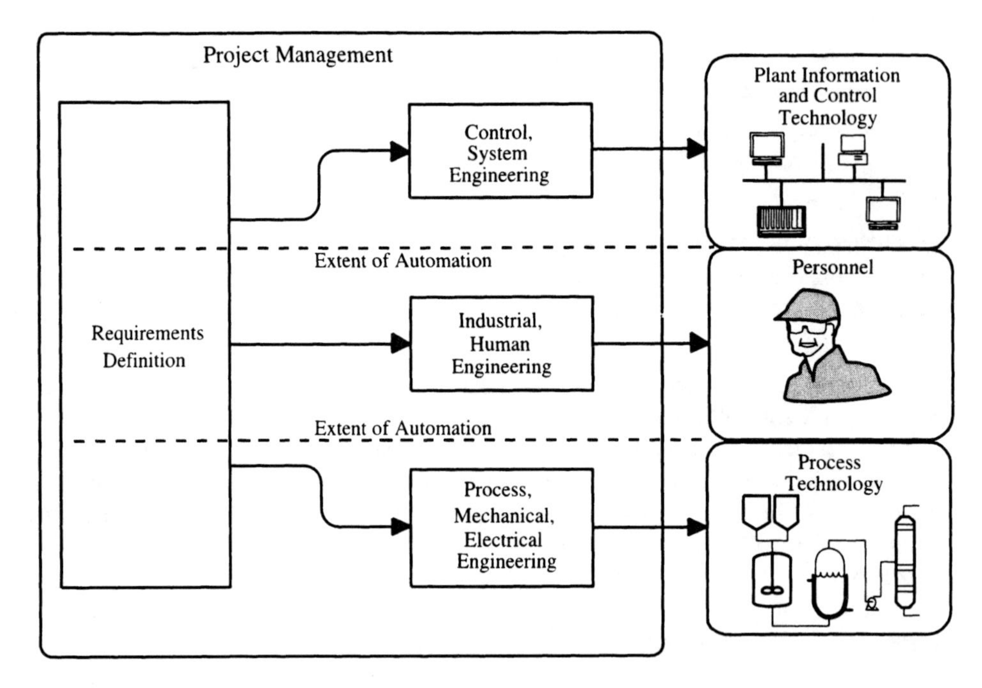


**Figure 2.1.** Plant capital project disciplines. (Reprinted by permission. Copyright © 1996, Automation and Control Technologies, Inc.)

The process technology and the control and information technology support the plant operating personnel. The technology also provides the mental and physical leverage to operate the plant effectively. The user interface lies at the boundary between the personnel and the other two major areas. Thus, the user interface is a critical consideration in capital project execution.

The three aspects of a plant are highly interactive and require an integrated project management approach to execute a capital project. Keeping this broad picture in mind proves valuable in achieving overall objectives during a capital project.

The overall flow on a capital project begins with the definition of the project requirements. A key factor in the failure of any project is missing this important first step. A multidisciplinary team approach is essential. The team should include plant operation and engineering experts familiar with the application and with the plant's current standards (equipment, documentation, etc.).

Once the requirements are clearly stated, they can be distributed to the appropriate engineering disciplines. These disciplines perform the detailed design and implementation of the required technology. Control engineers are involved in meeting the requirements for the control and information technology. Industrial and human engineering experts handle the plant organization requirements. Of course, the process technology is the responsibility of the process engineering experts. All this is orchestrated by the project management contingent.

During the lifecycle of a plant there are numerous capital projects. They range from small equipment modification projects to large, greenfield (grassroots) projects; from process debottlenecks to new product introduction projects. To understand the role of control engineering on a capital project, a plant lifecycle view must be considered. To this end, the authors have selected a plant lifecycle model that has industrywide acceptance.

*STEP Process Plant Lifecycle.* As part of the process industry's STEP standardization mentioned in Chapter 1, the participants developed a definition of a process plant lifecycle. This definition was developed to facilitate engineering data exchange in capital projects.

In Figure 2.2 the lifecycle is depicted in the form of an IDEF0 diagram (National Institute of Standards and Technology, 1993). It represents the activities during the life of a plant, from initial conception to final disposal. The time span for this lifecycle is many years. The focus of the model is to identify the information flow in performing plant design engineering. This focus is appropriate in the discussion of plantwide control.

As depicted in Figure 2.2, the major activities identified by the industry are to:

- Manage and plan the project.
- Design the plant.
- Procure the components.
- Construct and commission the plant.
- Manage, operate, and maintain the plant.
- Decommission and dispose of the plant.

These activities are performed concurrently throughout the lifecycle of the plant. The diagram does not represent a "once-through" cycle. An IDEF0 diagram should not be interpreted as a time sequence. At any point in time, there may be something going on in all the activities. One can be operating a plant, designing upgrades or retrofits, and procuring new equipment simultaneously. The IDEF0 diagram is not a schedule; it is a top-down view of the human activities and the information flow among the activities. Figure 2.2 is called the context level. Whether one is adding a new valve, upgrading an existing unit, building a new unit, or building a new plant, the philosophy in Figure 2.2 is the same, but the degree of activity will vary.

While there is general agreement among STEP participants at this level, there are two slightly different views of how the Design Plant (node A2 in Figure 2.2) activity at the context level is subdivided. Figure 2.3 is the process designer's viewpoint developed by the AP231 (pdXi) group (ISO, 1997b). This viewpoint divides node A2 into two subactivities:

- Define plant requirements.
- Define final plant design.

This corresponds to the idea expressed in Figure 2.1 where the requirement definition function precedes the detail engineering on a capital project.

The second view of the Design Plant activity is depicted in Figure 2.4. This latter view is the view of the architectural and engineering (A&E) contracting industry (AP221/227) (ISO 1994, 1997a). It divides the Design Plant activity into four subactivities:

- Conceptual process design
- Conceptual plant design
- Final process design
- Final plant design

The first activity of the AP221/227 view is roughly the same as the first activity in the AP231 activity model developed by the process engineering group. The AP221/227 view

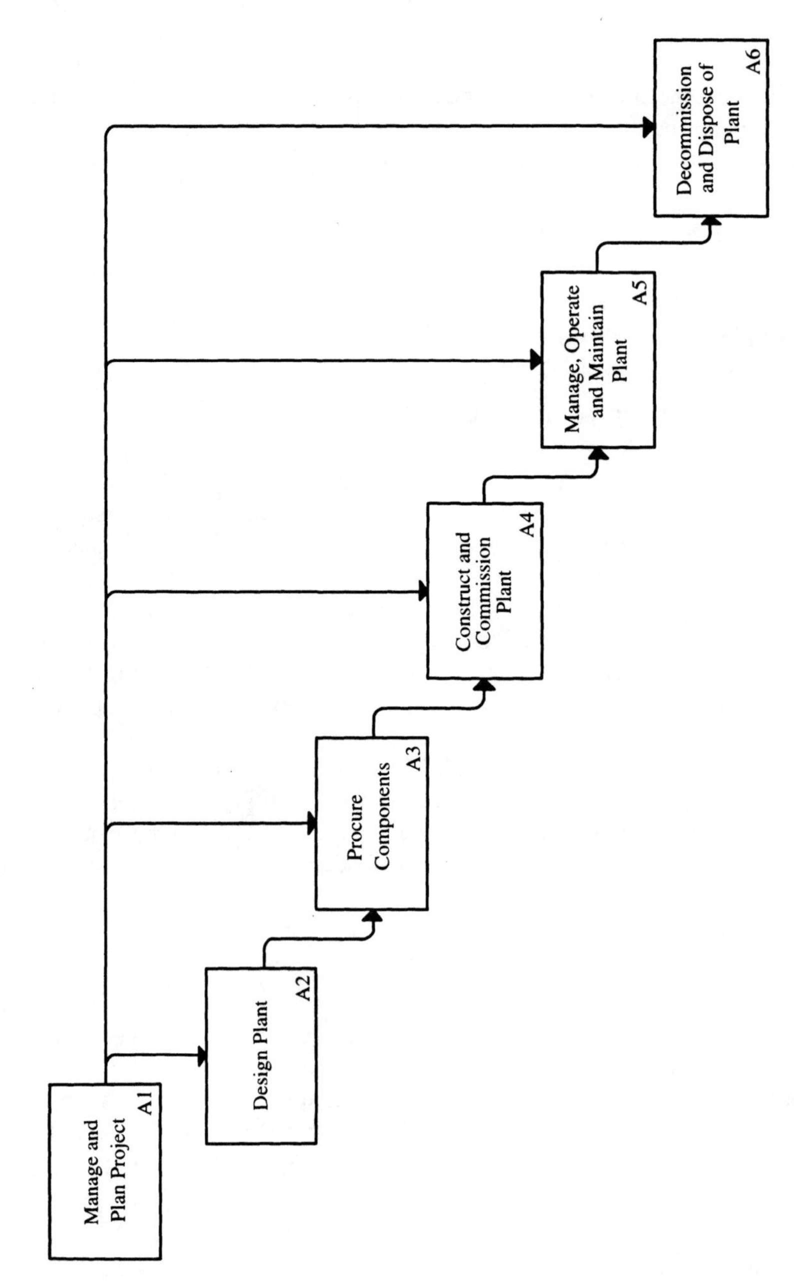


**Figure 2.2.** AP231 plant lifecycle IDEF0 diagram. (Reprinted by permission. Copyright © 1997, Automation and Control Technologies, Inc.)

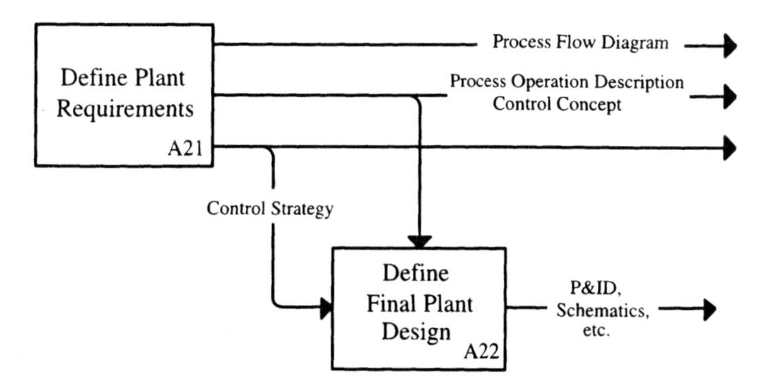


**FIGURE 2.3.** AP231 Design Plant activities. (Reprinted by permission. Copyright © 1997, Automation and Control Technologies, Inc.)

subdivides the AP231 Define Final Plant Design activity into three subactivities. This difference is interesting since it reflects the way the two groups see each other. Process designers tend to minimize the A&E effort on a project. The STEP effort has brought each of the roles in better focus for the other side to consider. It has the potential to improve cooperation and lead to more successful projects in the future while improving data exchange on capital projects.

The information flow depicted in Figure 2.2 for the AP231 activity model is limited to control engineering information packages. There are many information packages dealing with process design data that do not directly apply to the design of control and information technology. These are hidden to simplify and focus the discussion of activity modeling.

The control information package is collectively known as the initial control requirement definition. Eventually, the initial control requirement definition is detailed in subsequent design activities. This process adds increasing detail until it is ready to be procured, installed, and operated. The subsequent activities capture the detail on P&IDs and supporting documents.

The AP231 group developed a key design requirement for control technology, the control objectives. The control objectives are defined for both continuous and discrete control schemes. Furthermore, AP231 states that a control scheme may have different control objectives as a function of the operating states of a plant system, recognizing that plantwide control is a front-end design consideration. AP231 supports the idea of operating states and operating modules as introduced in Chapter 1. As defined by AP231, the control objectives are the basis to define the control requirements of the operating modules.

The process operation description (POD) is the design information package that defines the operating strategy and control objectives of the plant systems. It serves as the basis for implementing the control technology and the operating procedures for the manual activities on a project.

A companion to the POD is the Control Concept. This information package defines the extent of automation for the plant system. It is the statement of the control requirements needed to meet production objectives set for the capital project. For plantwide control, the Control Concept section is structured by the operating module definition in the POD. The reference model in Chapter 3 is a detailed statement of this structure used to establish the extent of automation for each operating module and to express the control requirements.

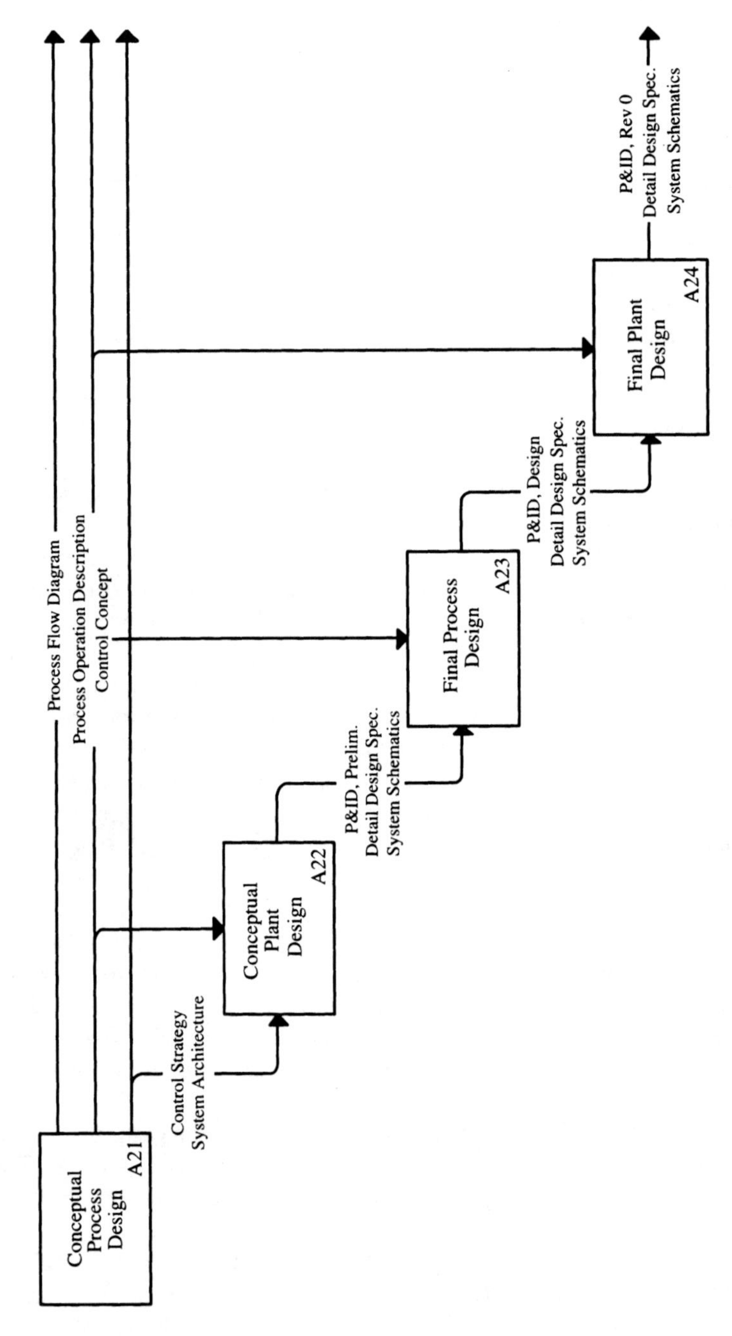


**FIGURE 2.4.** AP221/231 Design Plant activity model. (Reprinted by permission. Copyright © 1997, Automation and Control Technologies, Inc.)

AP231 also defines the data structure for basic control strategies, limited to discrete and regulatory control. It corresponds well with the object model described in Chapter 3. The AP231 control strategy data model defines the control strategy as part of the process description for the plant system. The control strategy is composed of continuous- and discrete-control schemes. The control schemes are composed of descriptive information that identifies the control type corresponding to the control types defined in Chapter 1.

The continuous-control schemes are implemented by control loops that are associated with process stream adjustments such as flows, temperatures, and concentrations. The discrete-control schemes are implemented by control logic associated with process stream events such as hazardous situations. AP231 defines the control strategy as the main input to the detail design activity.

In Figure 2.4, the AP221 and AP231 design plant activities are combined to represent the industry view of the information flow on capital projects. The detail design of plantwide systems starts with an initial control strategy defined in the process design. It is usually complemented by an initial system architecture. The initial control strategy and system architecture form the design basis for the capital project. This basis may change many times during the lifecycle of the control system.

Piping and instrument diagrams, detail specifications, and other schematics represent the detail design. In general, the control design information proceeds from the abstract expression of requirements to the concrete specification of the required control technology. The control technology specifications are used to procure the control technology hardware and program the application software.

The plantwide control portion of the STEP activity and data models is weak and limited to basic discrete and regulatory control types. This book presents a more comprehensive definition of the design information for plantwide control strategies and system architecture to augment the STEP definitions.

The capital project lifecycle used in this book, based on the AP231 and AP221 models with extensions for plantwide models, is outlined below.

**AP231: Design Plant**

- Preliminary design (define plant requirements)
- Detailed design (conceptual plant design, final process design, final plant design)

**AP231: Procure Components and Commission Plant**

- Implementation
- Installation
- Commissioning
- First production startup

**AP231: Manage, Operate, and Maintain Plant**

- System turnover

As defined, the AP231 Design Plant activity is subdivided into preliminary and detailed design. The preliminary design is assumed to be part of the AP231 Define Plant Requirements activity. The detailed design activity is performed concurrently with the

AP221 activities. This book is focused primarily on these two activities for they are the main activities of the control engineer on a capital project.

### 2.3.1 Preliminary Engineering

The preliminary engineering activity produces the POD, the Control Concept, and the initial control strategy and system architecture. These documents are collectively called the Control Requirements Definition (CRD). Ideally, the preliminary engineering is performed as part of the conceptual process design. Process experts are utilized most efficiently at this phase of a project.

The process engineering group conducts the conceptual process design. They are responsible for designing the process technology. Traditionally, the process design is the basis for the project scope throughout the lifecycle of the project.

As depicted in Figure 2.5, the process engineer's front-end activities are summarized as follows:

- Prepare the conceptual process design.
- Prepare the initial process technology budget.

The conceptual process design produces the design specifications for the physical plant equipment. It also includes data about any chemical unit operations and material/energy flow properties. The latter are specified at various operating points in the process for the various steady-state operating conditions.

The process design is usually represented by a process flow diagram (PFD) with an associated equipment item list for the major plant systems. The process technology capital budget is determined from this information. This is usually the first go/no-go decision in a project. It defines the financial constraints for executing a capital project.

Traditionally, the control technology for plantwide control is not part of the conceptual process design package. Traditionally, plantwide control is considered in the final plant design phase. This late entry usually means the budget for plantwide control is too low to realize any benefits. The design is already "cast in CAD" (computer-aided design), and it is too late to raise the level of control to accommodate plantwide schemes. Taking plantwide control into consideration in the early design phase forces the team to consider "design for operability," which can significantly reduce the lifecycle costs of a process plant.

It is significant, therefore, that AP231 acknowledges the need to begin control technology design earlier in the project. This acknowledgment alone is significant in the recognition of plantwide control as an essential technology in plant operations.

For the control aspects of a conceptual process design, the preliminary engineering activities for the project (Figure 2.5), in order, are as follows:

- Develop the POD.
- Develop the Control Concept.
- Define the preliminary automation and control strategy.
- Define the preliminary system architecture.
- Prepare the preliminary control technology budget.

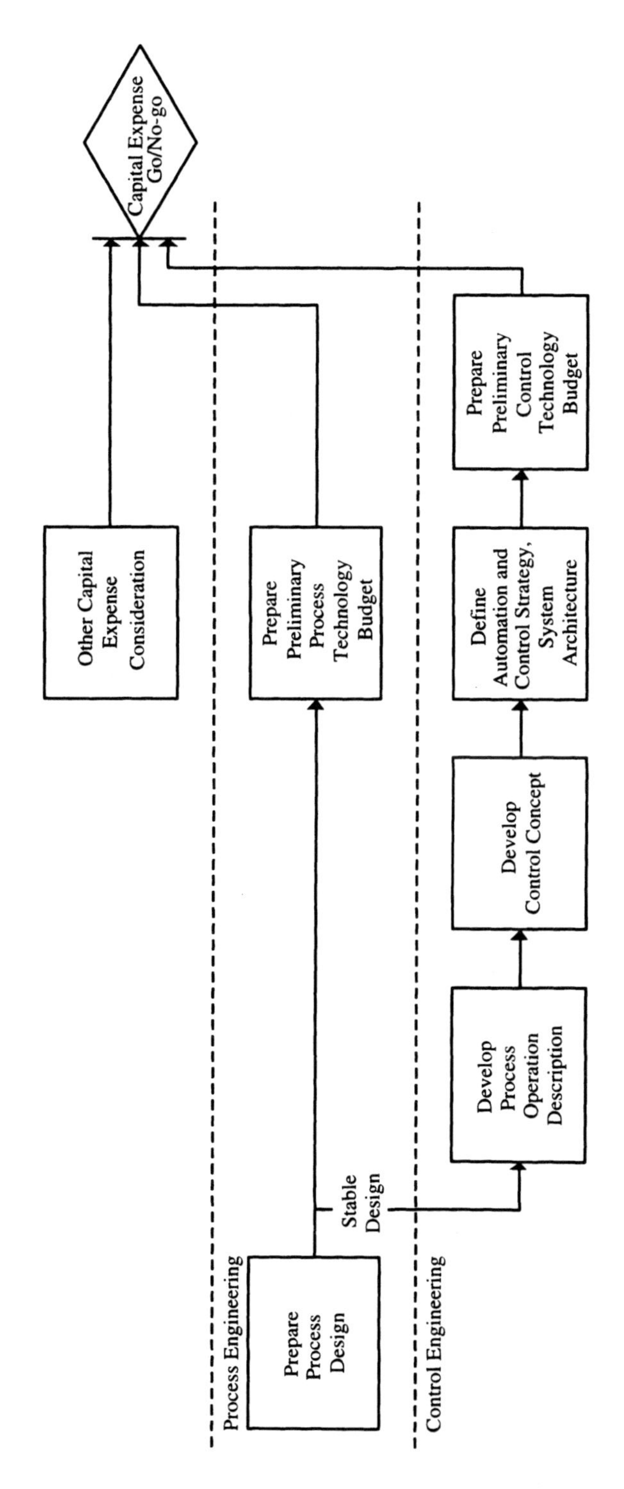


**FIGURE 2.5.** Preliminary engineering activities. (Reprinted by permission. Copyright © 1997, Automation and Control Technologies, Inc.)

During preliminary engineering, the primary objective is to first understand how to operate the process—implementation neutral—and then define the control requirements to support the operations.

The preliminary engineering design is an iterative process. Several iterations are required at various review levels before the requirements for the control system emerge. This process ensures that the control design is consistent with the operating philosophy of the plant. Only after the control requirements are well understood should the control strategy and system architecture be attempted.

The first step is to develop a POD. The POD defines the operating strategy for the proposed plant design. As such, operating personnel are key participants in its development. The POD includes a description of the operating modules and defines the boundaries of the process areas, cells, and units to be operated. It also defines the operating states and the activities to be performed for each. If the activities contain steps, the steps are listed, and their sequence is indicated. The control objectives are identified as part of the operating state description. The POD is implementation neutral; that is, all operating activities and criteria are described whether or not they will be automated.

In preparing a POD, the typical questions to be addressed include:

- What are the basic steps or divisions within the process operation? How is each operated?
- What are the significant process conditions and/or events that impact quality assurance, production, or protection of equipment, people, and the environment?
- What are the routine, special, and emergency response operations in each operating state?
- What types of process information must be collected and saved for performance monitoring (reports, data archiving, etc.)?
- What data must be supplied to other systems, including business data systems?

The next step after developing the POD is to decide on the amount of control to be implemented, captured as the Control Concept. The Control Concept states the requirements to be met by the control and information technology. It determines the portions of the POD that are to be automated. This statement defines the extent of automation required to operate the plant and achieve enterprise objectives. The POD and Control Concept statements are complementary to the PFD developed by process engineering.

The Control Concept is the basis for the control system design. It defines the extent of automation in terms of control requirements for the operating modules. The Control Concept is also used to verify the delivered system during system testing and commissioning. It should be kept technology neutral (e.g., a DCS or PLC) and especially vendor neutral. In other words, it should be prepared without embedded implementation.

In preparing the Control Concept, the typical questions to be considered include:

- What operating module activities and control objectives will be manual and what should be automated (i.e., the extent of automation)?
- How flexible must the automation be (e.g., multiple products, frequent equipment changes)?
- How do control activities interact (e.g., concurrent operation, shared resources)?
- How do control activities and operating personnel interact (e.g., data presentation, equipment selections, control commands, data entry)?

Within the Control Concept, control activities are the control types that will be used to automate the process operations to the extent required. They effect changes in the process as commanded by the operator and include loops, sequential control, interlocks, and data collection.

The level of abstraction for the control types should be the same as that for the PFDs. A natural tendency arises to consider details such as the control logic of individual valves, blowers, pumps, and agitators. While important details are often unearthed during the "discovery" process, the details should be kept to a minimum. It is proper to capture them on the side for later use. However, the overall design process is less efficient if too much detail is covered too soon (particularly if the process design is still unstable). Control details should be left for consideration during the detailed design activity.

After the control requirements are understood and documented, the project team can develop the preliminary control strategy and system architecture. The control strategy defines the control types and interrelations to fulfill the control requirements. The system architecture defines the hardware and software platform that will be required to execute the control strategies.

The Control Concept and control strategy are developed using the reference model described in Chapter 3. Operating modules as defined by the POD help segment these documents in order to relate them to the process design. The reference model provides the framework to ensure there is a consistent evolution from initial design through commissioning. The value of the right frame of reference is critical in a successful control engineering project.

A preliminary CRD package containing a first pass at the POD and the Control Concept should be produced as soon as practical. The level of detail is obviously limited to the information available at the time. Certain items, for instance, might simply be noted as TBD (to be determined).

Likewise, the initial control strategy and system architecture, based on the Control Concept, would be helpful in reviewing various Control Concept candidates early in the project. For both the control strategy and system architecture, the control algorithms and hardware/software requirements are very high level or conceptual. This philosophy provides flexibility in the budgeting activity.

The CRD sections are valuable inputs in developing the preliminary P&ID in the detail design effort. Later, as the process design becomes stable and the detail design is finalized, the CRD package is updated to produce a final version. All TBD references, for instance, should be resolved. This final version becomes the final project control documentation, an important project deliverable.

The CRD documentation is only the beginning of the control design package. It is used as a design basis for preparing the other follow-on project deliverables. For example, in the detailed design phase, the control strategy must be further described and detailed prior to beginning implementation. Similarly, the control system schematics may also be influenced by items mentioned in the documents.

A CRD has many benefits, both direct and indirect. The primary benefit is the early involvement of those persons most familiar with the plant's operation in developing the control system requirements. This method forces the participants to contemplate and describe the exact process area operation. Many issues relating to basic operations will surface *prior* to the selection of the control system equipment.

There are also several secondary benefits from the process. These include:

- Identification of the extent of automation early in a project, which provides a better baseline for budget control
- Initial definition of alarm management requirements to avoid unnecessary alarms in inactive operating states
- Initial definition of operator interface span of vision and control

A preliminary engineering effort is a valuable activity on a capital project when plantwide control is indicated.

### 2.3.2 Detailed Engineering

The detailed design begins after project approval, either near or at the end of the preliminary engineering activity. This phase requires many times the effort spent in preliminary engineering. Usually, some detail design is considered when developing the Control Concept in the preliminary engineering phase. It is impractical to develop the requirements without a reality check on whether or not the requirements can be implemented.

In the early days of control engineering when control type selection was limited to basic discrete and regulatory control, it was easy to say, "No, you cannot do that." With modern computer control capabilities it is virtually impossible to say "No." The only constraint is the control objective measurement. If the conditions to meet the control objective cannot be measured or calculated reliably, it cannot be automated. The control engineer may have to settle for control strategies that provide indications and statuses to the operator. In some cases, this strategy is perfectly adequate and desirable.

Detail design adds sufficient detail to the initial control strategy and system architecture to provide a firm basis for implementation. It is important that the detail specifications fulfill the control requirements as defined by the Control Concept. Detail design also includes defining the system test, education, and documentation requirements for the system.

The detail design activity develops the following deliverables:

- Measurement and final control functional specifications
- Safety instrumented system functional specifications
- Discrete and regulatory control functional specifications
- Procedural control functional specifications
- Process information data models
- Run management functional specifications
- User interface functional specifications

These specifications include the control technology specifications for the system hardware, embedded software, and application software programming. These specifications should address the size, performance, reliability, and environmental conditions to be met by the technology in this application.

Many projects must also address the process and control technology considerations of package process units. These units are preengineered packages that supply both the process and control technology for a plant system. These systems should be required to meet the same level of detail design specifications as the project-engineered systems. Of primary importance to the project control engineering team is how these systems are interfaced with the rest of the control technology.

In plantwide control systems, another interface consideration is the sharing of data between the control system and the Manufacturing Execution System (MES). The MES includes systems such as:

- Laboratory information management systems
- Order processing
- Inventory management
- Purchasing
- Shipping/receiving

A detail design is represented by several design information packages. These individual packages tend to be highly interrelated. The types of design documentation include:

- P&ID
- Loop diagrams
- Electrical schematics
- Panel layouts
- Sequence control description
- Graphical user interface (GUI) descriptions
- Implementation standards

Traditionally, the P&ID contains the functional specification of the plant controls. Today, control strategies are too complex and their representation on P&IDs is hard to read and, more importantly, hard to understand. The P&ID today can still represent the measurement and final control systems, but many projects stop here and use other representations for the higher level controls. Tools that include computer spreadsheets and databases capture these representations.

The measurement and final control systems are still largely distinct components that must be wired to the controllers. Loop diagrams and electrical schematics provide the interconnect details for implementation and installation.

If procedural control automation is required, a sequence description is prepared. This sequence description provides the instructions for programming the unit controllers to provide the required level of automation.

The user interface is defined by panel layout drawings if field operation is required. In current computer-based systems, a description of the GUI is prepared. The GUI description is a critical but often overlooked design statement. The user interface is the most important aspect of a control system design. The best control strategy will fail if the user interface is not utilized.

The emphasis in contemporary computer-based systems is to minimize "re-inventing" the wheel. In other words, if there is a design that has proven itself, do not re-engineer it. More importantly, it is important to develop a system of capturing reusable strategies that can be shared in an enterprise. The computer plays a valuable role in achieving this goal.

### 2.3.3 Implementation

Implementation is the capital project activity that includes the procurement and construction activities of the STEP activity model. In the case of the control system, it involves the procurement of the measurement and final control hardware and the control system hardware with its embedded software. It also includes the configuration and programming of the application software according to the detail design. This activity is complete after the system is verified in system testing.

Sometimes, the functions described in the detail design specifications require an implementation design. This design is necessary when the specific control type has not been previously implemented or a new type of system was procured for this application, which is often the case with application software.

The implementation of the system architecture requires hardware considerations such as power, grounding, wiring, and protection. The degree of concern is a function of the actual site environment of a process plant.

Software implementation considerations include the proper use of techniques and programming languages to convert the detailed logic diagrams into application software. There should be an implementation guideline to establish the proper methods of programming the control algorithms.

The use of staged system configurations is gaining wide acceptance. A staged system includes a process-side simulation to facilitate system logic and performance testing. It is usually located out of the way of actual site preparation and construction. Plant operators are often brought to the staged system to assist in the testing.

The final step in the implementation activity is to verify the correct operation of the system. System verification compares the system operation to the Control Concept to ensure that the required level of automation has been produced. It begins with module testing followed by preliminary system testing by the implementation team. The final test to verify the system functionality and performance is witnessed by the customer.

### 2.3.4 Installation

Installation is the activity that situates and powers the system at the plant site. This activity includes the interconnection wiring and system interconnection check-out. Each measurement and final control device is checked. The computer control system is set up and made operational to help in this effort. It provides a very powerful visual aid and logging system.

After measurement and final control system wring-out, the system hardware and embedded software performance is checked to make sure communication over the network is acceptable. Once system operation is checked and the installation is verified, the system is ready for commissioning.

### 2.3.5 Commissioning

This activity is usually done in conjunction with verifying the process technology. The combination is commonly called water testing or first chemical testing. As each plant system is brought on-line, the system is checked to make sure it works as required to support the operation of the process technology. The initial control settings are established and the system is readied for the first production run. At this point, the system is ready to make product.

### 2.3.6 First Production Startup and Turnover

The control engineering team has demonstrated to operating personnel that the control requirements have been met and the system is ready to be used. It is important to stay with the system as the first production begins. The transition is critical to the successful completion of the project. After the operating personnel take ownership, the project can be declared finished. The only activity left is to make sure the design information package is updated to an as-built state and turned over to production.

### 2.3.7 Training

During the latter stages of the implementation activity, training of the plant personnel that must use the system should start. Training activities proceed from this point to the end of the project and may extend beyond the end of the project as new personnel join the organization. Engineers, operators, supervisors, management, and maintenance personnel all need to be trained to use the new system. Of course, the type of training differs according to the position and duties of the personnel. The staged system often plays an important role in training operators since an emergency situation can be simulated on the staged system without fear of actual equipment damage or personnel injury if the operator makes a mistake.

## 2.4 PROJECT TEAM CONSIDERATIONS

An important consideration in conducting any capital project is assembling the right team members and assigning the responsibilities of the project team. Along with the appropriate disciplines, an important step is assigning clear responsibilities to each team member. The responsibilities of the team can be defined by using a roles-and-responsibilities practice and captured on RACI charts. This practice addresses the questions of "*Who* is responsible to do *what*?" and "*Who* can approve the *outcome*?" It also addresses who should review and be made aware of project information. This is especially important in the execution of the various control engineering tasks, many of which are shared among different project groups or disciplines.

The project team roles defined by the RACI chart are as follows:

- **Responsible (R)** The person (or persons) who carries out all or part of the task. There should be at least one "R" for each task.
- **Accountable (A)** The person who ensures that the task is done. There should be only one "A" for each task.
- **Consult (C)** A person (or persons) who must be consulted before the task is performed and should be informed afterward. This person's input may influence how the task is performed.
- **Inform (I)** A person (or persons) who must be informed after the task has been completed.

Some major tasks may need to be further subdivided prior to assigning the above roles. For example, the POD describes how the process area is operated. A process engineer familiar with how to operate the plant may lead this effort. The better lead-role person for the Control Concept development is usually the control engineer.

The assignment of these roles and responsibilities will vary from project to project and is not always obvious. The project team should negotiate a formal (or informal) RACI chart for each major task or activity. A simple matrix format is usually sufficient: Each task is described by a matrix row and each individual or organization is defined within the row by matrix columns labeled R, A, C, and I, respectively.

## REFERENCES

International Standards Organization, *ISO 10303: Industrial Automation Systems and Integration, Product Data Representation and Exchange—Part 227: Plant Spatial Data* (also known as AP227), working project draft, December 9, Geneva, Switzerland, 1994.

International Standards Organization, *ISO 10303: Industrial Automation Systems and Integration, Product Data Representation and Exchange—Part 221: Functional Data and Their Schematic Representation for Process Plant* (also known as AP221), working project draft, February 23, Geneva, Switzerland, 1997a.

International Standards Organization, *ISO 10303: Industrial Automation Systems and Integration, Product Data Representation and Exchange—Part 231: Application Protocol: Process Engineering Data: Process Design and Process Specification of Major Equipment* (also known as AP231), committee draft, July 17, Geneva, Switzerland, 1997b.

National Institute of Standards and Technology, *Integration Definition for Function Modeling (IDEF0)*, Federal Information Processing Standards Publication 183, December 21, Gaithersburg, MD, 1993.

# 3 A Practitioner's Model for Automation and Control

**Chapter Topics**

- Models for automation and control: process, physical, control
- Operating state
- User interface
- Control system technology
- Example models

*Scenario:* An understanding of the overall automation system is important to the success of an automation capital project.

There is the poem of six blind men from Indostan and their encounter with an elephant (Saxe, 1936, pp. 111–112):

*It was six men of Indostan*
*To learning much inclined,*
*Who went to see the Elephant*
*(Though all of them were blind)*
*That each by observation*
*Might satisfy his mind.*

*The first (touching its side) "Is nothing but a wall!"*
*The second (feeling its tusk) "Is very like a spear!"*
*The third (taking its trunk) "Is very like a snake!"*
*The fourth (feeling its knee) "Is very like a tree!"*
*The fifth (touching its ear) "Is very like a fan!"*
*The sixth (seizing its tail) "Is very like a rope!"*

*And so these men of Indostan*
*Disputed loud and long,*
*Each in his own opinion*
*Exceeding stiff and strong,*
*Though each was partly in the right,*
*And all were in the wrong!*

Automation and control systems, in many ways, are like the elephant. An individual usually only grasps one aspect of an entire automation system. One person may be intimately familiar with various pieces of equipment, their idiosyncrasies, and failure modes. Another understands the product produced by the automated system, the expected features and performance, and what to tweak to change the product. A third person understands the individual unit operations and the theory of operation. A fourth person understands the operational viewpoint. None of these views explains the entire picture of the overall plant operation. The purpose of this chapter is to present a framework, or model, that can encompass each of these views and yet present an overall picture of automation. After all, none of us would want to be embarrassed by insisting that the elephant is just a tree!

However, such a thing happened to our young control engineer. Our young engineer was directed to study the feasibility of a new advanced control algorithm. He enthusiastically worked with another engineer and the two of them proceeded to "rip out" the PID from a standard controller, replacing it with this new algorithm. They also developed an operator interface for a personal computer that would display the new process information this controller generated. Initial tests had shown great promise for this controller. However, field tests of this new controller were not so enthusiastically received. Some operational aspects of this controller came to light and the operational personnel at the plants where this controller was tried were uncomfortable with some aspects of the controller. Consequently, this controller never became a product. At the end of the project, the young engineer realized that he had not understood the entire picture and how this controller fit in. He was embarrassed because he was focusing on the algorithm (the tree) and not on other aspects of an overall automation system.

## 3.1 INTRODUCTION

As illustrated by the previous scenarios, it is beneficial to place a framework around the automation and control activities. This model, called an enterprise plant model, is used to assist in segmenting the automation system requirements into manageable pieces. This chapter defines the enterprise plant model and terminology for defining the control requirements of a plant. The S88.01 (batch control) international standard (ISA, 1995) is used as a starting point and expanded to include continuous manufacturing systems:

1. This model applies to continuous, batch, or discrete-parts manufacturing processes.
2. It emphasizes good practices for the design and operation of the plant.
3. It can be applied regardless of the extent of automation.

A process is a sequence of chemical, physical, or biological activities for the conversion, transport, or storage of material or energy (ISA, 1995). Industrial manufacturing processes are generally classified as continuous, batch, or discrete-parts manufacturing. This classification stems from the way the output appears: as a continuous flow (continuous); in finite quantities of material (batch); or in finite quantities of parts (discrete-parts manufacturing).

*Continuous Process.* In a continuous process, material passes in a continuous stream through the processing equipment. Once the process has established a steady operating state, the nature of the process does not depend on the length of time the process is operating (ISA, 1995). Commodity chemical manufacturing typically falls into this category. A fluid catalytic cracking unit (FCCU) in a petroleum refinery, simplified in

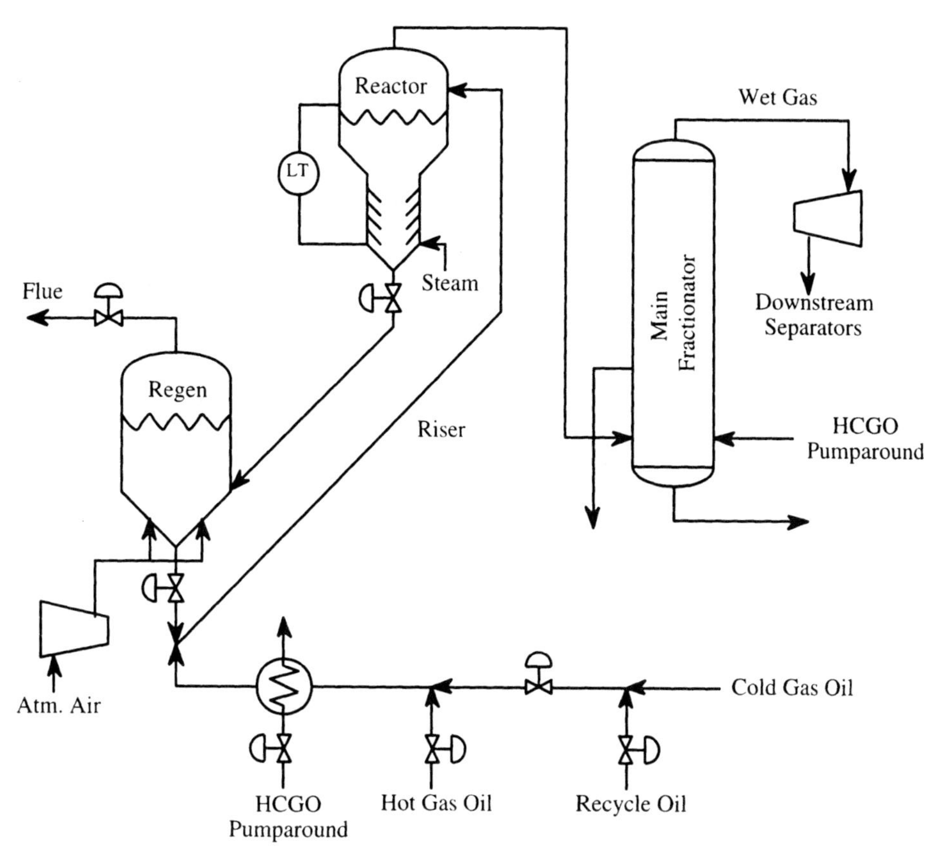


**FIGURE 3.1.** Fluid catalytic cracking unit. (Reprinted by permission of the AACC from *Proceedings of the 1992 American Control Conference*. Copyright © 1992, American Automatic Control Council.)

Figure 3.1, is an example continuous process (Grosdidier et al., 1992). A FCCU converts low-market-value petroleum chemicals into more profitable products, such as gasoline. Another example is a hot-strip-steel finishing mill, shown in Figure 3.2. The control system manipulates the force at each stand and the torque at each looper in order to control the tension between stands and the thickness of the steel at the outlet.

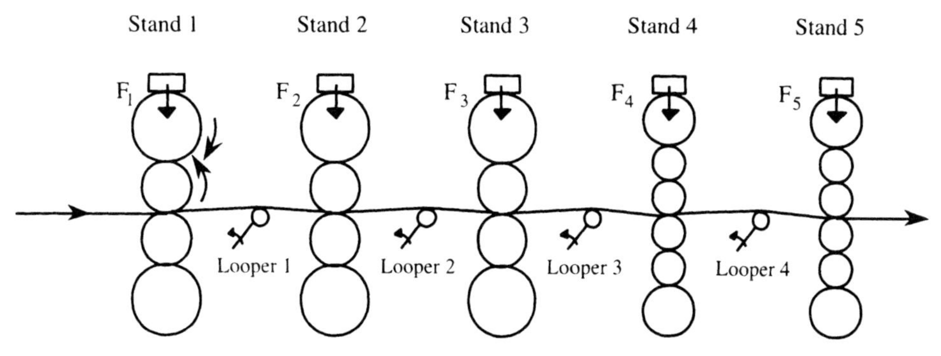


**FIGURE 3.2.** Steel-rolling mill.

*Batch Process.* In a batch process, finite quantities (batches) of material are produced by subjecting quantities of input materials to a defined order of processing actions using one or more pieces of equipment. Batch processes are discontinuous processes from a material flow standpoint. Batch processes are neither discrete nor continuous, though they have characteristics of both (ISA, 1995). Food, beverage, pharmaceutical, and specialty chemical processes are usually encompassed by this category. An example (small) batch process is illustrated in Figure 3.3. This process consists of a single tank reactor with two ingredients, and the product is emptied into another tank to be carried to another part of the plant. A more complicated example with several tanks and several types of products that can be manufactured is shown in Figure 3.4.

*Discrete-Parts Manufacturing Process.* In a discrete-parts manufacturing process, a specified quantity of material moves as a unit (part or group of parts) between workstations, and each unit maintains its unique identity (ISA, 1995). At a workstation, a unit may modified (drilled, machined, painted, etc.) or may be combined with one or more

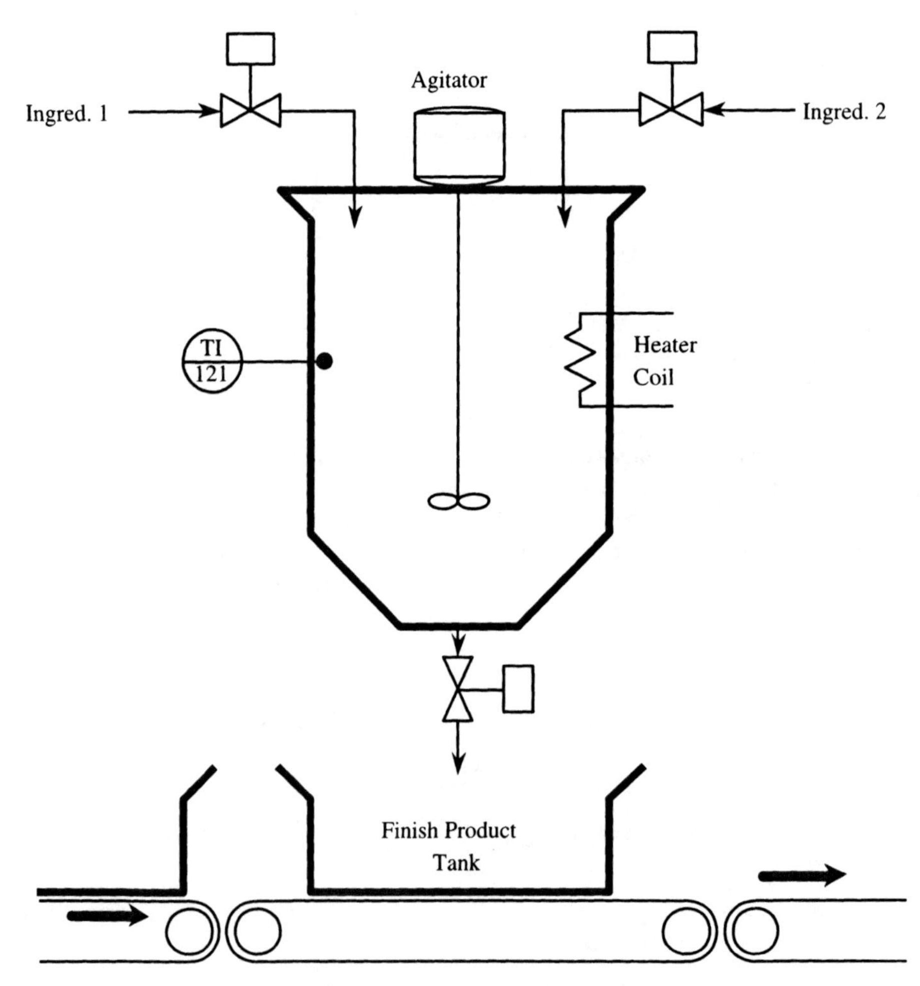


**FIGURE 3.3.** Simple batch process.

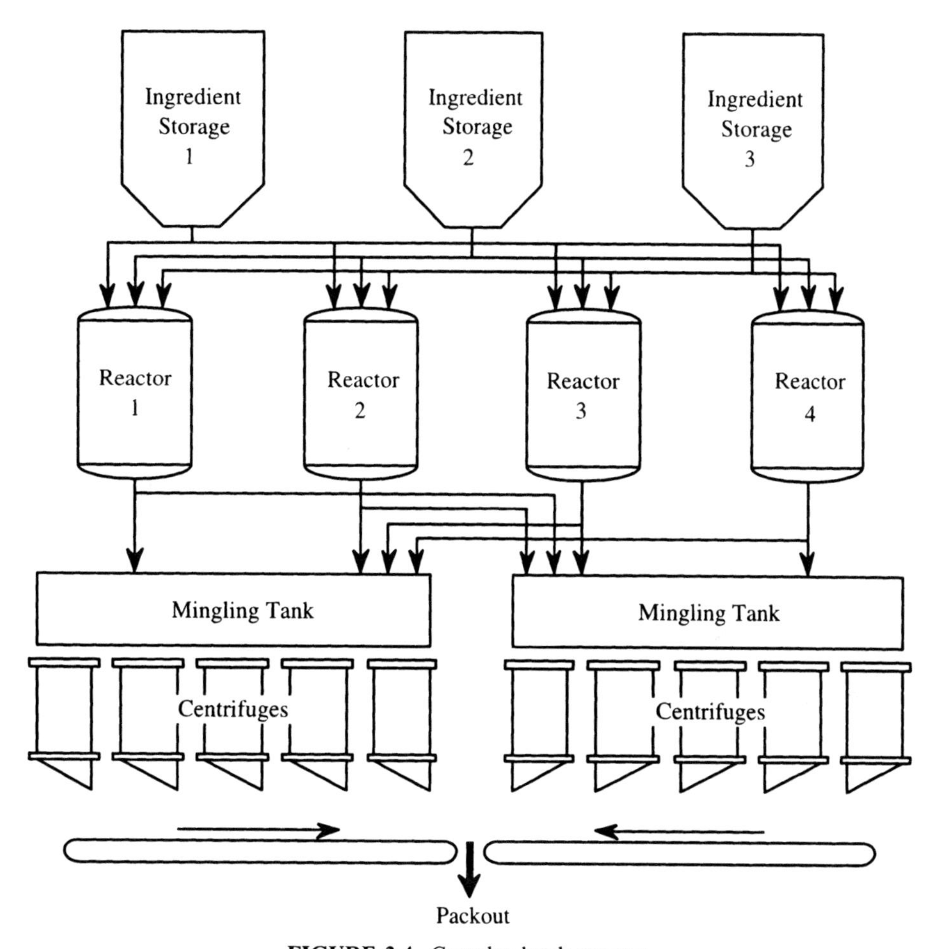


**FIGURE 3.4.** Complex batch process.

other parts (assembly). Packaging operations also fall into this category. An example discrete-parts workstation is shown in Figure 3.5. Incoming parts are received on the input conveyor, moved to the drilling station where a few holes are drilled, moved to the computer numerical control (CNC) station where the part is shaped, and then moved to the output conveyor. A simple packaging operation is shown in simplified form in Figure 3.6.

Viewed from a plantwide level, processes are often primarily one of the three main types, though they often contain at least one of the other types. For example, the output of many batch processes is packaged in containers smaller than a batch. The packaging operation is a discrete-parts operation. Even a discrete-parts manufacturing plant, for example, a small gasoline engine manufacturing facility, has waste recycling or treatment facilities that usually contain continuous or batch processes.

As explained in Chapter 1, a plant may be viewed from an equipment, product, or operations viewpoint. A practitioner's model of a process must incorporate all viewpoints. The S88.01 standard defines three different models of a process: a process model, a physical model, and a control model. These models and their relationship are presented and then extended to encompass non–batch control processes. Basically, the physical

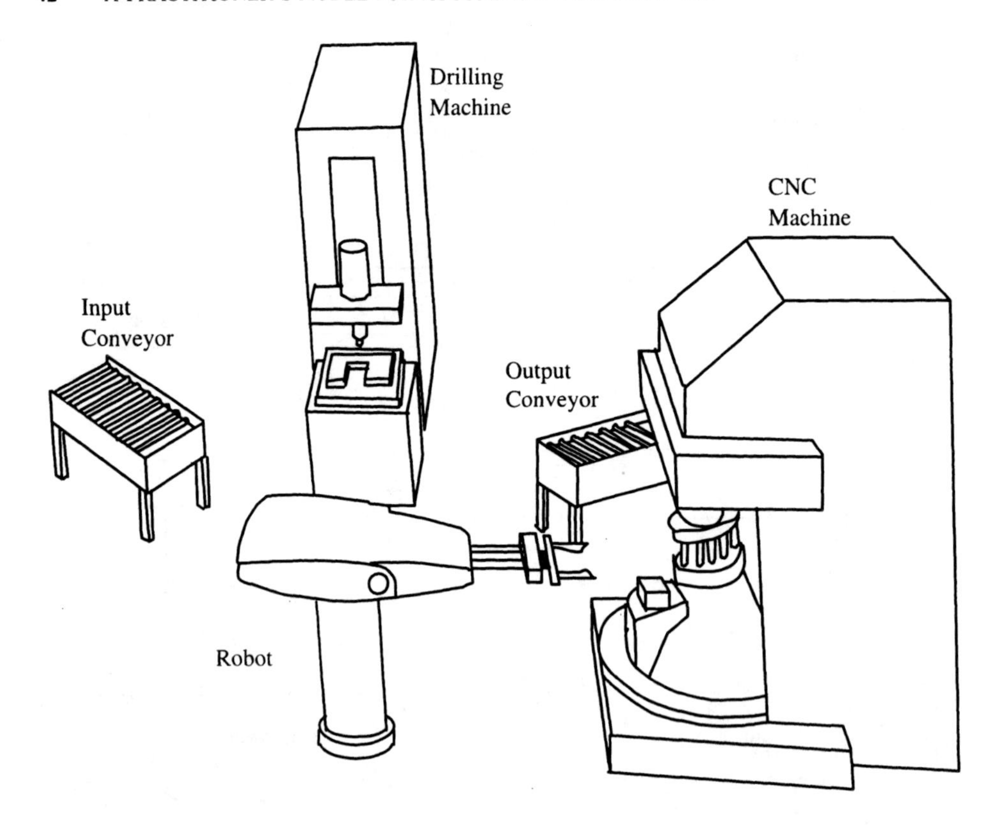


**FIGURE 3.5.** Discrete-parts manufacturing work cell. (Reprinted by permission. Copyright © 1996, Kelvin T. Erickson.)

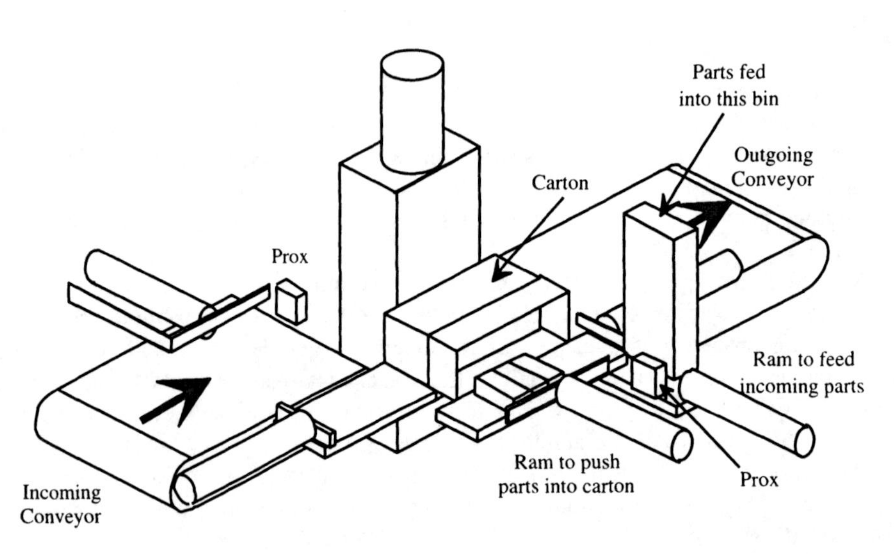


**FIGURE 3.6.** Example packaging operation. (Reprinted by permission. Copyright © 1996, Kelvin T. Erickson.)

model divides the physical assets of a plant into a series of hierarchical levels. Corresponding to each physical layer, there is a corresponding operation and control procedure. This overall model is explained and illustrated with examples. Next, the concept of a state is explained. Equipment entities have an operating state and control entities possess a control state. Following model development, aspects that affect all layers of the model are discussed: information management, user interface, and control system technology.

## 3.2 S88.01 MODELS

The S88.01 standard (ISA, 1995) defines three models: a process model, a physical model, and a procedure control model. All of these models are organized hierarchically. The process model is organized as shown in Figure 3.7 and describes the processing actions required to convert the raw materials into finished product. Starting from the top, the overall process is divided into stages, which are major processing activities needed to produce the finished goods. Stages are divided into operations, which are groupings of the minor processing activities.

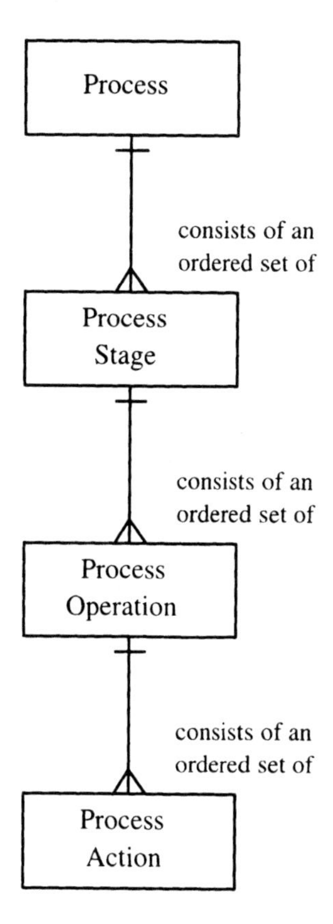


**FIGURE 3.7.** S88.01 process model. (Copyright © 1995, ISA, all rights reserved. Reprinted by permission from S88.01.)

The physical assets of an enterprise are organized in a hierarchical manner, as shown in Figure 3.8. Lower level groupings are combined to form higher levels in the hierarchy. In the case of the bottom two levels, a grouping in the level may be incorporated into another grouping at that same level. The S88.01 standard only addresses the lower four layers. The lower four layers refer to specific equipment types. An equipment type, shown in Figure 3.8, is a collection of physical processing and control equipment grouped for a specific purpose. During engineering activities on a project, the equipment at a lower level is grouped together to form a higher level equipment grouping. This grouping simplifies

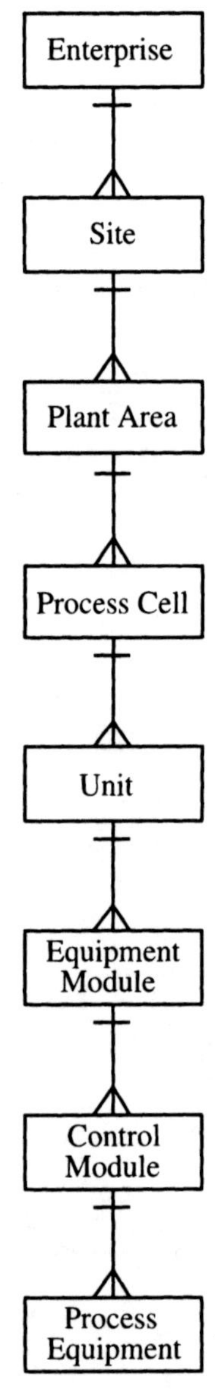


**FIGURE 3.8.** Hierarchical layers of S88.01 physical model. (Copyright © 1995, ISA, all rights reserved. Reprinted by permission with changes from S88.01.)

operation of that equipment and treats it as a single larger piece of equipment. Once created, the equipment cannot be split up except by reengineering the equipment at that level. An example hierarchical organization of a cell is shown later in Figure 3.16.

The S88.01 standard defines three types of control: basic control, procedural control, and coordination control. Basic control establishes and maintains a specific state of the equipment or process. It includes regulatory control, discrete control, sequential control, interlocking, monitoring, and exception handling. In S88.01, the procedural control model is hierarchical, as shown in Figure 3.9, and is definitely biased toward batch control. Coordination control initiates, directs, and/or modifies the execution of procedural control.

The physical and control models are important to the development of the Control Requirements Definition (CRD) document, as outlined in Chapter 2. The Process Operation Description (POD) sections of the CRD are organized according to the physical model. As expected, the Control Concept and control strategy sections of the CRD are organized according to the control model. The case study in Chapter 10 illustrates this organization.

After explaining the various parts of the process, physical, and control models, two example enterprises are used to illustrate these models. The first is a paper production

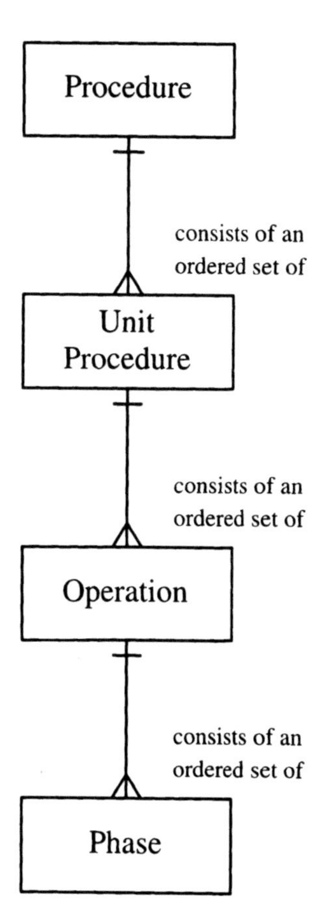


**FIGURE 3.9.** S88.01 procedural control model. (Copyright © 1995, ISA, all rights reserved. Reprinted by permission from S88.01.)

enterprise and is largely a continuous process. The second is a pharmaceutical enterprise and is largely a batch process. These two examples are not explored in detail, but enough information is given to illustrate the application of these process models.

### 3.2.1 Process Model

According to S88.01, a process consists of one or more process stages (Figure 3.7), which can be serial, parallel, or both: "A process stage is a part of a process that usually operates independently from other process stages. It usually results in a planned sequence of chemical or physical changes in the material being processed" (p. 20). The key concept for a process is that it transforms a material flow. For a continuous process, example stages are:

Reaction
Separation (e.g., distillation)
Chemical recovery

For a batch process, example stages are:

Reaction
Polymerization
Crystallization
Filtration
Drying
Packing-out

For discrete-parts manufacturing, example stages are:

Machining raw stock
Subassembly
Final assembly
Packaging

Each process stage consists of one or more process operations: "Process operations represent major processing activities. A process operation usually results in a chemical or physical change in the material being processed" (ISA, 1995, p. 20). Example operations for a *reaction* stage are:

Preparing reactor
Charging reactor
Reaction
Discharging reactor
Cleaning out reactor
Shutdown

Example operations for a distillation stage are:

Startup
Recycling
Process feed
Shutdown

Each process operation is further subdivided into one or more process actions. According to S88.01, "Process actions describe minor processing activities that are combined to make up a process operation" (ISA, 1995, p. 20). Example actions for the *prepare reactor* operation are:

Bring all valves to safe starting state
Initiate water jacket temperature control system
Heat reactor jacket to 40°C
Hold jacket temperature for 4 min

Example actions for the *startup* operation for a distillation stage are:

Fill bottoms with initial amount of material
Initiate bottoms heat exchanger temperature control
Wait for bottoms heat exchanger to reach minimum temperature
Initiate condenser temperature control
Wait for condenser temperature to reach maximum temperature
Increase bottoms temperature setpoint
Wait for column pressure to reach operating point

### 3.2.2 Enterprise Physical Model

The physical assets of an enterprise are usually grouped on a geographical basis, rather than in a strict hierarchical manner, as in S88.01. Also, in order to accommodate nonbatch processes, the lower two layers of the S88.01 model are only shown as parts of a unit. The physical model describes the physical assets of an enterprise in terms of enterprises, sites, areas, process cells, units, equipment modules, and control modules. The physical assets of an enterprise are organized geographically as shown in Figure 3.10. Hierarchically, they would be shown as in Figure 3.11. Practically, geographical divisions apply at the higher levels (enterprise, site, and area). Groupings at the lower levels (cell, unit, equipment module, and control module) are based on functionality.

***3.2.2.1 Enterprise.*** An enterprise coordinates the operation of one or more sites. It is responsible for determining what products will be manufactured and in general how they will be manufactured. An enterprise can be confined to one country or may be global in scope. A generic enterprise model showing layers from the enterprise level down to the process cell level is shown in Figure 3.12.

***3.2.2.2 Site.*** A site is a geographical grouping determined by the enterprise. All physical equipment in a site share common meteorological and geographic data. A site is typically

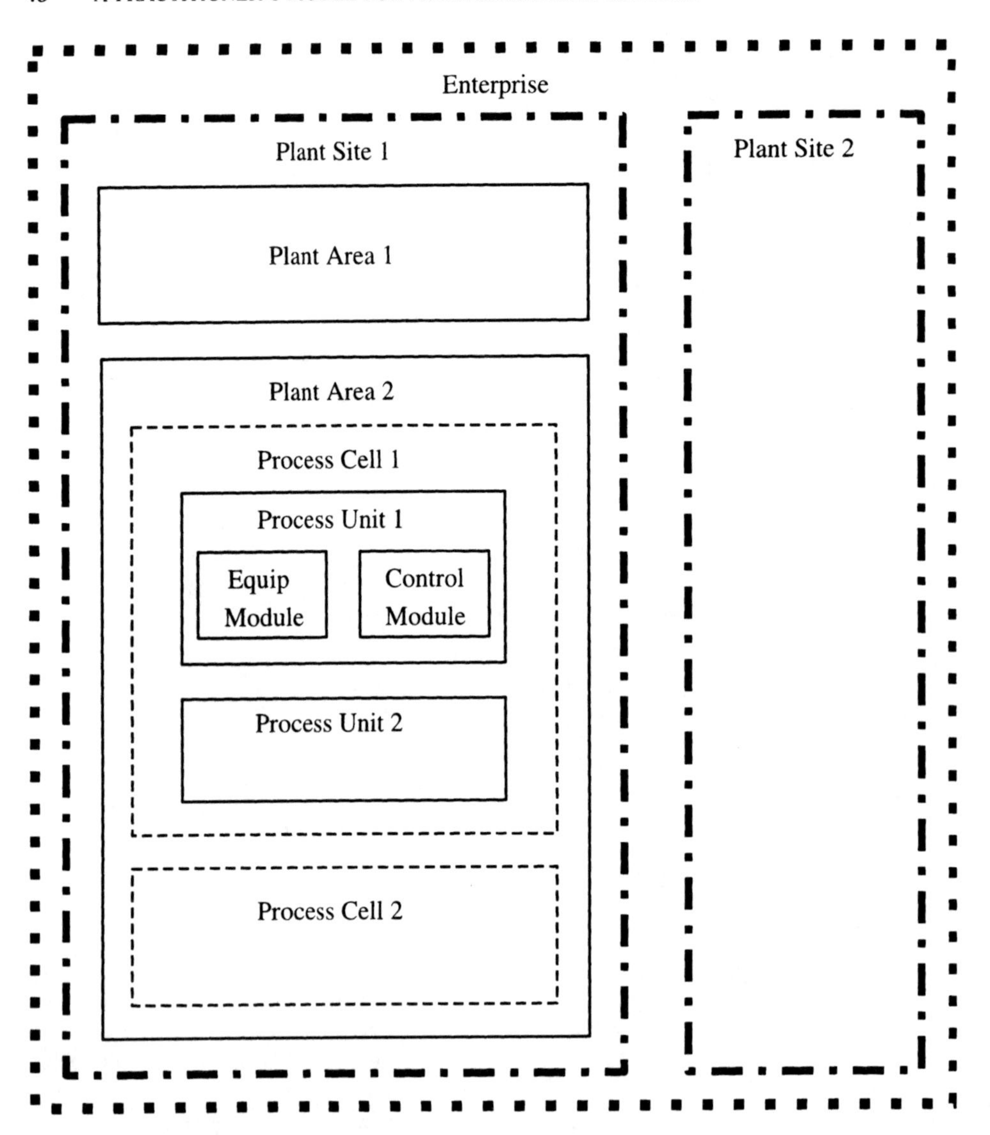


**FIGURE 3.10.** Geographical organization of physical model.

called a "plant," although enterprises sometimes divide a site into multiple "plants." Because of the confusion over this terminology, the word "plant" by itself is avoided in this model. A site consists of plant areas.

*3.2.2.3 Area.* An area is a geographical grouping smaller than a site. An area is generally a process or a utility, but it could be something else, for example, a warehouse. An area is a physical, logical grouping of process equipment that performs the desired cell operations (thermodynamic, chemical, biological) to make a product. Plant areas are often set by enterprise policy such as operator jurisdiction, product, or other criteria. The plant areas are sometimes called plant sections. A plant area is composed of process cells.

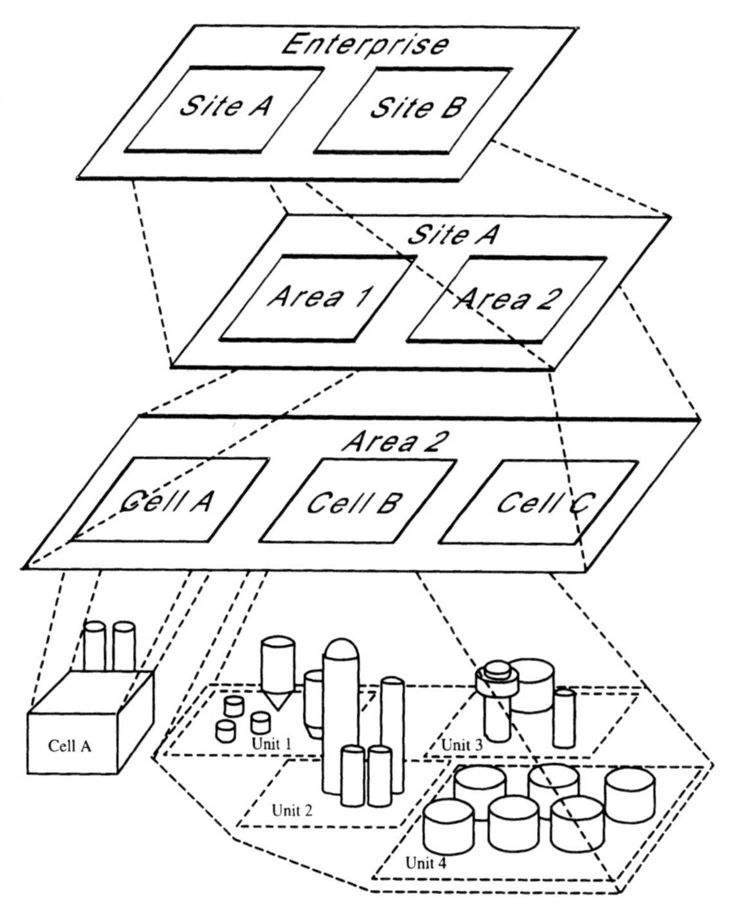


**FIGURE 3.11.** Hierarchical view of process.

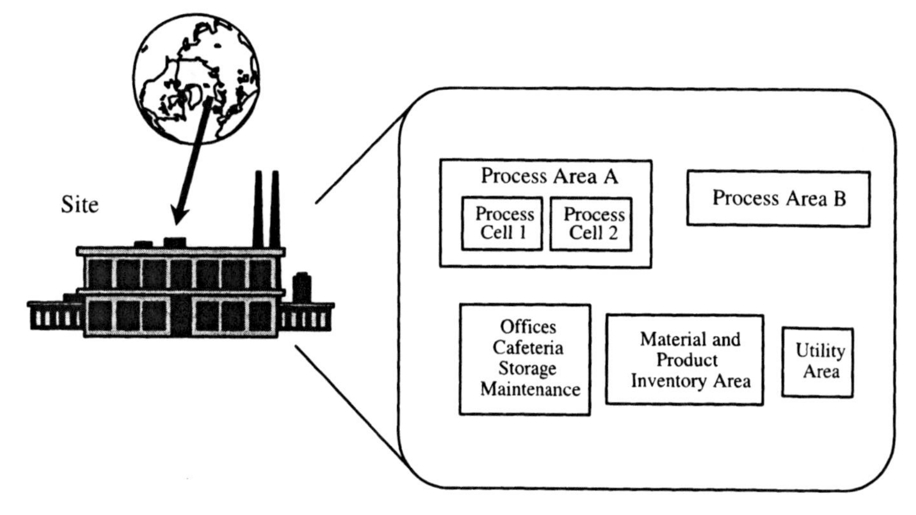


**FIGURE 3.12.** Enterprise model. (Reprinted by permission. Copyright © 1996, Automation and Control Technologies, Inc.)

***3.2.2.4 Process Cell.*** A process cell is a set of cooperating units. Typically, it is a logical grouping of equipment required to process one stream or manufacture one product or group of products. In a typical continuous process or discrete-parts manufacturing process, a cell makes one product or processes one stream. In a typical batch process, a cell produces more than one product. Sometimes, an area is composed of two or more cells that operate in parallel and concurrently in making the same or different products. In this case, each cell is referred to as a process train or a process line. In many batch processes, a cell is composed of a series of units. In this case, each unit acts on a batch of product and each unit is called a batch stage. In some batch processes, each batch product may follow a path through the units based on the selected recipe. Depending on the recipe, not all of the equipment in the area may be used to produce a batch.

An example area consisting of two cells is shown in Figure 3.13. In this example, one cell is batch and the other is continuous. Some other example process cells are:

Chemical recovery
Nylon fiber manufacture (batch or continuous)
Packaging
Corn starch production
Specialty chemical production

***3.2.2.5 Unit.*** A unit is a set of equipment modules and control modules; from an operating viewpoint, it is the smallest subdivision of the process plant equipment in an area or cell. A unit is usually centered on a major piece of equipment, such as a mixing tank, reactor, or distillation column. One or more major processing activities—such as reacting, separating, and making a solution—can be conducted in a unit. The unit combines all necessary physical processing and control equipment required to perform those activities.

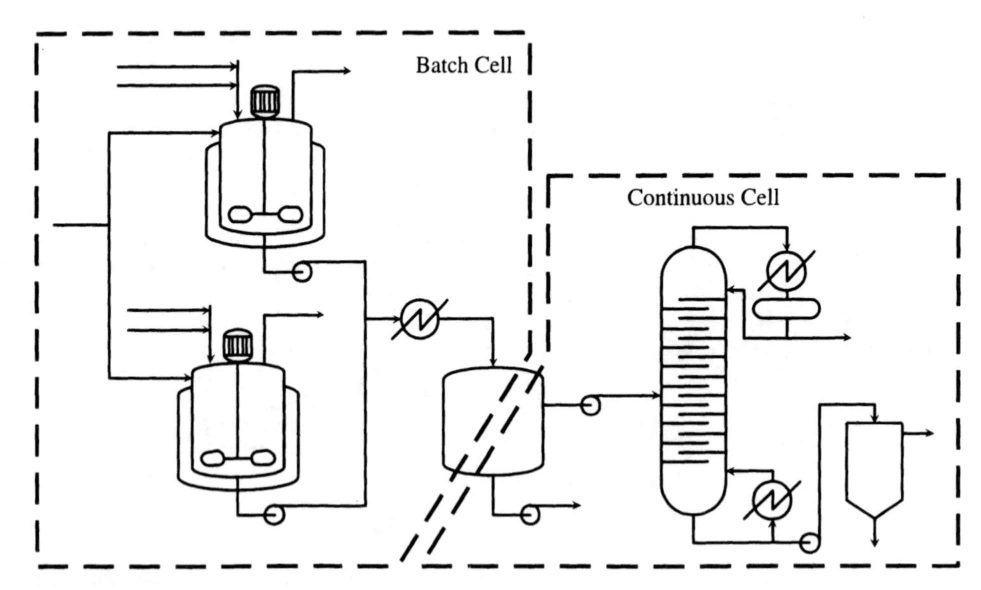


**FIGURE 3.13.** Plant area consisting of two cells. (Reprinted by permission. Copyright © 1996, Automation and Control Technologies, Inc.)

Units often operate relatively independently of each other. Units may also be shared by more than one cell.

As shown in Figure 3.14, the two cells of Figure 3.13 are further subdivided into units. The batch cell is divided into two batch reactors and the continuous cell is divided into separator and distillation column units. Note that the two reactor units share a heat exchanger and that the cells share a decanter unit.

Some examples of units are:

Distillation column
Continuously stirred tank reactor
Crystallizer
Evaporator
Fermentor
Decanter
Ion exchange column

***3.2.2.6 Equipment Module and Control Module.*** A unit consists of equipment and/or control modules. There is a clear distinction between an equipment module and a control module. An equipment module is associated with procedural (sequence or batch) control and a control module is not associated with procedural control. An equipment module is a plant item whose control scheme is a phase that coordinates other plant items whose function is discrete or regulatory control. In contrast, a control scheme for a control module is generally a discrete or regulating loop.

Equipment and control modules may be shared among units. For example, in Figure 3.14, the two batch reactors share a heat exchanger to cool the product before it enters the decanter.

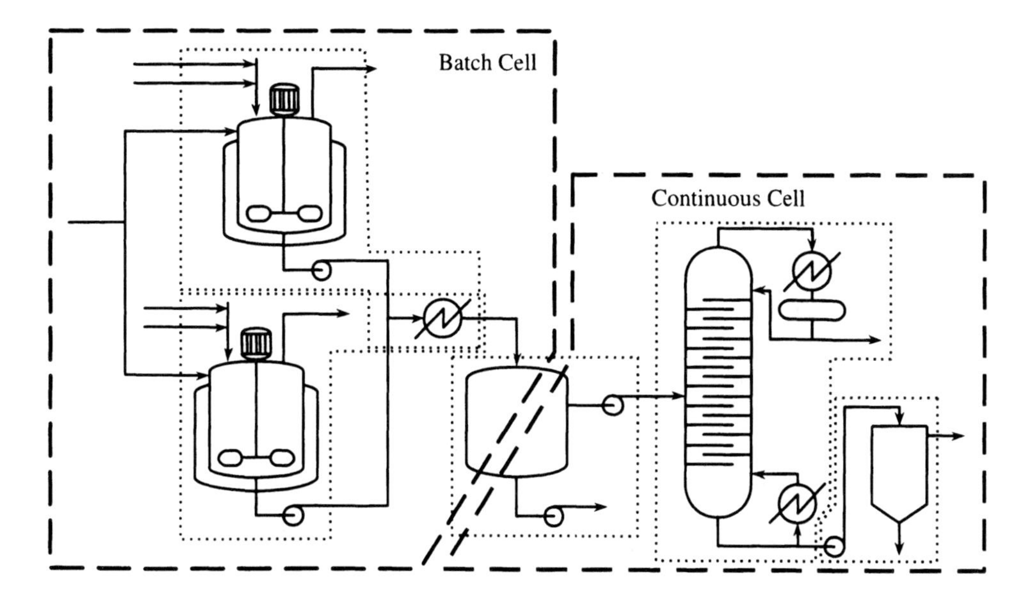


**FIGURE 3.14.** Division of process cell into units.

Physically, the equipment module may consist of subordinate equipment modules. An equipment module may be part of a unit or a stand-alone equipment grouping within a cell. A stand-alone equipment grouping can be an exclusive-use resource or a shared-use resource. An equipment module carries out a finite number of specific minor processing activities, such as weighing, mixing, and ratioing of feed streams. The equipment module contains all necessary physical processing and control equipment required to perform those activities. The scope of the equipment module is defined by the finite tasks it is designed to perform.

As an example of units containing equipment modules, consider the cell in Figure 3.15 that mixes two feedstock materials, then filters, dries, and stores the resulting product to provide an inventory for product pack-out. The cell is divided into two units: process tank and process drying train. The process tank contains two equipment modules: mix tank and agitator. The drying train consists of the equipment modules: exchanger, dryer, delumper, and product bin.

Some other examples of equipment modules are:

Compressor
Filter
Pump
Splitter
Scrubber
Tank

A control module is a collection of sensors, actuators, and associated processing equipment that is operated as a single entity. A control module can also consist of other control modules. For example, a header control module could be defined as a combination of several automatic valve control modules.

As an example of units containing control modules, again consider the cell in Figure 3.15. The process tank unit contains three control modules: two flow controllers and a mixer speed control. The process drying train has a dryer belt speed and delumper speed control modules.

Some examples of control modules are:

A regulating loop consisting of a transmitter, a controller, and a control valve that is operated via the controller setpoint

A regulating loop consisting of a speed transmitter, a controller, a variable speed drive, and an agitator

A state-oriented device that consists of an on–off automatic block valve with position feedback switches that is operated via the setpoint on the device

A header that contains several on–off automatic block valves and that coordinates the valves to direct flow to one or several destinations based on the setpoint directed to the header control module

A position sensor, motion controller, stepper motors, and associated mechanical components that are used to apply a label to a bag as it comes down a conveyor

For the cell of Figure 3.15, the units, equipment modules, and control modules are arranged hierarchically in Figure 3.16. This figure is an expansion of the S88.01 symbology of Figure 3.8.

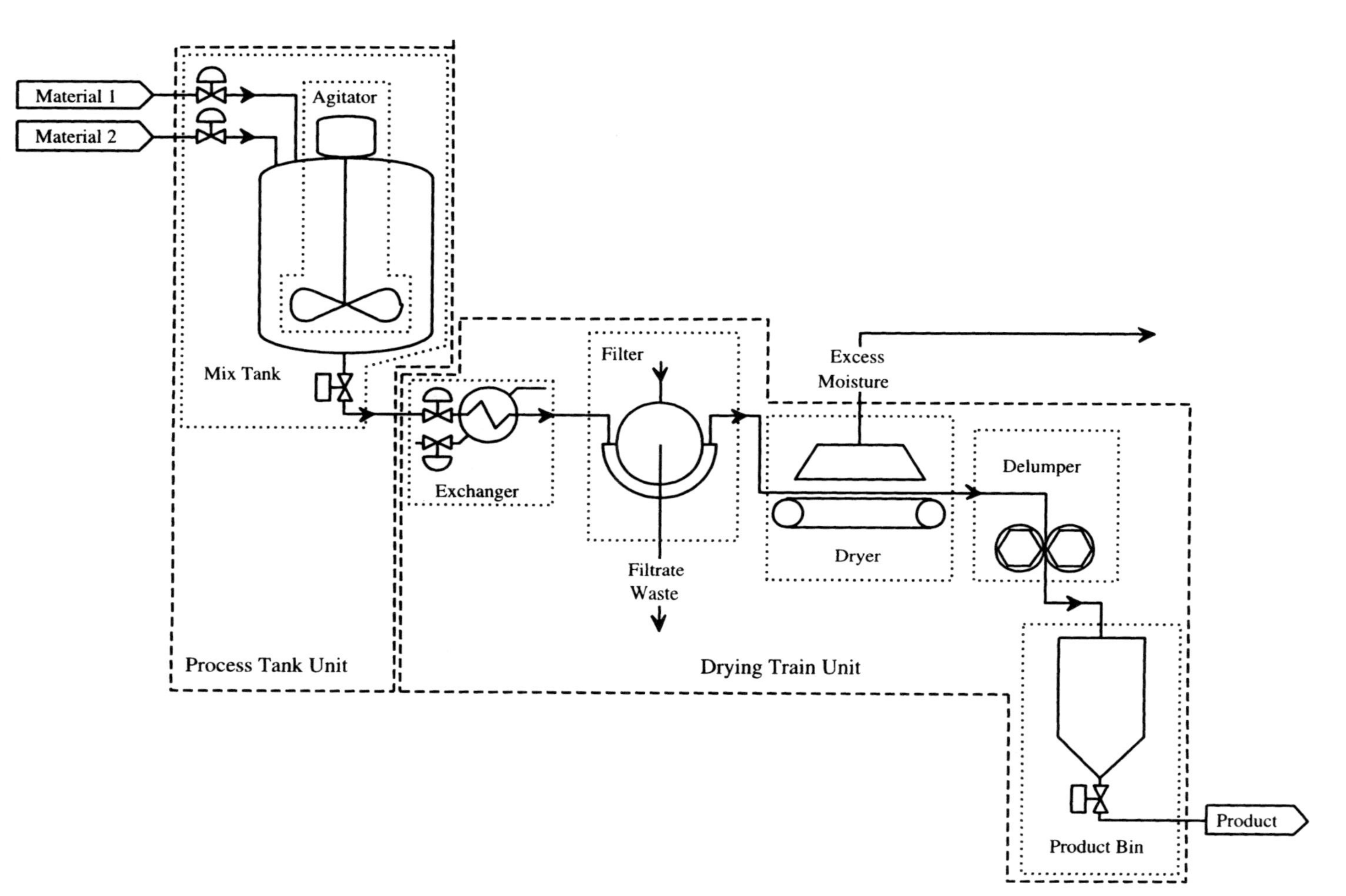


**FIGURE 3.15.** Process cell divided into units and equipment and control modules.

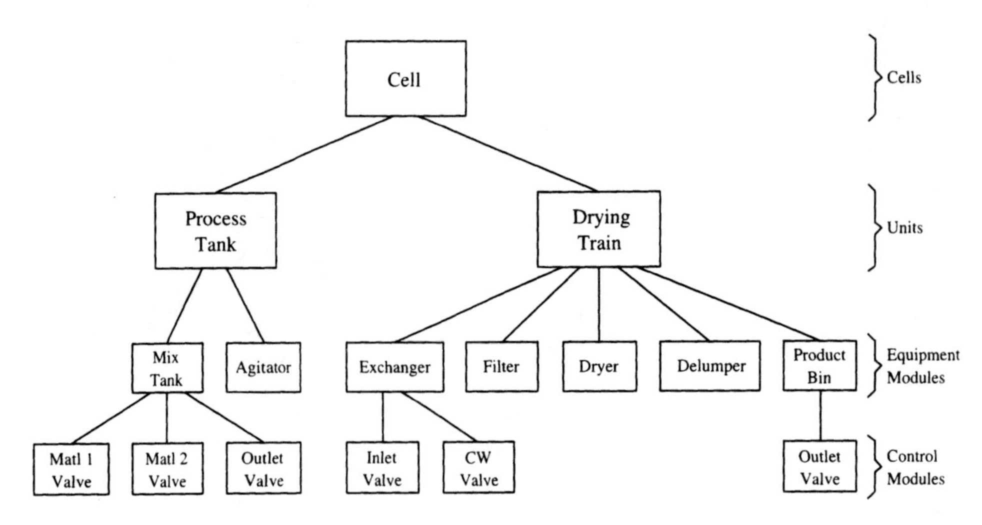


**FIGURE 3.16.** Process cell physical model arranged hierarchically.

### 3.2.3 Control Model

Each level of the physical model corresponds to a level in the control activity model. As for the physical model, the control procedures can be organized hierarchically, as shown in Figure 3.17. Smaller groupings are combined to form larger entities in the model. The control model describes the control procedures of an enterprise in terms of enterprise management, site management, plant area management, process management, unit supervision, process control, and safety protection. The control procedures may also be organized geographically in the same manner as the physical model (Figure 3.11). However, below the unit supervision level, it may be difficult to enforce a strict geographical division when some process control objects are shared among units.

The control activity model (Figure 3.18) provides an overall perspective of the major activities involved in plantwide process control and their relationships. This model is focused toward the lower layers in the hierarchical control model (Figure 3.17) because these layers are the typical domain of the control engineer. The relationships between the activities are defined by the information flowing between them and will be defined as the activities are discussed.

***3.2.3.1 Enterprise Management.*** Enterprise management is responsible for coordinating the management of all sites. It is concerned with such issues as allocation of production among similar sites, coordination of material transfer between sites, and maintaining enterprise recipes.

***3.2.3.2 Site Management.*** Site management is responsible for coordinating the management of the plant areas. It is concerned with such issues as coordinating production of plant areas, scheduling production of products based on the time of year, coordination of material transferred between areas, and maintaining site recipes.

***3.2.3.3 Plant Area Management.*** Plant area management comprises the control activities that coordinate the process cell operations. Typical Manufacturing Execution

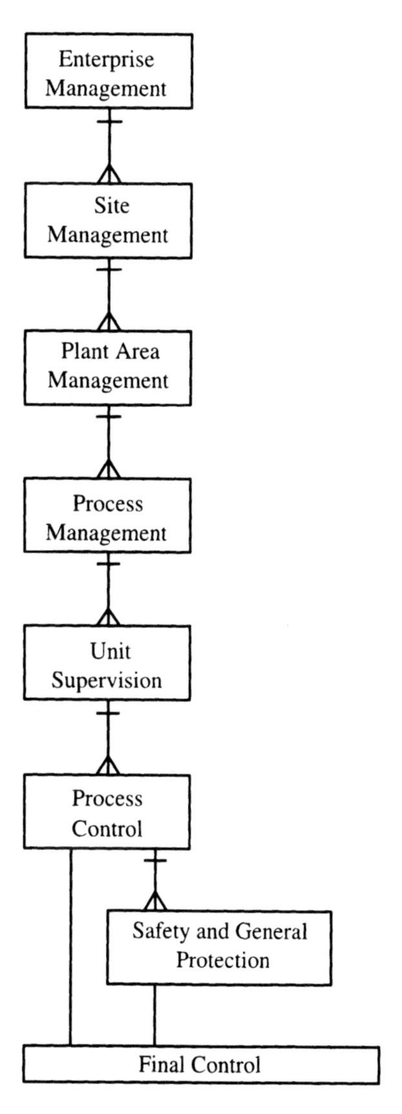


**FIGURE 3.17.** Hierarchical layers of control model.

System (MES) tasks are done at this level: production planning, coordination, and scheduling; production information management; and recipe management. Figure 3.19 depicts the production planning and scheduling activities of this layer, which take a production run and translate it into the commands of process management layer activities. Here, a production run causes two batch processes to be run to generate a portion of the raw material needed by three continuous production lines.

***3.2.3.4 Process Management.*** The process management layer is concerned with cellwide control and unit operation coordination. Figure 3.20 depicts an overall view of process management functions. An area schedule manager handles any inter-unit coordination, especially when it involves sharing a common resource, for example,

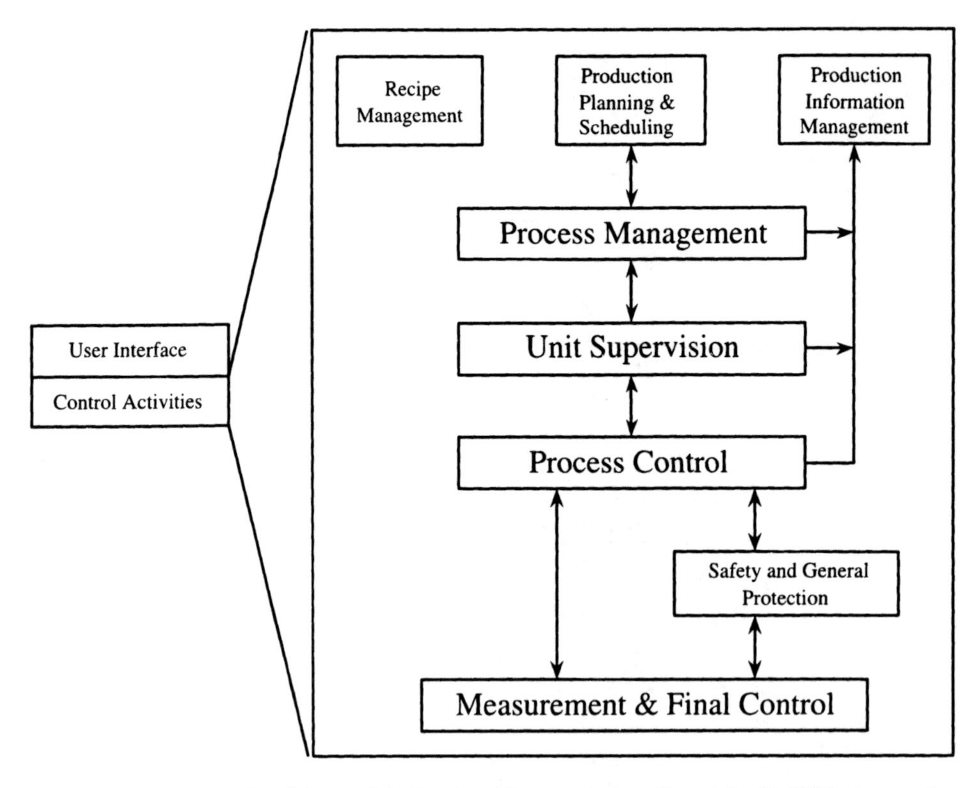


**FIGURE 3.18.** Control activity model. (Reprinted by permission. Copyright © 1996, Automation and Control Technologies, Inc.)

reactor clean-in-place equipment or common chemical tanks used for the regeneration of multiple ion beds. The inevitable coordination that must take place during process startups and shutdowns are handled at this level. In batch systems, the master recipe is stored and executed at this level and transferred to the appropriate unit supervision activity. Path selection may be done at this level. Any journaling of the batch operation is also done here. For both continuous and batch processes, trending and event information is initially kept at this level before it is transferred to more permanent storage. A procedure is the procedural control element at the process management level. For a batch process, a batch manager executes the procedure to make a batch at the cell level. For a continuous cell, the process management functions are often handled by a supervisory control manager. Depending on the control scheme, multivariable control may be handled at this level.

***3.2.3.5 Unit Supervision.*** Automation of unit supervision is at the heart of any good control strategy. For batch control systems, the unit procedure executes the unit recipe and is concerned with such issues as resource coordination and allocation. For continuous processes, the various single-loop controllers are integrated into a functional unit. Multivariable and advanced control algorithms are generally executed at this level. An overall view of unit supervision is shown in Figure 3.21. The unit manager coordinates the functions of the process control objects: equipment module control, loops, devices, indicators, and statuses. Each object in the unit, including the unit manager, has a state, as explained in Section 3.2.4.

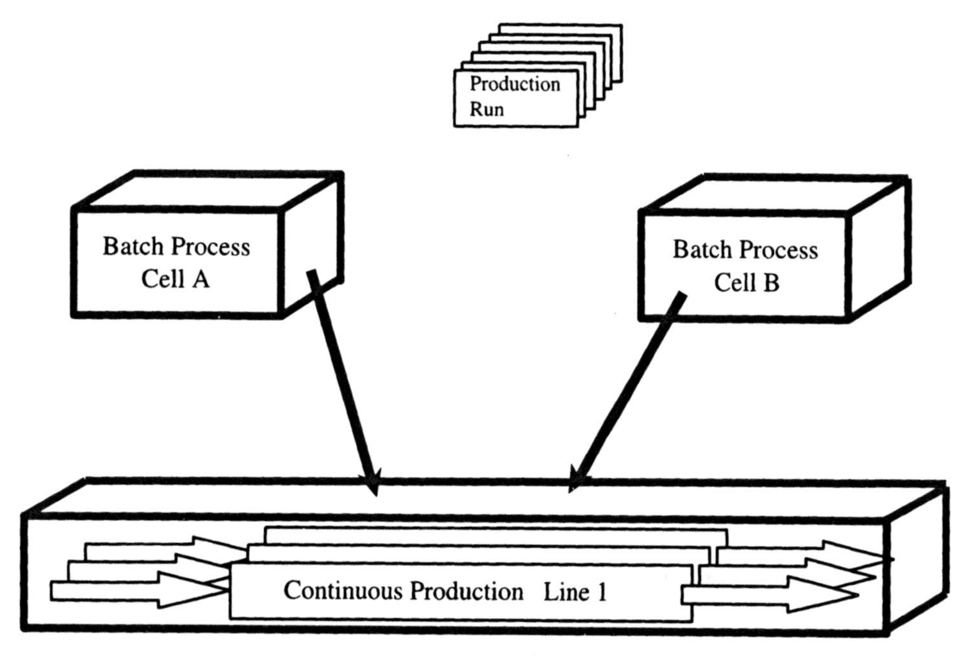


**FIGURE 3.19.** Production planning at plant area management level. (Reprinted by permission. Copyright © 1996, Automation and Control Technologies, Inc.)

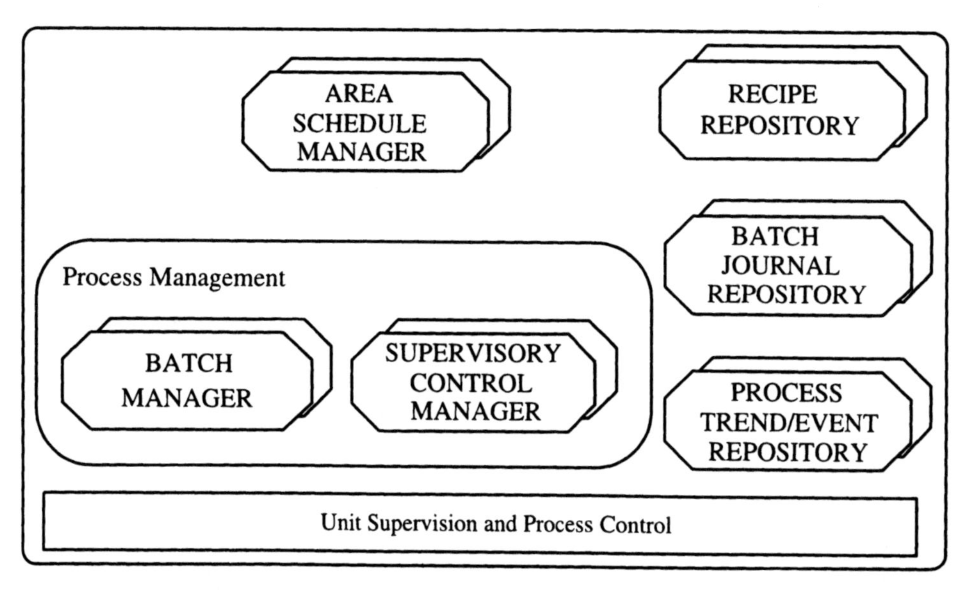


**FIGURE 3.20.** Process management and its relation to unit supervision. (Reprinted by permission. Copyright © 1996, Automation and Control Technologies, Inc.)

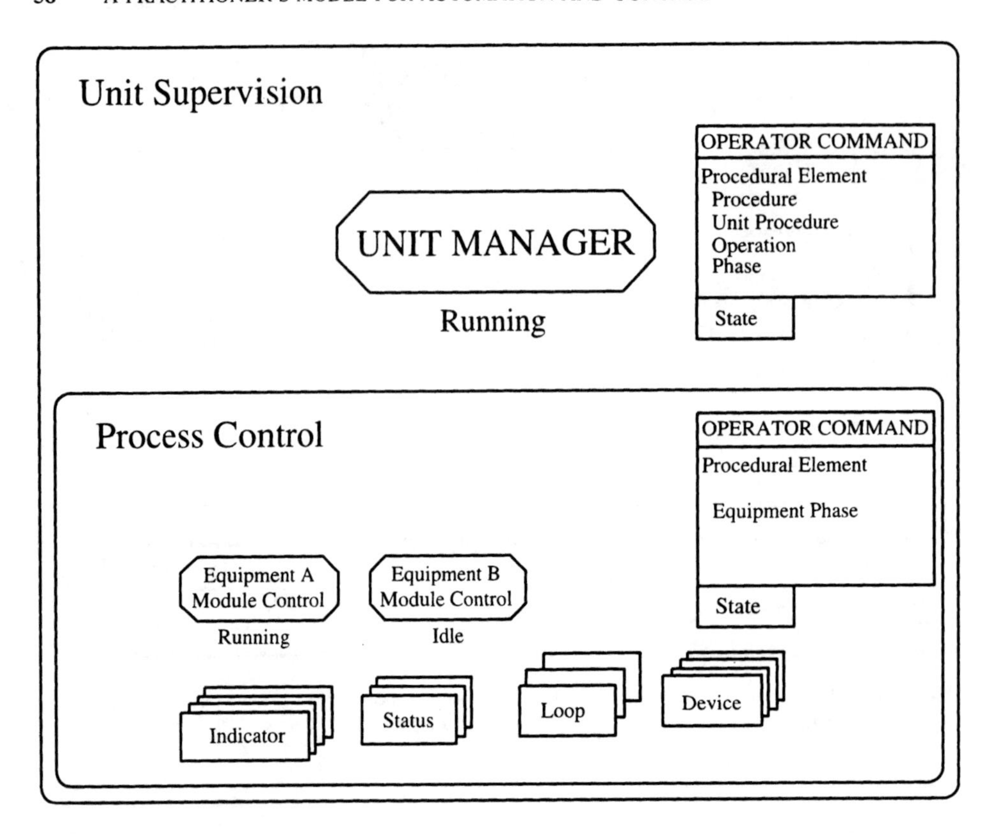


**FIGURE 3.21.** Unit supervision and its relation to process control layer objects. (Reprinted by permission. Copyright © 1996, Automation and Control Technologies, Inc.)

Unit supervision has procedural elements: unit procedure, operation, and phases (ISA, 1995). Even though these terms were originally defined for use in batch control, they are also applicable to continuous control. The procedural control aspects of process control are detailed in Chapter 9 and are only summarized here. The lowest level, a phase, is divided into steps and represented as a function chart. Phases are grouped into an operation, and operations are part of a procedure. The main distinction between operations and phases is that only one operation can be running at once while multiple phases may be running simultaneously. A unit procedure, in batch control terminology, is the strategy needed to make a batch at the unit level. Continuous processes will generally not have a procedure at the cell level. The main operator interface is usually through the unit supervisor, although lower level maintenance operator functions are provided at the process control layer.

***3.2.3.6 Process Control.*** The process control layer encompasses the basic discrete, regulatory, and equipment module procedural control elements, as shown in Figure 3.22. At this level, procedural control is very basic. For example, turn on the sump pump when the water level is at the high limit switch and keep it on until the low-level limit switch is deactivated, at which point the pump is turned off. Process control encompasses the types of objects that can be supervised at the unit level:

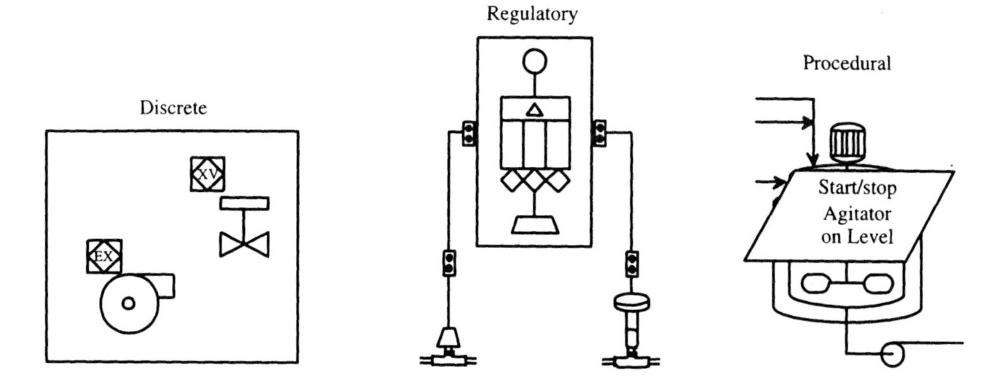


**FIGURE 3.22.** Entities at process control level. (Reprinted by permission. Copyright © 1996, Automation and Control Technologies, Inc.)

*Equipment Module Control.* Batch or sequential control of an equipment module. Discrete manufacturing and batch control systems use this type of control object extensively.

*Loop.* Typically a single-input, single-output PID controller that monitors a transmitter and manipulates a control valve position in order to force a process variable to the desired setpoint. Cascade control is also considered a loop, as well as other types of single-input, single-output algorithms.

*Device.* Has a single output with two possible values (e.g., on–off, open–close). A typical device has one or two inputs.

*Indicator.* A single analog value, typically in engineering units.

*Status.* A single discrete value.

These types of objects can also generically be called *control modules*. Loops and devices can be composite modules consisting of other loops and devices. A control module can also be a strategy and consist of multiple loops. A typical unit will have at most one equipment module control procedure and one or more other type of object. The process control objects for the sample units of Figure 3.15 are shown in Figures 3.23 and 3.24. In the format of Figure 3.16, the control objects are arranged hierarchically as in Figure 3.25.

Equipment module control is concerned with one phase. An equipment module is a plant item whose control scheme is a phase that coordinates other plant items whose function is discrete or regulatory control. Equipment module control is associated with a shared system, transfer system, or a continuous system that has a small sequence. Examples include:

Demineralizer
Clean-in-place
Blend out
Transfer system between two units

*3.2.3.7* ***Safety and General Protection.*** This layer includes those control functions that specifically protect personnel, equipment, and product from damage. Interlocks can be

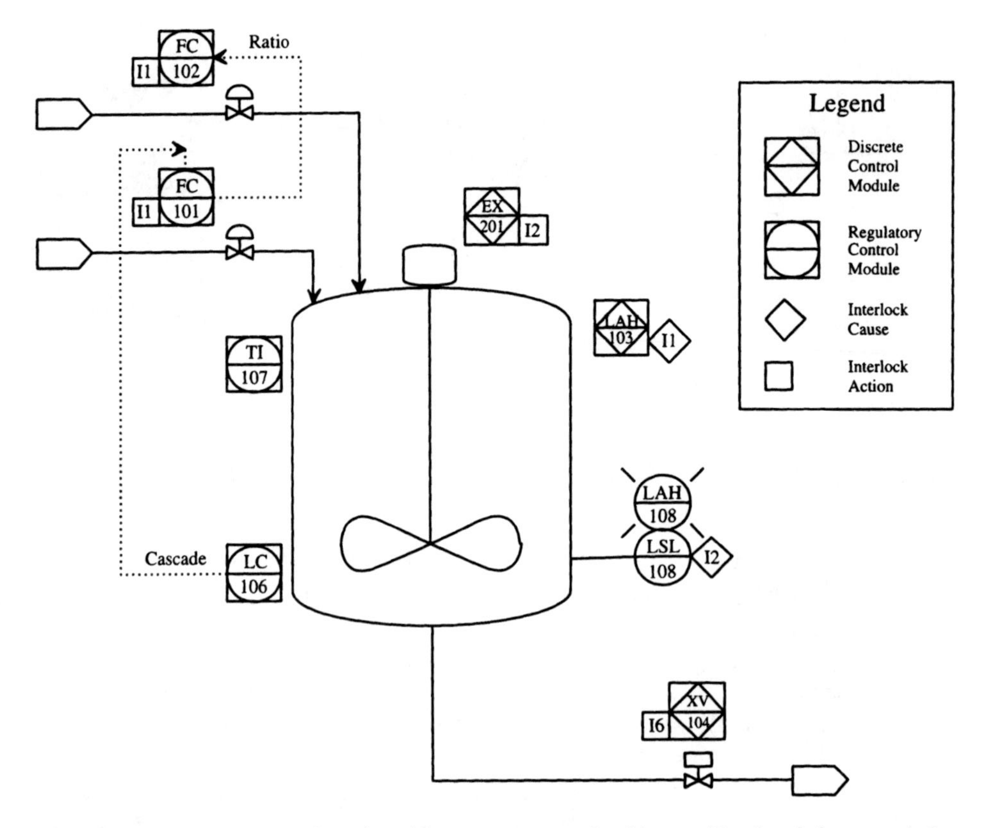


**FIGURE 3.23.** Example tank unit with process control objects. (Reprinted by permission. Copyright © 1996, Automation and Control Technologies, Inc.)

mechanical, hard wired, or programmed into the electronic control system. Many interlocks prevent the operator from commanding actions that are unsafe. Other interlocks detect a process malfunction or dangerous condition and then drive the system into a safe condition. Interlocks are treated in more detail in Chapter 8.

***3.2.3.8 Measurement and Final Control*** This layer includes the measurement sensors and final control elements that interface directly to the physical elements of the process:

Analog sensors and transmitters
Control valves and actuators
Analytical instrumentation
Motor drives
Digital sensors
Digital actuators
Vision systems
Motion detectors

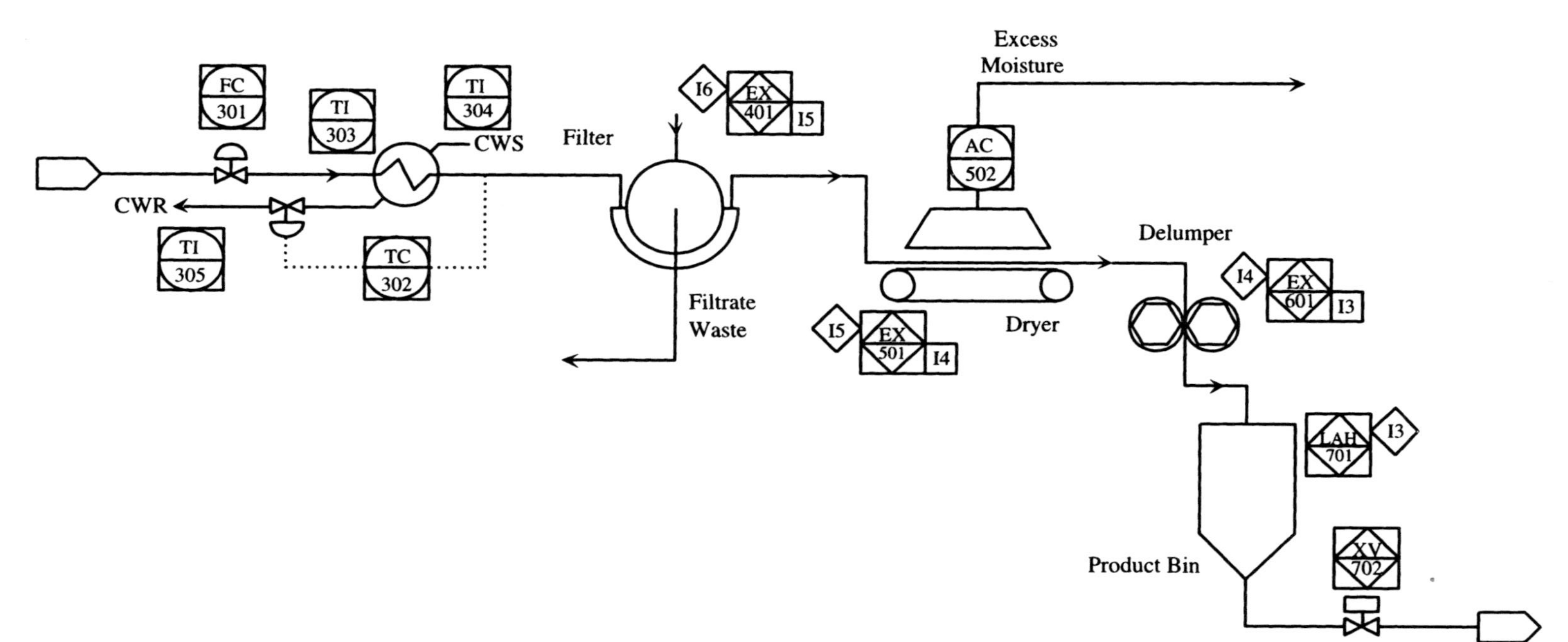


**FIGURE 3.24.** Example drying train unit with process control objects. (Reprinted by permission. Copyright © 1996, Automation and Control Technologies, Inc.)

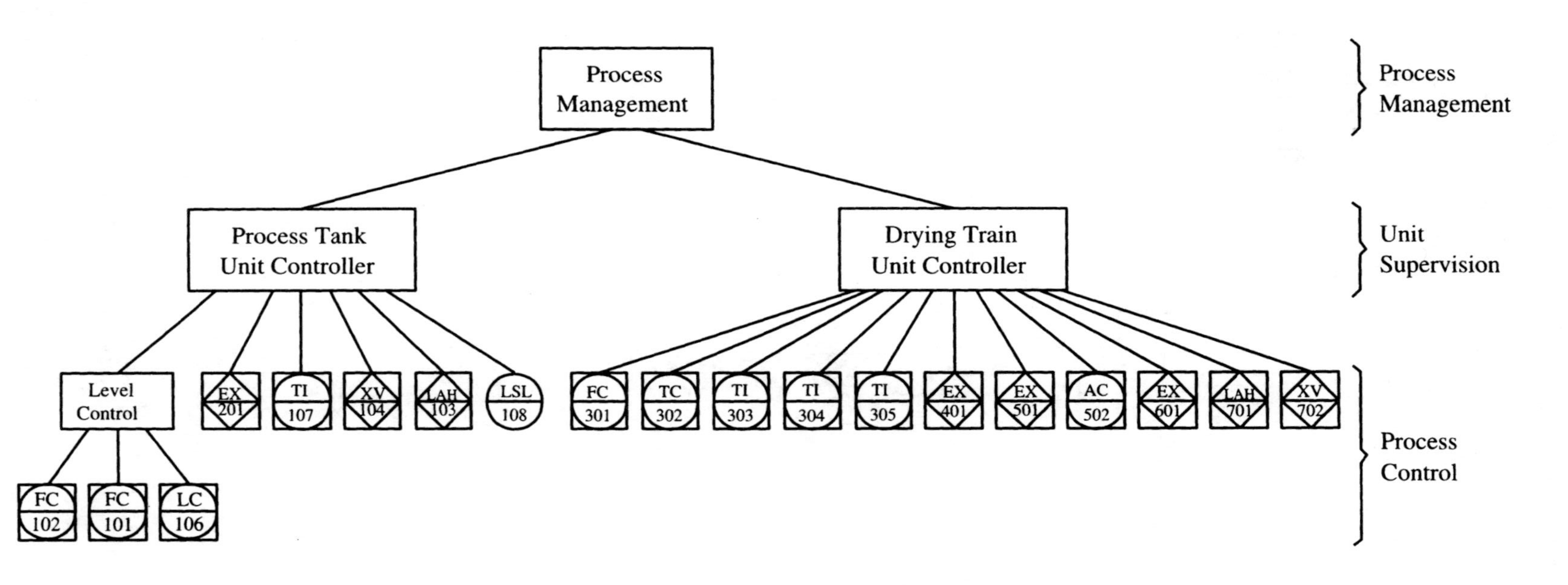


**FIGURE 3.25.** Process cell control objects arranged hierarchically.

### 3.2.4 Operating State and Control State

Each entity of the physical and control models possesses a state. The state specifies the current condition of the entity and defines how the entity will operate and how it will respond to commands. Equipment entities have an operating state and control entities possess a control state. Example operating states for physical entities include On, Off, Open, Closed, 20% open, Tripped, Failed, and Available. Example control states for control entities include RUNNING, IDLE, HOLDING, PAUSED, STOPPED, ABORTED, AUTOTUNING, and BUMPLESS TRANSFER. The operating state of an equipment entity and the control state of the corresponding control entity are related. Often, the operating state of an equipment entity is dependent on the control state of the control entity controlling it.

As an example of operating states in a process cell, consider the cell of Figure 3.15, called the A Process. This process is a single process cell with two units. The states of this cell and its associated units are illustrated in Figure 3.26. Normal operation is continuous to maintain an adequate inventory of product for packout. Tank 100 startup is performed by first filling the tank with the desired ratio. The mixture is then agitated while laboratory

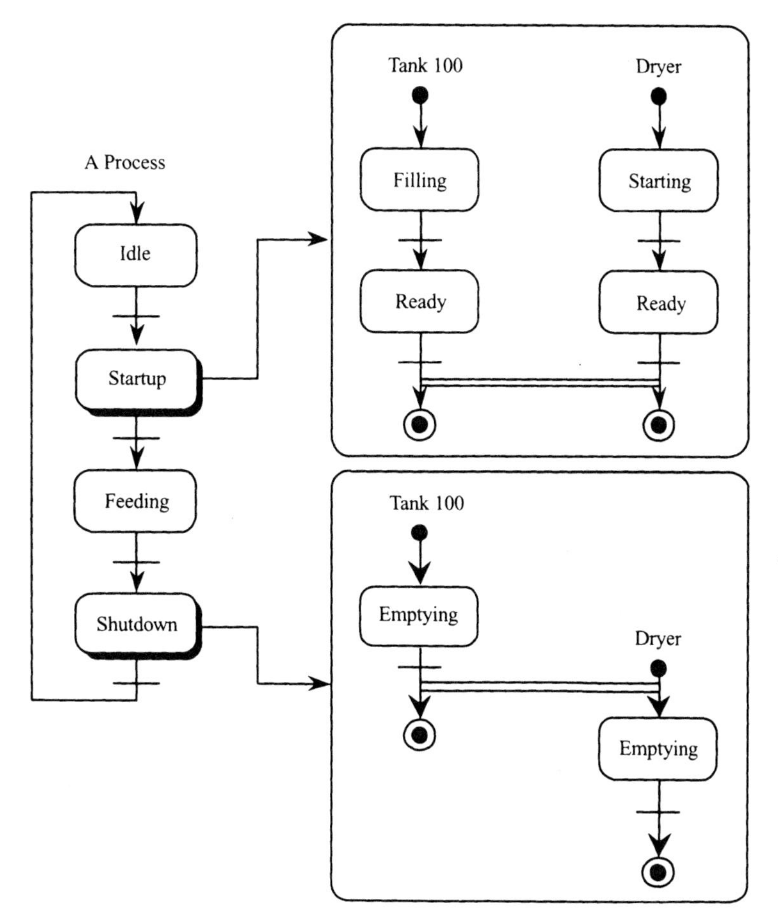


**FIGURE 3.26.** Operating state diagram for the A process. (Reprinted by permission. Copyright © 1996, Automation and Control Technologies, Inc.)

tests are performed. Once the laboratory tests are complete and the mixture meets product quality standards, the unit is ready for operation. Meanwhile, the dryer is not started until all equipment is running and ready for feeding. When the material is approved and the dryer is operational, the feeding is initiated. Feeding continues until the bin is full or the operator decides to shut down the process. Tank 100 is stopped by stopping the feedstock and continuing the feed until the tank is empty. Once it is empty, the material in the dryer is purged from the conveying system. The units are then considered idle and ready for the next run. The states are described as follows.

**Operating State Name: Idle**

*Reference.* A Process Tank 100.

*Routine Activities*

Clean tank.

Perform nightly inspection.

*Exception Handling.* None.

*Primary Control Objectives.* None.

*Performance Information.* None.

*State End Conditions.*

Filling started.

**Operating State Name: Filling**

*Reference.* A Process Tank 100.

*Routine Activities*

Charge material 2 and material 1 in a 2 : 1 ratio.

Begin agitating mixture during charging, after tank level is above the lower level limit.

Stop material flows when tank has reached designed fill level.

Extract sample and perform laboratory analysis.

*Exception Handling*

Loss of material 1

Loss of material 2

High level detected

Low level detected

Discharge valve failure

*Primary Control Objectives*

Concentration of mixture (material flow ratio)

Material amount for proper mixing (residence time based on level in tank)

*Performance Information.*

Mixture quality standards.

*State End Conditions.*

Acceptable laboratory sample.

**Operating State Name: Ready**

*Reference.* A Process Tank 100.

*Routine Activities*

Agitate.

Take periodic laboratory sample (hourly).

*Exception Handling*

Agitator failure.

Material amount in tank is less than required for proper mixing.

*Primary Control Objectives.* None.

*Performance Information.*

Mixture quality standards.

*State End Conditions.*

Feeding started.

**Operating State Name: Starting**

*Reference.* A Process Drying Train.

*Routine Activities*

Start delumper.

Start dryer.

Start filter.

Check cooling water availability.

*Exception Handling*

Loss of cooling water.

Loss of drying.

Delumper clogged.

Filter clogged.

Product bin full.

*Primary Control Objectives.* None.

*Performance Information.* None.

*State End Conditions.*

All equipment is operating.

**Operating State Name: Ready**

*Reference.* A Process Drying Train.

*Routine Activities:*

Monitor equipment continuously.

Check cooling water availability.

*Exception Handling.*

Any equipment in drying train fails.

*Primary Control Objective.* None.

*Performance Information.* None.

*State End Conditions.*

Feeding started.

**Operating State Name: Feeding.**

*Reference.* A Process Cell.

*Routine Activities*

Restart flow of material 1 and material 2 in 2 : 1 ratio to the tank.

Open discharge valve.

Cool product in the exchanger.

Filter product.

Dry product.

Delump product.

Deliver product to bin.

Take laboratory sample each hour that process remains in feeding state.

Generate production/shift reports.

*Exception Handling*

Material 1 or 2 valve failure.

Agitator failure.

Drain valve failure.

Cooling water loss.

Filter clog.

Delumper clog.

Loss of drying.

Material amount for optimum mixing not maintained in tank.

Product bin overflow.

*Primary Control Objectives*

Concentration (material 1 and 2 ratio).

Amount of material (liquid level in tank).

Separation (manipulated by temperature of cooler).

Moisture content (moisture level of dryer).

*Performance Information*

Quality (concentration, moisture levels).

Production by shift/day.

*State End Conditions*

Production run complete.

**Operating State Name: Emptying**

*Reference.* A Process Tank 100.

*Routine Activities*

Shut off flows of material 1 and material 2.

Drain Tank 100 until empty.

*Exception Handling*

Material 1 valve failure.

Material 2 valve failure.

Discharge valve failure.

*Primary Control Objectives.* None.

*Performance Information.* None.

*State End Conditions.*

Tank is drained.

**Operating State Name: Emptying**

*Reference:* A Process Drying Train.

*Routine Activities*

Close product inlet valve.

Wait filter purge time.

Shut down filter.

Stop temperature control on exchanger.

Wait time determined for product to pass through dryer.

Shut down dryer.

Stop moisture control on dryer.

Wait to allow material to pass through delumper.

Shut down delumper.

*Exception Handling*

Inlet valve failure.

Dryer conveyor failure.

Dryer clogged.

Filter clogged.

*Primary Control Objectives.* None.

*Performance Information.* None.

*State End Conditions.*

All materials purged from drying train.

Concerning the control state, the S88.01 standard follows traditional process control in making a distinction between modes and control states. Traditionally, the state of a control entity, for example, a PID loop, is called the "mode." Typical modes are "manual," "automatic," "semi-automatic," "cascade," "computer manual," and "computer automatic." The S88.01 standard makes a distinction between modes and states. Both modes and states describe the status of physical entities and procedural elements. States specify

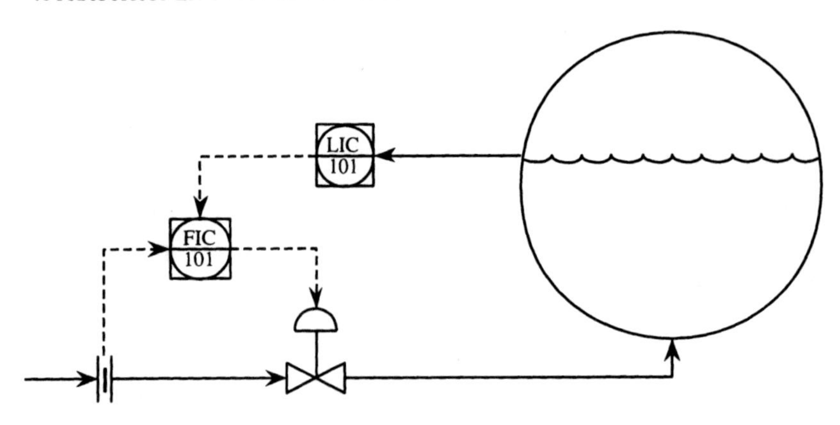


**FIGURE 3.27.** Boiler feedwater control cascade loop.

their current status and modes specify the manner in which transitions take place between the states. This distinction between modes and states is unnecessary and only serves to cloud the real control issues. In addition, in many control systems the traditional notion of a mode is often difficult to explain and leads to confusion. An example is used to show the relation between a control state and the traditional concept of a mode.

Consider the boiler control system in Figure 3.27. This type of arrangement of controllers is commonly called a cascade control system. There are two measurements: boiler level and feed flow. Only one physical variable is manipulated—the valve position. In traditional control, the mode of the controllers is set, depending on what one wishes to control. If one wants no control, both controllers are set to manual mode and the operator manipulates the valve position directly. To control the feed flow, FIC101 is set to automatic mode and LIC101 is set to manual mode and the operator adjusts the setpoint of the feed flow controller, FIC101. If one wants to control the boiler level, FIC101 is set to cascade mode, LIC101 is set to automatic mode, and the setpoint of FIC101 is connected to the output of LIC101. The operator now manipulates the setpoint of LIC101. If the system is controlling the boiler level and the mode of FIC101 is switched to manual, then LIC101 should automatically be forced to the manual mode. When viewing this system as a control module, it has three basic operating states: Manual, Flow, and Level (Figure 3.28). The interface to the operator is thus simplified, because one does not need to be concerned about the individual controller modes. Using the concept of a control state, the faceplate for the controller can appear as in Figure 3.29. It has five values on it: current level, level setpoint, feedwater flow, feedwater flow setpoint, and percent output. The current level and feedwater flow are only indications. The current state (Level, Flow, or Manual) is shown in

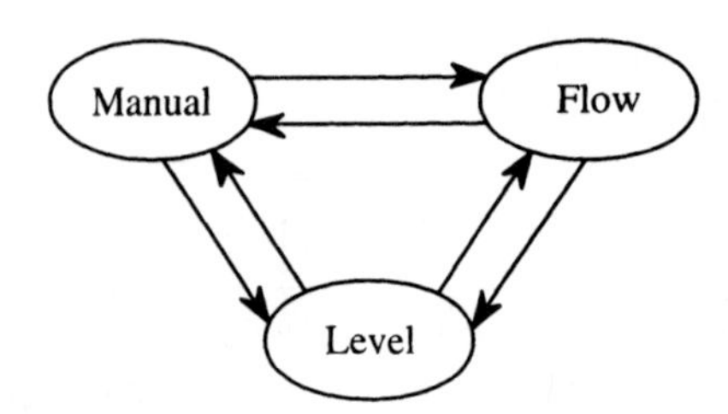


**FIGURE 3.28.** Control states for boiler feedwater control.

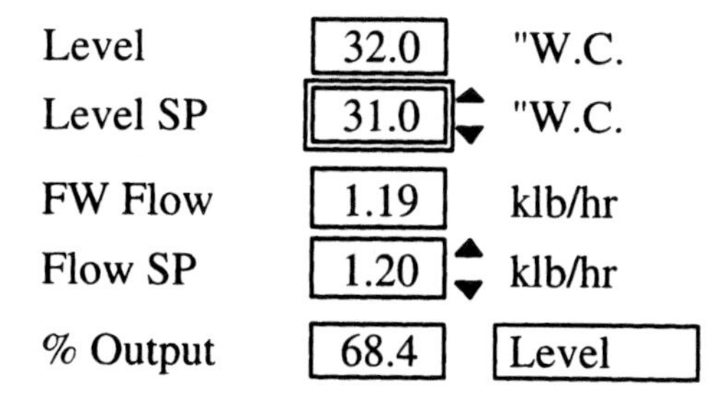


**FIGURE 3.29.** Feedwater loop user interface.

a separate field to the right of the percent output value. The particular setpoint or percent output that may be manipulated by the operator depends on the state. The figure shows the control module in the Level state and only the boiler level setpoint can be manipulated by the operator, which is indicated in the display by placing a box around the level setpoint. Color is also used to differentiate which variable is being controlled.

As another example of control states, consider the following override control scheme to regulate steam header pressure (Stephanopoulos, 1984), shown in Figure 3.30. The high-pressure steam is "let down" to the lower pressure level. The pressure in the low-pressure steam line is controlled by PIC101. To prevent the high-pressure line from excessive pressures, an override control system with a high-signal selector (HSS) transfers control

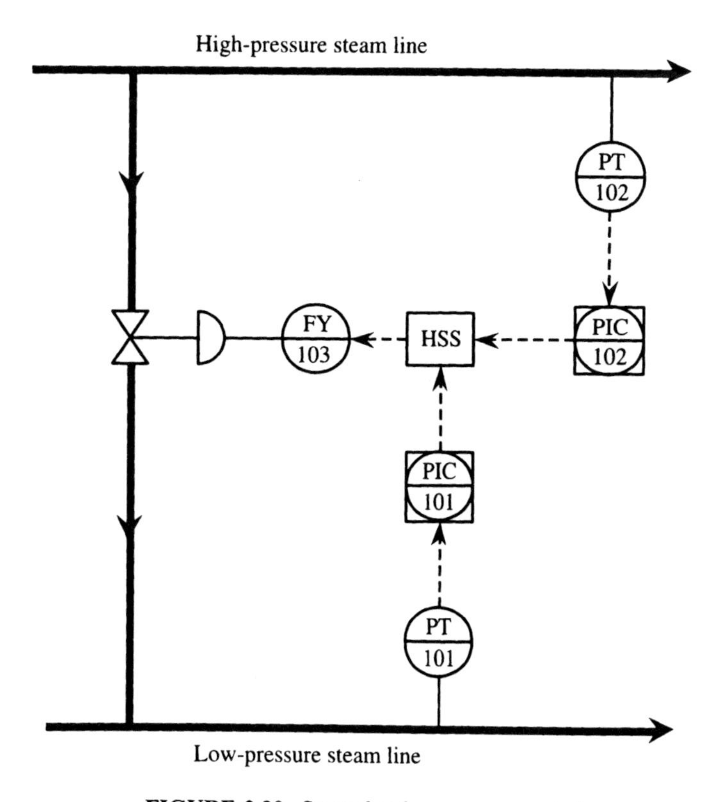


**FIGURE 3.30.** Steam header pressure control.

action from PIC101 to PIC102 when the pressure in the high-pressure line exceeds an upper limit. From a state perspective, this system has three control states: Manual, Low-pressure Control, and High-pressure Control.

From the two previous examples, cascade control and override are seen as schemes of interaction among objects that the system manipulates. The operation should be transparent to the operator. State transitions now are important and serve to appropriately describe the system operation.

As a more complicated example of control states, consider the state transition diagram for a typical batch or continuous procedural element (procedure, operation, or phase) in Figure 3.31. This diagram is a modification from S88.01 (ISA, 1995). Certain control states and transitions usually apply only to either a batch control system or a continuous control system but not both. Generally, the STARTING state applies to continuous processes and the COMPLETE state applies to batch processes. The process normally starts in the IDLE state. The shaded states are final states. The valid states are described as:

*IDLE.* The procedural element is waiting for a Start command that will cause a transition to the STARTING state.

*STARTING.* The procedural element executes startup logic. Once complete, the procedural element automatically transitions to the RUNNING state. This state is generally only for continuous-control systems. In most batch systems, the procedural element proceeds immediately to the RUNNING state.

*RUNNING.* Normal operation. In batch control systems, the operations, phases, and procedures are executing. In purely continuous control systems, the procedural

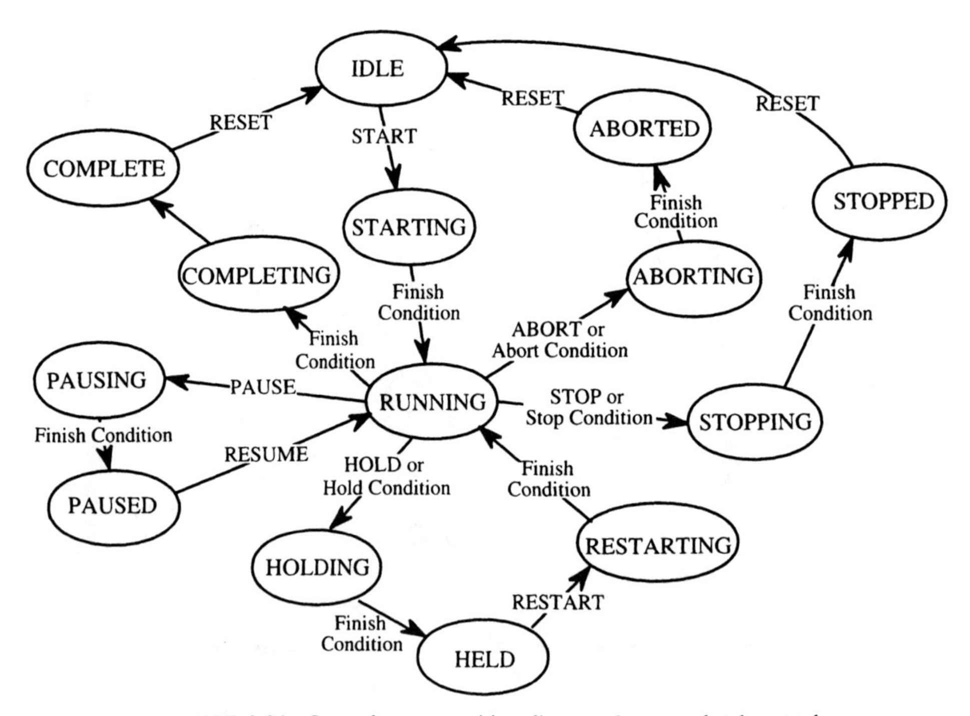


**FIGURE 3.31.** General state transition diagram for procedural control.

element remains in this state until the operator or an abnormal condition forces it to another state.

*COMPLETING.* For a batch control system, the normal operation has run to completion and any special logic needed to bring the process to a shutdown state is executed. Many continuous-control systems do not run to completion and so do not transition to this state. Once finished, the procedural element automatically transitions to the COMPLETE state.

*COMPLETE.* Once the procedural element has finished any special completion logic, the state changes to COMPLETE. The procedural element is waiting for a Reset command that will cause a transition to IDLE.

*PAUSING.* The procedural element or equipment entity has received a Pause command. The procedural element stops at the *next* defined state or stable stop location in its normal RUNNING logic. Once stopped, the state automatically transitions to PAUSED.

*PAUSED.* Once the procedural element has paused at the defined stop location, the state changes to PAUSED. This state is normally used for short-term stops. A Resume command causes a transition to the RUNNING state, resuming normal operation immediately following the defined stop location.

*HOLDING.* The procedural element receives a Hold command and is executing its HOLDING logic to place the procedural element or equipment entity into a known state. If no sequencing is required, then the procedural element or equipment entity transitions immediately to the HELD state.

*HELD.* Once the procedural element has completed its HOLDING logic and is at the known state, the state changes to HELD. This state is normally used for a long-term stop. A Restart, Stop, or Abort command causes a transition out of this state.

*RESTARTING.* The procedural element has received a Restart command while in the HELD state. If no sequencing is required, the procedural element transitions immediately to the RUNNING state; otherwise restart logic is executed.

*STOPPING.* The procedural element has received a Stop command and is executing its STOPPING logic, which facilitates a controlled shutdown. If no sequencing is required, then the procedural element or equipment entity transitions immediately to the STOPPED state.

*STOPPED.* The procedural element has completed its STOPPING logic and is waiting for a Reset command to transition to IDLE.

*ABORTING.* The procedural element has received an Abort command and is executing its ABORTING logic, which facilitates a quicker, and not generally controlled, emergency shutdown. Generally, no sequencing is required, and the procedural element or equipment entity transitions immediately to the ABORTED state.

*ABORTED.* The procedural element has completed its ABORTING logic and is waiting for a Reset command to transition to IDLE.

In an actual process, the RUNNING state may be broken down further. For example, ion exchange beds typically have Ready, On-line, and Regenerating states.

## 3.3 INFORMATION MANAGEMENT

Information is exchanged at all levels of the enterprise model. Figure 3.32 shows the exchange of information with other enterprise functions. Due to the nature of this text, information management that concerns control is emphasized. In Figure 3.32, the control functions handled at the cell and lower levels are represented by the inner circle. The functions of the site and area levels are indicated between the two circles. Functions at the enterprise level are shown outside the circles. These three divisions also correspond with the implementation technologies available today: Enterprise Resource Planning Systems (ERPSs), Manufacturing Execution Systems (MESs), and control systems. These technologies are related to the layers of the enterprise model in Figure 3.33. Typical control system technology handles the control functions at the unit and lower levels. Typical MES packages handle the control functions at the cell and area level, and ERPS packages are used at the enterprise level.

## 3.4 USER INTERFACE

The user interface is a window to the process and occurs at each control level. The relationship between the user interface and the control model is shown in Figure 3.34. In this section, the general functions of a user interface are reviewed. Its specific role and

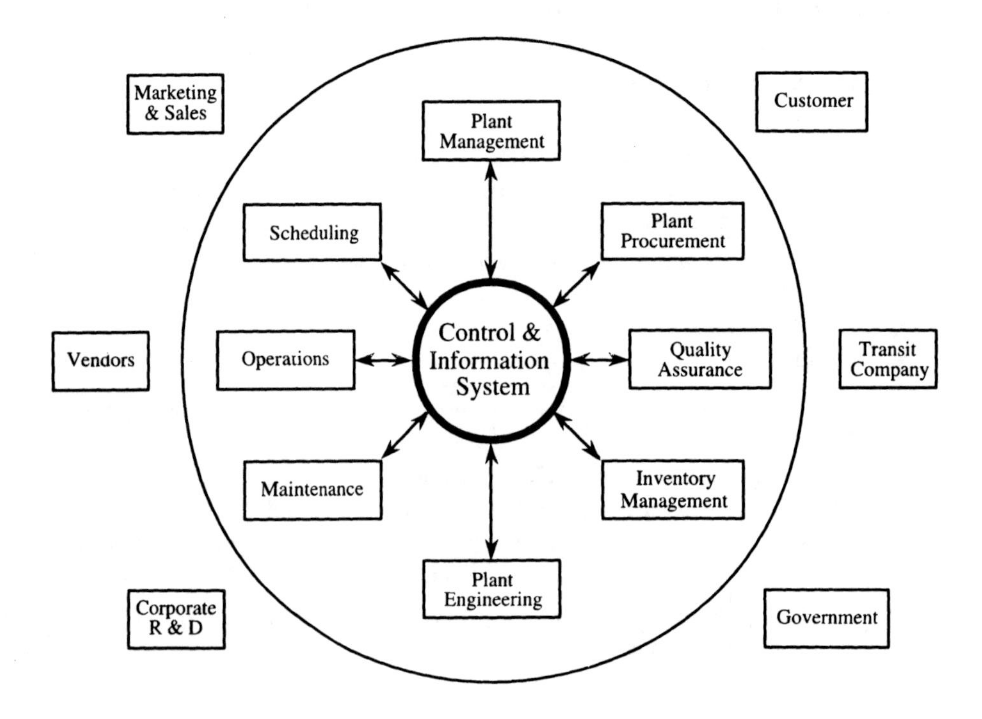


**FIGURE 3.32.** Information exchange with other enterprise functions. (Reprinted by permission. Copyright © 1996, Automation and Control Technologies, Inc.)

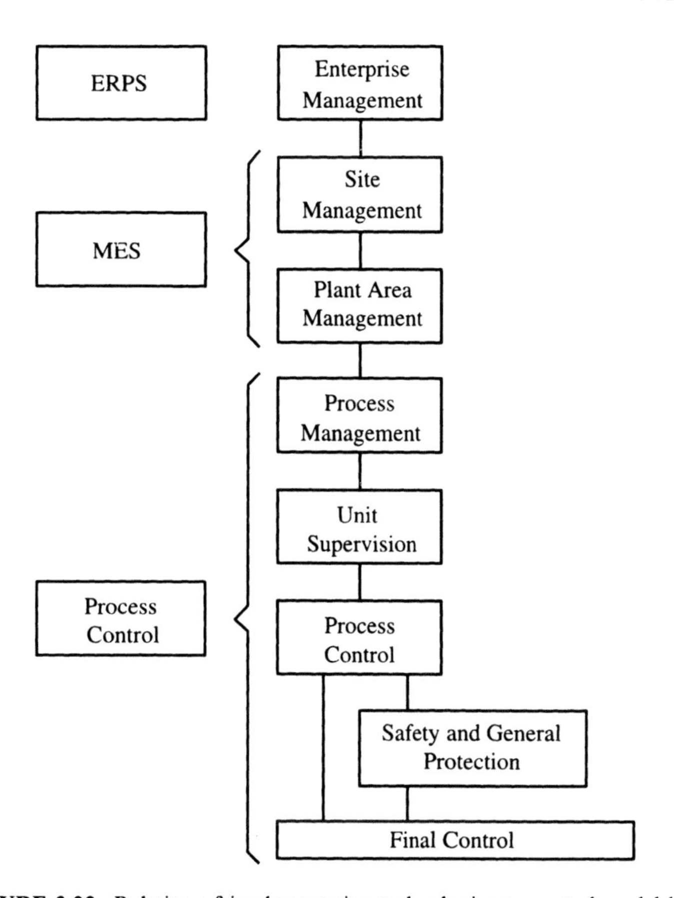


**FIGURE 3.33.** Relation of implementation technologies to control model layers.

appearance are discussed in conjunction with the various parts of the control system in succeeding chapters.

In the not-too-distant past, the typical user interface to a process control system consisted of a panel along one or more walls of a room. The panel had a graphical representation of the process and a multitude of industrial-sized switches, push buttons, gauges, PID controllers, and strip chart recorders. All of these devices were hard wired to the field devices, transmitters, and actuators. With the advent of digital computers, an alternative display device became available: a computer screen with process graphics and operator commands entered via a keyboard. Technology has progressed to the point where a typical user interface is a color graphic touch screen and the keyboard and mouse are only used by the engineer making changes to the screens. Panel displays have not disappeared entirely, though. They are sometimes still used as a hot backup if the computer system fails. In addition, some equipment has local operator control panels. For example, a motor or pump may have a hand–off–auto (HOA) switch that allows maintenance personnel to override the action of the control system.

The type of user interface is often very different for the various levels of the control hierarchy. At the lowest level—measurement and final control—the user interface is very simple and will either be hard-wired physical switches and indicator displays or a touch

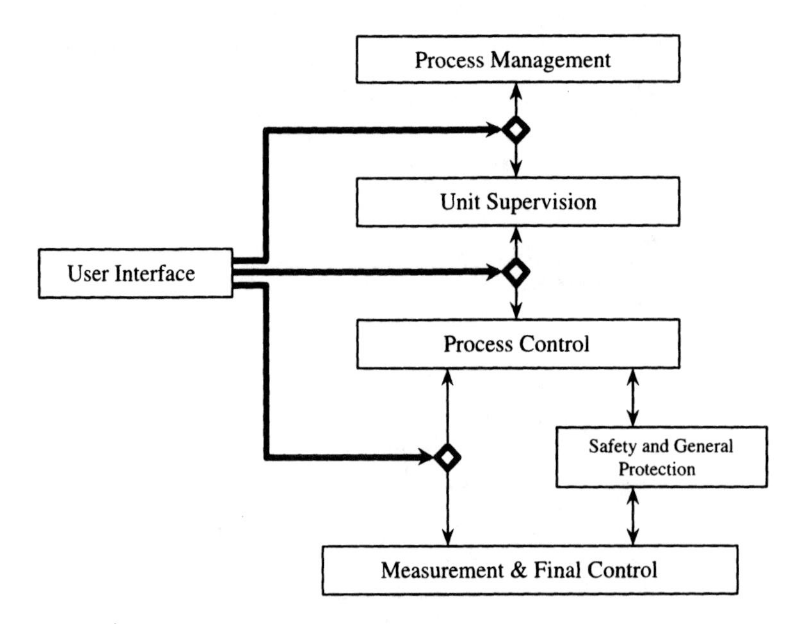


**FIGURE 3.34.** User interface and the control hierarchy. (Reprinted by permission. Copyright © 1996, Automation and Control Technologies, Inc.)

screen with graphical representations of the switches and indicators. At the process control level, the typical interface for loops is called a faceplate display that appears similar to that of the single-loop digital or analog controller. A device will also have a faceplate, but it will mimic a small hard-wired control panel. Generally, loops and devices have tuning and/or configuration parameters on a detail display that is accessed from the faceplate. At the unit supervision and process management levels, a graphical representation of the process presents overall status information about the unit or cell. Detailed information about objects is easily accessed. The displays are generally organized in a hierarchical manner, roughly corresponding to the hierarchical control model. For example, a plant area management screen shows the overall status and material flow between cells. A simple touch on an individual cell will bring up a graphic display of the cell with information about each unit. Touching a unit brings up a unit display graphically showing the process with device, loop, and other object information. Touching a loop device brings up a controller faceplate with information about that loop.

The basic types of user interface displays are as follows:

1. *Operational Summary.* These are the displays one uses in monitoring a process. They present summary information and usually display some graphic representation of the process. The amount and level of detail depend on the level in the hierarchy that one finds the display. For example, a reactor unit display has an indication for every temperature measurement device attached to the reactor: inlet, outlet, upper, middle and lower wall, cooling jacket inlet, and cooling jacket outlet. The cell display that includes that reactor unit has only one temperature, the average wall temperature. Alarms—indications of abnormal conditions—are displayed as indicators. For example, a failed valve may be indicated by changing its color to red. Detailed alarm information is usually presented on the alarm summary screen.

2. *Configuration/Setup.* Detailed process command parameters and parameters used to set up and/or configure control modules occur on this type of display. For example, PID loop tuning parameters are displayed and modified on this type of display. For batch control systems, the recipe is viewed and/or modified on this type of screen. This type of display tends to be more textual. Also, there are usually security measures attached to this screen so that only lead operators or engineers can change the parameters.

3. *Alarm Summary.* The alarm summary presents a complete list of time-stamped active alarms. It includes alarms that have been acknowledged by the operator as well as those that have not been acknowledged by an operator. An alarm remains in the alarm summary until the condition causing the alarm has cleared and an operator has acknowledged it. The three possible states of an alarm in this summary are thus (1) active, not acknowledged; (2) active, acknowledged; and (3) not active, not acknowledged. Generally, a different text/background color or blinking/not blinking is used to distinguish the three categories of alarms. Alarms may also be prioritized. Also, the alarm summary list may be displayed in priority order (highest at top), in time stamp order (most recent at top), in acknowledgment order (unacknowledged alarms at top), or in some combination of these.

4. *Event History.* An event history presents a time-stamped list of all significant events that have occurred in the process. Besides alarms, events such as valves opening/closing, operators logging in/out, batch start/stop, and changes in configuration parameters are recorded. The type of information that may be logged depends on the user interface vendor. Generally, the event history is maintained in more permanent storage and eventually copied to an archival device.

5. *Trend.* Values of pertinent process variables, such as flow, temperature, and pressure, over a period of time are shown by this type of display. This type of screen provides the ability to chart the progress of the process in real time. Generally, more than one variable may be plotted on the same screen. This type of display replaces the strip chart recorder. Trended data kept for archival purposes is often called *historical trend* data.

6. *Manual Control.* Displays used to manually control a device are generally only available to maintenance personnel. Depending on the process, they may be accessed only from a screen near the equipment rather than from a centrally located control room. These screens are meant to bypass the automatic control system when conditions preclude the control system from working. Generally, these displays are used cautiously, and permission is granted only when the process is in a safe state.

7. *Diagnostic.* These screens are used by maintenance personnel to diagnose equipment failures and usually contain information more detailed than alarms. For example, most variable-frequency motor drives (controllers) have a wealth of diagnostic information available. A diagnostic display could be set up to display this information when requested. For example, with this display an operator could diagnose a drive fault as an overcurrent fault on acceleration without needing to walk to the physical drive and access the information from the drive front panel.

Security issues are often overlooked when designing a user interface. The computer-based user interface allows access to a large amount of process information and parameters. Thus, it is necessary to restrict access to certain functions. Typically, a user logs on to the system with a *user name* and *password,* which grants him or her an *access level.* A higher access level permits greater access to the restricted functions. For example, an operator generally is not permitted to change recipes or PID tuning parameters. However, an engineer logging onto the same system can access these functions provided his or her access level is high enough.

## 3.5 CONTROL SYSTEM TECHNOLOGY

The typical architecture of a process control system is shown in Figure 3.35. This figure emphasizes the lower layers of the control model and is intended to be generic. The particular architecture depends on the control system vendor. The technology used to implement control generally depends on the particular layer. Critical interlocks may be hardwired or implemented in special safety protection controllers. Other interlocks will be programmed in a discrete or regulatory controller. Connections to the physical process transducers and actuators are typically accomplished using input–output (I/O) panels or files that are remotely mounted. There is intense activity in the interconnection of field devices to standard remote I/O networks (e.g., Foundation Fieldbus, Interbus-S, Ethernet, and Control Area Network) in order to save wiring costs. At the process control level there has traditionally been two types of controllers. Traditionally, discrete-parts

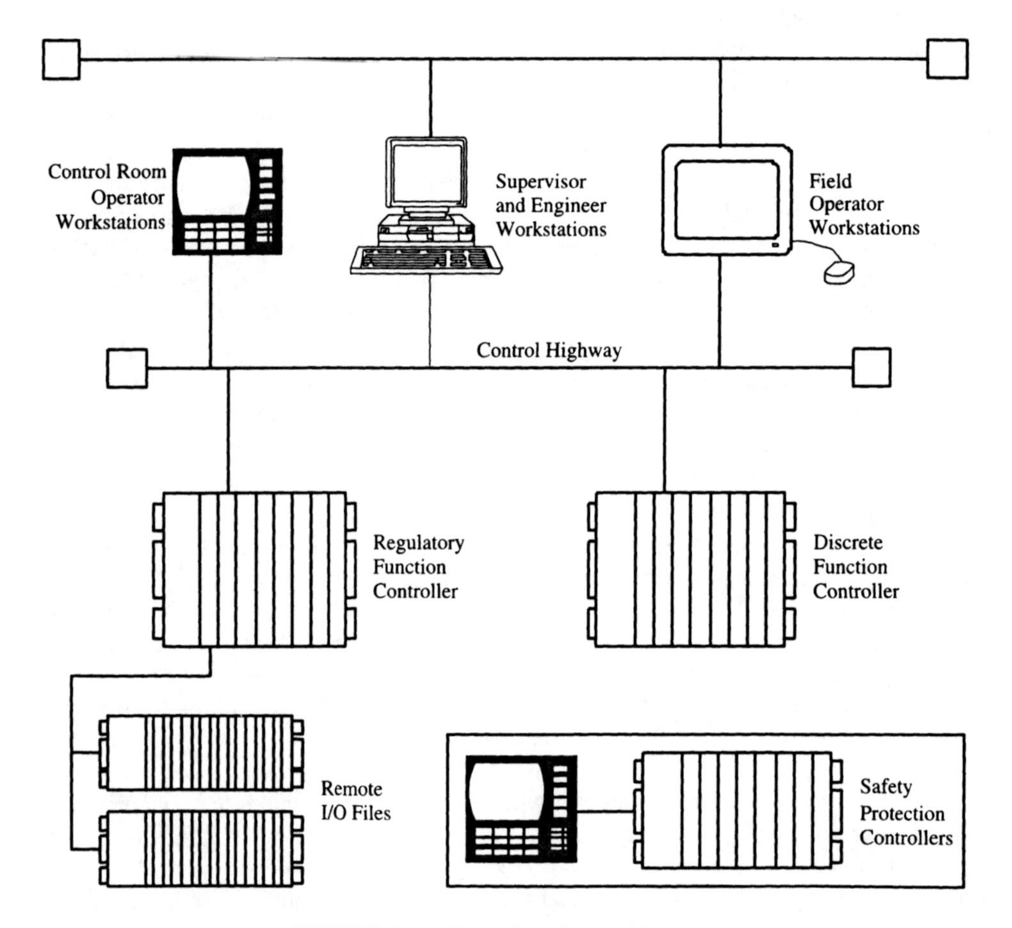


**FIGURE 3.35.** General system architecture.

manufacturing processes have been controlled using a programmable logic controller (PLC) and continuous systems have been controlled with a distributed control system (DCS). Batch control uses either or both. Recently, DCS and PLC systems have been evolving toward each other. With the advent of the IEC 1131 (International Electrotechnical Commission, 1993) standard programming languages (ladder logic, function block, sequential function chart, structured text, and statement list), both DCS and PLC functionality can reside in the same device. Because of this DCS/PLC trend toward a device with the same functionality, the authors will use the term programmable electronic system (PES) to refer to any system that implements the control functions in the process management layer or lower control layers.

All of the PES controllers are typically interconnected with a proprietary network called a control highway. However, the trend is away from these proprietary networks and toward standard protocols, like Ethernet [IEEE 802.3 (IEEE, 1996)] or the IEEE 802.4 (IEEE, 1995) token bus protocol of the Insitute of Electrical and Electronics Engineers. The control highway also connects the workstations that implement the operator, engineer, and supervisor user interfaces. Typically, there is a plant information network, generally Ethernet, that connects the plant control and information system with the functions of the area and site enterprise functions. Typically, the area and site control functions reside in a host computer.

## 3.6 EXAMPLE MODELS

Two examples from the public domain are used to illustrate the automation and control models presented in this chapter. Neither example is developed fully, though. The physical model of the first example, a paper production site, is shown down to the unit level. This process is largely a continuous-control example. The functions of the unit supervisory control schemes within a particular recovery cell are used as illustrations of control modules. The second example, a batch process, is a pharmaceutical manufacturing process. A particular cell process is divided into stages and operations. The physical model and procedural control is developed in detail for one unit.

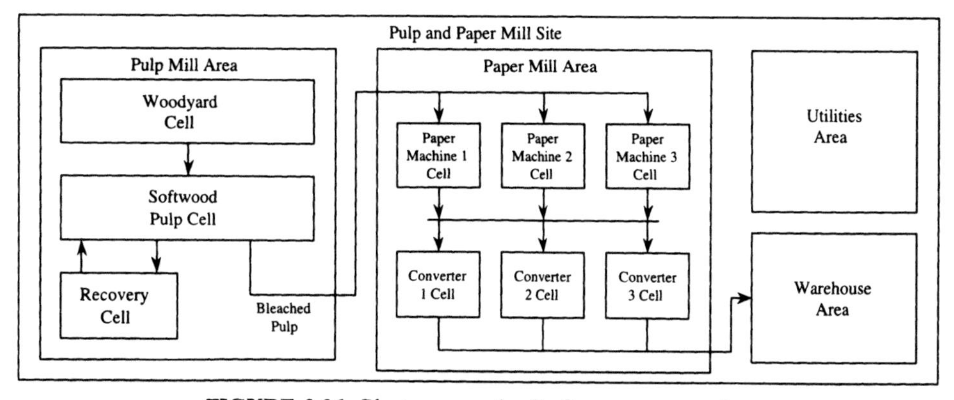


**FIGURE 3.36.** Plant areas and cells for paper enterprise.

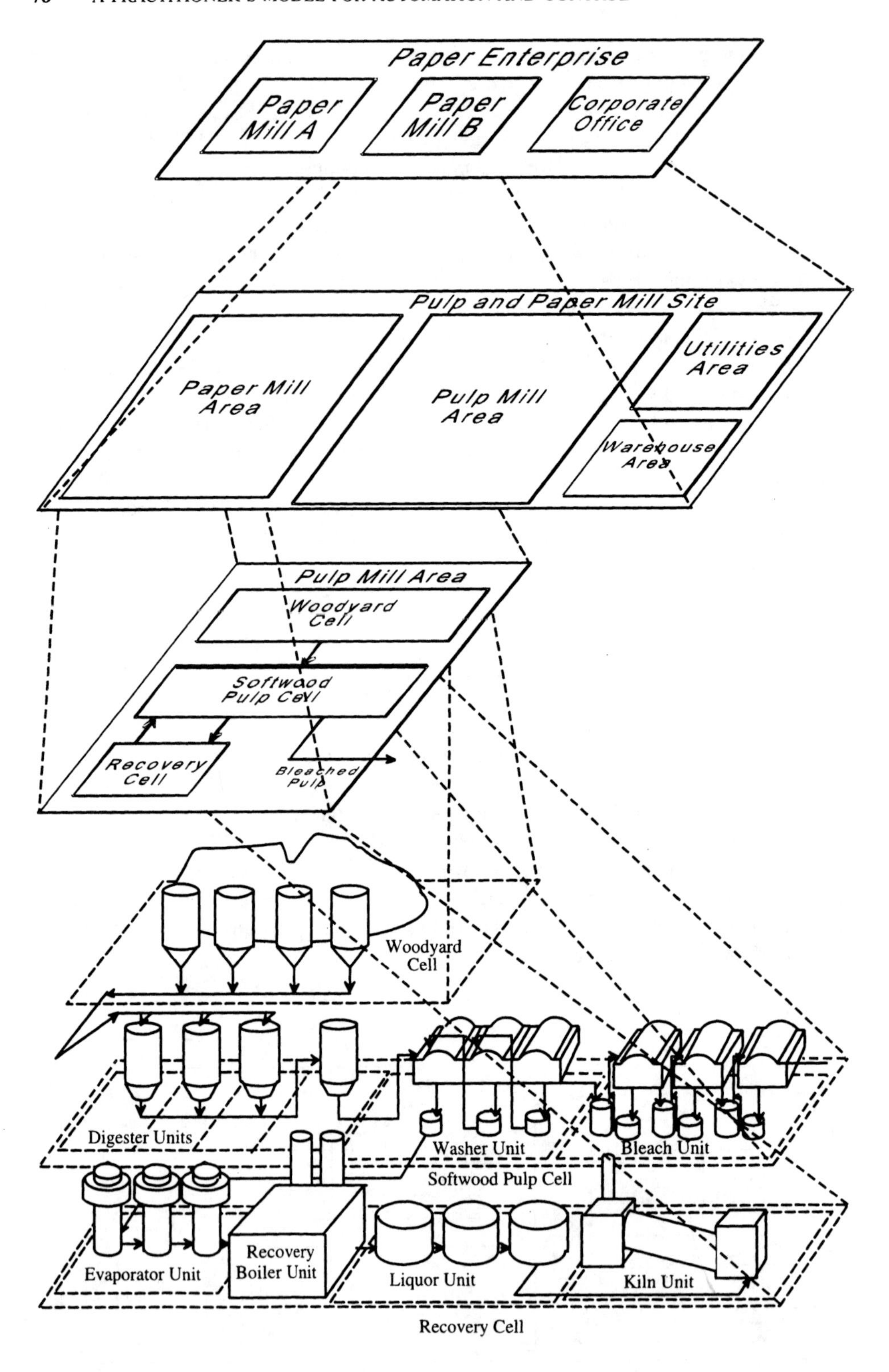


**FIGURE 3.37.** Hierarchical view of paper enterprise.

### 3.6.1 Paper Production Enterprise

A typical site of a paper production enterprise is shown in Figure 3.36 broken down into plant areas and process cells (Biermann, 1996; MacDonald and Franklin, 1969, 1970). The site is divided into four plant areas: the pulp mill, the paper mill, the warehouse area, and the utilities area. In addition, the pulp mill area is subdivided into the process cells of woodyard, softwood pulp, and recovery. The paper mill area is subdivided into multiple paper machine cells and multiple converter cells. The enterprise, site, and plant area levels are shown hierarchically in Figure 3.37. On this diagram, the recovery cell is further broken down into units.

An alternative model of a paper production enterprise is shown in the case study of Chapter 10. Both models are equally valid and illustrate that there is no unique model of a plant. In this section, the emphasis is on the physical model of the plant. In Chapter 10, the emphasis is on the control aspects of the plant.

The site is broken down into areas, cells, and units, as shown in Table 3.1. The pulp mill process cells for a representative kraft process are shown in Figures 3.38–3.41. As shown in Figure 3.38, the woodyard consists of a barker unit that removes the bark from each log, a chipper unit that mechanically breaks down the wood into chips, a screener to classify the chip size, and chip storage. A simplified schematic of the softwood pulp cell is depicted in Figure 3.39. Batch digesters are shown here, with each one considered a separate unit. Continuous digesters may also be present in a pulp mill. In a typical sequence, the batch digester is charged with softwood chips, white liquor, and black liquor. The mixture is circulated and chips are added as the contents settle. The digester is then sealed and heated with steam. After the cooking temperature is reached, it is maintained for 20–45 min. When the cook is completed, the contents are discharged to one of the blow tank units. A

**TABLE 3.1 Paper Production Site Areas, Cells, and Units**

| Plant Area | Cell | Unit |
|---|---|---|
| Pulp mill | Woodyard | Barker<br>Chipper<br>Ship storage |
| | Softwood pulp | Digester<br>Refiner<br>Brown stock washer<br>Bleach |
| | Recovery | Evaporator<br>Recovery boiler<br>Liquor processing<br>Lime kiln |
| Paper mill | Paper machine | Stock preparation<br>Wet end<br>Dryer<br>Calender |
| | Converter | Coater<br>Slitter |

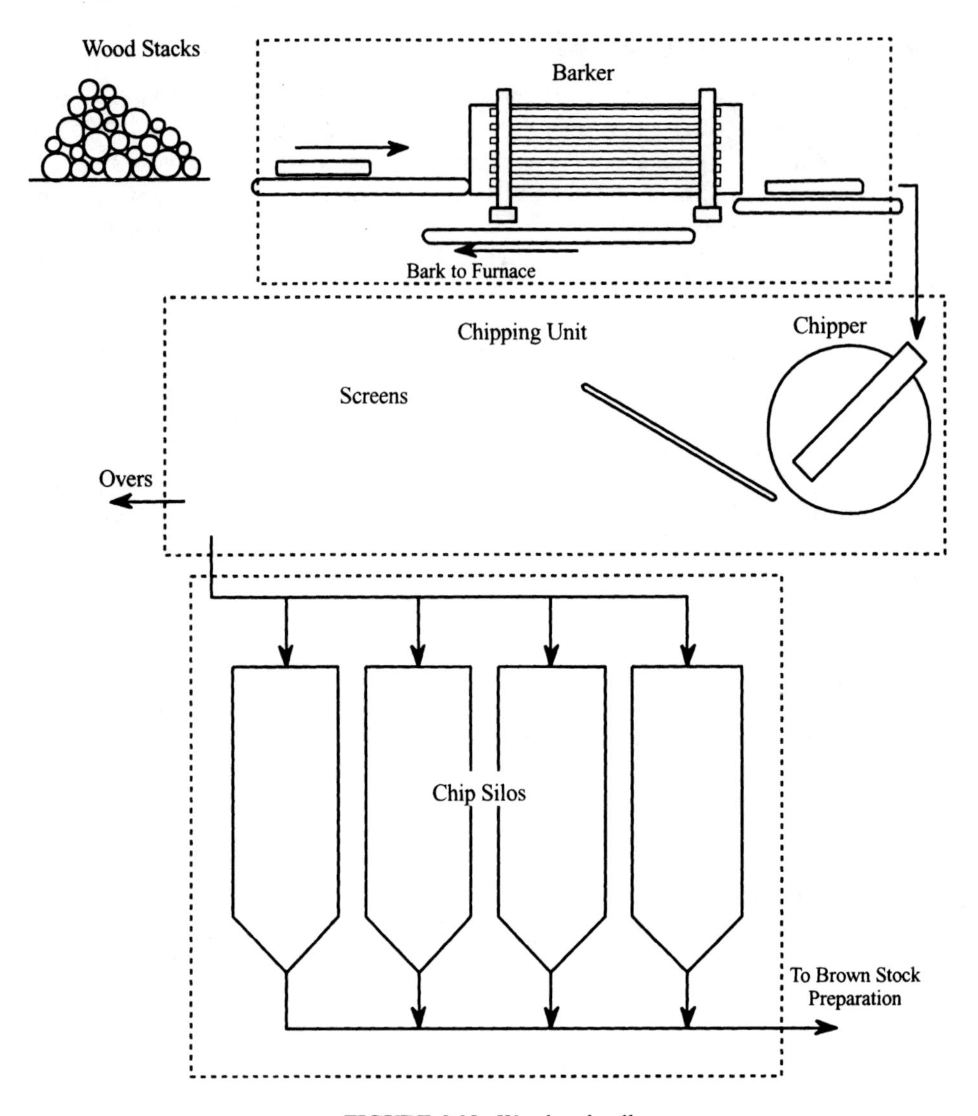


**FIGURE 3.38.** Woodyard cell.

blow tank receives the hot pulp from a digester, mixes it with weak black liquor, and cools the pulp to about 100°C. The hot stock refining equipment module follows the blow tanks. The fibrilizer separates the wood fibers prior to washing and the knotter removes large rejects. The brown stock washer recovers process chemicals in the form of weak black liquor. The washed pulp proceeds to the bleach unit. In a typical bleaching unit, shown in Figure 3.40, the pulp is treated in a series of chemical stages. Each stage typically consists of a mixing unit to mix the chemical with the pulp, a retention tower to provide time for the bleaching chemical to react with the pulp and a washer to remove the bleaching chemicals. The recovery cell is shown schematically in Figure 3.41. The evaporator unit concentrates the weak black liquor that is sprayed into the recovery furnace. The sulfur- and sodium-based inorganic materials are liberated and recovered as a liquid smelt. This smelt is

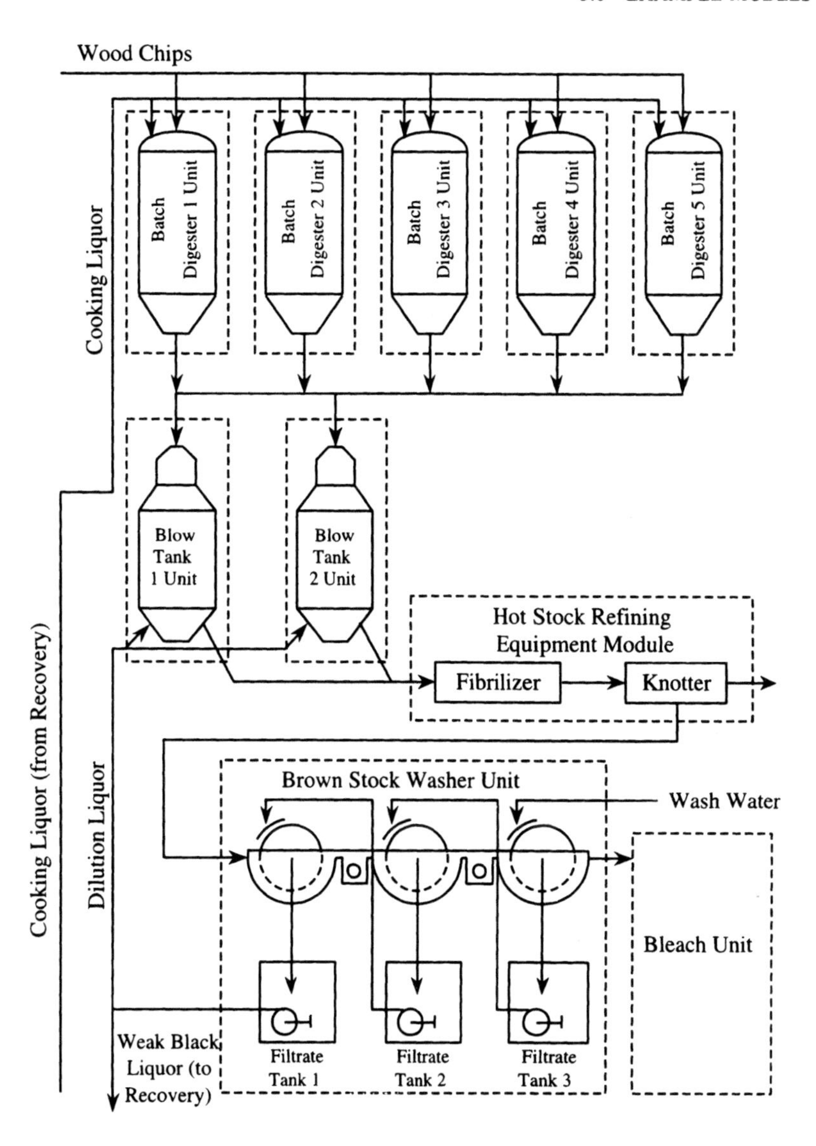


**FIGURE 3.39.** Softwood pulp cell.

dissolved in water to form green liquor. The white cooking liquor is formed by converting the sodium carbonate in the green liquor to sodium hydroxide using calcium hydroxide. The precipitate from the white liquor clarifier (calcium carbonate), is heated in a kiln and converted to calcium oxide (lime), which is then used to convert more green liquor.

Figures 3.42–3.44 show the cells for the paper mill area divided into units. The units of a paper machine cell are shown in Figures 3.42 and 3.43. The stock preparation unit is shown in Figure 3.42. The stock proportioner blends the appropriate raw pulp and additives into the proper recipe for the paper machine. The machine chest acts as a buffer between stock preparation and the paper machine. Contaminants are removed from the pulp slurry with a series of vortex cleaners. A high percentage of usable fiber occurs in the

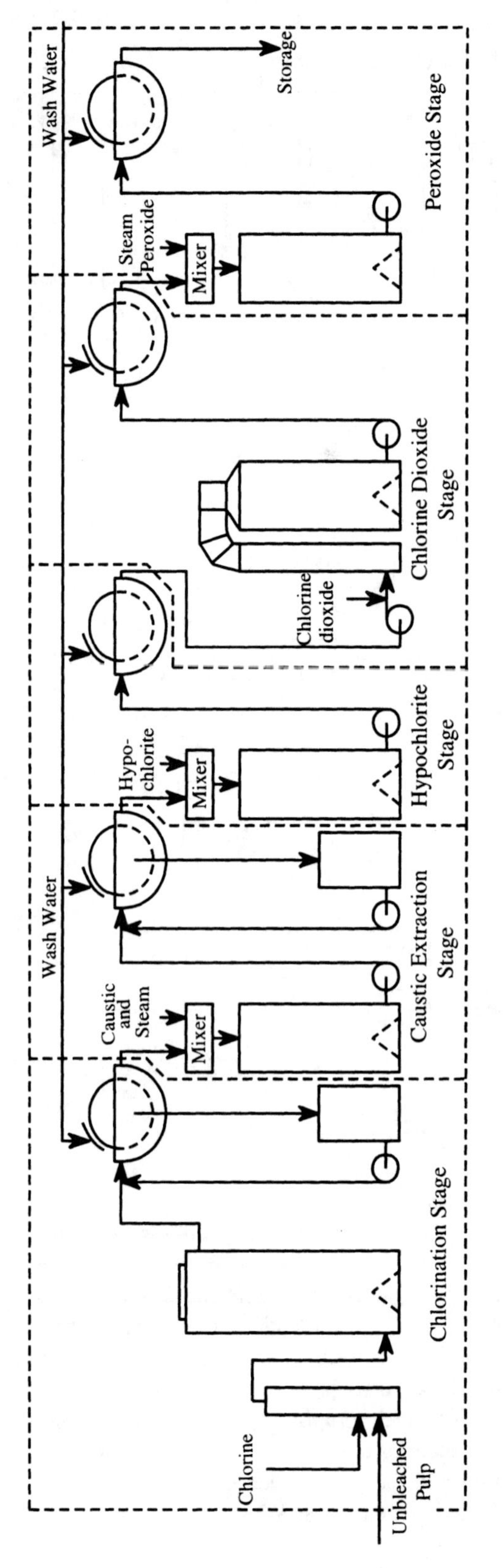


**FIGURE 3.40.** Bleach unit of pulp cell.

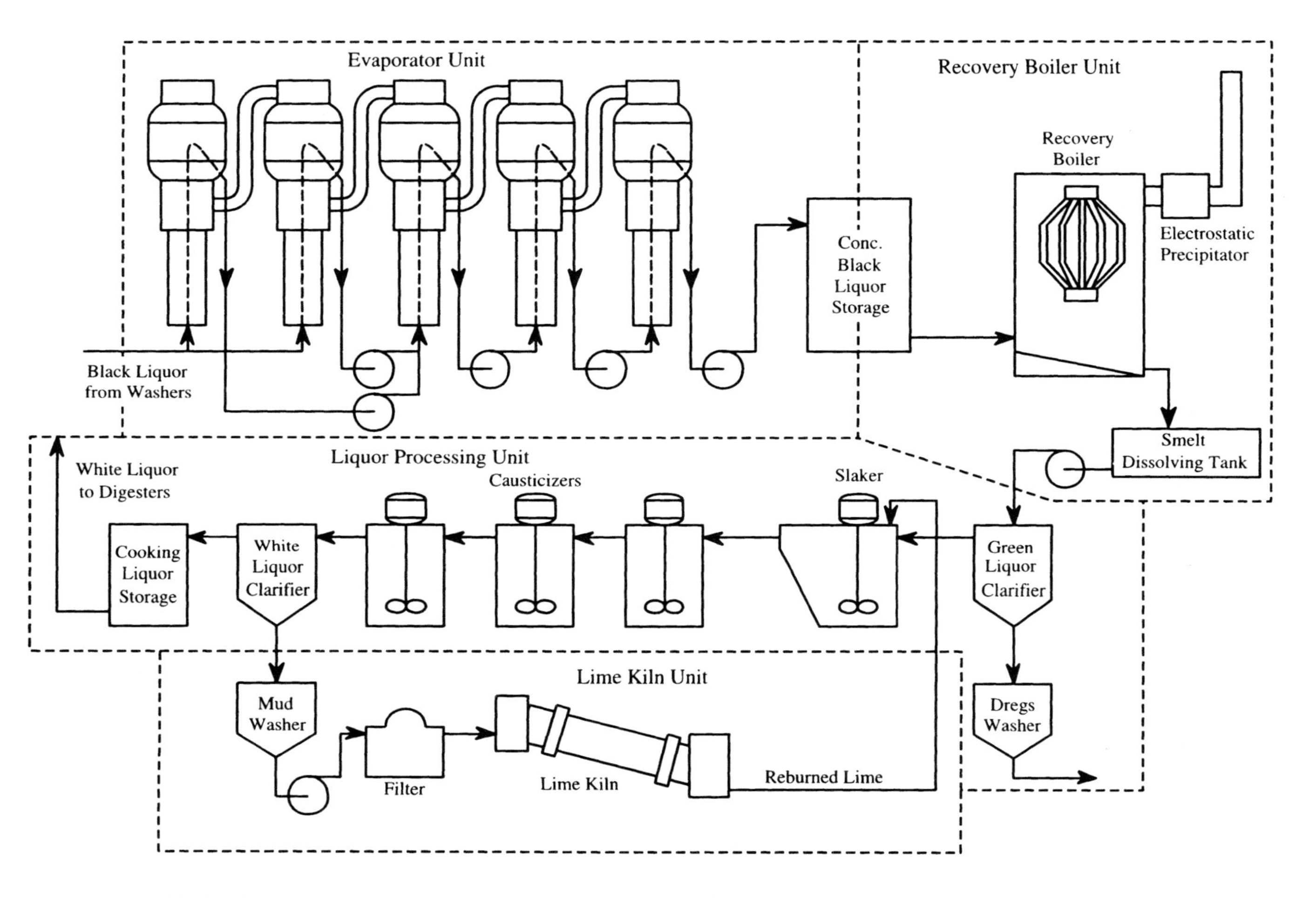


**FIGURE 3.41.** Recovery cell. (Reprinted by permission, with changes. Copyright © 1988, M. A. Keyes.)

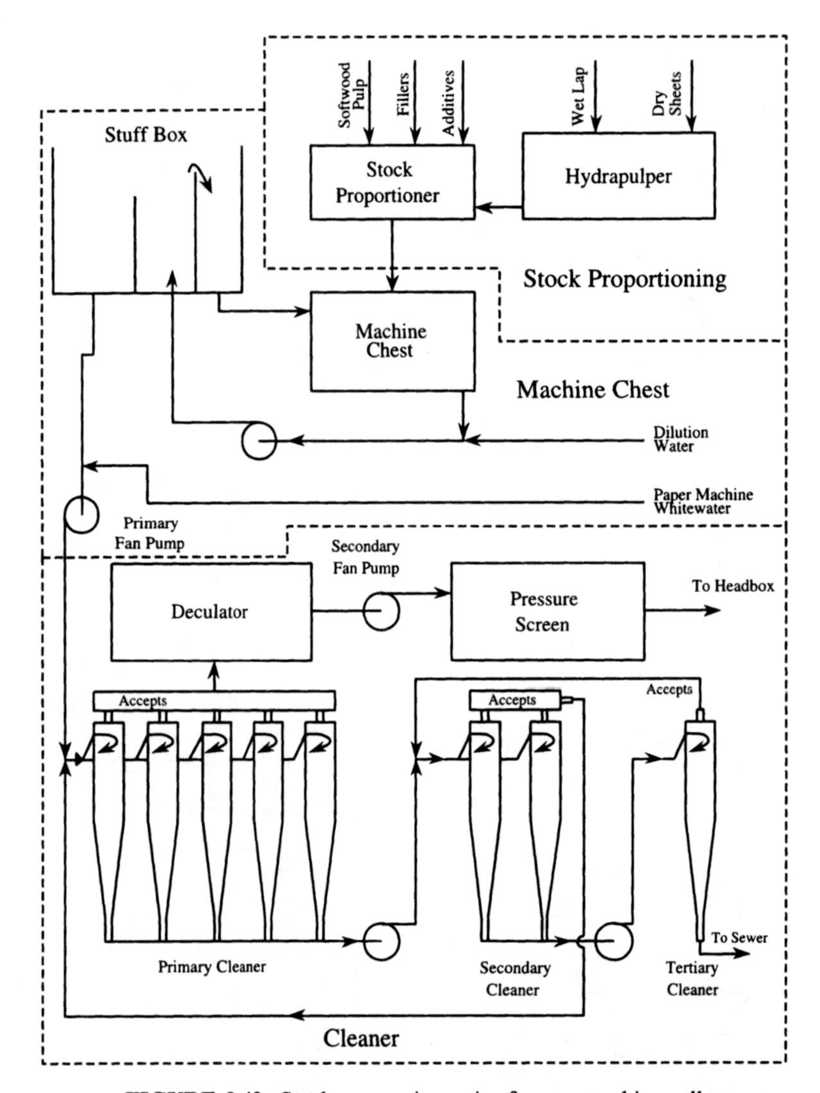


**FIGURE 3.42.** Stock preparation unit of paper machine cell.

reject stream from the vortex cleaners; hence the rejects from the primary cleaners are further treated with secondary cleaners and the rejects from the secondary cleaners are treated with a tertiary cleaner. The deculator removes entrapped air in the slurry and the pressure screen removes dirt and other large particles from the stock. The other units of the paper machine cell are shown in Figure 3.43. The wet end unit contains the head box, which meters the slurry onto a horizontal, moving, fine-mesh woven wire cloth. The wire cloth runs over the table rolls, suction boxes, and then over the couch roll, where the web of fibers leaves the Fourdrinier table. The web proceeds through one or more presses that mechanically remove water and compress the sheet and then through a set of steam-heated dryer rolls that removes the water by evaporation. The dryer also contains a size press that applies sizing, which improves the water resistance of the paper. The calender improves the

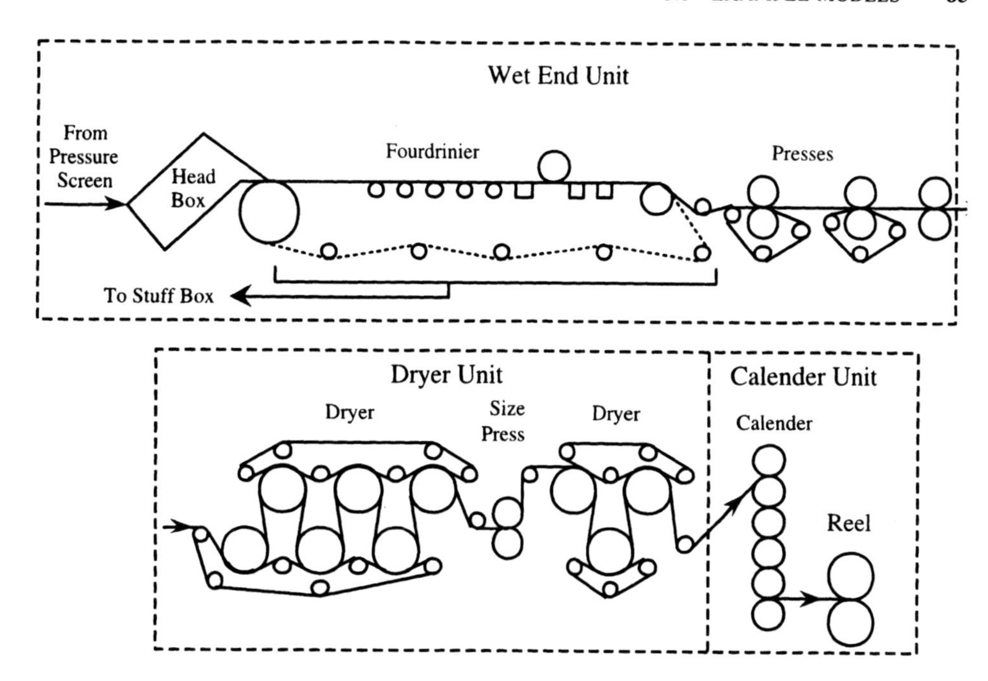


**FIGURE 3.43.** Other units of paper machine cell.

smoothness of the paper. Finally, the paper is collected on a reel. Conversion is a general term for any operation that follows the paper machine cell. Figure 3.44 depicts a converter cell that coats the paper and then cuts the paper into narrower rolls during rewinding.

As an example of unit supervision control, consider the recovery cell in Figure 3.41. Keyes and Kaya (1988) outline advanced control strategies for the liquor processing and lime kiln units of a pulp mill. The overall goal of these two unit controls is to provide efficient operation of the lime circulation system (causticizing and calcining) in the pulping recovery area.

The control for the liquor processing unit accounts for the inherent complexities of the recausticizing process by using an inferential controller without an actual measurement of lime feed rate. The equipment and instrumentation for one control strategy of this unit are

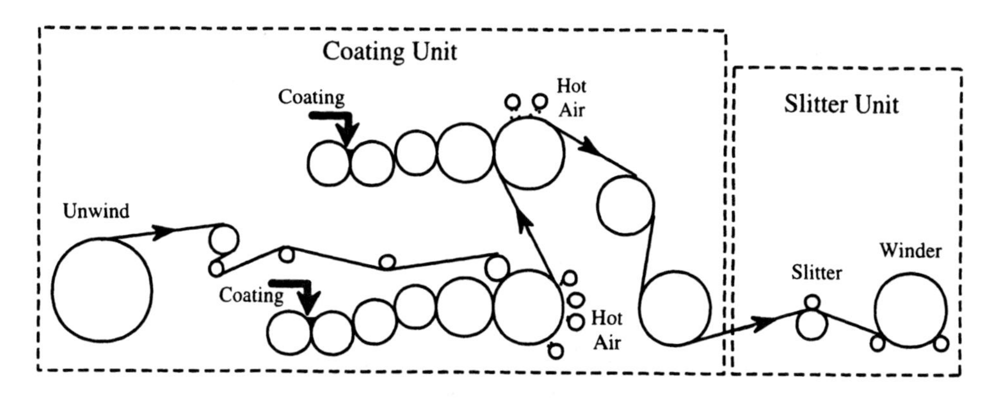


**FIGURE 3.44.** Example converter cell.

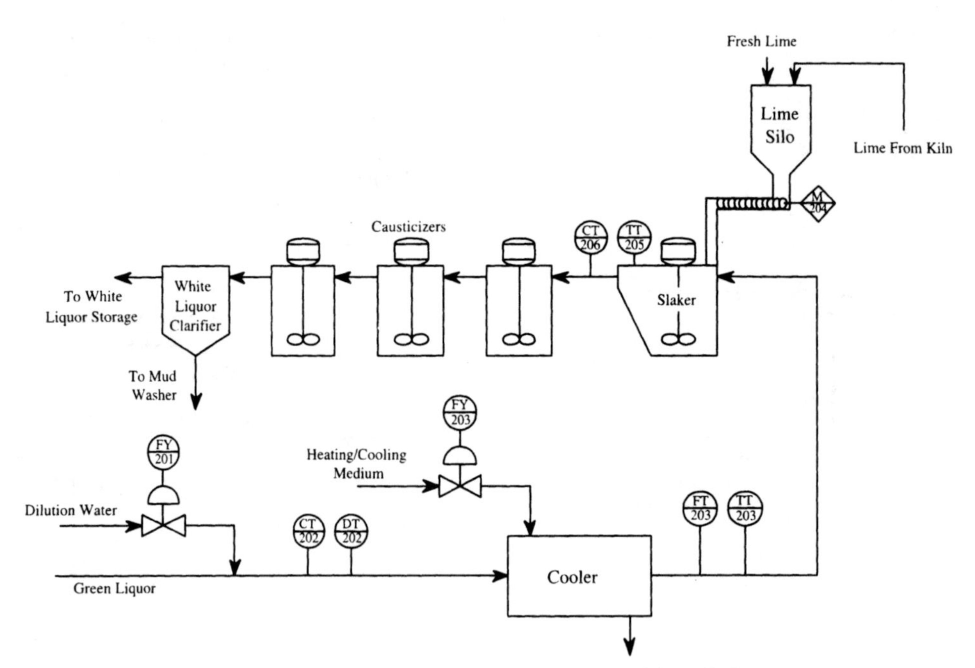


**FIGURE 3.45.** Liquor processing unit control installation.

shown in Figure 3.45 and a simplified representation of the unit supervision dealing with the recausticizing control is shown in Figure 3.46. The entire unit control strategy also includes control of the causticizer agitators and control of the white liquor clarifier, which is not shown. The control strategy automatically maximizes conversion of $Na_2CO_3$ to NaOH without overliming. The unit controller calculates the predicted lime–liquor ratio, automatically adjusts this ratio based on the NaOH concentration at the slaker discharge, inferentially controls the lime feed rate, and includes boil-over safety features. Advanced controller inputs include standard process measurements and periodic operator entry of laboratory test results from both the green liquor and the liquor discharge from the slaker. The two advanced controller outputs are the remote setpoint for the lime feeder speed control and an override signal to the green liquor temperature control valve.

The recausticizing control strategy broken down into control modules is shown in Figure 3.46. The [NaOH] feedback control, inferential lime feed rate control, and slaker temperature override control (shown as hatched boxes) can all be considered loop control modules. The latter two could be classified as parallel acting since they both control the lime feeder speed control. However, [NaOH] feedback control can also be considered part of a larger control module—the advanced recausticizing controller—that generates the setpoint to the inferential lime feed rate control.

Effective control of the lime kiln is crucial to the overall economics of the pulping process. Underburned or overburned lime can be costly by reducing pulping efficiency and increasing energy costs. The goal of the lime kiln control strategy is to stabilize lime quality and minimize kiln fuel consumption. The equipment and instrumentation for the lime kiln unit is shown in Figure 3.47. The temperature of the flue gas indicates the temperature at the entrance of the kiln and an optical pyrometer measures the temperature of the lime leaving the kiln. The unit control strategy is shown in Figure 3.48. The hot-end and cold-end temperatures are stabilized using dynamic decoupling and lime mud flow rate

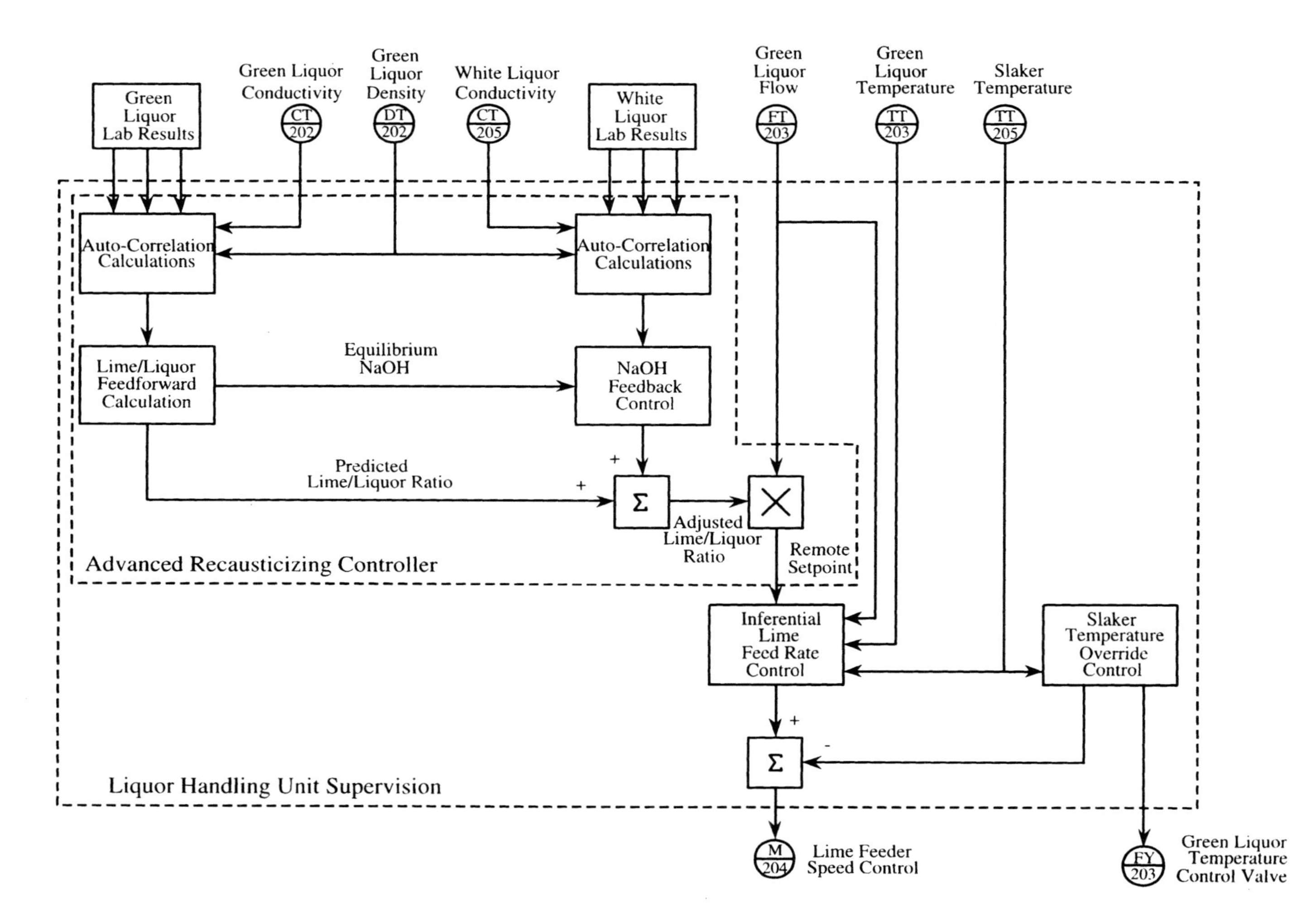


**FIGURE 3.46.** Recausticising control part of liquor processing unit. (Reprinted by permission, with changes. Copyright © 1988, M. A. Keyes.)

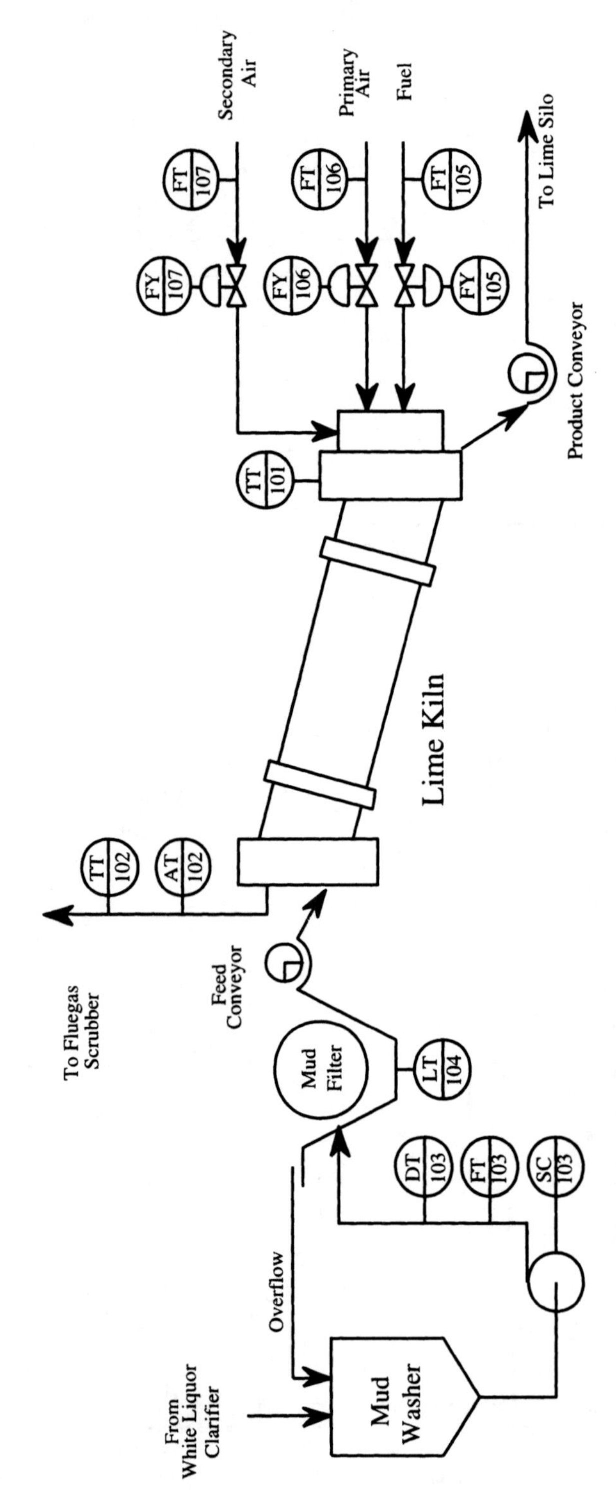


**FIGURE 3.47.** Lime kiln unit installation. (Reprinted by permission, with changes. Copyright © 1988, M. A. Keyes.)

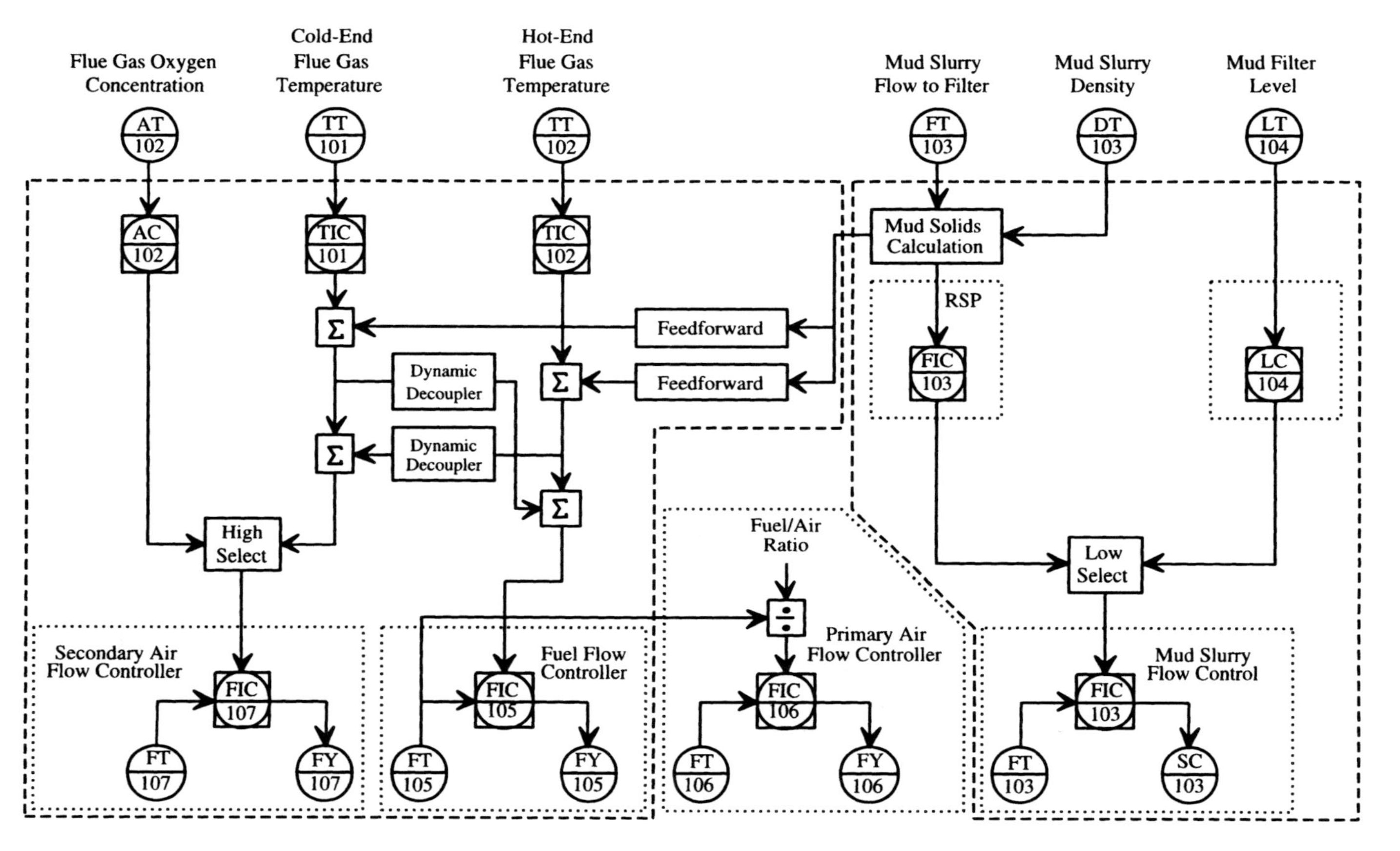


**FIGURE 3.48.** Lime kiln control part of liquor processing unit. (Reprinted by permission, with changes. Copyright © 1988, M. A. Keyes.)

feedforward. The operator determines the kiln speed and temperature setpoints that optimize lime availability and fuel consumption, but these setpoints could be determined by a cell controller that calculates this optimization on a period basis. The unit controller provides remote setpoints to the air and fuel flow controllers and the mud slurry flow rate controller. When the mud filter vat overflows, the measured slurry flow does not indicate actual kiln feed rate, so an override control is added to prevent this condition. Also, there is a flue gas override control to prevent unsafe kiln operation when the secondary air flow is reduced to maintain the cold-end temperature.

This unit is also shown divided into control modules. The secondary and fuel flow controllers are loop control modules. However, these two control modules are part of another control module, a multivariable kiln temperature controller. The entire mud slurry flow control can be considered a single control module. But it is formed from three loop control modules: mud solid flow control, filter level control, and mud slurry flow control.

### 3.6.2 Pharmaceutical Manufacturing Enterprise

A pharmaceutical manufacturing site is shown in Figure 3.49. The site is divided into five plant areas: solid dosage manufacturing, liquid manufacturing, inhalation manufacturing, finished-goods area, and offices area. A typical solid dosage manufacturing cell (SDMC), which does bulk tablet manufacturing, is described by Cole (1995). The process overview is shown in Figure 3.50. The process, which is called tablet manufacturing, is broken up into stages and operations, as shown in Table 3.2.

The physical model layers of a SDMC cell are organized into the following units:

Material reception
Bulk dispensing
Manual dispensing
Granulation
Blending
Compression
Coating
Packing
Dispatch
Quality assurance

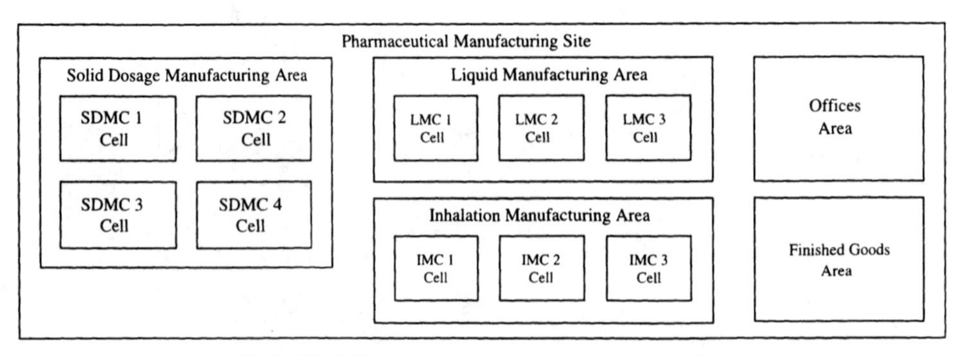


**FIGURE 3.49.** Pharmaceutical manufacturing site.

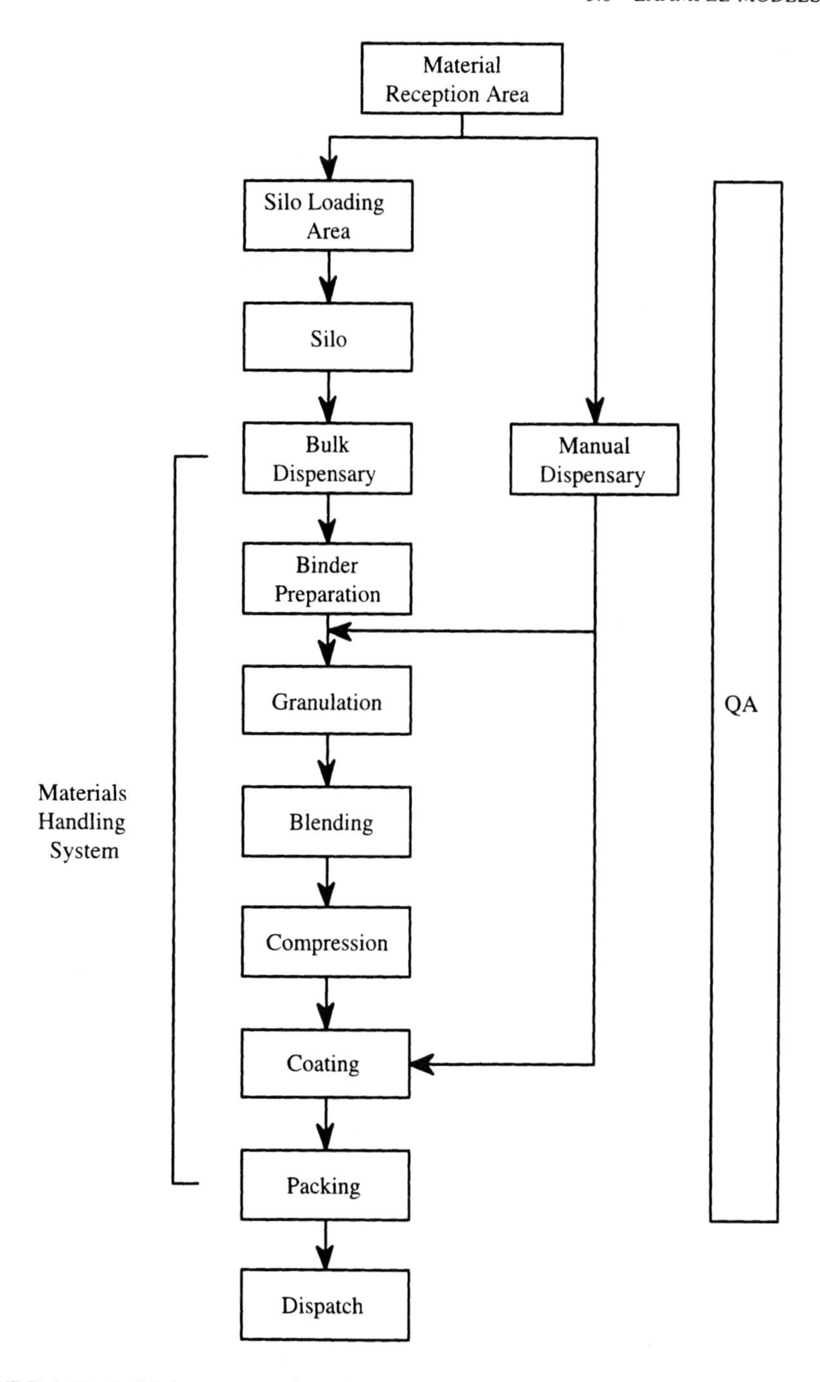


**FIGURE 3.50.** Solid dosage manufacturing cell process overview. [This figure was originally presented in the World Batch Forum (WBF) publication *SP88 Batch Automation in the Process Industries*, vol. 1. It is reproduced here by arrangement with the World Batch Forum which holds the copyright.]

**TABLE 3.2 Tablet Manufacturing Stages and Operations**

| Process Stage | Process Operation |
|---|---|
| Raw-materials handling | Material arrives in SDMC<br>Quality assurance check<br>Sieving<br>Load silo<br>Transfer to holding area |
| Manual dispensing | Equipment check<br>Material check<br>Booth dispensing<br>Isolator dispensing<br>Reconciliation<br>Sampling<br>Sieving |
| Bulk dispensing | Equipment check<br>Material check<br>Bulk dispense<br>Transfer manually dispensed<br>Material<br>Discharge to bin<br>Reconciliation |
| Binder preparation | Water addition<br>Solid addition<br>Heat<br>Slurry preparation<br>Slurry addition<br>Heat<br>Agitate |
| Granulation | Equipment check<br>Load mixer<br>Dry mix<br>Solvent addition<br>Wet mix<br>Mill/transfer to dryer<br>Drying<br>Mill/discharge to bin<br>Clean-in-place |
| Blending | Transfer to blend station<br>Blend<br>Transfer to holding area |
| Compression | Equipment check<br>Material check<br>Manual tablet compression<br>Compression run<br>Process tests<br>Layering press<br>Discharge<br>Clean-in-place |

**TABLE 3.2** (*continued*)

| Process Stage | Process Operation |
|---|---|
| Coating | Equipment check<br>Material check<br>Solution preparation<br>Charge coater<br>Coat<br>Drying<br>Discharge<br>Clean-in-place |
| Packing | Line clearance<br>Material check<br>Machine set up<br>Material sorting<br>Initial run<br>Main run<br>Carton<br>Sampling<br>Casing<br>Overwrapping<br>Bulk packing<br>Granule packing |

**TABLE 3.3 Granulation Unit Physical Model Organization**

| Unit | Sub unit | Equipment Module | Control Module |
|---|---|---|---|
| Granulation | Granulator | Loading system | Transfer valve<br>Hopper contact<br>Chute feedback |
| | | Solvent addition | Load cell<br>Transfer valve |
| | | Binder addition | Load cell<br>Transfer valve |
| | | Granulator | Chopper motor<br>Impeller motor<br>Impeller load |
| | | Transfer | Mill feedback<br>Discharge valve<br>Sieve size |
| | Dryer | Dryer | Air flow<br>Temperature<br>Bowl pressure drop<br>Filter pressure drop |
| | | Discharge system | Dry mill<br>Hopper feedback |

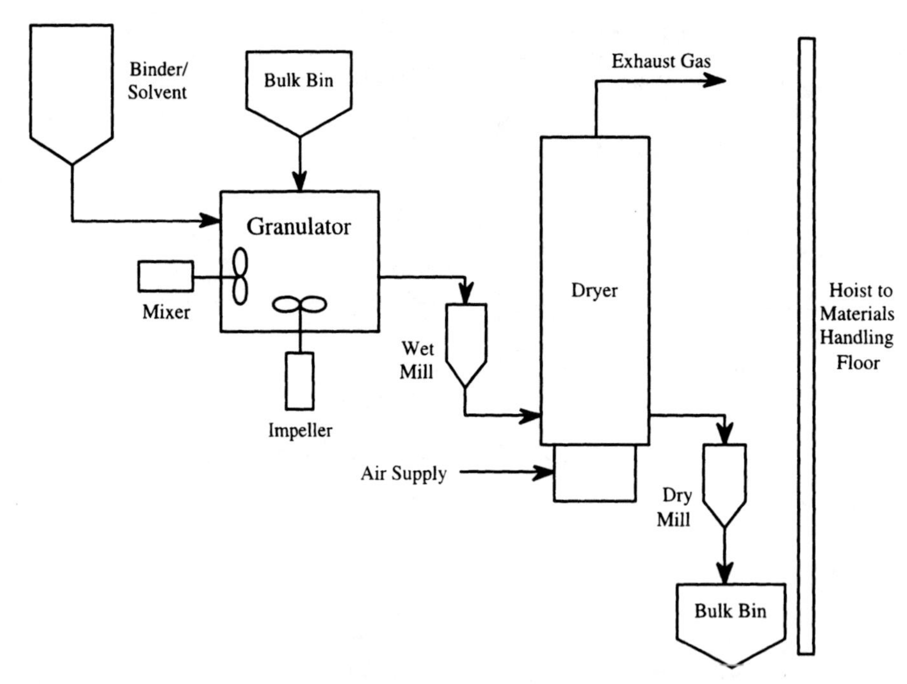


**FIGURE 3.51.** Granulation unit. [This figure was originally presented in the World Batch Forum (WBF) publication *SP88 Batch Automation in the Process Industries*, vol. 1. It is reproduced here by arrangement with the World Batch Forum which holds the copyright.]

**TABLE 3.4 Overall Procedure for Making Tablets**

| Procedure | Unit Procedure | Operation |
|---|---|---|
| Make tablets | Receive materials | |
| | Bulk dispense | |
| | Manual dispense | |
| | Granulate | Preparation<br>Load<br>Dry mix<br>Binder addition<br>Wet mix<br>Transfer (wet mill)<br>Dry<br>Discharge (dry mill)<br>Clean-in-place |
| | Blend | |
| | Compress | |
| | Coat | |
| | Pack | |
| | Dispatch | |
| | Quality assurance | |

**TABLE 3.5 Granulate Unit Procedure Expanded into Phases**

| Operation | Phase |
|---|---|
| Preparation | Area clean check<br>Clothing challenge<br>Ambient condition check<br>Spray arm check<br>Binder vessel level<br>Prime binder pipe<br>Spray arm position<br>Through-hole feeder<br>Preheat dryer |
| Load mixer | Check bin<br>Open valve<br>Transfer<br>Close valve |
| Dry mix | Start impeller<br>Start chopper<br>Start timer |
| Binder addition | Impeller speed check<br>Chopper speed check<br>Start pump<br>Run timer<br>Stop pump |
| Wet mix | Impeller speed check<br>Chopper speed check<br>Run timer<br>Stop impeller<br>Stop chopper |
| Transfer (wet mill) | Fit screen<br>Open discharge valve<br>Open dryer feed valve<br>Transfer |
| Drying | Run dryer<br>Record results |
| Transfer (dry mill) | Fit screen<br>Locate product bin<br>Transfer<br>Material balance |

The granulation unit is depicted in Figure 3.51, and the organization of the physical model into subunits, equipment modules, and control modules is shown in Table 3.3.

The procedural control structure for the SDMC cell is outlined in accordance with S88.01 and consists of a procedure divided into unit procedures, operations, and phases. The entire procedure for the cell is described, and then the procedural control for the granulation unit is broken down into operations and phases. The overall procedure for making tablets is divided as shown in Table 3.4. The granulate unit procedure is expanded into phases in Table 3.5. Note that not all of the phases are automated.

## References

Biermann, C. J., *Handbook of Pulping and Papermaking*, 2nd ed., Academic, San Diego, CA, 1996.

Cole, N., "SP88 Batch Control Models and Terminology: An Example Application Based on a General Purpose Solid Dosage Manufacturing Centre," *SP88 Batch Automation in the Process Industries*, World Batch Forum, Phoenix, AZ, 1995.

Grosdidier, P., A. Mason, A. Aitolahti, P. Heinonen, and V. Vanhamäki, "FCC Unit Reactor-Regenerator Control," *Proceedings of the 1992 American Control Conference*, American Automatic Control Council, Evanston, IL, 1992, pp. 117–121.

IEC, *IEC 1131-3: Programmable Logic Controllers—Part 3: Programming Languages*, International Electrotechnical Commission, Geneva, Switzerland, 1993.

IEEE, *Local Area Networks—Part 4. Token-Passing Bus Access Method and Physical Layer Specifications [ANSI/IEEE Std 802.4-1990 (R1995)]*, Institute of Electrical and Electronics Engineers, New York, 1995.

IEEE, *Information Technology—Telecommunications and Information Exchange between Systems—Local and Metropolitan Area Networks—Specific Requirements—Part 3. Carrier Sense Multiple Access with Collision Detection (CSMA/CD) Access Method and Physical Layer Specifications [ANSI/IEEE Std 802.3; 1996 edition]*, Institute of Electrical and Electronics Engineers, New York, 1996.

ISA, *ISA-S88.01, Batch Control, Part 1: Models and Terminology*, Instrument Society of America, Research Triangle Park, NC, 1995.

Keyes, M. A., and A. Kaya, "A Perspective on Advanced Control of Lime and Mud Processing," presented at the 1988 American Control Conference, Atlanta, GA, June 15–17, 1988.

MacDonald, R. G., and J. N. Franklin, Eds., *Pulp and Paper Manufacture, Vol. 1: The Pulping of Wood*, 2nd ed., McGraw-Hill, New York, 1969.

MacDonald, R. G., and J. N. Franklin, Eds., *Pulp and Paper Manufacture, Vol. 3: Papermaking and Paperboard Making*, 2nd ed., McGraw-Hill, New York, 1970.

Saxe, J. G., excerpted from "The Blind Men and the Elephant," in H. Felleman, *The Best Loved Poems of the American People*, Doubleday, New York, 1936.

Stephanopoulos, G., *Chemical Process Control*, Prentice-Hall, Englewood Cliffs, NJ, 1984.

# 4 Process Modeling

**Chapter Topics**

- Dynamic models based on first principles
- Linearization
- Discrete-time models
- Empirical model identification

*Scenario:* Modeling plays a central role in process control and is a key element in the successful application of control.

Once our young engineer had solved the air/fuel controllability problem outlined in the Chapter 1 scenario, the normal mode of control became automatic and the operators appreciated the help. Operator acceptance opened the door to more advanced control possibilities. Operating personnel started asking, "Can you do this or that?" The young engineer felt a sense of accomplishment. While investigating the operator's requests, he noticed that operators were trying to adjust the firing rate to maintain melter glass temperature. He discovered that this was difficult for them to do and was a source of quality problems. Upon further investigation, he began to appreciate what was happening. The difficulty was in the dynamics of melter temperature control. It had an 8-hr time constant with significant deadtime. Also, it was affected by the pull rate changes on the melter caused by production run changes in the forming machines. Could this be improved? Maybe, if he could understand the relation between temperature and firing rate. The challenge was in how to use glass temperature to adjust the firing rate. After a few months of unsuccessful attempts he was exposed to a control concept called model-based control. He learned that he needed a dynamic computer model that could run in real time. By chance, his employer was involved with a high-technology university that was conducting research on glass processing. He decided to visit the campus and find out if they could help. At first he thought he had found the model. They had developed a computer model of a glass melter/furnace that addressed energy and material flow. Upon further investigation, he discovered that they had not considered the coupling of the two from an energy exchange standpoint. He was told that it was too complex. In other words, the energy exchange between the firing rate and glass melt was not coupled. The model was not useful in his quest to solve the glass melting control problem. Later, he discussed the problem with a long-time supervisor at the plant. The engineer asked a simple question, "When the pull rate on the glass melter changes, what do you do?" The supervisor said he adjusted the furnace temperature up/down 2°F for every ton/day increase/decrease in the feed rate. The young engineer immediately saw a practical model that he could use. The final model turned out to be more complex than he first thought, but with persistence he developed a glass temperature control scheme that was based on the supervisor's method. This model eventually resulted in improved glass melt quality and reduced fuel costs. The engineer's first adventure into model-based control was successful. He also learned the value of understanding the process by representing it as a mathematical model and simulating it in real time.

As illustrated by our young engineer, modeling plays a central role in process control and is a key element in the successful application of control. Without a model, the engineer can spend much time in a fruitless attempt to do the control in an ad hoc way. The type of model depends on its use. Generally models at the lower levels of the control system tend to be more complicated. Models do not need to be complex. For many systems, a simple empirical model that describes the gross behavior of the system is more than sufficient to design the control system illustrated by the glass furnace problem.

## 4.1 INTRODUCTION

Suppose our young engineer is told to develop a control system for a continuously stirred tank reactor (CSTR; see Figure 4.2 later). Faced with this new challenge, the first impulse ought to be to develop a model of the process either by first principles or from empirical data. The process model should answer the following questions:

1. *What are the controlled variables?* Proper regulation of these variables is the ultimate goal of the control system. These variables may not be directly measurable. For example, the temperature of molten glass in a furnace is usually not directly measurable but is inferred from other measurements. In the glass furnace example, the quality of the glass is the important controlled variable. Ultimately, the goal of any control system is to regulate the key variables in the process so that the enterprise makes a profit.
2. *What are the measured variables*? These variables are those that can be directly measured and thus used by the control algorithm. Usually, they are quantities like flow, pressure, temperature, and concentration. These variables are somehow related to the controlled variables. For the glass furnace, the measured variable is the furnace temperature even though the controlled variable is the glass temperature.
3. *Which variables can be manipulated*? Control inherently involves variables that can be adjusted in order to affect a change in the measured variables. For the glass furnace, one of the manipulated variables is the fuel input to the furnace.
4. *What is the expected effectiveness of the control system*? In order to implement the control properly, the model must determine the following aspects of the process:
   a. *Sign and magnitude changes*: When a manipulated variable is increased, does the controlled variable increase or decrease? For the glass furnace, an increase in the furnace firing rate increases the glass temperature.
   b. *Speed of response*: Does the controlled variable change rapidly or slowly when the given manipulated variable changes? The answer to this question determines what kind of controller manipulations can be tolerated by the process. For example, the glass furnace has a very sluggish response to a change in the firing rate. An aggressive controller response to a change in the furnace temperature will probably cause the system response to oscillate, which is not acceptable.
5. *How sensitive is the system to changes in the operating point*? Industrial control systems often experience changes in the operating conditions and equipment

performance. For example, the glass furnace performance will depend on the type of fuel, which may change depending on the time of year. A sensitivity analysis at the expected operating points of the process has to occur as a part of modeling.

This chapter reviews the procedures needed to obtain the models necessary for control system design. Many types of models are available to the control system designer, and these types are compared and contrasted.

1. *Mathematical Models.* According to Denn (1986, p. 1), "A mathematical model of a process is a system of equations whose solution, given specific input data, is representative of the response of the process to a corresponding set of inputs." Mathematical models are used in this text. These models are simple or complex, as dictated by the intended use.

2. *Fundamental and Empirical Models.* Fundamental models are based on fundamental concepts such as the conservation of material and/or energy. These models can provide great perception into the process operation but can be very complex and costly to develop. Therefore, empirical models based on experimental data are often developed and are usually sufficient for most control system design. Both types are treated in this chapter.

3. *Dynamic and Steady-State Models.* Both steady-state and dynamic models are covered in this chapter.

4. *Lumped and Distributed Models.* A lumped system is one in which the system properties do not change with position within the system and the model involves algebraic and differential equations. A distributed system is one in which the properties are dependent on position and the model involves partial differential equations. The models in this text involve lumped systems, with the exception of transport delay. For example, though the glass melt temperature changes as it moves through the furnace (a distributed system), the model developed by our young engineer treated the furnace as a lumped system where only the exit temperature is important.

5. *Continuous- and Discrete-Time Models.* Most real systems have continuous and discrete elements. For example, the glass furnace will have a continuous control valve to regulate the fuel flow but also has a series of solenoid valves to direct the proper type of fuel (natural gas, propane, etc.) to the fuel control valve. The continuous control valves have a range over which the position can be varied, whereas the solenoid valves are discrete, having only two states: on and off. The discrete model for a device like a solenoid valve is not treated in this chapter since it is usually very simple. However, the discrete-time model is considered since many control algorithms are implemented by digital controllers. The discrete-time model used by digital controllers arises because the signals are discrete-time signals. In addition, the accuracy of the discrete-time signal values is finite. However, the accuracy of the discrete-time signals is smaller than the accuracy of the sensors and actuators, and so the accuracy of discrete-time signals is considered to be identical to the accuracy of continuous signals. Both discrete-time and continuous models are treated in this text.

The type of model one uses generally depends on the context in which it is used. In general, fundamental dynamic models are preferred when designing control systems. However, these types of models are often expensive to build and many processes are not well known. For these reasons, empirical and steady-state models are often preferred. In addition, for control systems at the cell level and higher, the models are generally simple because detailed process dynamics are often not important at the higher control system levels.

At each level of the control model introduced in the previous chapter, a different type of process model is usually utilized.

*Process Control.* All model types may be used at this level, depending on the level of detail needed to design the control system. If a system is difficult to control, a fundamental dynamic model is preferred.

*Unit Supervision.* All mathematical models are also used at this level, though fundamental dynamic models are generally used only for designing and testing multivariable control systems. Otherwise, empirical steady-state models are quite sufficient.

*Process Management.* Steady-state empirical models are used at this level if modeling is needed at all.

*Plant Area Management.* Extremely simple empirical models are used at this level.

*Site Management.* If any models are used, they are very simple empirical models.

This chapter is a broad overview of the types of models that we will use in the design of control systems. Models based on fundamental principles are treated first and are taken from the differential equations through the linearization and methods of obtaining an analytic solution. As an alternative, the types of simple empirical models are introduced. Discrete-time models, which will be needed for model identification and digital control, are reviewed, followed by a review of numerical simulation of the system as an alternative to finding the analytic solution. The chapter is concluded by reviewing identification methods to obtain the empirical model parameters.

## 4.2 DYNAMIC MODELS BASED ON FUNDAMENTAL PRINCIPLES

The equations for dynamic models based on first principles must be formulated in terms of fundamental quantities. In chemical engineering, these quantities are mass, energy, and momentum. Under assumptions that are generally valid in chemical engineering systems, these quantities obey the principle of conservation, which is generally stated as

$$\text{Accumulation} = \text{In} - \text{Out} + \text{Generation} - \text{Consumption}$$

When the accumulation is zero, the balance results in an algebraic equation. For a nonzero accumulation, this balance results in a differential equation, which is generally written as

$$\frac{\begin{pmatrix}\text{accumulation of}\\ X\text{ within system}\end{pmatrix}}{\text{time period}} = \frac{\begin{pmatrix}\text{flow of }X\\ \text{into system}\end{pmatrix}}{\text{time period}} - \frac{\begin{pmatrix}\text{flow of }X\\ \text{out of system}\end{pmatrix}}{\text{time period}} + \frac{\begin{pmatrix}\text{amount of }X\\ \text{generated in system}\end{pmatrix}}{\text{time period}} - \frac{\begin{pmatrix}\text{amount of }X\\ \text{consumed in system}\end{pmatrix}}{\text{time period}} \tag{4.1}$$

where $X$ is one of the following fundamental quantities:

Total mass
Mass of a chemical component
Energy
Momentum

Other constitutive equations may be used to completely determine the model, such as

$$Q = hA(\Delta T) \qquad \text{(heat transfer)}$$
$$r_A = k_0 e^{-E/RT} c_A \qquad \text{(chemical reaction rate)}$$
$$PV = nRT$$
$$F = c_v \sqrt{\frac{\Delta P}{\rho}} \qquad \text{(fluid flow}$$
$$y_i = k_i x_i \qquad \text{(phase equilibrium)}$$

The model has the proper number of equations when the behavior of the system can be predicted from the model. A correctly formulated model has no degrees of freedom. The concept of degrees of freedom is expressed as

$$\text{DOF} = \text{NV} - \text{NE}$$

where DOF = degrees of freedom, NV = number of variables, and NE = number of equations. Note that NV represents the number of variables in the system and does not include the constant parameters or external stimuli. If NV is greater than NE, then the system is underspecified and the model must be corrected either by including more appropriate equations or by correctly designating a variable as a specified parameter or external stimuli. If NE is greater than NV, then the system is overspecified and in general no unique solution exists. In this situation, there are one or more dependent equations or constant parameters that ought to be designated as variables.

However, a system model expressed as a set of (usually nonlinear) differential equations is not immediately useful for control system design. One could use a nonlinear control algorithm, but these are difficult to develop and are usually overly complicated. In order to design a control system, it is generally useful to model the system as a set of linear differential equations. The control system designer has three options:

1. Simulate the nonlinear system on a computer and numerically compute its solution.
2. Develop a linear model that approximates the dynamic behavior of the system in the neighborhood of a specified operating point.
3. Transform the nonlinear system into a linear system by an approximate transformation of variables.

The third option can be done in a few cases. The first two options are generally always feasible. However, the second method is most useful for control system design, and it is the approach used by most control system designers. The first option is still useful for control system design testing and checkout before trying it on the real process. Both of the first two options are outlined in this chapter.

Developing a linear model that approximates the dynamic behavior of the system in the neighborhood of a specified operating point involves the following steps:

1. Formulate system differential equations based on first principles.
2. Linearize the differential equations about the operating point.
3. Laplace transform these equations.
4. Express as a transfer function.

### 4.2.1 Formulate Model Based on First Principles

In this section, first-principle models will be developed to illustrate the general technique outlined in the previous section. This section is only intended as an introduction to the subject. More information may be found in texts (Luyben, 1990; Marlin, 1995; Stephanopoulos, 1984).

In any modeling effort, simplifying assumptions are usually made in order to make the problem manageable.

The general principle of Equation (4.1) is expressed more concisely as follows:

Total mass balance:

$$\frac{d(\rho V)}{dt} = \sum_{i=1}^{\#\text{ of inlets}} \rho_i F_i - \sum_{j=1}^{\#\text{ of outlets}} \rho_j F_j \tag{4.2}$$

Mass balance on individual component $A$:

$$\frac{d(n_A)}{dt} = \frac{d(c_A V)}{dt} = \sum_{i=1}^{\#\text{ of inlets}} c_{A_i} F_i - \sum_{j=1}^{\#\text{ of outlets}} c_{A_j} F_j \pm r_A V \tag{4.3}$$

Total energy balance:

$$\frac{d(E)}{dt} = \frac{d(U+K+P)}{dt} = \sum_{i=1}^{\#\text{ of inlets}} \rho_i F_i h_i - \sum_{j=1}^{\#\text{ of outlets}} \rho_j F_j h_j + Q + W_S \tag{4.4}$$

where the variables in (4.2)–(4.4) are

$\rho$ = density of material in the system

$V$ = total volume of material in the system

$\rho_i, \rho_j$ = density of the material in the $i$th input stream and the $j$th output stream, respectively

$F_i, F_j$ = volumetric flow rate of the $i$th input stream and the $j$th output stream, respectively

$n_A$ = number of moles of component $A$ in the system

$c_A$ = molar concentration of $A$ in the system

$c_{A_i}, c_{A_j}$ = molar concentration of $A$ in the $i$th input stream and the $j$th output stream, respectively

$r_A$ = reaction rate per unit volume for component $A$ in the system. A positive reaction rate indicates that $A$ is produced; a negative $r_A$ indicates that $A$ is being consumed.

$U, K, P$ = internal, kinetic, and potential energies of the system, respectively

$h_i, h_j$ = specific enthalpy of the material in the $i$th input stream and the $j$th output stream, respectively

$Q$ = amount of heat flowing into the system from its surroundings. A negative sign before this term indicates that heat is flowing out of the system.

$W_S$ = shaft work done by the surroundings on the system. A negative sign on this term indicates the system is performing shaft work on the surroundings.

**Example 4.1.** For the first process, consider the continuous-flow stirred tank reactor (CSTR) of Figure 4.1. It consists of a tank with an inlet and outlet flow. This process is a very simple reactor where chemical $A$ in the inlet stream reacts to form chemical $B$. This reaction generates no heat and the volume of the liquid is constant due to the nature of the outlet. In order to keep the model simple, the following assumptions are made:

1. Densities of inlet and outlet streams are identical and constant.
2. The potential and kinetic energies of the inlet and outlet streams are zero.
3. The heat capacities of the inlet and outlet streams are identical and constant.
4. The shaft work done by the surroundings on the system is zero.
5. The heat of reaction is negligible and therefore ignored.
6. The reactor volume is perfectly mixed.
7. The chemical reaction is first order.

***Solution.*** There is no energy transfer in this system and so the differential equations are developed based on the mass balances:

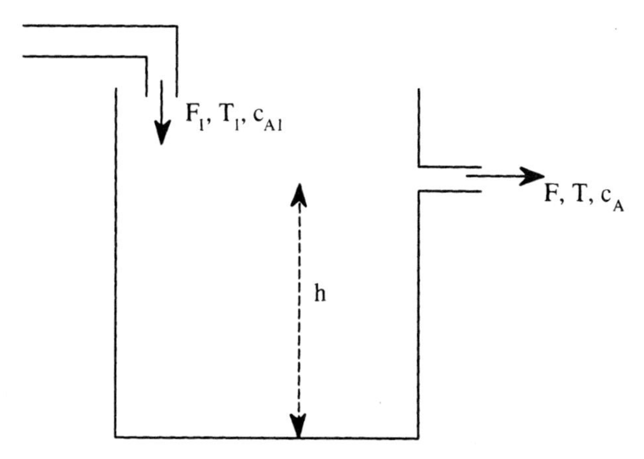


**FIGURE 4.1.** Isothermal CSTR reactor.

Overall mass balance:

$$\frac{d(\rho V)}{dt} = \frac{d(\rho A h)}{dt} = \rho F_1 - \rho F$$

$$0 = \rho F_1 - \rho F$$

$$F_1 = F \tag{4.5}$$

Component mass balance:

$$\frac{d(n_A)}{dt} = \frac{d(c_A V)}{dt} = c_{A_1} F_1 - c_A F - r_A V$$

$$\frac{d(c_A V)}{dt} = c_A \frac{dV}{dt} + V \frac{dc_A}{dt} = c_{A_1} F_1 - c_A F - k_0 e^{-E/RT_1} c_A V \tag{4.6}$$

where $V$ is the reactor volume. Using (4.5), Equation (4.6) becomes

$$\frac{dc_A}{dt} = \frac{F_1}{V}(c_{A1} - c_A) - k_0 e^{-E/RT_1} c_A \tag{4.7}$$

Therefore this system consists of one algebraic equation (4.5) and one differential equation (4.7). This system has two variables, $c_A$ and $F$, and two equations. The quantity $c_{A1}$ is an external stimulus. All other symbols represent specified constants and depend on the chemicals involved and reactor size. Thus this system has no degrees of freedom and is completely specified.

**Example 4.2.** For the second process, consider the CSTR of Figure 4.2. It consists of a tank with an inlet and outlet flow and a heat flow out of the tank. This process is a very simple reactor where chemical $A$ in the inlet stream reacts to form chemical $B$. However, unlike Example 4.1, this reaction generates heat and a cooling jacket is used to remove the heat of reaction. In addition, the height of the liquid is allowed to vary. First, the system

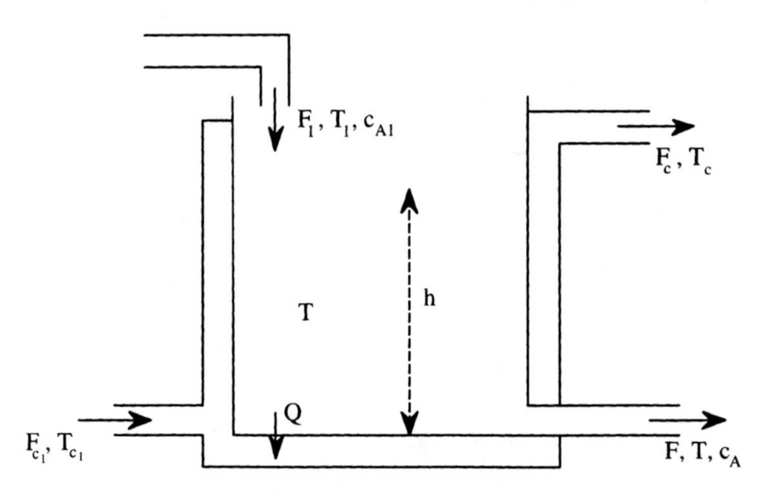


**FIGURE 4.2.** Nonisothermal CSTR reactor.

will be considered without the reaction and then modified to include the reaction. In order to keep the model simple, the following assumptions are made:

1. Densities of inlet and outlet streams are identical and constant.
2. The potential and kinetic energies of the inlet and outlet streams are zero.
3. The heat capacities of the inlet and outlet streams are identical and constant.
4. The shaft work done by the surroundings on the system is zero.
5. There is negligible mass of the reactor walls and bottom.
6. Jacket and reactor volume are perfectly mixed.
7. The chemical reaction is first order.

***Solution.*** The differential equations are developed based on the mass and energy balances:

Overall mass balance:

$$\frac{d(\rho V)}{dt} = \frac{d(\rho A h)}{dt} = \rho F_1 - \rho F$$

$$\rho A \frac{dh}{dt} = \rho F_1 - \rho F$$

$$A \frac{dh}{dt} = F_1 - F \tag{4.8}$$

Energy balance:

$$\frac{dH}{dt} = \{H_{\text{in}}\} - \{H_{\text{out}}\} + Q$$

$$\rho A C_\rho \frac{d(hT)}{dt} = \rho C_\rho F_1[T_1 - T_{\text{ref}}] - \rho C_\rho F[T - T_{\text{ref}}] + Q$$

Letting $T_{\text{ref}} = 0$ and dividing both sides by $\rho C_\rho$,

$$A \frac{d(hT)}{dt} = F_1 T_1 - FT + \frac{Q}{\rho C_\rho}$$

$$A \frac{d(hT)}{dt} = Ah \frac{dT}{dt} + AT \frac{dh}{dt} = Ah \frac{dT}{dt} + T(F_1 - F) = F_1 T_1 - FT + \frac{Q}{\rho C_\rho}$$

$$\frac{dT}{dt} = \frac{F_1 T_1}{Ah} - \frac{F_1 T}{Ah} + \frac{Q}{\rho C_\rho A h} \tag{4.9}$$

To eliminate $Q$, the heat transferred between the process at temperature $T$, and cooling water at a temperature $T_c$,

$$Q = UA_H(T - T_c) \tag{4.10}$$

where

$U$ = overall heat transfer coefficient, Btu/sec-ft$^2$-°R, assumed constant

$A_H$ = heat transfer area

Now, from Equation (4.9), the reactor total energy is

$$\frac{dT}{dt} = \frac{F_1 T_1}{Ah} - \frac{F_1 T}{Ah} - \frac{UA_H(T - T_c)}{\rho C_\rho Ah} \tag{4.11}$$

Adding the jacket energy equation yields

$$\frac{dT_c}{dt} = \frac{F_c(T_{c1} - T_c)}{V_c} + \frac{UA_H(T - T_c)}{\rho_c C_c Ah} \tag{4.12}$$

where $V_c$ is the jacket volume. The quantity $A_H$ varies with $h$, the height of the reactor fluid. If the reactor is a flat-bottomed cylinder with the jacket only around the outside and with diameter $D$,

$$A_H = h\pi D = h\pi 2\sqrt{\frac{A}{\pi}} = 2h\sqrt{\pi A} \tag{4.13}$$

and (4.12) becomes

$$\frac{dT_c}{dt} = \frac{F_c(T_{c1} - T_c)}{V_c} + \frac{2U\sqrt{\pi}(T - T_c)}{\rho_c C_c \sqrt{A}} \tag{4.14}$$

When the chemical reaction is added, an equation expressing the mass balance on component $A$ is added and the energy balance equation is modified.

Component mass balance:

$$\frac{d(n_A)}{dt} = \frac{d(c_A V)}{dt} = c_{A1}F_1 - c_A F - r_A V$$

$$\frac{d(c_A V)}{dt} = c_A \frac{dV}{dt} + V\frac{dc_A}{dt} = c_{A1}F_1 - c_A F - k_0 e^{-E/RT} c_A V \tag{4.15}$$

Using (4.8) and manipulating (4.15), the component mass balance can be expressed as

$$\frac{dc_A}{dt} = \frac{F_1}{Ah}(c_{A1} - c_A) - k_0 e^{-E/RT} c_A \tag{4.16}$$

Adding the heat of reaction to the energy balance equation (4.11) (Stephanopoulos, 1984) and using (4.13), the reactor energy balance is

$$\frac{dT}{dt} = \frac{F_1 T_1}{Ah} - \frac{F_1 T}{Ah} + Jk_0 e^{-E/RT} c_A - \frac{2U\sqrt{\pi}(T - T_c)}{\rho C_\rho \sqrt{A}} \tag{4.17}$$

where $J = -\Delta H_r / \rho C_\rho$ and $-\Delta H_r$ is the heat of reaction.

Therefore the equations for the system model are (4.8), (4.14), (4.16), and (4.17). This system has four variables and four equations. The quantities $F_1$, $T_1$, $c_{A1}$, $F_c$, and $T_c$ are all external stimuli. All other symbols represent specified constants and depend on the chemicals involved, reactor size, and reactor material. Thus this system has no degrees of freedom and is completely specified.

One more important modeling item must be introduced. In a chemical processing system, transport delay arises because of the inevitable finite time that is required to move material from one point to another. For example, suppose we have a viscous, clear liquid flowing through the pipe shown in Figure 4.3*a*. If dye is constantly injected at point $A$ in the pipe starting at a given time, how long will it take for the dye to reach point $B$? If we

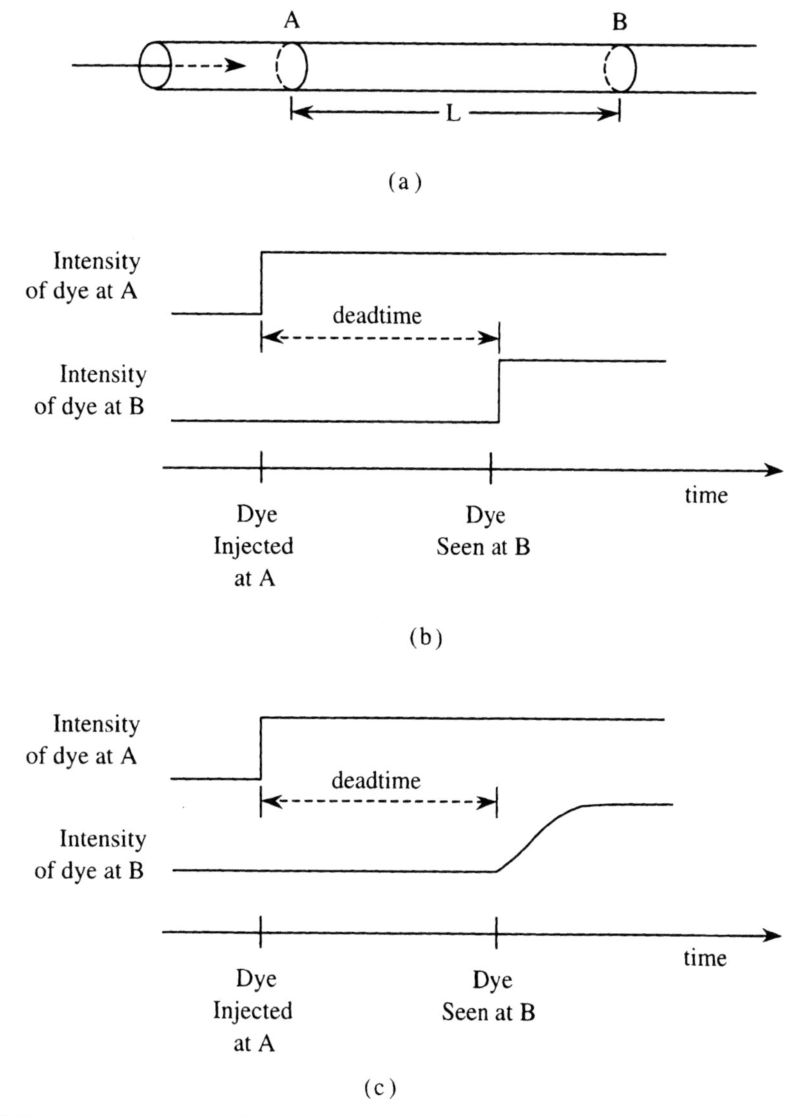


**FIGURE 4.3.** Illustration of deadtime: (*a*) pipe; (*b*) ideal response of dye; (*c*) actual response of dye.

assume that the dye does not disperse as it travels down the pipe, then the time required to travel between points $A$ and $B$ is

$$\text{Time} = \frac{\text{volume between } A \text{ and } B}{\text{volumetric flow rate}} = \frac{AL}{Av} = \frac{L}{v}$$

where $v$ is the velocity of the fluid. In this case, one expects a graph of the intensity of the dye at points $A$ and $B$ to be like Figure 4.3*b*. In reality, the graph of this example looks more like Figure 4.3*c* because the dye disperses as it travels down the pipe. The time during which no effect of the manipulated variable can be observed at the measurement is called *transport delay*. Transport delay, also called *transport lag*, or *deadtime*, is difficult to mathematically represent and analyze. However, it is often an important part of many real chemical process control systems and cannot be ignored in the control system design if it is significant. Significant deadtime is detrimental to control in real process systems, and thus the deadtime is reduced as much as possible.

### 4.2.2 Linearization About Operating Point

In order to analyze a nonlinear system with linear analysis tools, the set of nonlinear differential equations must be linearized about an operating point. Linearization about an operating point is consistent with control practice. Many continuous process control systems operate at an operating point for long lengths of time. For example, the operating point of the main fractionating column in an oil refinery is constant for long periods of time and depends on the time of year, that is, whether more gasoline or fuel oil needs to be produced. First, linearization of a system with one variable will be explained and illustrated. Then the procedure will be extended to multivariable systems.

Consider the general nonlinear function $f(x)$ as shown in Figure 4.4:

$$y = f(x) \tag{4.18}$$

The operating point is $(x_0, y_0)$. The linear approximation is the tangent to the function at the point $(x_0, y_0)$. In mathematical terms, the nonlinear function $f(x)$ is expanded into a Taylor series about the operating point $(x_0, y_0)$,

$$\begin{aligned} f(x) = f(x_0) &+ \left(\frac{df}{dx}\right)_{x_0} \frac{x - x_0}{1!} + \left(\frac{d^2f}{dx^2}\right)_{x_0} \frac{(x - x_0)^2}{2!} + \cdots \\ &+ \left(\frac{d^nf}{dx^n}\right)_{x_0} \frac{(x - x_0)^n}{n!} + \cdots \end{aligned} \tag{4.19}$$

Neglecting all but the first two terms, $f(x)$ is approximated as

$$f(x) \approx f(x_0) + \left(\frac{df}{dx}\right)_{x_0} (x - x_0) \tag{4.20}$$

So the linearized approximation of the nonlinear equation (4.18) is

$$y = f(x_0) + \left(\frac{df}{dx}\right)_{x_0} (x - x_0) \tag{4.21}$$

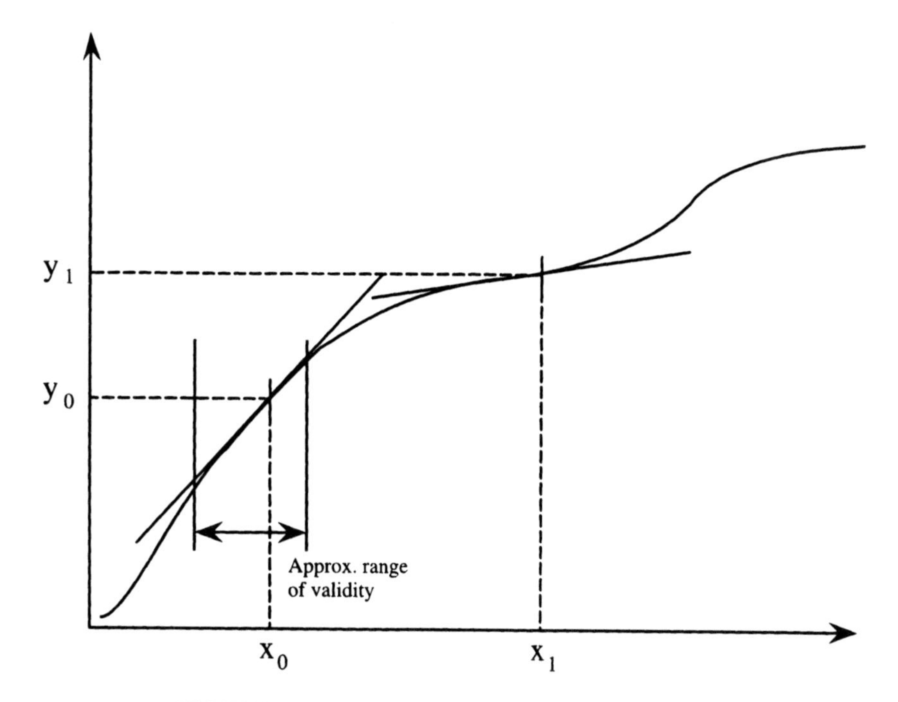


**FIGURE 4.4.** General nonlinear system with one variable.

Since $y_0 = f(x_0)$, Equation (4.21) can be rewritten as

$$y - y_0 = \left(\frac{df}{dx}\right)_{x_0} (x - x_0) \tag{4.22}$$

By defining the deviation variables $\bar{x} = x - x_0$ and $\bar{y} = y - y_0$, (4.22) becomes

$$\bar{y} = \left(\frac{df}{dx}\right)_{x_0} \bar{x} \tag{4.23}$$

This approximation is only valid when $x$ is close to $x_0$. The validity of the approximation is shown in Figure 4.4. The linear approximation depends on the location of the point $x_0$ around which the Taylor series is expanded. As shown in Figure 4.4, the linear approximation at the point $x_1$ is not the same as the approximation at the point $x_0$.

**Example 4.3.** Find a linear approximation to the orifice flow sensor shown in Figure 4.5. The volumetric flow rate $F$ is related to the pressure drop across the orifice $\Delta P$ in the following manner:

$$F = K\sqrt{\frac{\Delta P}{\rho}} \tag{4.24}$$

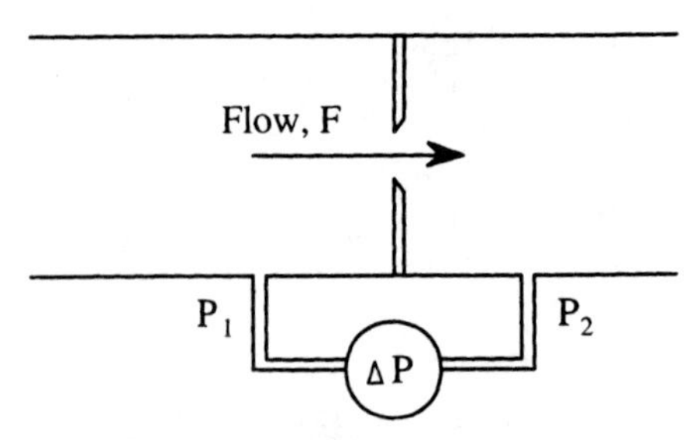


**FIGURE 4.5.** Orifice flow sensor.

where $K$ is the sensor constant, related to the size of the orifice opening and pipe diameter, and $\rho$ is density. If we let $K' = K/\sqrt{\rho}$, (4.24) can be written as

$$F = f(\Delta P) = K'\sqrt{\Delta P} \tag{4.25}$$

The graph of $\Delta P$ versus $F$ is shown in Figure 4.6. Find a linear approximation to this characteristic.

***Solution.*** Using (4.22), the linear approximation is

$$F - F_0 = \left(\frac{d}{dx}(K'\sqrt{\Delta P})\right)_{\Delta P_0}(\Delta P - \Delta P_0)$$

$$F - F_0 = \left(\frac{\frac{1}{2}K'}{\sqrt{\Delta P}}\right)_{(\Delta P)_0}(\Delta P - \Delta P_0) = \tfrac{1}{2}K'\left(\frac{1}{\sqrt{\Delta P_0}}\right)(\Delta P - \Delta P_0) \tag{4.26}$$

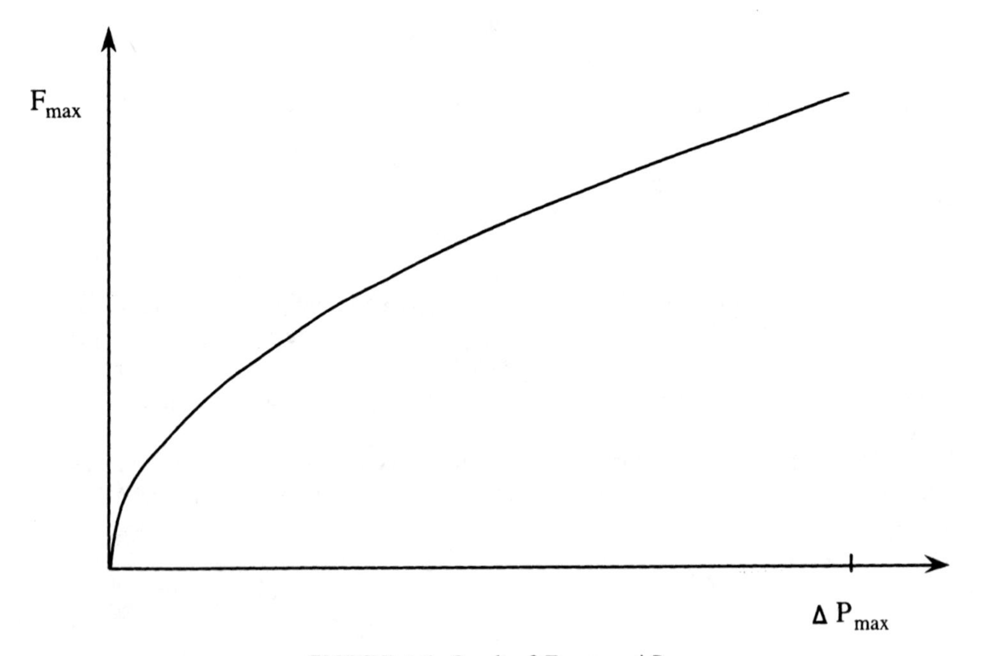


**FIGURE 4.6.** Graph of $F$ versus $\Delta P$.

In terms of deviation variables,

$$\bar{F} = \tfrac{1}{2}K'\left(\frac{1}{\sqrt{\Delta P}}\right)\overline{\Delta P} \tag{4.27}$$

As an example, let the orifice be sized so that $\Delta P_{max} = 10$ psi for $F_{max} = 25$ gpm. For this example, $K' = 7.91$. The linearized models for some example operating points are

| | | |
|---|---|---|
| $F_0 = 5$ gpm | $\Delta P_0 = 0.40$ psi | $\bar{F} = 6.25\overline{\Delta P}$ |
| $F_0 = 12$ gpm | $\Delta P_0 = 2.30$ psi | $\bar{F} = 2.60\overline{\Delta P}$ |
| $F_0 = 20$ gpm | $\Delta P_0 = 6.40$ psi | $\bar{F} = 1.56\overline{\Delta P}$ |

For multivariable systems, where the function depends on more than one variable, the linearization approach is basically the same as for a system of one variable. Consider the general nonlinear function

$$y = f(x_1, x_2, x_3, \ldots, x_n) \tag{4.28}$$

The operating point is $(x_{10}, x_{20}, x_{30}, \ldots, x_{n0}, y_0)$. The Taylor series expansion of the nonlinear function about the operating point is

$$\begin{aligned} y = {} & f(x_{10}, \ldots, x_{n0}) \\ & + (x_1 - x_{10})\left(\frac{\partial f}{\partial x_1}\right)_{(x_{10},\ldots,x_{n0})} + (x_2 - x_{20})\left(\frac{\partial f}{\partial x_2}\right)_{(x_{10},\ldots,x_{n0})} \\ & + \cdots + (x_n - x_{n0})\left(\frac{\partial f}{\partial x_n}\right)_{(x_{10},\ldots,x_{n0})} + \text{higher order terms} \end{aligned} \tag{4.29}$$

Neglecting all of the higher order terms, the linearized approximation of (4.28) may be written as

$$\begin{aligned} y \approx {} & f(x_{10}, \ldots, x_{n0}) \\ & + (x_1 - x_{10})\left(\frac{\partial f}{\partial x_1}\right)_{(x_{10},\ldots,x_{n0})} + (x_2 - x_{20})\left(\frac{\partial f}{\partial x_2}\right)_{(x_{10},\ldots,x_{n0})} \\ & + \cdots + (x_n - x_{n0})\left(\frac{\partial f}{\partial x_n}\right)_{(x_{10},\ldots,x_{n0})} \end{aligned} \tag{4.30}$$

Since $y_0 = f(x_{10}, \ldots, x_{n0})$, Equation (4.30) can be rewritten as

$$\begin{aligned} y - y_0 = {} & (x_1 - x_{10})\left(\frac{\partial f}{\partial x_1}\right)_{(x_{10},\ldots,x_{n0})} + (x_2 - x_{20})\left(\frac{\partial f}{\partial x_2}\right)_{(x_{10},\ldots,x_{n0})} + \cdots \\ & + (x_n - x_{n0})\left(\frac{\partial f}{\partial x_n}\right)_{(x_{10},\ldots,x_{n0})} \end{aligned} \tag{4.31}$$

By defining the deviation variables $\bar{x} = x - x_0$ and $\bar{y} = y - y_0$, (4.31) becomes

$$\bar{y} = \bar{x}_1\left(\frac{\partial f}{\partial x_1}\right)_{\text{OP}} + \bar{x}_2\left(\frac{\partial f}{\partial x_2}\right)_{\text{OP}} + \cdots + \bar{x}_n\left(\frac{\partial f}{\partial x_n}\right)_{\text{OP}} \tag{4.32}$$

where OP means operating point.

**Example 4.4.** Linearize the CSTR model of Example 4.2. This model has one obvious nonlinear element, the reaction rate part of the equation. The model constants are defined as: $A = 2\ \text{m}^2$, $C_p = C_c = 1$ cal/g K, $\rho = \rho_c = 10^6\ \text{g/m}^3$, $k_0 = 1.0 \times 10^{10}\ \text{min}^{-1}$, $E/R = 8330.1$ K, $-\Delta H_r = 130 \times 10^6$ cal/kmol, $U = 5 \times 10^5$, and $V_c = 7\ \text{m}^3$. The operating point inlet stream data is defined as $h_o = 5$ m, $F_{1o} = 1\ \text{m}^3/\text{min}$, $T_{1o} = 340$ K, $c_{A1o} = 2.0\ \text{kmol/m}^3$, $F_{c1o} = 3\ \text{m}^3/\text{min}$, and $T_{c1o} = 293$ K.

***Solution.*** At the operating point (equilibrium) all of the derivatives (4.8), (4.14), (4.16), and (4.17) are zero. Using a nonlinear equation solution package or simulating the system until it reaches steady state, the dependent variables are found to be $F_o = 1\ \text{m}^3/\text{min}$, $T_o = 359.94$ K, $c_{Ao} = 0.2021\ \text{kmol/m}^3$, and $T_{co} = 342.88$ K. Linearizing the equation for the mass balance on component $A$ (4.16) yields

$$\frac{dc_A}{dt} = f(h, F_1, T, c_{A1}, c_A)$$

$$\frac{\overline{dc_A}}{dt} = \bar{h}\left(\frac{\partial f}{\partial h}\right)_{\text{OP}} + \bar{F}\left(\frac{\partial f}{\partial F}\right)_{\text{OP}} + \bar{T}\left(\frac{\partial f}{\partial T}\right)_{\text{OP}} + \bar{c}_{A1}\left(\frac{\partial f}{\partial c_{A1}}\right)_{\text{OP}} + \bar{c}_A\left(\frac{\partial f}{\partial c_A}\right)_{\text{OP}}$$

$$\begin{aligned}\frac{\overline{dc_A}}{dt} &= \bar{h}\left(-\frac{F_{1o}}{h_o^2 A}(c_{A1o} - c_{Ao})\right) + \bar{F}\left(\frac{c_{A1o} - c_{Ao}}{Ah_o}\right) + \bar{T}\left(\left(\frac{E}{RT_o^2}\right)(-k_o e^{-e/RT_o})c_{Ao}\right)\\ &\quad + \bar{c}_{A1}\left(\frac{F_{1o}}{Ah_o}\right) + \bar{c}_A\left(-\frac{F_{1o}}{Ah_o} - k_o e^{-E/RT_o}\right)\end{aligned} \tag{4.33}$$

In a similar manner, the other linearized equations for this system are:

$$\frac{\overline{dh}}{dt} = \bar{F}_1\left(\frac{1}{A}\right) - \bar{F}\left(\frac{1}{A}\right) \tag{4.34}$$

$$\begin{aligned}\frac{\overline{dT}}{dt} &= \bar{h}\left(-\frac{F_{1o}}{h_o^2 A}(T_{1o} - T_o)\right) + \bar{T}_1\left(\frac{F_{1o}}{Ah_o}\right) + \bar{T}_c\left(\frac{2U\sqrt{\pi}}{\rho_c C_c \sqrt{A}}\right) + \bar{c}_A\left(Jk_o e^{-E/RT_o}\right)\\ &\quad + \bar{T}\left(-\frac{F_{10}}{Ah_o} + \left(\frac{E}{RT_o^2}\right)(Jk_o e^{-E/RT_o})c_{Ao} - \frac{2U\sqrt{\pi}}{\rho_c C_c \sqrt{A}}\right)\end{aligned} \tag{4.35}$$

$$\frac{\overline{dT_c}}{dt} = \bar{T}\left(\frac{2U\sqrt{\pi}}{\rho_c C_c \sqrt{A}}\right) + \bar{F}_c\left(\frac{(T_{c1o} - T_{co})}{V_c}\right) + \bar{T}_{c1}\left(\frac{F_{co}}{V_c}\right) + \bar{T}_c\left(-\frac{F_{co}}{V_c} - \frac{2U\sqrt{\pi}}{\rho_c C_c \sqrt{a}}\right) \tag{4.36}$$

Substituting in the values at the operating point, the linearized model equations are

$$\frac{\overline{dh}}{dt} = 0.50\bar{F}_1 - 0.50\bar{F} \tag{4.37}$$

$$\frac{\overline{dT}}{dt} = 0.3988\bar{h} + 0.10\bar{T}_1 + 0.1492\bar{T} + 1.2533\bar{T}_c + 115.63\bar{c}_A \tag{4.38}$$

$$\frac{\overline{dT_c}}{dt} = 1.2533\bar{T} - 7.126\bar{F}_c + 0.4286\bar{T}_{c1} - 1.6819\bar{T}_c \tag{4.39}$$

$$\frac{\overline{dc_A}}{dt} = 0.03596\bar{h} + 0.1798\bar{F} - 0.01156\bar{T} + 0.10\bar{c}_{A1} - 0.9894\bar{c}_A \tag{4.40}$$

### 4.2.3 Analytic Solution of Model Equations

Given a linear approximate representation of a physical system, the *Laplace transformation* allows one to analyze the system and design a suitable control system. The Laplace transform changes the differential equations describing the system into a set of algebraic equations, which are easier to solve and analyze. The Laplace transform is primarily used in control systems in order to obtain the time response solution to the differential equation(s) that describe the system and to obtain the transfer function of the system, which describes the system dynamics. The review here is brief and only contains those items used later in the text. The interested reader is referred to Carlson (1994) or Ziemer et al. (1989) for more information.

The Laplace transform exists for a function $f(t)$ that is sufficiently well behaved so that $e^{-\sigma t}f(t)$ is absolutely integrable if $\sigma$ is sufficiently large. Fortunately, physically realizable signals always have a Laplace transform. The Laplace transformation[1] for a function, $f(t)$ is

$$F(s) = \int_{0^-}^{\infty} f(t)e^{-st}\,dt = \mathscr{L}\{f(t)\} \tag{4.41}$$

The *inverse Laplace transform* is written as

$$f(t) = \frac{1}{2\pi j}\int_{\sigma-j\infty}^{\sigma+j\infty} F(s)e^{+st}\,ds \tag{4.42}$$

These two transformation integrals are rarely used to transform a function. Instead, one ordinarily uses a table of Laplace transforms to do the transforms and inverse transforms. Table 4.1 lists some important Laplace transform pairs.

The Laplace variable $s$ can also be considered to be the differential operator so that

$$s \equiv \frac{d}{dt} \tag{4.43}$$

[1] Technically, (4.41) defines the one-sided Laplace transform (Ziemer et al., 1989). Signals will be assumed to exist for positive time only and the one-sided and two-sided Laplace transforms give the same result.

**TABLE 4.1 Table of Laplace Transforms**

| $F(s)$ | $f(t), t \geq 0$ |
|---|---|
| $1$ | $\delta(t)$, Dirac delta function |
| $\frac{1}{s}$ | $(t)$, unit step function |
| $\frac{1}{s^2}$ | $t$ |
| $\frac{2!}{s^3}$ | $t^2$ |
| $\frac{1}{s+a}$ | $e^{-at}$ |
| $\frac{1}{(s+a)^2}$ | $te^{-at}$ |
| $\frac{a}{s(s+a)}$ | $1 - e^{-at}$ |
| $\frac{a}{s^2(s+a)}$ | $\frac{1}{a}(at - 1 + e^{-at})$ |
| $\frac{a}{s^2+a^2}$ | $\sin at$ |
| $\frac{s}{s^2+a^2}$ | $\cos at$ |
| $\frac{a}{(s+b)^2+a^2}$ | $e^{-bt}\sin at$ |
| $\frac{s+b}{(s+b)^2+a^2}$ | $e^{-bt}\cos at$ |
| $\frac{s+\alpha}{(s+b)^2+a^2}$ | $\frac{\sqrt{(\alpha-b)^2+a^2}}{a}e^{-bt}\sin(at+\phi),\ \phi = \tan^{-1}\frac{a}{\alpha-b}$ |

The integral operator is

$$\frac{1}{s} = \int_{0^-}^{t} dt \tag{4.44}$$

The inverse Laplace transformation is usually obtained by doing a partial fraction expansion on the equation in terms of the variable $s$, inverse transforming each term, and combining the results to get the answer. This approach is often useful for control system analysis and design because the time response due to each of the denominator polynomial roots can clearly be observed.

Two Laplace transform properties will be useful in control system analysis. The first is the *time delay property*. Given

$$\mathscr{L}\{f_1(t)\} = F_1(s)$$

If the original function is delayed by $\theta_d > 0$ units of time, the Laplace transform of the delayed function is

$$\mathscr{L}\{f_1(t-\theta_d)\} = e^{-\theta_d s}F_1(s) \tag{4.45}$$

The second useful property is the *final-value theorem*,

$$\lim_{t\to\infty} f_1(t) = \lim_{s\to 0} sF_1(s) \tag{4.46}$$

provided the limit on the right side of (4.46) exists.

The *transfer function* of a linear system is defined as the ratio of the Laplace transform of the output variable to the Laplace transform of the input variable, with all initial conditions assumed to be zero. A transfer function can only be defined for a linear, stationary (constant parameter) system. A transfer function is an input–output description of the system behavior and thus does not include any information about the internal structure of the system.

**Example 4.5.** Use the Laplace transform to determine the response of the isothermal CSTR of Example 4.1 to a +0.2 step change in the inlet concentration $c_{A1}$. Also, determine the transfer function of the system where $c_{A1}$ is the input and $c_A$ is the output.

***Solution.*** Since we are considering changes in concentration ($F_1$ and $T$ constant), Equation (4.7) does not need to be linearized. So, given $F_1$ and $T$, the response to a *change* in concentration will be the same regardless of the initial concentration assuming the concentration is always positive. Without loss of generality, the initial concentration will be assumed to be zero. This assumption will allow a transfer function to be obtained for the system. The Laplace transform of the differential equation (4.7) is

$$sC_A(s) = \frac{F_1}{V}C_{A1}(s) - \left(\frac{F_1}{V} + k_0 e^{-E/RT_1}\right)C_A(s)$$

$$C_A(s) = \frac{F_1/V}{s + F_1/V + k_0 e^{-E/RT_1}} C_{A1}(s) \tag{4.47}$$

Assume the following conditions for the system: $T_1 = 340$ K, $k_0 = 1.0 \times 10^{10}$ min$^{-1}$, $E/R = 8330.1$ K, $V = 10$ m$^3$, $F_1 = 1$ m$^3$/min, and $c_{A1} = 0.2$ kmol/m$^3$. Now (4.47) becomes

$$C_A(s) = \frac{0.1}{s + 0.9894} C_{A1}(s) \tag{4.48}$$

Now $c_{A1}(t) = 0.2u(t)$ kmoles per cubic meter, so

$$C_{A1}(s) = \frac{0.2}{s}$$

and from (4.48),

$$C_A(s) = \frac{0.02}{s(s + 0.9894)} \tag{4.49}$$

Doing a partial fraction expansion on $C_A(s)$ and taking the inverse Laplace transform, the time response is obtained as

$$C_A(s) = \frac{0.02}{s} + \frac{-0.02}{s + 0.9894}$$

$$c_A(t) = 0.02u(t) - 0.02e^{-0.9894t}u(t) \tag{4.50}$$

The time response (4.50) is plotted in Figure 4.7. The *final value* of the response is 0.02. The final value can also be obtained without plotting the time response by using the final-value theorem (4.46) on Equation (4.49),

$$\lim_{t\to\infty} c_A(t) = \lim_{s\to 0} sC_A(s) = \lim_{s\to 0} s\left[\frac{0.02}{s(s+0.9894)}\right] = 0.02$$

The *transfer function* of this system is obtained from (4.48) as the ratio of the Laplace transform of the output $C_A(s)$ to the Laplace transform of the input $C_{A1}(s)$:

$$G(s) = \frac{C_A(s)}{C_{A1}(s)} = \frac{0.1}{s + 0.9894} \tag{4.51}$$

Remember that since this system is linear for changes in concentration, a different change in initial concentration will only scale the time response. An initial concentration other than zero will simply bias the time response curve. So, if the initial concentration is $c_{A1} = 1.0\ \text{kmol/m}^3$ and the change is 0.1, the shape of the response is the same as shown in Figure 4.7 but shifted up 1 unit and scaled by 0.5.

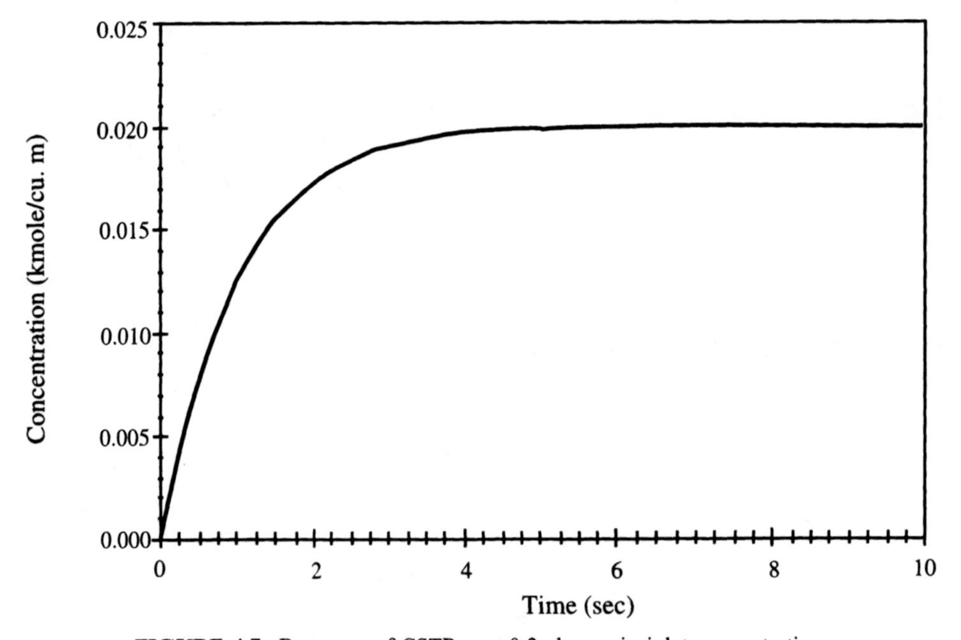


**FIGURE 4.7.** Response of CSTR to +0.2 change in inlet concentration.

Many times one is not interested in the dynamic response of the output, only the steady-state effect of a step change in the input. The *gain* of the system is obtained by assuming a unit step input signal and using the final-value theorem to obtain the steady-state response,

$$\text{Gain} = \lim_{s\to 0} s\left[\frac{1}{s}G(s)\right] = \lim_{s\to 0} G(s) \tag{4.52}$$

So the gain can be simply found as the limit of the transfer function when $s$ is allowed to approach zero. Of course, the limit must exist. For Example 4.5, the gain is 0.1.

With a transfer function description of a system, a block diagram can be used to describe a system. For example, Example 4.5 can be shown as the block diagram in Figure 4.8. If a controller to regulate the concentration is introduced into the system, the block diagram will appear similar to Figure 4.9. Note that in a real system the concentration is treated as a disturbance and the controller manipulates the inlet flow.

**Example 4.6.** Use the Laplace transform to determine the response of the linearized CSTR of Example 4.4 to a +10% step change in the inlet concentration $c_{A1}$.

***Solution.*** Taking the Laplace transform of the linearized differential equations (4.33)–(4.36),

$$s\bar{H}(s) = 0.50\bar{F}_1(s) - 0.50\bar{F}(s)$$
$$s\bar{T}(s) = 0.3988\bar{H}(s) + 0.10\bar{T}_1(s) + 0.1492\bar{T}(s) + 1.2533\bar{T}_c(s) + 115.63\bar{c}_A(s)$$
$$s\bar{T}_c(s) = 1.2533\bar{T}(s) - 7.126\bar{F}_c(s) + 0.4286\bar{T}_{c1}(s) - 1.6819\bar{T}_c(s)$$
$$s\bar{c}_A(s) = 0.0359\bar{H}(s) + 0.1798\bar{F}(s) - 0.01156\bar{T}(s) + 0.10\bar{C}_{A1}(s) - 0.9894\bar{C}_A(s)$$

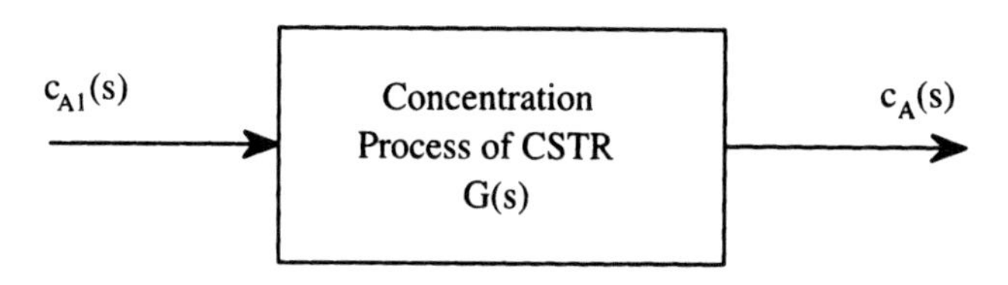


**FIGURE 4.8.** Block diagram of CSTR system.

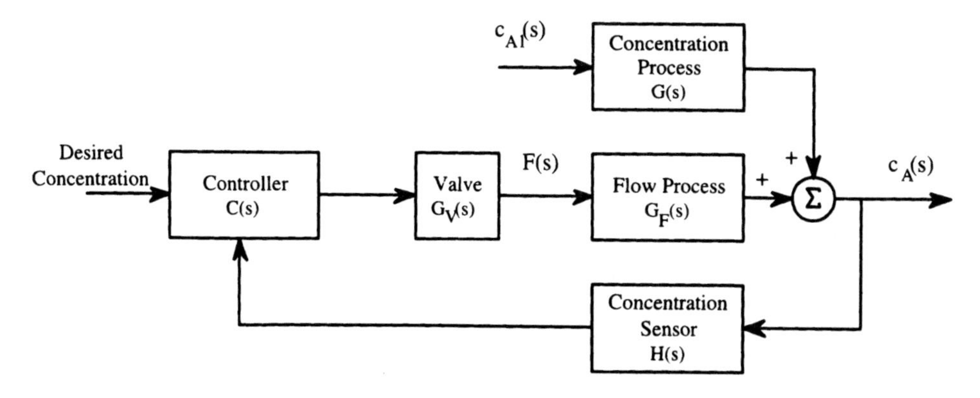


**FIGURE 4.9.** Block diagram of CSTR system with controller.

Rearranging the equations,

$$\bar{H}(s) = \frac{0.50}{s}\bar{F}_1(s) - \frac{0.50}{s}\bar{F}(s) \tag{4.53}$$

$$\bar{T}(s) = \frac{0.3988}{s - 0.1492}\bar{H}(s) + \frac{0.10}{s - 0.1492}\bar{T}_1(s) + \frac{1.2533}{s - 0.1492}\bar{T}_c(s) + \frac{115.63}{s - 0.1492}\bar{c}_A(s) \tag{4.54}$$

$$\bar{T}_c(s) = \frac{1.2533}{s + 1.6819}\bar{T}(s) - \frac{7.126}{s + 1.6819}\bar{F}_c(s) + \frac{0.4286}{s + 1.6819}\bar{T}_{c1}(s) \tag{4.55}$$

$$\bar{c}_A(s) = \frac{0.03596}{s + 0.9894}\bar{H}(s) + \frac{0.1798}{s + 0.9894}\bar{F}(s) - \frac{0.01156}{s + 0.9894}\bar{T}(s) + \frac{0.10}{s + 0.9894}\bar{C}_{A1}(s) \tag{4.56}$$

Eliminating all dependent variables from the right side of the equations can be done algebraically but is very tedious due to the coupling between the temperatures and concentrations. A much simpler approach is to define individual transfer functions for (4.53)–(4.56),

$$\bar{H}(s) = G_1\bar{F}_1(s) - G_1\bar{F}(s) \tag{4.57}$$

$$\bar{T}(s) = G_2\bar{H}(s) + G_3\bar{T}_1(s) + G_4\bar{T}_c(s) + G_5\bar{c}_A(s) \tag{4.58}$$

$$\bar{T}_c(s) = G_6\bar{T}(s) + G_7\bar{F}_c(s) + G_8\bar{T}_{c1}(s) \tag{4.59}$$

$$\bar{c}_A(s) = G_9\bar{H}(s) + G_{10}\bar{F}(s) + G_{11}\bar{T}(s) + G_{12}\bar{C}_{A1}(s) \tag{4.60}$$

An overall block diagram of the system is shown in Figure 4.10, and overall transfer functions between each of the inputs and each output may be obtained using Mason's loop rule (Dorf and Bishop, 1995). It can be shown that the four outputs in terms of the six inputs are

$$\bar{H}(s) = G_1\bar{F}_1(s) - G_1\bar{F}(s) \tag{4.61}$$

$$\begin{aligned}\bar{T}(s) = [&G_1G_2\bar{F}_1(s) + (G_5G_{10} - G_1G_2)\bar{F}(s) + G_3\bar{T}_1(s) + G_4G_7\bar{F}_c(s)\\ &+ G_4G_8\bar{T}_{c1}(s) + G_5G_{12}\bar{c}_{A1}(s)]/\Delta\end{aligned} \tag{4.62}$$

$$\begin{aligned}\bar{T}_c(s) = [&G_1G_2G_6\bar{F}_1(s) + (G_5G_{10} - G_1G_2)G_6\bar{F}(s) + G_3G_6\bar{T}_1(s) + G_7\bar{F}_c(s)\\ &+ G_8\bar{T}_{c1}(s) + G_5G_6G_{12}\bar{c}_{A1}(s)]/\Delta\end{aligned} \tag{4.63}$$

$$\begin{aligned}c_A(s) = [&G_1G_2G_{11}\bar{F}_1(s) + (G_{10} - G_1G_2G_4)\bar{F}(s) + G_3G_{11}\bar{T}_1(s) + G_4G_7G_{11}\bar{F}_c(s)\\ &+ G_4G_8G_{11}\bar{T}_{c1}(s) + G_{12}(1 - G_4G_6)\bar{c}_{A1}(s)]/\Delta\end{aligned} \tag{4.64}$$

where

$$\Delta = 1 - G_5G_{11} - G_4G_6$$

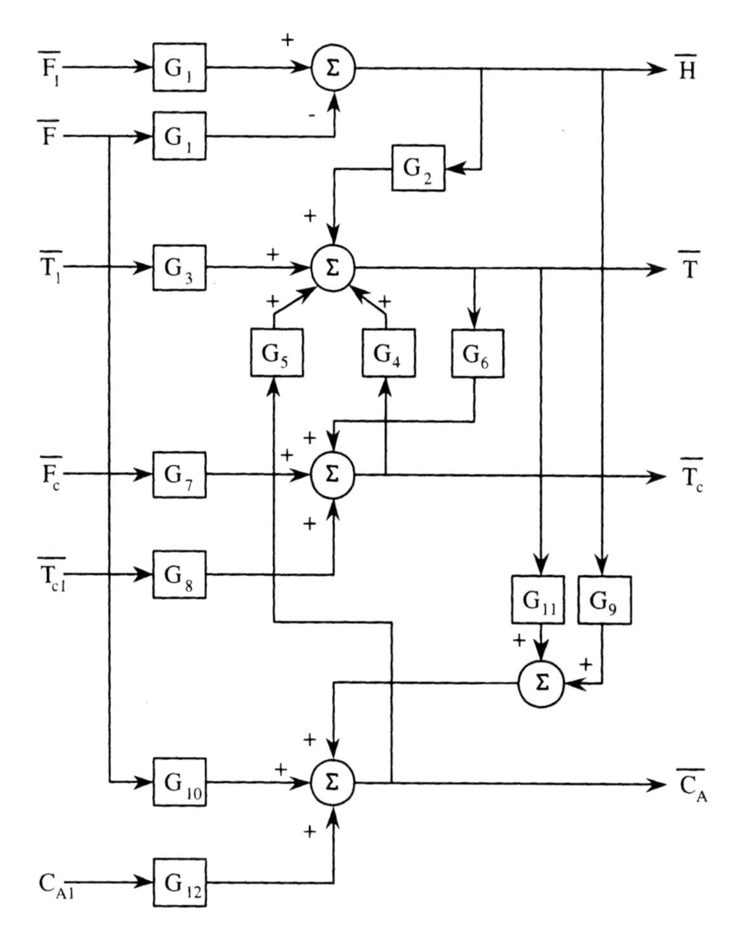


**FIGURE 4.10.** Block diagram of linearized CSTR (Example 4.6).

For the step change in the inlet concentration,

$$\bar{c}_{A1}(s) = \frac{0.20}{s}$$

All other inputs are zero. Therefore, the Laplace transform of the responses for the height, reactor temperature, coolant temperature, and concentration of chemical $A$ can be shown to be

$$\bar{H}(s) = 0$$

$$\bar{T}(s) = \frac{2.3125(s + 1.6819)}{s(s + 2.1371)(s^2 + 0.3850s + 0.2086)}$$

$$\bar{T}_c(s) = \frac{2.898}{s(s + 2.1371)(s^2 + 0.3850s + 0.2086)}$$

$$\bar{C}_A(s) = \frac{0.02(s - 0.7857)(s + 2.3184)}{s(s + 2.1371)(s^2 + 0.38505s + 0.2086)}$$

The partial fraction expansions of the above equations are

$$\bar{T}(s) = \frac{8.73}{s} + \frac{0.1246}{s+2.1371} - \frac{8.8499s + 3.1412}{s^2 + 0.3850s + 0.2086}$$

$$\bar{T}_c(s) = \frac{6.50}{s} - \frac{0.3431}{s+2.1371} - \frac{6.1579s + 3.1041}{s^2 + 0.3850s + 0.2086}$$

$$\bar{C}_A(s) = -\frac{0.0817}{s} + \frac{0.001255}{s+2.1371} + \frac{0.0805s + 0.0537}{s^2 + 0.3850s + 0.2086}$$

The time responses of the deviations from the operating point are

$$\bar{T}(t) = [8.73 + 0.1246e^{-2.1371t} - 9.5185e^{-0.1925t}\sin(0.4142t + 1.197)]u(t)$$

$$\bar{T}_c(t) = [6.50 - 0.3431e^{-2.1371t} - 7.7058e^{-0.1925t}\sin(0.4142t + 0.926)]u(t)$$

$$\bar{C}_A(t) = [-0.0817 + 0.00125e^{-2.1371t} + 0.122e^{-0.1925t}\sin(0.4142t + 0.717)]u(t)$$

The responses are shown in Figure 4.11 in terms of the temperatures and concentrations, not just the deviations. Generally, for systems that are at least as complex as this one, one would simulate the system and obtain a response rather than determine the exact analytical expression for the response. However, it is relatively easy to find the steady-state values. Using the final-value theorem (4.46), the steady-state values of the deviations are

$$\bar{T}_{ss} = \lim_{t\to\infty} \bar{T}(t) = \lim_{s\to 0} s\bar{T}(s) = \frac{2.3125(1.6819)}{(2.1371)(0.2086)} = 8.73$$

$$\bar{T}_{c,ss} = \lim_{s\to 0} s\bar{T}_c(s) = \frac{2.898}{(2.1371)(0.2086)} = 6.50$$

$$\bar{C}_{A,ss} = \lim_{s\to 0} s\bar{C}_A(s) = \frac{0.02(-0.7851)(2.3184)}{(2.1371)(0.2086)} = -0.082$$

and the final values of the responses are $T = 368.15$ K, $c_A = 0.120\ \text{kmol/m}^3$, and $T_c = 349.38$ K. Unfortunately, when substituting these values into the original differential equations (4.14), (4.16), and (4.17), they do not define an equilibrium point. Thus, one can see in this example that the linearized equations are not exact when deviating from the equilibrium point. However, as will be shown in Example 4.8, the error is relatively minor.

## 4.3 EMPIRICAL MODELS

In many cases, detailed dynamic models based on first principles are not needed or are prohibitively expensive to develop. In these cases, simple empirical models are used. This section outlines the various types of these models. Identification of the model parameters is explained in Section 4.6.

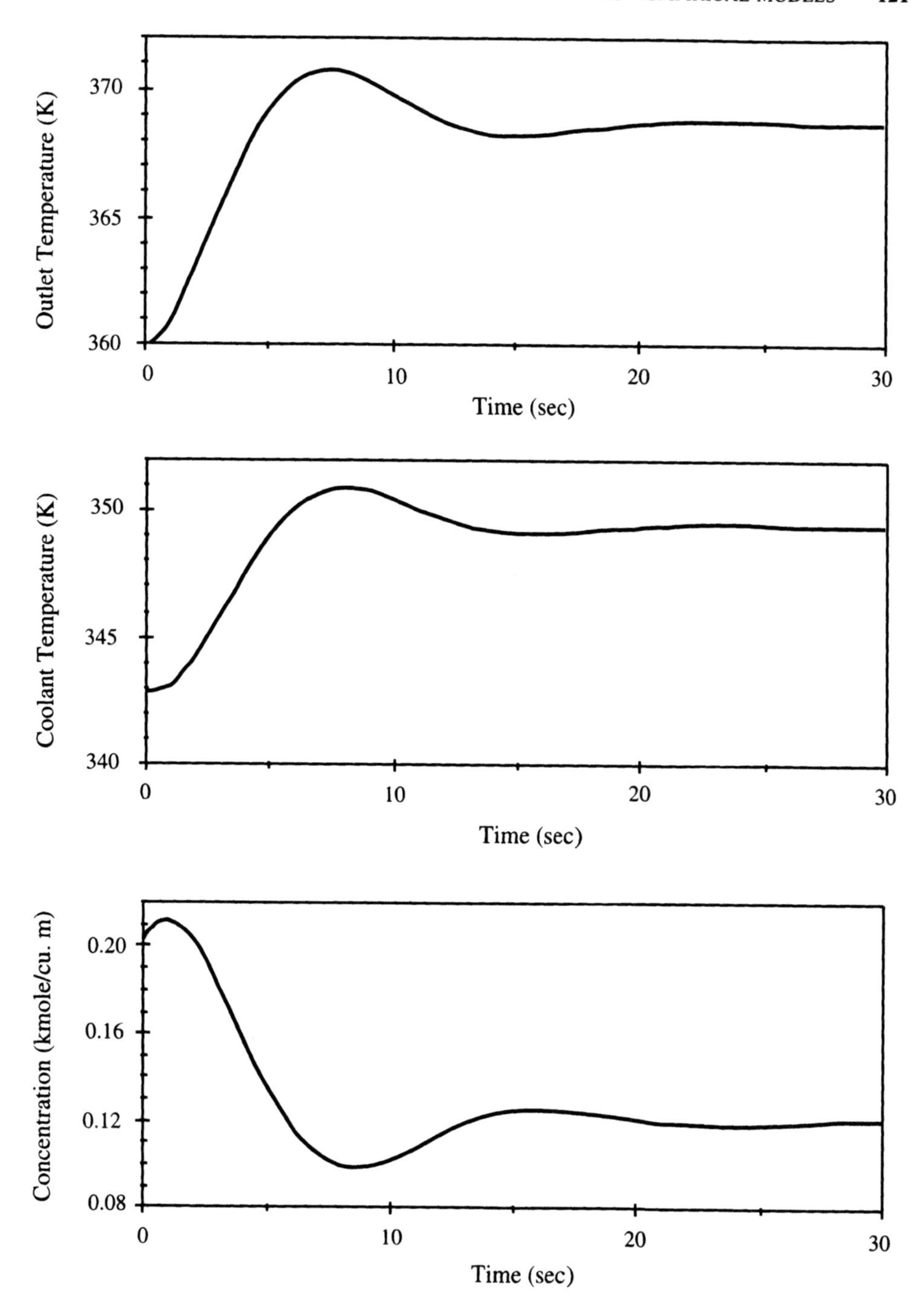


**FIGURE 4.11.** Responses of CSTR outputs to change in inlet concentration.

*Gain-Only Process.* When the response of the process variable to a change in the manipulated variable is essentially instantaneous (Figure 4.12), the process is called a *gain-only* process. Example processes that are gain-only are:

Speed control of centrifugal compressors
Liquid flow control with a fast actuator

The transfer function of a gain-only process is

$$G_G(s) = K \tag{4.65}$$

Depending on the location of the measuring sensor relative to the manipulating device (e.g., position of flow sensor relative to manipulating valve), there may be deadtime in the controlled variable response. In this case, the process is called a *gain-plus-deadtime* process. The transfer function of a gain-plus-deadtime process is

$$G_{GDT}(s) = Ke^{-s\theta_D} \tag{4.66}$$

*First-Order Process.* When the response of the process variable to a step change in the manipulated variable is represented as in Figure 4.13, the process is called a *first-order lag process*. Example processes are:

Gas pressure control
Liquid flow control with slow actuators
Concentration control

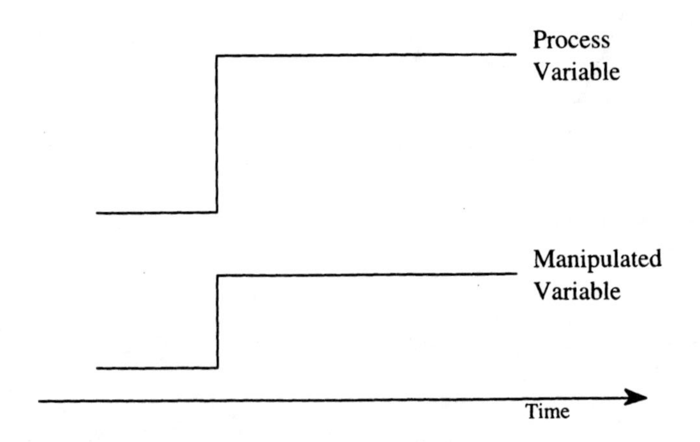


**FIGURE 4.12.** Response of a gain-only process.

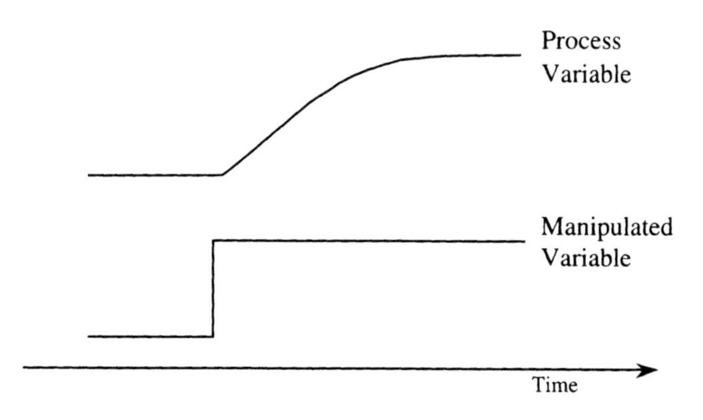


**FIGURE 4.13.** Response of a first-order lag process.

The CSTR of Example 4.1 is a first-order process when considering $c_{A1}$ as the manipulated variable and $c_A$ as the process variable. The transfer function of a first-order lag process is

$$G_{\mathrm{FO}}(s) = \frac{K}{\tau_1 s + 1} \tag{4.67}$$

where $K$ is the gain and $\tau_1$ is the time constant of the process. If the first-order lag process contains deadtime, then it is called a *first-order-plus-deadtime* (FODT) process and the transfer function is

$$G_{\mathrm{FODT}}(s) = \frac{Ke^{-s\theta_D}}{\tau_1 s + 1} \tag{4.68}$$

where $\theta_D$ is the process deadtime.

*Second-Order Overdamped Process.* When the response of the process variable to a step change in the manipulated variable is represented by the curve in Figure 4.14, the system may be a second-order overdamped process. The difference in the response, compared to a first-order process, is seen in the initial change away from the initial value. A first-order process makes an abrupt change at a line of nearly constant slope. A second-order response more gradually makes its initial change. For a higher order process, the initial change becomes even more gradual. In practice, it is often difficult to distinguish higher order processes from a second-order process by examining the step response. Example second-order processes are:

Temperature process, where the lag of the temperature sensor and thermowell is significant when compared to the process lag

Heat exchanger

Cascaded tanks

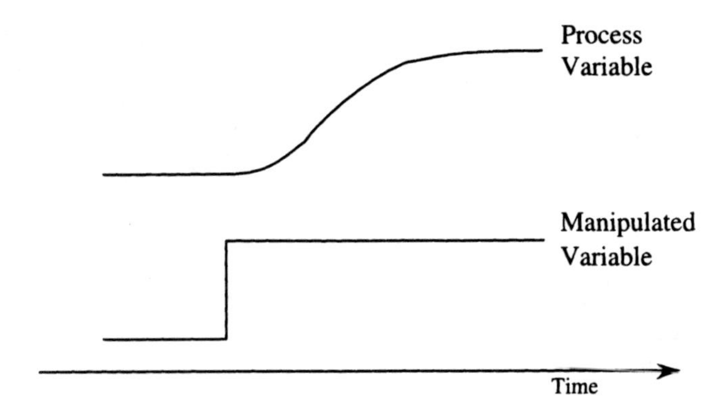


**FIGURE 4.14.** Response of a second-order lag process.

The transfer function of a second-order overdamped process with deadtime is

$$G_{\mathrm{SODT}}(s) = \frac{Ke^{-s\theta_D}}{(\tau_1 s + 1)(\tau_2 s + 1)} \tag{4.69}$$

where $K$ is the gain, $\tau_1$ and $\tau_2$ are the time constants, and $\theta_D$ is the process deadtime.

*Second-Order Underdamped (Sinusoidal) Response.* A second-order underdamped response takes a form similar to that of Figure 4.15. An underdamped response often arises when doing cascade control, covered in Chapter 6. In this case, when tuning the outer loop, the "process" includes the feedback combination of the inner loop process and a controller. Often, the inner controller is tuned so that the inner loop exhibits an

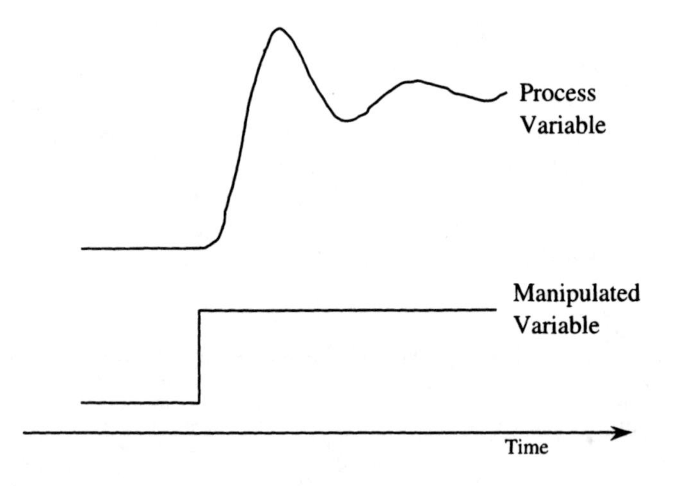


**FIGURE 4.15.** Response of an underdamped second-order process.

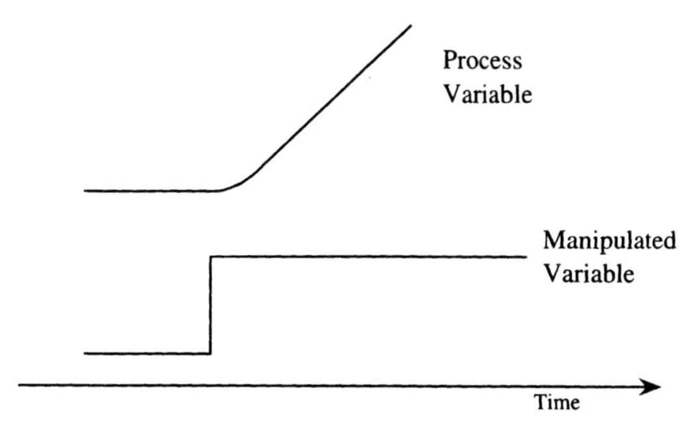


**FIGURE 4.16.** Response of an integrating process.

underdamped response. The general form of the transfer function of a second-order underdamped response is

$$G_{\mathrm{SUDT}}(s) = \frac{Ke^{-s\theta_D}}{(s/\omega_n)^2 + 2\xi(s/\omega_n) + 1} \qquad (\xi < 1) \tag{4.70}$$

where $K$ is the gain, $\omega_n$ is the undamped natural frequency, $\xi$ is the damping ratio, and $\theta_D$ is the process deadtime.

*Integrating Process.* If the process contains an integrator, then its response will look like Figure 4.16. This type of process is not self-regulating; that is, it does not reach a finite value in steady state when the input is a step function. The response of Figure 4.16 represents a first-order process with an integrator. If the process was pure integration, then the response would be a pure ramp. An example integrating process is:

Tank level control by regulation of the outflow or inflow

The transfer function of an integrating process with first-order dynamics is

$$G_{\mathrm{IFO}}(s) = \frac{K_S}{s(\tau_1 s + 1)} \tag{4.71}$$

where $K_S$ is the integrating constant and $\tau_1$ is the time constant.

## 4.4 DISCRETE-TIME MODELS

Most control algorithms are executed with digital computers, and hence discrete-time models must be used to design these types of controllers. Also, most model identification techniques directly identify discrete-time models.

Discrete-time models are based on the need to sample the analog process signals. A digital computer samples, that is, at periodic intervals takes a snapshot of the measurement signals rather than continuously monitoring the signals. For digital control systems, the $Z$-transform is used to analyze the system in much the same way as the Laplace transform is used in continuous systems.

The $Z$-transform is developed from the Laplace transform of a sampled signal. Consider a signal $f(t)$ sampled at a sampling period of $T$ seconds with a sampler as in Figure 4.17$a$. The original signal is shown in Figure 4.17$b$, and the sampled signal is shown in Figure 4.17$c$. The sampled signal can be written as

$$f^*(t) = f(0)\delta(t) + f(T)\delta(t-T) + f(2T)\delta(t-2T) + \cdots \tag{4.72}$$

where $\delta(t)$ is the Dirac delta function and $\delta(t-T)$ is the Dirac delta function delayed by $T$ seconds. Equation (4.72) can be written as an infinite sum,

$$f^*(t) = \sum_{k=0}^{\infty} f(kT)\delta(t-kT) \tag{4.73}$$

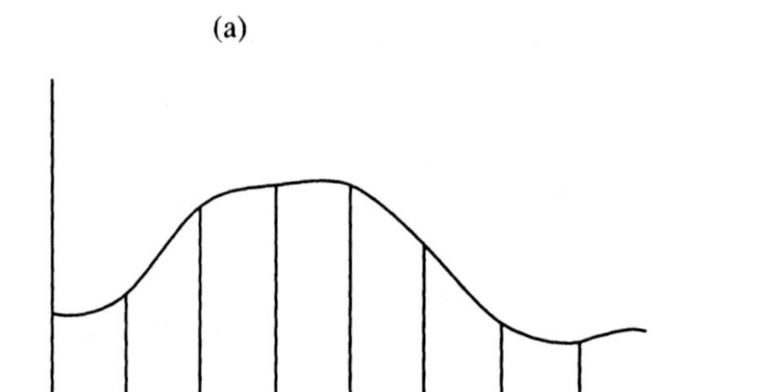


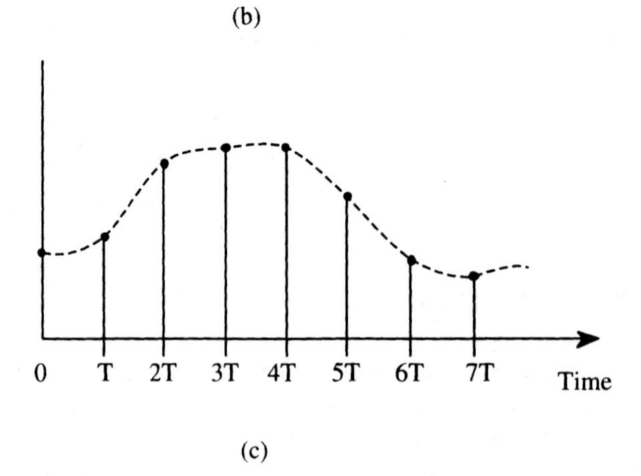


**FIGURE 4.17.** Process of sampling: ($a$) sampler; ($b$) original signal; ($c$) sampled signal.

Taking the Laplace transform of the sampled signal,

$$\mathscr{L}[f^*(t)] = \sum_{k=0}^{\infty} f(kT)e^{-skT} \tag{4.74}$$

and definining $z = e^{sT}$,

$$\mathscr{L}[f^*(t)] = \sum_{k=0}^{\infty} f(kT)z^{-k} \equiv \mathscr{Z}[f(t)] = F(z) \tag{4.75}$$

This sequence converges for a region $|z| < r$. Often, a constant sampling period of $T$ seconds that generates a sequence of values $\{e(k)\}$ is assumed, and the $Z$-transform is written as

$$F(z) = \mathscr{Z}[\{f(k)\}] = \sum_{k=0}^{\infty} f(k)z^{-k} \tag{4.76}$$

As in the case of the Laplace transform, the sampled signal is assumed to exist for positive time only, so (4.76) defines the single-sided $Z$-transform.

The $Z$-transform exists for a sampled signal that is well-behaved; that is, samples are not at discontinuities. If $e(t)$ is discontinuous at $t = kT$, where $k$ is an integer, then the sample $f^*(kT)$ is taken to be $f^*(kT^+)$, where $f^*(kT^+) = f^*(kT + \varepsilon)$, $\varepsilon$ being arbitrarily small.

The inverse $Z$-transform is defined to be

$$f(kT) = \frac{1}{2\pi j} \oint F(z)z^{k-1}\, dz \tag{4.77}$$

where the contour of integration encloses all singularities of $F(z)$. As in the case of the Laplace transform, one rarely uses the transformation summation to transform a signal. Instead, one uses a table of $Z$-transforms (Table 4.2). The inverse $Z$-transform of an arbitrary function $F(z)$ is obtained by the following procedure:

1. Divide $F(z)$ by $z$.
2. Do a partial fraction expansion on $F(z)/z$.
3. Multiply both sides of the equation by $z$.
4. Inverse transform each term.
5. Combine to get the final $f(k)$ result.

Steps 1 and 3 ensure that terms contain a $z$ in the numerator and the table can be directly used.

Two $Z$-transform properties will be useful in control system analysis. The first is the *time delay property*. Given

$$\mathscr{Z}[f_1(k)\}] = F_1(z)$$

**TABLE 4.2 Table of Z-Transforms**

| $F(z)$ | $f(t), t \geq 0$ |
|---|---|
| $1$ | $\delta(t)$, Dirac delta function |
| $\dfrac{z}{z-1}$ | $u(t)$, unit step function |
| $\dfrac{Tz}{(z-1)^2}$ | $t$ |
| $\dfrac{2!Tz}{(z-1)^3}$ | $t^2$ |
| $\dfrac{z}{z-e^{-aT}}$ | $e^{-at}$ |
| $\dfrac{(1-e^{-aT})z}{(z-1)(z-e^{-aT})}$ | $1-e^{-at}$ |
| $\dfrac{z \sin aT}{z^2-2z\cos aT+1}$ | $\sin at$ |
| $\dfrac{z(z-\cos aT)}{z^2-2z\cos aT+1}$ | $\cos at$ |
| $\dfrac{ze^{-aT}\sin aT}{z^2-2ze^{-aT}\cos aT+e^{-2aT}}$ | $e^{-bt}\sin at$ |
| $\dfrac{z^2-ze^{-aT}\cos aT}{z^2-2ze^{-aT}\cos aT+e^{-2aT}}$ | $e^{-bt}\cos at$ |

If the samples of the original function are delayed by $n$ samples, the $Z$-transform of the delayed function is

$$\mathscr{Z}[\{f_1(k-n)\}] = z^{-n}F_1(z) \tag{4.78}$$

The second useful property is the *final-value theorem*,

$$\lim_{k\to\infty} f_1(k) = \lim_{z\to 1}(z-1)F_1(z) \tag{4.79}$$

provided the limit on the right side of (4.79) exists.

The *transfer function* of a linear discrete-time system is defined as the ratio of the $Z$-transform of the output samples to the $Z$-transform of the input samples, with all initial conditions assumed to be zero. A transfer function can only be defined for a linear, stationary (constant parameter) system. A transfer function is an input–output description of the system behavior and thus does not include any information about the internal structure of the sytem.

In a digital control system, the process is still continuous, even though the controller is digital. As shown in Figure 4.18, the digital controller samples the continuous system output signal, which has already been discussed. However, the output of the digital controller is also a sampled signal that must be transformed into a continuous signal which is the input to the continuous-control system. This process is called *data reconstruction*

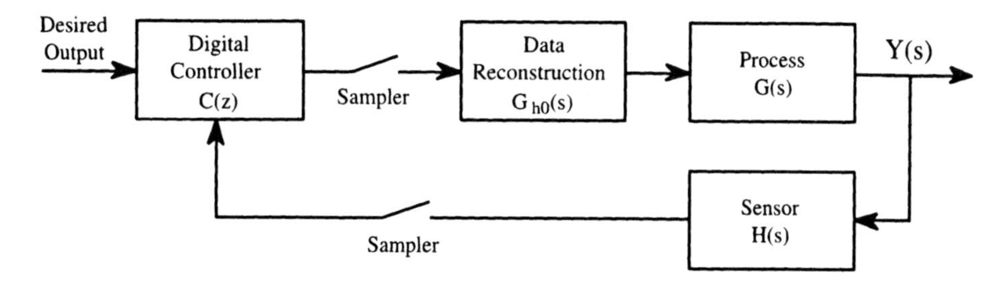


**FIGURE 4.18.** Discrete-control system block diagram.

because the continuous signal is reconstructed from the samples. The discrete-time model of the process includes the data reconstruction. First, the data reconstruction will be devised and then the process of coverting a continuous model and the data reconstruction device into an equivalent discrete-time model will be formulated.

The most popular data reconstruction device is called a *zero-order hold* (ZOH). A ZOH clamps the output signal to the value of the input sample at the sample instant. A typical input to a ZOH is shown in Figure 4.19*a*, and the corresponding output is shown in Figure

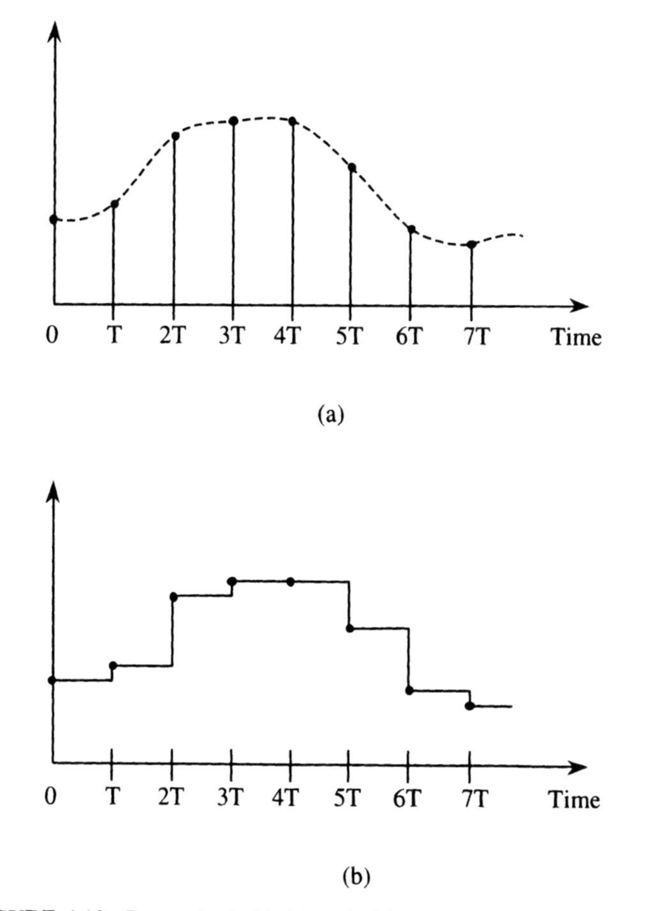


**FIGURE 4.19.** Zero-order hold: (*a*) typical input signal; (*b*) output signal.

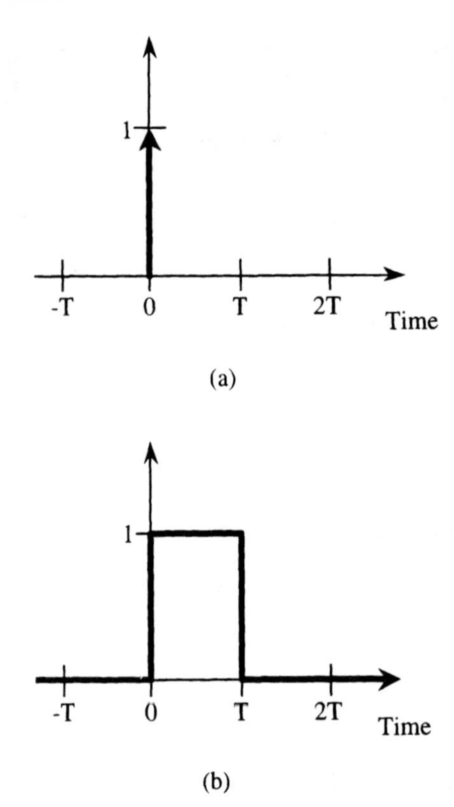


**FIGURE 4.20.** (*a*) Unit impulse input to ZOH. (*b*) Output of ZOH.

4.19*b*. The transfer function of the ZOH can be easily developed by considering the input to the ZOH to be a unit impulse (sample a signal of value 1 at the zeroth instant); Figure 4.20*a*. The output of the ZOH is as shown in Figure 4.20*b*. The output signal is

$$y(t) = u(t) - u(t - T)$$

$$Y(s) = \frac{1}{s} - \frac{1}{s}e^{-Ts} = \frac{1}{s}(1 - e^{-Ts}) \tag{4.80}$$

Since the Laplace transform of the impulse input is 1, the transfer function is the same as $Y(s)$ in (4.80). Therefore, the transfer function of a ZOH, $G_{h0}(s)$, is

$$G_{h0}(s) = \frac{1}{s}(1 - e^{-Ts}) \tag{4.81}$$

To develop a discrete-time equivalent of the ZOH and $G(s)$ (Figure 4.21) it is necessary to find

$$G_{zeq}(z) = \mathscr{Z}\{G_{\mathrm{ZOH}}(s)G(s)\}$$

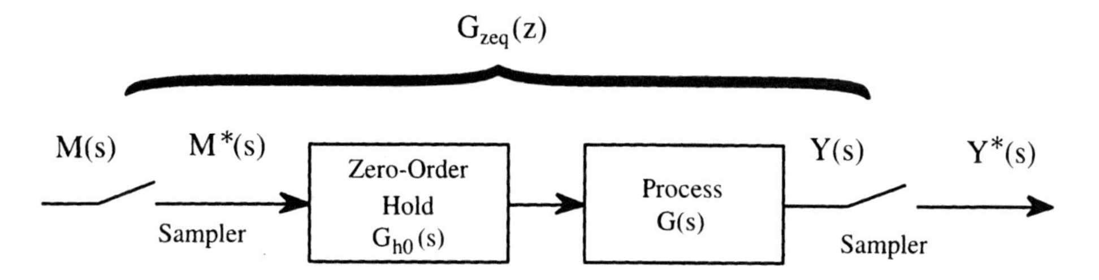


**FIGURE 4.21.** Discrete equivalent of a continuous transfer function.

which is interpreted to mean the $Z$-transform of the inverse Laplace transform of the transfer function, $G_{ZOH}(s)G(s)$. The development is as follows:

$$\begin{aligned} G_{zeq}(z) &= \mathscr{Z}\left\{\frac{1}{s}(1 - e^{-Ts})G(s)\right\} \\ &= \mathscr{Z}\left\{\frac{G(s)}{s} - e^{-Ts}\frac{G(s)}{s}\right\} \\ &= \mathscr{Z}\left\{\frac{G(s)}{s}\right\} - \mathscr{Z}\left\{e^{-Ts}\frac{G(s)}{s}\right\} \end{aligned}$$

On the right side of the equation, the second term is simply the first term delayed by one sample; thus

$$\begin{aligned} G_{zeq}(z) &= \mathscr{Z}\left\{\frac{G(s)}{s}\right\} - z^{-1}\mathscr{Z}\left\{\frac{G(s)}{s}\right\} \\ &= (1 - z^{-1})\mathscr{Z}\left\{\frac{G(s)}{s}\right\} = \frac{z-1}{z}\mathscr{Z}\left\{\frac{G(s)}{s}\right\} \end{aligned}$$

As an example, consider

$$G(s) = \frac{1}{s/a + 1} = \frac{a}{s + a}$$

The discrete-time equivalent $G_{zeq}(z)$ is developed as

$$\begin{aligned} G_{zeq}(z) &= \frac{z-1}{z}\mathscr{Z}\left\{\frac{a}{s(s+a)}\right\} \\ &= \frac{z-1}{z}\mathscr{Z}\{1 - e^{-at}\} \\ &= \frac{z-1}{z}\frac{(1 - e^{-aT})z}{(z-1)(z - e^{-aT})} \\ &= \frac{1 - e^{-aT}}{z - e^{-aT}} \end{aligned} \tag{4.82}$$

In a similar manner, the discrete-time equivalent of a general overdamped second-order process,

$$G(s) = \frac{1}{(s/a+1)(s/b+1)} = \frac{ab}{(s+a)(s+b)}$$

is

$$G_{zeq}(z) = \frac{b_1 z^{-1} + b_2 z^{-2}}{(1 - e^{-aT}z^{-1})(1 - e^{-bT}z^{-1})} \tag{4.83}$$

where

$$b_1 = \frac{1}{b-a}[b(1 - e^{-aT}) - a(1 - e^{-bT})]$$
$$b_2 = \frac{1}{b-a}[a(1 - e^{-bT})e^{-aT} - b(1 - e^{-aT})e^{-bT}]$$

**Example 4.7.** Derive the discrete-time equivalent transfer function of the isothermal CSTR of Example 4.5 and compare the response of the continuous system and the discrete-time equivalent system to a step input of 0.02 units. The sampling period is 0.25 s.

***Solution.*** The transfer function of the system is

$$G(s) = \frac{C_A(s)}{C_{A1}(s)} = \frac{0.1}{s + 0.9894}$$

Using (4.82), the discrete-time equivalent transfer function is

$$G_{zeq}(z) = \frac{0.02215}{z - 0.7809}$$

The $Z$-transform of the input signal is $0.02z/(z-1)$ and so the samples of the output are

$$C_A(z) = \frac{0.000443z}{(z-1)(z-0.7809)} = \frac{0.02z}{z-1} - \frac{0.02z}{(z-0.7809)}$$
$$c_A(kT) = 0.02 - 0.02e^{-0.9894kT} \qquad k \geq 0$$

In Figure 4.22, the samples are compared with the response from Example 4.5. The sampling period is a little longer than what one would normally use but illustrates that the response of a discrete-time equivalent exactly represents the response of the continuous system to a step signal *at the sample instants.*

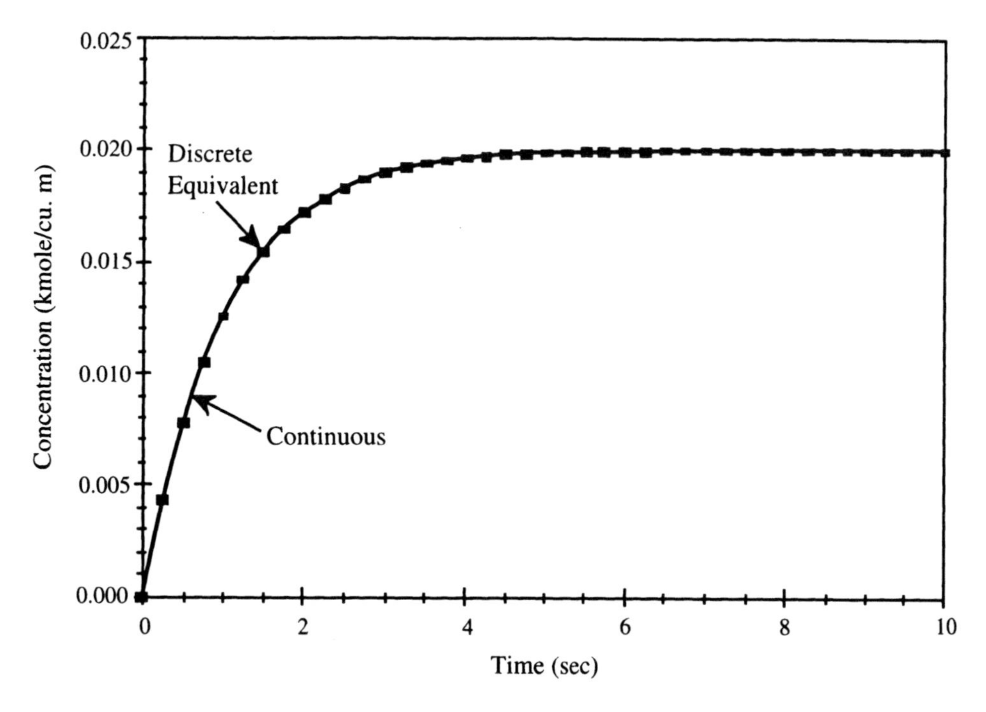


**FIGURE 4.22.** Comparison of continuous and discrete equivalent.

## 4.5 SIMULATION

For a simple linear system that has only one input and one output, it is relatively easy to derive an analytic expression for the output given a particular type of input signal. However, as demonstrated by the CSTR example, it is often cumbersome to obtain an analytic expression for the response of a system linearized at its operating point when the system has more than one input and output. Also, it is generally not possible to derive an analytic expression for the response of a nonlinear system regardless of the operating point. For these reasons, the responses for many systems are obtained by simulating the system on a computer. The heart of simulation involves the numerical solution of the differential equations that describe a system. Numerical solutions of differential equations that describe a system is briefly surveyed in this section and illustrated by simulating the CSTR of Example 4.2. A more comprehensive treatment of this topic may be found elsewhere (e.g., Constantinides, 1987; Maron and Lopez, 1991).

Numerical methods do not obtain an analytic solution for the response of a system as done in Example 4.6. Instead, they provide a response that is "close" to the true solution of the differential equations. In addition, they provide a response that is like discrete-time control in that the response is obtained at specific time samples called time steps. Starting with initial estimates of the variables, an approximation of integraion is used to determine the values of the variables at the end of the time step.

Because integration is numerically better behaved than differentiation, the differential equations must be transformed to equations that use integration. This step is actually easy, since the input to an integrator is the derivative of its output. For example, the differential equation (4.17) is implemented by using the derivative as the input to an integrator. If one

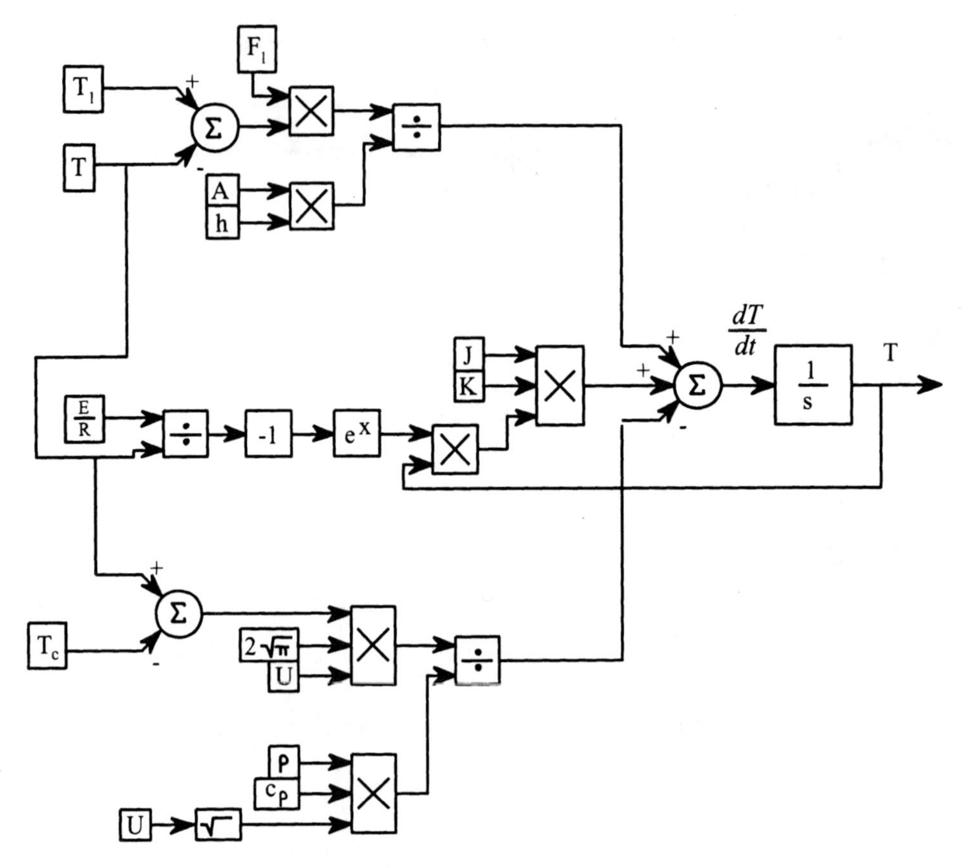


**FIGURE 4.23.** Example differential equation simulation.

is using a simulation package that builds the system using a block diagram approach, Equation (4.17) is implemented similarly to Figure 4.23, where the various blocks represent arithmetic operations.

There are many ways to approximate the integration operator. The three simplest methods are described here: forward rectangular, backward rectangular, and trapezoidal. These three methods of integration approximation are illustrated in Figure 4.24. The

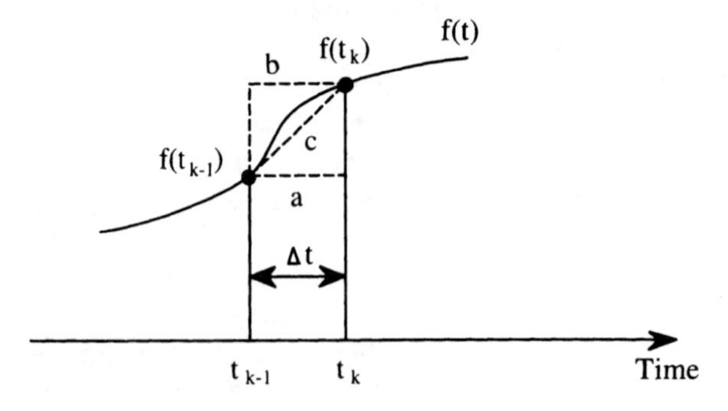


**FIGURE 4.24.** Integration approximations: (*a*) forward rectangular; (*b*) backward rectangular; (*c*) trapezoidal.

general idea is to approximate the integral at the current time step $y(t_k)$ given the value of the integral at the previous time step $y(t_{k-1})$ and the values of the input function at the current and/or previous time step. The integral of the input $f(t)$ between $t_{k-1}$ and $t_k$ can be approximated as:

Forward rectangular (explicit Euler):

$$y(t_k) \approx y(t_{k-1}) + \Delta t f(t_{k-1})$$

Backward rectangular (implicit Euler):

$$y(t_k) \approx y(t_{k-1}) + \Delta t f(t_k)$$

Trapezoidal:

$$y(t_k) \approx y(t_{k-1}) + \tfrac{1}{2}\Delta t[f(t_{k-1}) + f(t_k)]$$

Other methods have been derived to approximate integration using higher order terms. Among these methods, the Runge–Kutta methods are popular (Constantinides, 1987; Maron and Lopez, 1991). Other methods, such as the adaptive fifth-order Runge–Kutta, have a variable step size, taking small steps when the input function has discontinuities and large steps when the input function is smooth.

All numerical methods introduce some error at each step, and these errors accumulate as the simulation progresses. Higher order methods produce less error. For example, the rectangular methods produce accumulated error on the order of $\Delta t^2$, the trapezoidal accumulated error is on the order of $\Delta t^3$, and the fourth-order Runge–Kutta produces accumulated error on the order of $\Delta t^4$ (Constantinides, 1987). Thus, for a given accuracy, the Euler integration method requires a smaller step size than the fourth-order Runge–Kutta algorithm. However, the larger step size for the Runge–Kutta algorithm is partially offset by the larger number of calculations required per step. Stability considerations may also determine the maximum step size. For example, for most systems, the methods are unstable for larger step sizes, even if the step size meets accuracy considerations (Constantinides, 1987). Practically, there is also a lower limit on the step size. Obviously, for a smaller step size, more calculations will be required and thus the simulation will require more time to reach the final simulation time. Second, a step size that is very small results in a slight decrease in accuracy due to accumulated round-off error. However, this decrease in accuracy is generally very slight and usually unnoticed. With all of these considerations, an intermediate range of step sizes yields the best trade-off.

One rule of thumb used to choose the step size is to relate it to the dynamics of the process. The authors' rule of thumb is to choose the step size in the range of from 0.1 to 0.01 times the smallest time constant of the system. However, when simulating a closed-loop control system, the dynamics of the closed-loop system are generally faster than the original system without control. In addition, for a nonlinear system it may be impossible to determine the time constant ahead of time. So, another approach is to run the simulation at a number of step sizes and compare the results. An appropriate step size is one at which a smaller step size does not significantly change the results and that has sufficient accuracy. Some systems of differential equations are called *stiff*, which means that the time constants differ in orders of magnitude (e.g., $\tau_1 = 1$ sec and $\tau_2 = 6000$ sec $= 100$ min). In this case, the step size for the methods already cited, called *explicit* methods, require a step size small relative to the smallest time constant for stability and accuracy. One solution is to use an explicit numerical algorithm that works well with stiff systems, for example, the implicit Euler (Luyben, 1990). Another approach involves the use of *implicit* numerical methods that iteratively solve the equations at each time step (Maron and Lopez, 1991).

**Example 4.8.** Use a commercial simulation package to determine the response of the CSTR of Example 4.2 to a +10% change in the inlet concentration at the same operating point as used in Example 4.4.

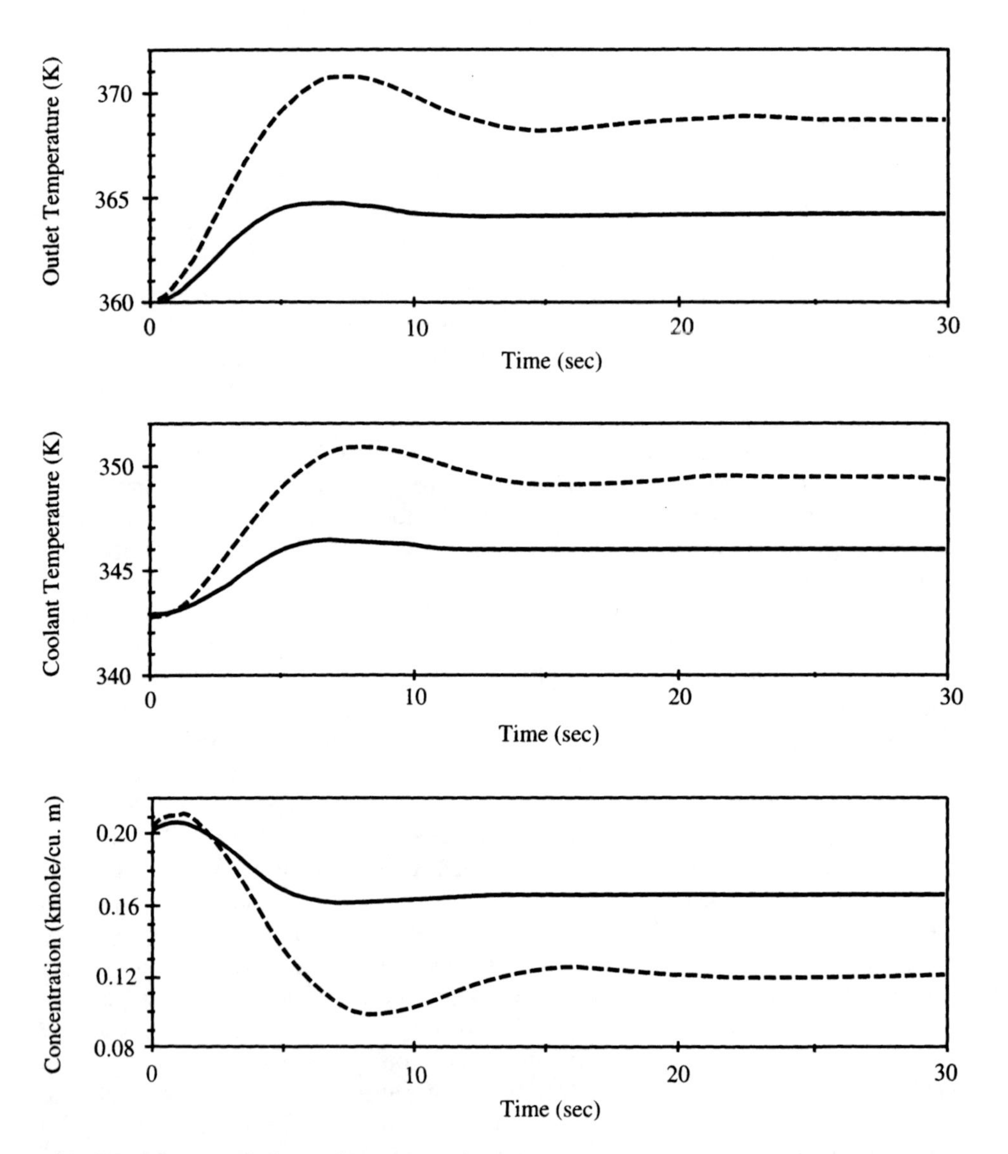


**FIGURE 4.25.** Comparison of response of continuous system with discrete equivalent. —— Nonlinear simulation; ----- linearized model.

***Solution.*** The simulated response will be compared to the responses obtained in Example 4.6, which is the linearized response. The VisSim or SIMULINK software package may be used to implement the simulation and obtain the responses in Figure 4.25. The details are not shown, but each differential equation is constructed as in Figure 4.23. The particular integration method chosen is Runge–Kutta second order with a time step of 0.05 sec. For this problem, the simulation results do not change much for time steps less than 0.1 sec. Note the comparison with the response of the linearized model of Example 4.6. For the nonlinear simulation, the final values of the responses are $T = 364.15$ K, $c_A = 0.166$ kmol/m$^3$, and $T_c = 346.02$ K. That the simulation of the differential equations is correct is verified by taking the final values and verifying that they satisfy the original differential equations (4.14), (4.16), and (4.17). Compared with the simulation results, the predicted final temperatures of the linearized model are only off by about 1%. The predicted final concentration of the linearized model deviates by almost 28%, but that is still small compared to the inlet concentration.

## 4.6 EMPIRICAL MODEL IDENTIFICATION

The parameters of the empirical models of Section 4.3 can be identified in two ways. One can generate a reaction curve, that is, make a step change in the manipulated variable and record (with a data collection device or strip chart recorder) the manipulated variable and the response of the process variable. The reaction curve is graphically analyzed in order to determine the model parameters. Since most data collection devices sample the signals, the response is represented by an array of samples that can be plotted and then graphically analyzed. An alternative approach is to collect samples of both the manipulated and process variables and then use least-squares estimation to estimate the discrete-time model parameters. The discrete-time model parameters may then be translated into the empirical model parameters if needed.

A disadvantage of the first method is that the process cannot be controlled when the samples of the reaction curve are being collected. Also, significant disturbances to the process that occur will corrupt the data and model. Noise on the data samples, whether introduced by the transducer or by the nature of the measurement (i.e., pressure), will need to be removed by filtering the data. However, if filtering needs to be severe, when the filter time constant is greater than $\frac{1}{10}$ any significant process time constant, then the model parameters will be corrupted. In contrast, least-squares identification will tolerate some noise on the data samples and can be used when the process is being controlled, though not aggressively (Gustavsson et al., 1977). Also, the second method will allow the manipulated variable to be a random signal with zero average value, causing smaller deviations on the process variable.

### 4.6.1 Reaction Curve Methods

The reaction curve is the response of the process variable to a step change in the manipulated variable. This response may be recorded by a pen-based device or by a data collection system that samples the response. In the latter case, a response is generated by plotting the samples. The empirical model parameters may be determined by graphically analyzing the reaction curve plot or by analyzing the samples of the process and manipulated variables. Graphical methods will be outlined and demonstrated in this section, although computer programs can easily be written to perform the analysis on data samples.

When generating the reaction curve, it is important that the plot (or samples) of the manipulated variable be obtained. This curve is needed to determine the gain and deadtime of the empirical model. Also, the process must be at steady state when the manipulated variable is changed and the process must be allowed to reach its new steady-state value before terminating data collection. If the process has a significant disturbance during data collection, then the data set must be discarded and a new data set obtained.

*Gain-Only Process.* Even if the process contains dynamics, a gain-only model can be obtained if the process has no integrators. Although the dynamics are neglected, just knowing the gain of the process is often the first step in determining information about the process. Regardless of whether the process has no or little dynamics (Figure 4.26*a*) or has dynamics (Figure 4.26*b*), the gain is determined simply as the ratio of the measured change in the process variable to the change in the manipulated variable,

$$G_G(s) = K = \frac{\Delta \mathrm{PV}}{\Delta \mathrm{MV}} = \frac{\mathrm{PV}_2 - \mathrm{PV}_1}{\mathrm{MV}_2 - \mathrm{MV}_1} \tag{4.84}$$

A gain-only process is also the most reliable to obtain if there is significant noise on the process variable measurement. Filtering also does not affect the process gain.

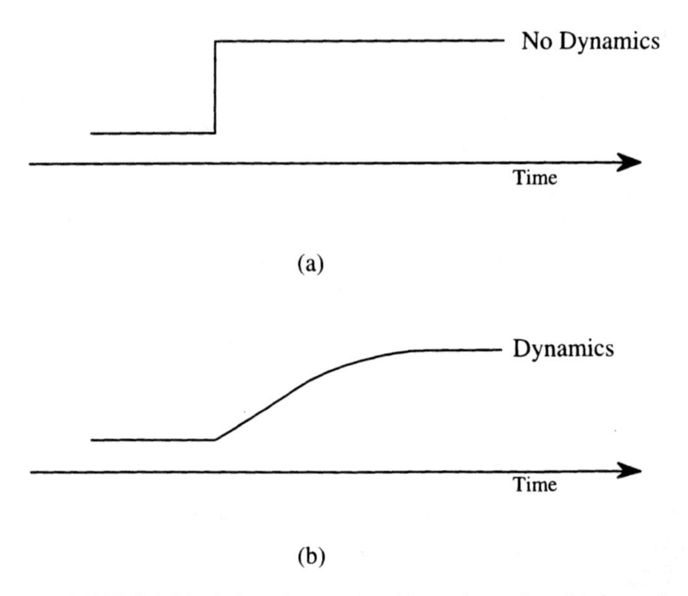


**FIGURE 4.26.** Gain-only process: (*a*) no dynamics; (*b*) dynamics.

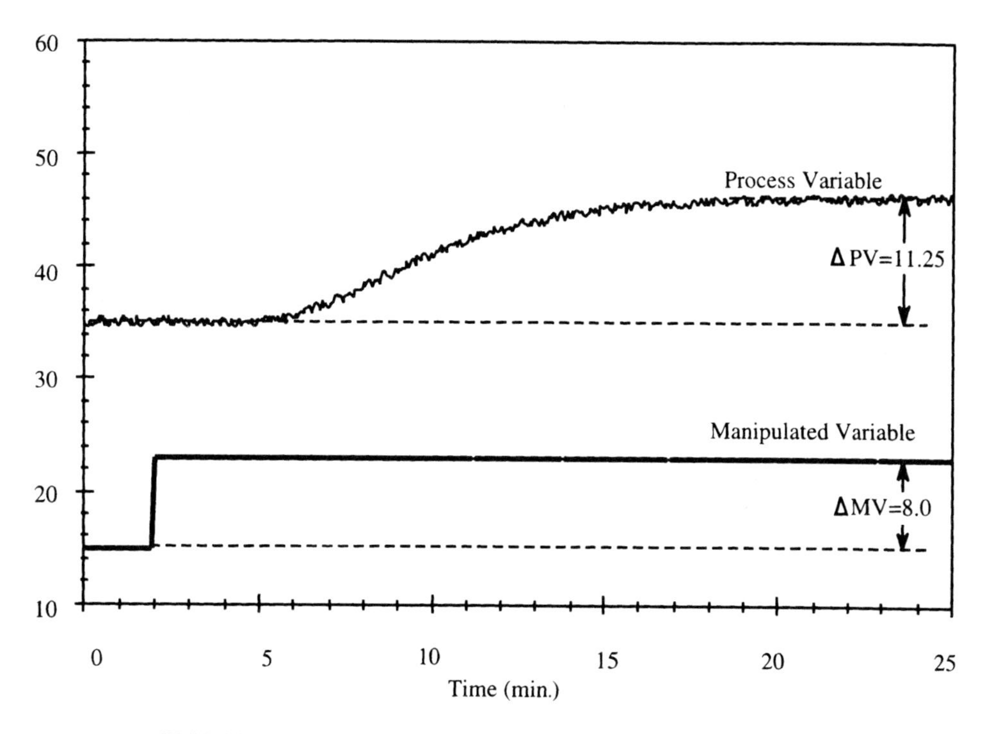


**FIGURE 4.27.** Example gain-only process model from graphical data.

**Example 4.9.** Find a gain-only process model of a process whose manipulated variable change and process variable response are as in Figure 4.27. The model is

$$G_G(s) = \frac{\Delta \text{PV}}{\Delta \text{MV}} = \frac{11.25}{8.0} = 1.41 \tag{4.85}$$

Note that the model (4.85) has ignored the deadtime and dynamics that are so obviously present in the response.

*First-Order-Plus-Deadtime (FODT) Model.* The transfer function of a general FODT model is

$$G_{\text{FODT}}(s) = \frac{Ke^{-s\theta_D}}{\tau_1 s + 1} \tag{4.86}$$

where

$$K = \text{process gain,}$$
$$\theta_D = \text{apparent process deadtime}$$
$$\tau_1 = \text{apparent process time constant}$$

Many processes are higher than first order (contain more than one lag term) and so any first-order model will only be an approximation. However, the approximation is sufficient in most cases and many of the PID tuning procedures use a FODT process model.

Two methods will be demonstrated. Both of these methods are based on features of the response of (4.86) to a step input of magnitude $A$,

$$y(t) = KA(1 - e^{-(t-\theta_D)/\tau_1}) \tag{4.87}$$

The response, (4.87), is evaluated for two values of the exponent, 1 ($t = \theta_D + \tau_1$) and $\frac{1}{3}$ ($t = \theta_D + \frac{1}{3}\tau_1$). When the exponent is 1, the value of the response is

$$y(t) = KA(1 - e^{-1}) = 0.632KA$$

In other words, the response reaches 63.2% of its change when $t = \theta_D + \tau_1$. Similarly, when the exponent is $\frac{1}{3}$, the value of the response is

$$y(t) = KA(1 - e^{-1/3}) = 0.283KA$$

In this case, the response reaches 28.3% of its change when $t = \theta_D + \frac{1}{3}\tau_1$.

The first graphical procedure, Method I, is based on Ziegler and Nichols (1942) and finds the gain, apparent deadtime, and apparent time constant as follows:

1. Draw a tangent to the process variable response curve at its point of maximum slope, the point of inflection (Figure 4.28). In practice, this step may be tricky, especially if the response contains noise.
2. The term $T_1$ is the time at which the step change in the manipulated variable was imposed; $T_2$ is the time at which the tangent drawn in the previous step intersects the time axis, *drawn horizontally from the beginning of the process variable response curve*. The apparent process deadtime $\theta_D = T_2 - T_1$.
3. The term $T_4$ is the time at which the process variable response has made 63% of its total change. The apparent process time constant $\tau_1 = T_4 - T_2$.

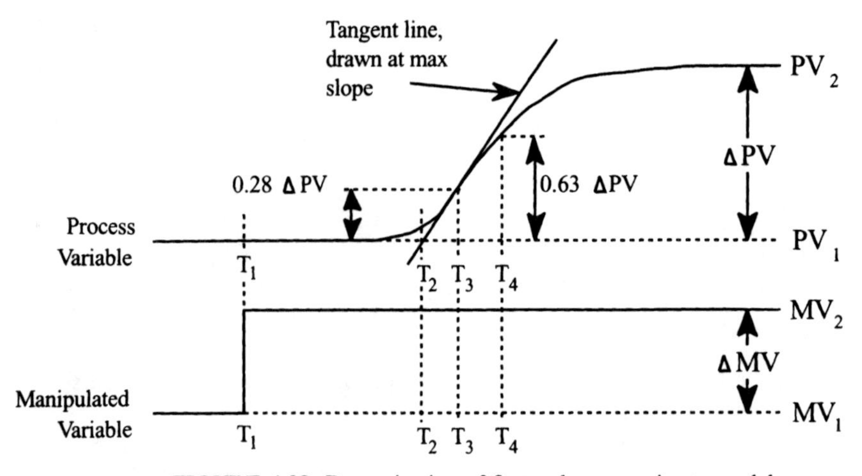


**FIGURE 4.28.** Determination of first-order approximate model.

4. The process gain is the change in the process variable divided by the change in the manipulated variable,

$$K = \frac{\Delta PV}{\Delta MV} = \frac{PV_2 - PV_1}{MV_2 - MV_1}$$

The second graphical procedure, Method II, is more suited to processes with significant measurement noise (Marlin, 1995) and is as follows (Figure 4.28):

1. The term $T_1$ is the time at which the step change in the manipulated variable was imposed; $T_3$ is the time at which the process variable response has made 28% of its total change ($T_3 = \theta_D + \frac{1}{3}\tau_1$); and $T_4$ is the time at which the process variable response has made 63% of its total change ($T_4 = \theta_D + \tau_1$).
2. The approximate process deadtime and time constant are

$$\tau_1 = 1.5(T_4 - T_3) \qquad \theta_D = T_4 - \tau_1 - T_1$$

3. The process gain is the change in the process variable divided by the change in the manipulated variable,

$$K = \frac{\Delta PV}{\Delta MV} = \frac{PV_2 - PV_1}{MV_2 - MV_1}$$

**Example 4.10.** Find a FODT approximation to a process, given the manipulated variable change and resulting process variable response in Figure 4.29. The actual process transfer function is

$$G(s) = \frac{1.4e^{-2.3s}}{(2s+1)^3}$$

with random noise on the measurement and a noise amplitude randomly distributed between ±0.5 units.

***Solution.*** The appropriate information needed for both methods is marked on the response in Figure 4.29. For Method I, the FODT model is

$$\tau_1 = T_4 - T_2 = 10 - 5.9 = 4.1 \text{ min}$$
$$\theta_D = T_2 - T_1 = 5.9 - 2.0 = 3.9 \text{ min}$$
$$K = \frac{\Delta PV}{\Delta MV} = \frac{11.25}{8.0} = 1.41$$

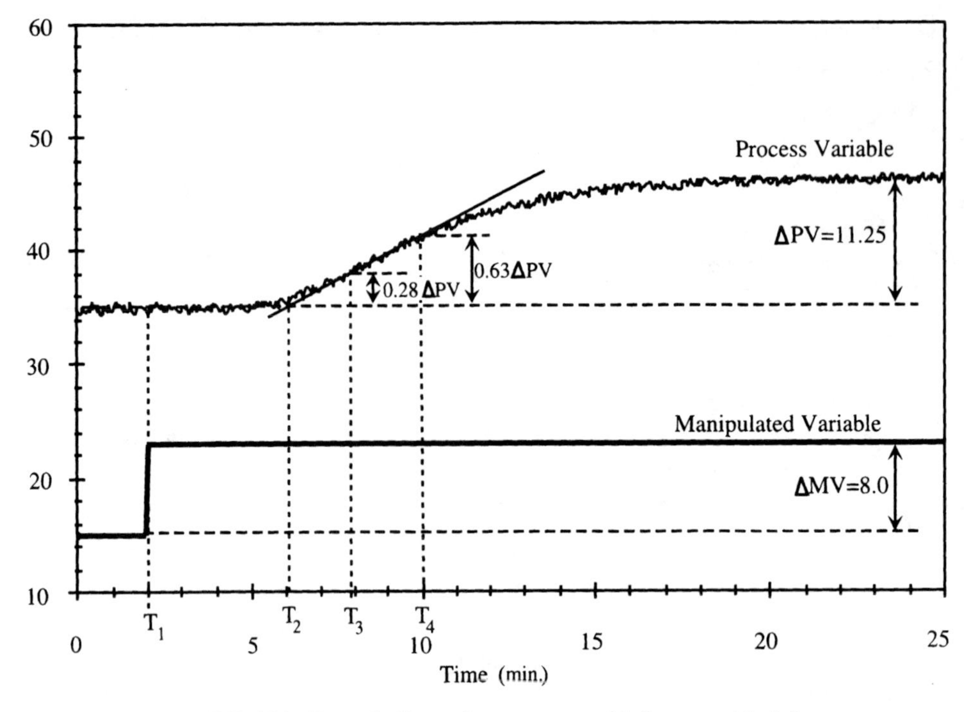


**FIGURE 4.29.** Example first-order process model from graphical data.

For Method II, the FODT model is

$$\tau_1 = 1.5(T_4 - T_3) = 1.5(10 - 7.9) = 3.15 \text{ min}$$

$$\theta_D = T_4 - \tau_1 - T_1 = 10 - 3.15 - 2 = 4.85 \text{ min}$$

$$K = \frac{\Delta \text{PV}}{\Delta \text{MV}} = \frac{11.25}{8.0} = 1.41$$

Comparing the response of these two approximate models with the plotted response (Figure 4.30) shows that the model obtained by Method I is a little closer to the real process response, especially since it seems to get a deadtime closer to the actual response. However, keep in mind that a good result for Method I depends on the accuracy of the tangent line drawn at the inflection point.

*Second-Order-Plus-Deadtime (SODT) Model.* The particular method used here to find the second-order model parameters is from Sundaresan et al. (1978). Depending on the process response, an overdamped or underdamped model may be more appropriate. An underdamped response, as shown by the sinusoidal response, may arise when the "process" is the inner loop of a cascade control scheme and one wants to determine an approximate process model in order to tune the outer loop.

The general overdamped process model is

$$G_{\text{SODT}}(s) = \frac{Ke^{-s\theta_D}}{(\tau_1 s + 1)(\tau_2 s + 1)} \tag{4.88}$$

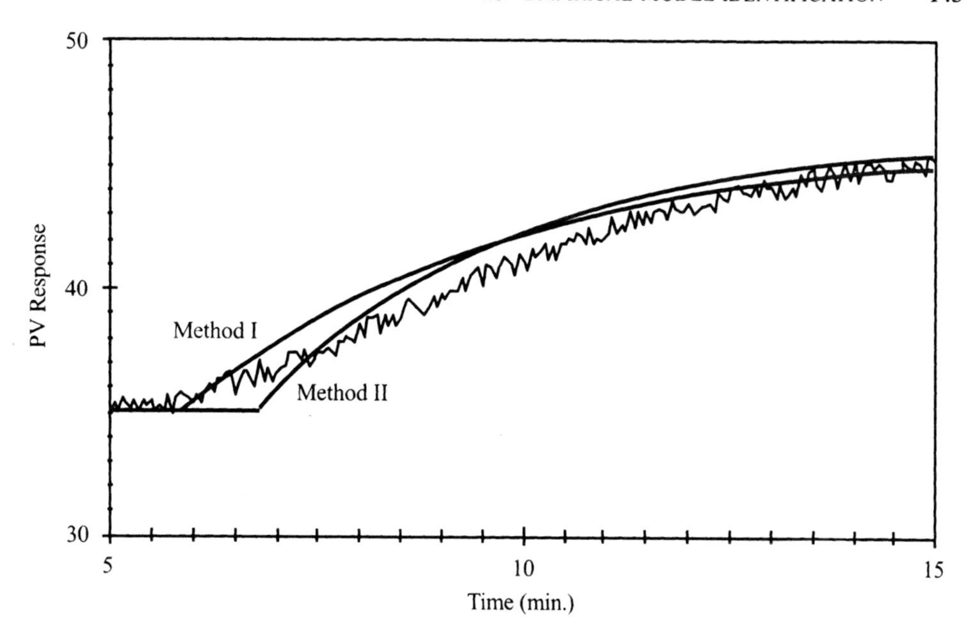


**FIGURE 4.30.** Comparison of Method I and II models with process data.

where $K$ is the gain, $\tau_1$ and $\tau_2$ are the time constants, and $\theta_D$ is the process deadtime. If $\tau_1 = \tau_2$, the system is called *critically damped.* Assume as given the general overdamped response in Figure 4.31, where the process variable response has been normalized to a maximum value of 1 and the time of the manipulated variable change is defined to be zero time. The gain of the process is determined in the same manner as for the gain-only and first-order models, namely,

$$K = \frac{\Delta \mathrm{PV}}{\Delta \mathrm{MV}}$$

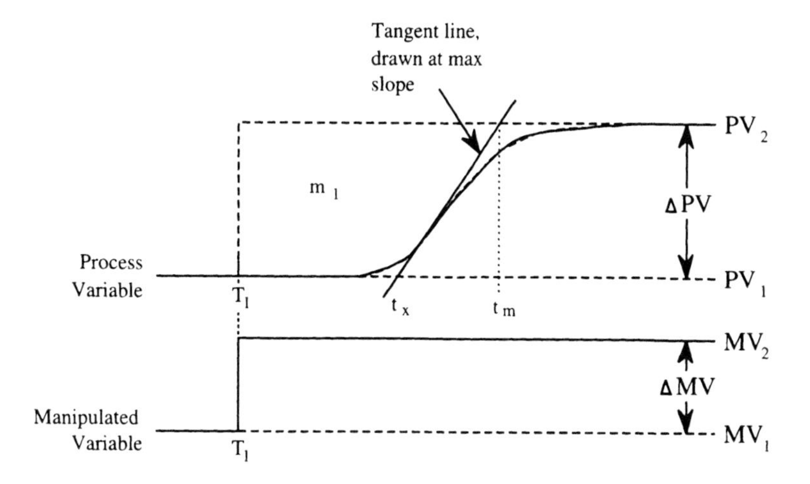


**FIGURE 4.31.** Determination of overdamped second-order approximate model.

It can be shown (Gibilaro and Waldram, 1972) that the first moment of the normalized response $c(t)$ is

$$m_1 = \int_0^\infty [1 - c(t)]\, dt \tag{4.89}$$

which is the shaded area of Figure 4.31. The first moment is also related to the sum of the parameters that need to be identified,

$$m_1 = \theta_D + \tau_1 + \tau_2 \tag{4.90}$$

Taking the second derivative of the time domain response of (4.88) and equating it to zero, the time at the point of inflection can be found as

$$t_i = \theta_D + \alpha \ln \eta \tag{4.91}$$

where

$$\eta = \frac{\tau_1}{\tau_2} \tag{4.92}$$

$$\alpha = \frac{\tau_1 \tau_2}{\tau_1 - \tau_2} \tag{4.93}$$

At the point of inflection, the slope of the tangent line $M_i$ is

$$M_i = \frac{(\eta)^{1/(1-\eta)}}{(\eta - 1)\alpha} \tag{4.94}$$

This tangent intersects the final value of $c(t)$ at time $t_m$, whose value is given by

$$t_m = \theta_D + \alpha\left(\ln \eta + \frac{\eta^2 - 1}{\eta}\right) \tag{4.95}$$

By combining Equations (4.90), (4.94), and (4.95), one obtains

$$(t_m - m_1)M_i = \frac{(\eta)^{1/(1-\eta)}}{(n - 1)} \ln \eta \tag{4.96}$$

If in Equation (4.96) the $\eta$ is changed to $1/\eta$, the right-hand side does not change. Therefore, it is sufficient to consider $0 \leq \eta \leq 1$. Equation (4.96) can be written in the form

$$\lambda = \chi e^{-\chi} \tag{4.97}$$

where

$$\lambda = (t_m - m_1)M_i \qquad \chi = \frac{\ln \eta}{\eta - 1}$$

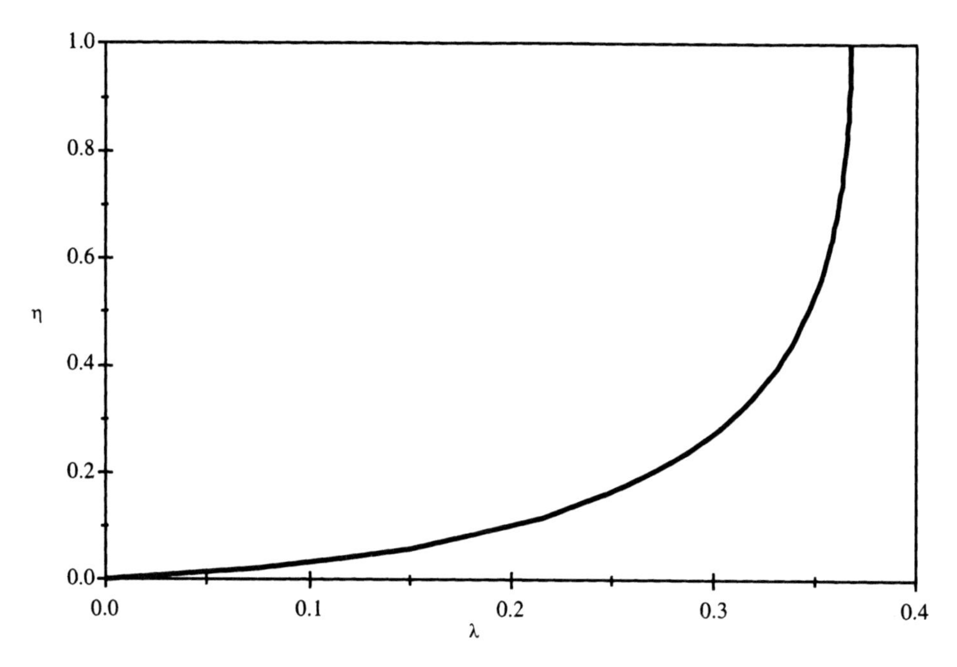


**FIGURE 4.32.** Plot of $\lambda$ versus $\eta$. (Reprinted with permission from K. R. Sundaresan et al., "Evaluating Process Parameters from Process Transients" *Ind. Eng. Chem. Process Des. Dev.*, **17**, 237–241 (1978). Copyright © 1978, American Chemical Society.)

It is evident from these equations that the maximum value of $\lambda$ is $e^{-1}$, which occurs when the system is critically damped at $\eta = 1$. When $\eta = 0$, the system is first order. Hence, the range of $\eta$ is $0 < \eta < 1$ and, consequently, $0 < \lambda < e^{-1}$. A plot of (4.97) for $0 < \eta < 1$ is shown in Figure 4.32 as a function of $\lambda$. Once $m_1$, $t_m$, and $M_i$ are measured and $\lambda$ is computed, the value of $\eta$ is determined from Figure 4.32 . Using (4.94), the value of $\alpha$ is determined. The values of $\tau_1$, $\tau_2$, and $\theta_D$ are determined by solving (4.90), (4.92), and (4.93).

Based on this development, the procedure to find the approximate gain, deadtime, and two time constants is as follows:

1. The process gain is the change in the process variable divided by the change in the manipulated variable, $K = \Delta\text{PV}/\Delta\text{MV}$.
2. Determine the value of the shaded area, where $\Delta$PV has been normalized to 1. For quick graphical results, approximate the integral by doing rectangular integration,

$$m_1 \approx \frac{T}{\Delta\text{PV}} \sum_{k=0}^{k=n} [\text{PV}_2 - \text{PV}(t_k)] \tag{4.98}$$

or trapezoidal integration,

$$m_1 \approx \frac{T}{\Delta\text{PV}} \left( \frac{1}{2}[\text{PV}_2 - \text{PV}(t_0)] + \sum_{k=1}^{k=n-1} [\text{PV}_2 - \text{PV}(t_k)] + \frac{1}{2}[\text{PV}_2 - \text{PV}(t_k)] \right) \tag{4.99}$$

where $t_0$ is the time of the manipulated variable change, $T$ is the uniform time between equidistant samples of the process variable response, and $PV_2$ is the final process variable response value. Ten to 20 samples are usually adequate.

3. Draw a tangent to the process variable response curve at its point of maximum slope, the point of inflection (Figure 4.31). In practice, this step may be tricky, especially if the response contains noise. Find the time at which the tangent intersects the final value of the process variable response and calculate the slope of this line as

$$M_i = \frac{1}{t_m - t_x}$$

Note that $\Delta PV$ has been normalized to 1.

4. Compute $\lambda = (t_m - T_1 - m_1)M_i$.
5. From the graph of $\lambda$ versus $\eta$ (Figure 4.32), determine $\eta$.
6. Solve for $\tau_1$, $\tau_2$, and $\theta_D$,

$$\tau_1 = \frac{\eta^{1/(1-\eta)}}{M_i} \qquad \tau_2 = \frac{\tau_1}{\eta} \qquad \theta_D = m_1 - \tau_1 - \tau_2$$

For critically damped systems, $\tau_1 = \tau_2 = 1/eM_i$, $\theta_D = m_1 - 2\tau_1$

The procedure for an underdamped system is only a modification of the previous procedure. The general underdamped process model is

$$G_{\text{SUDT}}(s) = \frac{Ke^{-s\theta_D}}{(s/\omega_n)^2 + 2\xi(s/\omega_n) + 1} \qquad (\xi < 1) \tag{4.100}$$

where $K$ is the gain, $\omega_n$ is the undamped natural frequency, $\xi$ is the damping ratio, and $\theta_D$ is the process deadtime. A general underdamped response is given in Figure 4.33. An

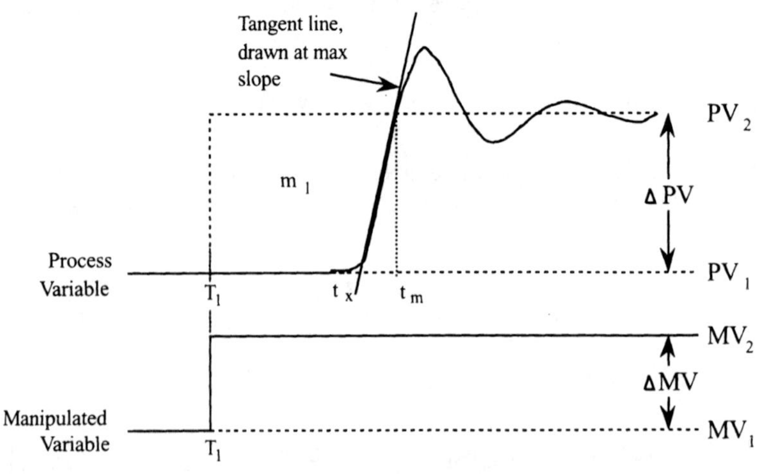


**FIGURE 4.33.** Determination of underdamped second-order approximate model.

underdamped response has more than one point of inflection. The first point of inflection is analogous to $t_i$ for the overdamped case. Starting with the expression for the time response of (4.100), a similar procedure to that of the overdamped case gives

$$\lambda = (t_m - T_1 - m_1)M_i = \frac{\cos^{-1}\xi}{\sqrt{1-\xi^2}}\exp\left(\frac{-\xi}{\sqrt{1-\xi^2}}\cos^{-1}\xi\right) \tag{4.101}$$

$$\omega_n = \frac{\cos^{-1}\xi}{\sqrt{1-\xi^2}}\frac{1}{t_m - T_1 - m_1} \tag{4.102}$$

$$\theta_D = m_1 - \frac{2\xi}{\omega_n} \tag{4.103}$$

In order to properly compute $m_1$, the area below the final value is considered positive and the area above the final value is negative. In the region of interest for the damping ratio, $0 \le \xi \le 1$, Figure 4.34 provides a plot of Equation (4.101) as a function of $\lambda$. The procedure for obtaining the apparent gain, deadtime, damping ratio, and natural frequency only differs from the overdamped case in the last two steps:

1. The process gain is the change in the process variable divided by the change in the manipulated variable, $K = \Delta PV/\Delta MV$.
2. Determine the value of the shaded area, where $\Delta PV$ has been normalized to 1. For quick graphical results, approximate the integral by doing a trapezoidal integration (4.99). This method requires a close estimate of the area and the rectangular

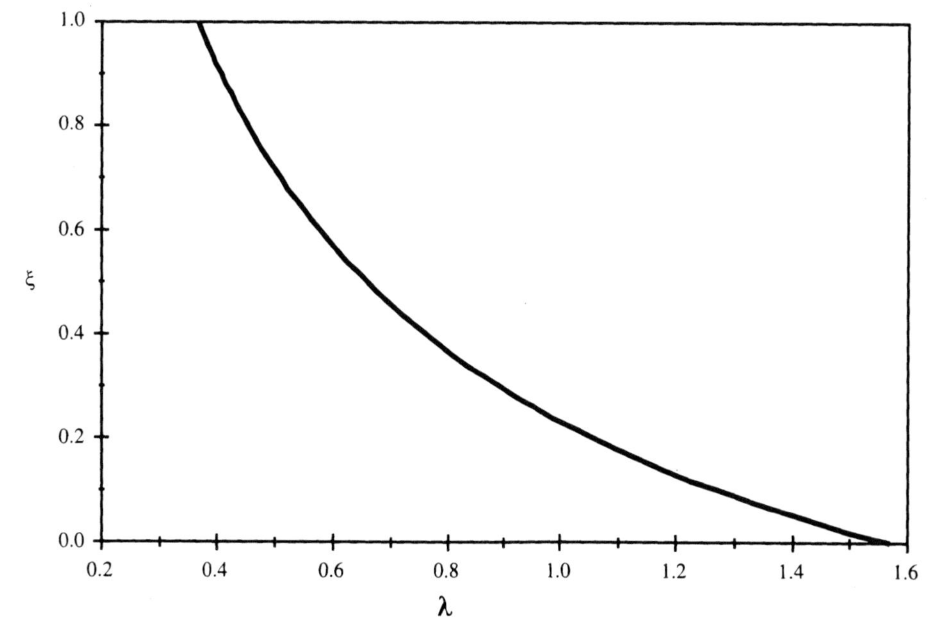


**FIGURE 4.34.** Plot $\lambda$ of versus $\xi$. (Reprinted with permission from K. R. Sundaresan et al., "Evaluating Process Parameters from Process Transients," *Ind. Eng. Chem. Process Des. Dev.*, **17**, 237–241 (1978). Copyright © 1978, American Chemical Society.)

integration is often not accurate enough. Any areas below the final value are positive and any areas above the final value are negative.

3. Draw a tangent to the process variable response curve at its point of maximum slope, the point of inflection. Find the time at which the tangent intersects the final value of the process variable response and calculate the slope of this line as

$$M_i = \frac{1}{t_m - t_x}$$

Note that ΔPV has been normalized to 1.

4. Compute $\lambda = (t_m - T_1 - m_1)M_i$.
5. From the graph of $\lambda$ versus $\xi$ (Figure 4.34), determine $\xi$.
6. Solve for $\omega_n$ and $\theta_D$ using (4.102) and (4.103).

**Example 4.11.** Find a second-order approximation to a process, given the manipulated variable change and resulting process variable response in Figure 4.35, which is the same response used in Example 4.10.

***Solution.*** The response is that of an overdamped process, so the overdamped process model (4.88) will be used. The tangent line is already marked in Figure 4.35. The process gain $K = 1.41$ is obtained in the same manner as for Example 4.10. The value of the shaded area is approximated with rectangular integration, (4.98), using 10 samples of the

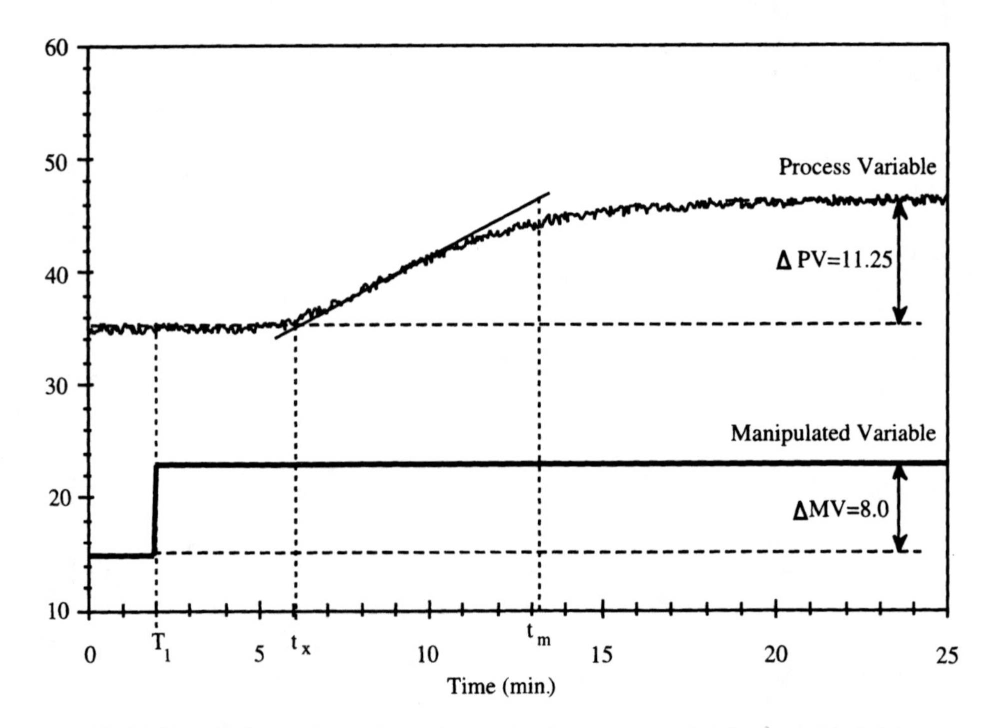


**FIGURE 4.35.** Example overdamped second-order process model from graphical data.

process variable response. The first sample is $t = 2$ min and the other samples are obtained at 2-min intervals. The approximate value of the area is 9.12. From Figure 4.35, the slope of the tangent is determined to be

$$M_i = \frac{1}{t_m - t_x} = \frac{1}{13.2 - 6.1} = 0.141$$

Further,

$$\lambda = (t_m - T_1 - m_1)M_i = (13.2 - 2 - 9.12)(0.141) = 0.293$$

Using Figure 4.32, $\eta = 0.26$. Therefore, the process parameters are

$$\tau_1 = \frac{\eta^{1/(1-\eta)}}{M_i} = \frac{0.26^{1/(1-0.26)}}{0.141} = 1.15 \qquad \tau_2 = \frac{\tau_1}{\eta} = \frac{1.15}{0.26} = 4.42$$

$$\theta_D = m_1 - \tau_1 - \tau_2 = 9.12 - 1.15 - 4.42 = 3.55$$

and the approximate process model is

$$G_{\mathrm{SODT}}(s) = \frac{1.41e^{-3.55s}}{(1.15s + 1)(4.42s + 1)}$$

Comparing the response of this approximate model with the plotted response (Figure 4.36) shows that the model is a little closer to the actual process than either of the first-order models (Figure 4.30), though there is not a significant difference.

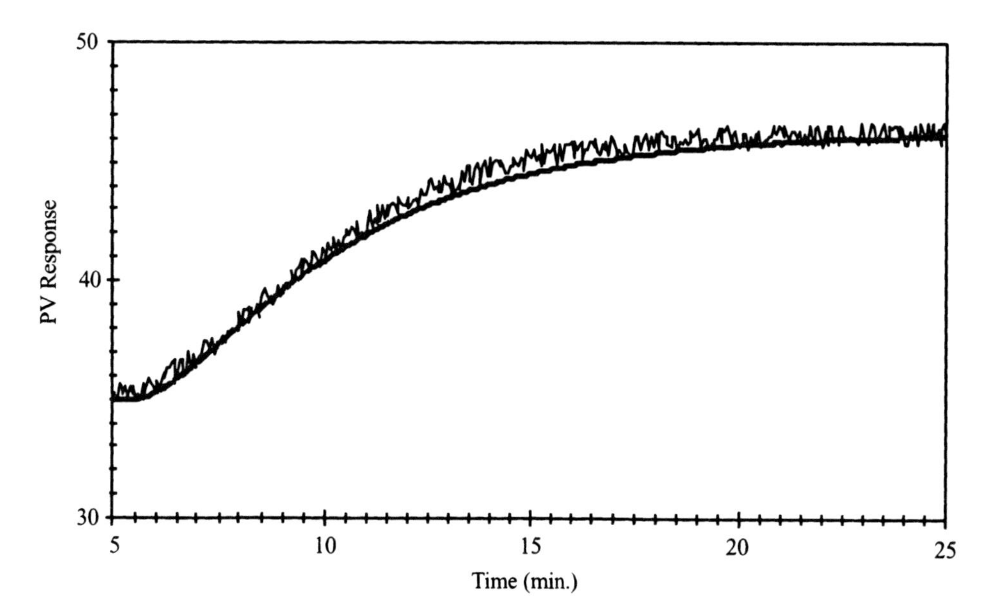


**FIGURE 4.36.** Comparison of overdamped second-order model with process data.

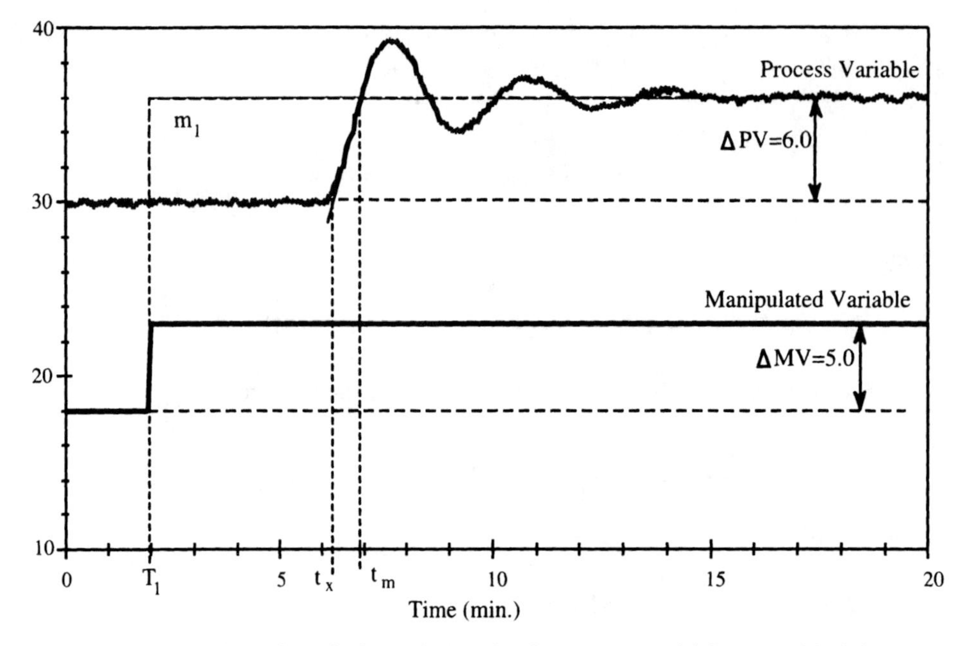


**FIGURE 4.37.** Example underdamped second-order process model from graphical data.

**Example 4.12.** Find a second-order approximation to a process, given the manipulated variable change and resulting process variable response in Figure 4.37. The actual process is

$$G(s) = \frac{1.2e^{-4s}}{(0.1s+1)[(s/2)^2 + 2(0.175)(s/2) + 1]}$$

with random noise on the measurement and noise amplitude randomly distributed between $\pm 0.5$ units.

***Solution.*** The response is that of an underdamped process, so the underdamped process model (4.100) will be assumed. The tangent line is already marked in Figure 4.37. The process gain $K = 1.2$. The value of the shaded area is approximated with trapezoidal integration, (4.99), using 14 samples of the process variable response. The first sample is at $t = 2$ min and the other samples are obtained at 1-min intervals. The approximate value of the area is 4.19. From Figure 4.37, the slope of the tangent is determined to be

$$M_i = \frac{1}{t_m - t_x} = \frac{1}{6.9 - 6.3} = 1.67$$

Further,

$$\lambda = (t_m - T_1 - m_1)M_i = (6.9 - 2 - 4.19)(1.67) = 1.19$$

Using Figure 4.34, $\xi = 0.14$. Therefore, the process parameters are

$$\omega_n = \frac{\cos^{-1}\xi}{\sqrt{1-\xi^2}}\frac{1}{t_m - T_1 - m_1} = \frac{\cos^{-1}(0.14)}{\sqrt{1-(0.14)^2}}\frac{1}{6.9-2-4.19} = 2.03$$

$$\theta_D = m_1 - \frac{2\xi}{\omega_n} = 4.19 - \frac{2(0.14)}{2.03} = 4.05$$

and the approximate process model is

$$G_{\text{SUDT}}(s) = \frac{1.2e^{-4.05s}}{(s/2.03)^2 + 2(0.14)(s/2.03) + 1}$$

Comparing the response of this approximate model with the plotted response (Figure 4.38) shows good agreement with the data.

*Integrator-Plus-Deadtime (ID) Model.* Since an integrating process is not self-regulating, the process will "run away" if a step change is imposed on the manipulated variable and held constant. Therefore, as soon as the reaction curve reaches a constant slope, data collection should be terminated and the controller should be placed in the Automatic state to restore control of the process. The basic integrating process transfer function with deadtime and no other dynamics is

$$G_{\text{ID}}(s) = \frac{K_S}{s}e^{-\theta_D s} \tag{4.104}$$

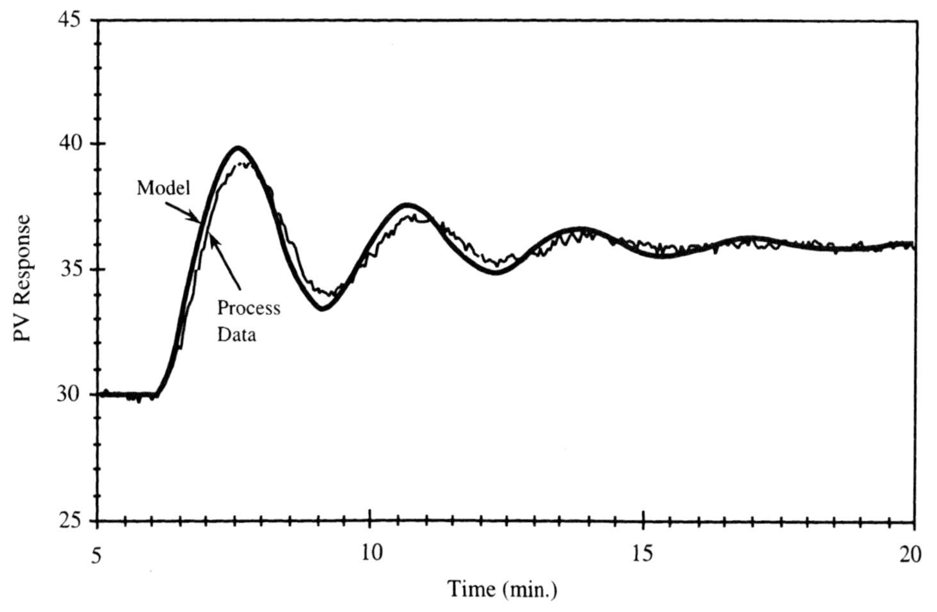


**FIGURE 4.38.** Comparison of underdamped second-order model with process data.

where

$$K_S = \text{apparent integrating constant}$$
$$\theta_D = \text{apparent process deadtime}$$

The graphical procedure to find the integrating constant and apparent deadtime is as follows:

1. Draw a tangent to the process variable response curve at its point of maximum slope (Figure 4.39). In practice, this step may be tricky, especially if the response contains noise.
2. The term $T_1$ is the time at which the step change in the manipulated variable was imposed; $T_2$ is the time at which the tangent drawn in the previous step intersects the initial condition from which the process variable response originates. The apparent process deadtime $\theta_D = T_2 - T_1$.
3. The integrating constant is the slope of the process variable response divided by the change in the manipulated variable,

$$K_S = \frac{\text{Slope of tangent}}{\Delta \text{MV}} = \frac{\text{Y/X}}{\text{MV}_2 - \text{MV}_1}$$

**Example 4.13.** Find an approximation to a process, given the manipulated variable change and resulting process variable response in Figure 4.40. In order to keep the manipulated variable from moving too far away from the operating point, as soon as the response seems to assume a constant slope, the manipulated variable is changed to a step in the negative direction for the same duration as the original change. This change brings the process back to its original operating point.

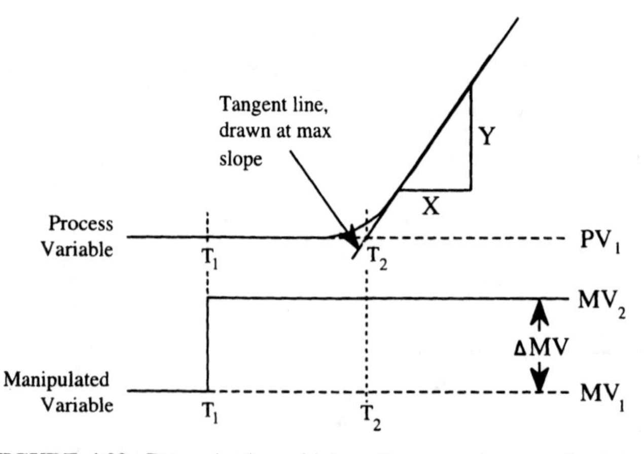


**FIGURE 4.39.** Determination of integrating process approximate model.

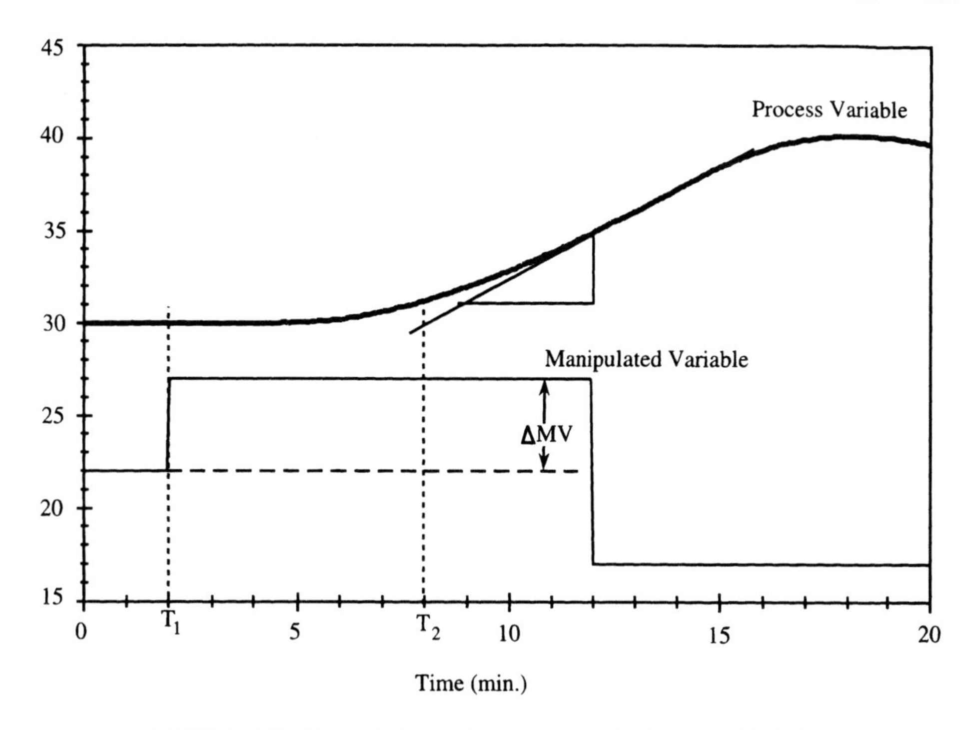


**FIGURE 4.40.** Example integrating process model from graphical data.

***Solution.*** The response is that of a process containing an integrator. The appropriate information needed to determine the approximate process is marked in Figure 4.40. The model parameters are thus

$$\theta_D = T_2 - T_1 = 8 - 2 = 6 \text{ min}$$
$$K_S = \frac{\text{slope of tangent}}{\Delta\text{MV}} = \frac{4/(12 - 8.8)}{5} = 0.25$$

So the approximate ID model is

$$G_{\text{ID}}(s) = \frac{0.25}{s} e^{-6s} \tag{4.105}$$

A comparison of the approximate model response with the actual process is shown in Figure 4.41. The comparison is extended to show the effect of moving the manipulated variable to bring the system back to the operating point. The model does an adequate job of predicting the constant slope of the first change, but because it does not accurately reflect the entire process dynamics, there is significant modeling error for the second manipulated variable move.

*Integrating FODT Model.* If one wants to recover the dynamics of a process containing an integrator, the integrator must be "removed" from the reaction curve by taking its derivative and then analyzing this derivative curve in the same manner as for a first- or second-order process with deadtime. An example is used to show this technique.

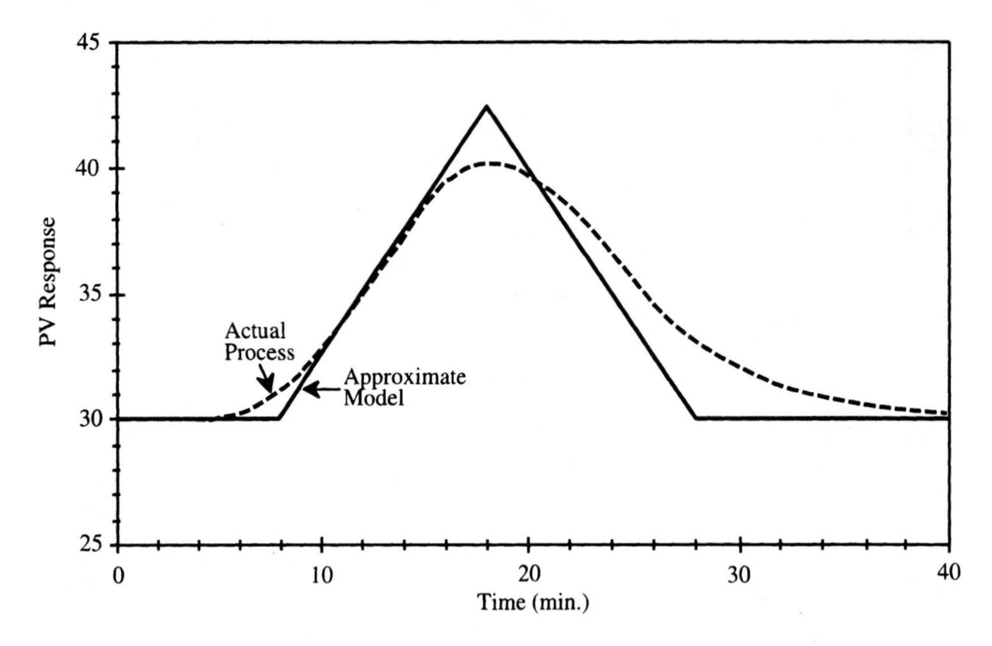


**FIGURE 4.41.** Comparison of integrator-plus-deadtime model with process data.

**Example 4.14.** Find the dynamics of the integrating process whose manipulated variable change and resulting process variable response are shown in Figure 4.40. The actual process transfer function is

$$G(s) = \frac{0.3e^{-2s}}{s(5s+1)(s+1)}$$

with random noise on the measurement and a noise amplitude randomly distributed between $\pm 0.03$ units.

***Solution.*** When the integrator is removed from the process variable response by taking its derivative, the response is as in Figure 4.42. Note that the small amount of noise on the process variable measurement is magnified by the derivative. Nevertheless, it is still possible to obtain an approximate model. Because of the noise in the response curve, a second-order process is difficult to reliably estimate, although filtering the data does help. The unfiltered points are presented here to illustrate it is still possible to obtain a good approximation with noisy data. The information needed to obtain a first-order approximate model is also shown in Figure 4.42. The time of the manipulated variable move $T_1$ is 2 min and its magnitude is 5. Using first-order method I, the FODT model parameters are

$$\tau_1 \approx T_4 - T_2 = 8.6 - 4.7 = 3.9 \text{ min}$$

$$\theta_D \approx T_2 - T_1 = 4.7 - 2.0 = 2.7 \text{ min}$$

$$K = \frac{\Delta \text{PV}}{\Delta \text{MV}} = \frac{1.25}{5.0} = 0.25$$

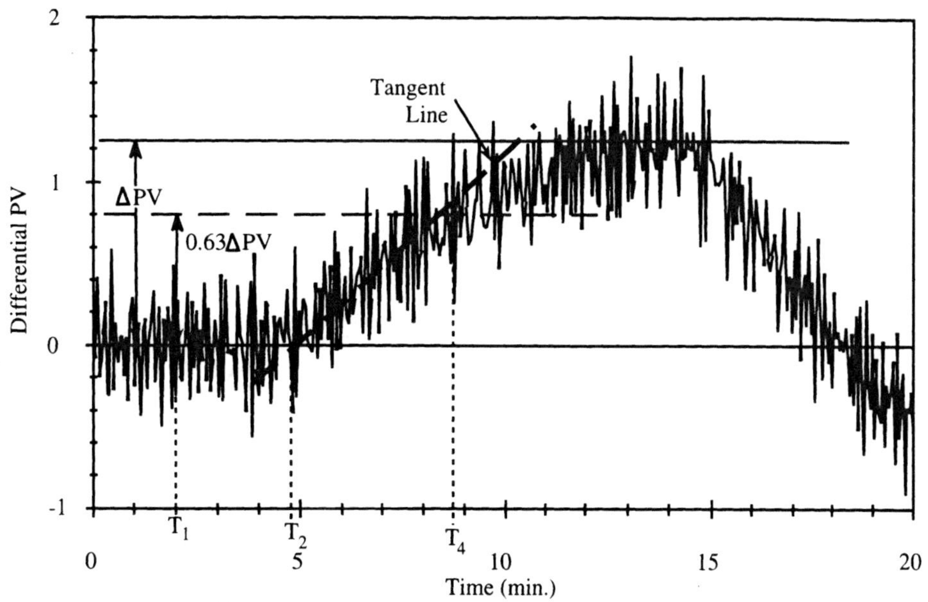


**FIGURE 4.42.** Differentiated data used to determine approximate dynamics of process with integrator.

Combining these parameters with the integrator, the approximate process model is

$$G_{\text{ID}}(s) = \frac{0.25e^{-2.7s}}{s(3.9s + 1)} \tag{4.106}$$

A comparison of the approximate model response with the actual process is shown in Figure 4.43. As in the previous example, the comparison is extended to show the effect of moving the manipulated variable to bring the system back to the operating point. Comparing with the model of the previous example (Figure 4.41), the model that includes first-order dynamics does better at predicting the process response. However, as for the previous example, there is significant modeling error for the second manipulated variable move.

### 4.6.2. Least-Squares Estimation

In the previous section, the process model was determined by using the process step response to estimate the parameters of a continuous-time transfer function process model. In contrast, least-squares estimation finds the parameters of a discrete-time transfer function model. In many cases, it is possible to transform the discrete-time model parameters into the continuous transfer function parameters (4.82) and (4.83) and thus produce the same models generated in the previous section. However, many of the newer control algorithms (e.g., model predictive control) use the discrete-time model directly in the control algorithm and eliminate the transformation step.

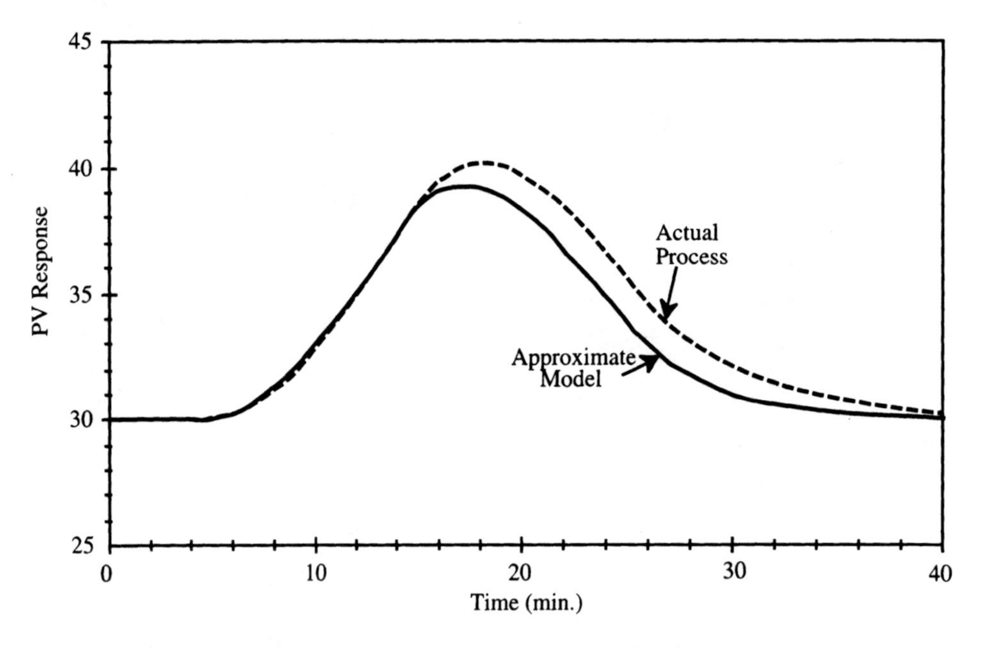


**FIGURE 4.43.** Comparison of integrator and first-order dynamics with process data.

The first choice one needs to make is the form of the general model. Two types of models are considered here, the autoregressive moving-average (ARMA) model and the moving-average (MA) model. The ARMA model has the general form

$$G(z) = \frac{Y(z)}{U(z)} = \frac{b_1 z^{-1} + b_2 z^{-2} + \cdots + b_{n-1} z^{-n+1} + b_n z^{-n}}{1 - a_1 z^{-1} - a_2 z^{-2} - \cdots - a_{n-1} z^{-n+1} - a_n z^{-n}} \tag{4.107}$$

where $U(z)$ is the input and $Y(z)$ is the output. The difference equation describing the system is

$$\begin{aligned} y(k) &= a_1 y(k-1) + a_2 y(k-2) + \cdots + a_n y(k-n) \\ &\quad + b_1 u(k-1) + b_2 u(k-2) + \cdots + b_n u(k-n) + e(t) \end{aligned} \tag{4.108}$$

where $e(t)$ represents the modeling error inherent in the procedure to find a model. The coefficient vector containing the parameters to be estimated is

$$\theta = [a_1 \quad a_2 \quad \cdots \quad a_n \quad b_1 \quad b_2 \quad \cdots \quad b_n]^{\mathrm{T}} \tag{4.109}$$

The MA model represents the model as one polynomial, rather than the ratio of two polynomials, in the form

$$G(z) = H(z) = h_1 z^{-1} + h_2 z^{-2} + \cdots + h_{n-1} z^{-n+1} + h_n z^{-n} \tag{4.110}$$

The difference equation describing the system is

$$y(k) = h_1 u(k-1) + h_2 u(k-2) + \cdots + h_n u(k-n) + e(t) \tag{4.111}$$

The coefficient vector is

$$\theta = [h_1 \quad h_2 \quad \cdots \quad h_{n-1} \quad h_n]^{\mathrm{T}} \tag{4.112}$$

Note that the MA model may be obtained from the ARMA model by polynomial division. In general, the MA model contains more parameters than the ARMA model, and thus one would assume that more data samples are required. However, it does have the advantages that unusual dynamics are handled easily and the step response, which has intuitive appeal, is easily obtained by numerically integrating the MA model. Because of these advantages, the MA model is used in many model predictive control algorithms.

The general idea of identification is to calculate the "best" value of the coefficient vector given samples of the input $u(k)$ and the output $y(k)$. Of course, some criterion must be used to get the best estimate. A popular method minimizes the sum of squared errors, where the error is defined as the actual model output minus the predicted model output for each data sample. Hence the term *least-squares estimation*. Many other forms of estimation are also possible, but discussion of these is beyond the scope of this text. The interested reader is referred to Ljung and Söderström (1983) or Goodwin and Sin (1984). For simplicity, the MA model is used in the initial development and is extended to the ARMA model. Formally, consider the $n$th-order MA model

$$y(k) = h_1 u(k-1) + h_2 u(k-2) + \cdots + h_n u(k-n) + e(t)$$

and a set of $N$ data measurements

$$\{u(1), y(1)\}, \{u(2), y(2)\}, \{u(3), y(3)\}, \ldots, \{u(N), y(N)\}$$

where $N > n$. The coefficient vector is

$$\theta = [h_1 \quad h_2 \quad \cdots \quad h_{n-1} \quad h_n]^{\mathrm{T}}$$

Define the regression vector as

$$\phi(k) = [u(k-1) \quad u(k-2) \quad \cdots \quad u(k-n+1) \quad u(k-n)]^{\mathrm{T}}$$

The predicted output at sample $k$ given previous samples of the input and the estimated parameters is

$$\hat{y}(k|\theta) = \phi^{\mathrm{T}}(k)\theta$$

The function to be minimized is the sum of the squared errors, defined as

$$V_N(\theta) = \frac{1}{N}\sum_{k=1}^{N} e^2(k) \tag{4.113}$$

where

$$e(k) = y(k) - \hat{y}(k|\theta)$$

Define the $n \times N$ regressor matrix

$$\Phi_N = [\phi(1) \quad \phi(2) \quad \phi(3) \quad \cdots \quad \phi(N)]$$

and the vector of output measurements

$$\psi = [y(1) \quad y(2) \quad \cdots \quad y(N)]^{\mathrm{T}}$$

It can be shown that the value of the parameter vector that minimizes the sum of the squared errors (4.113) is

$$\hat{\Theta}_N^{\mathrm{LS}} = (\Phi_N \Phi_N^T)^{-1} \Phi_N \psi \tag{4.114}$$

where $(\cdot)^{-1}$ denotes matrix inversion. The expression in (4.114) requires the multiplication of two large matrices. An equivalent formulation that involves smaller matrices is the following (Ljung and Söderström, 1983):

$$\hat{\Theta}_N^{\mathrm{LS}} = R^{-1}(N) f(N) \tag{4.115}$$

where

$$R(N) = \frac{1}{N} \sum_{k=1}^{N} \phi(k) \phi^{\mathrm{T}}(k)$$

$$f(N) = \frac{1}{N} \sum_{k=1}^{N} \phi(k) y(k)$$

If identifying an ARMA model, the only change in the development is that the coefficient vector is

$$\theta = [a_1 \quad a_2 \quad \cdots \quad a_n \quad b_1 \quad \cdots \quad b_n]^{\mathrm{T}}$$

and the regression vector is

$$\phi(k) = [y(k-1) \quad y(k-2) \quad \cdots \quad y(k-n) \quad u(k-1) \quad \cdots \quad u(k-n)]^{\mathrm{T}}$$

Recursive versions of the least-squares algorithm are used for on-line identification (Ljung and Söderström, 1983). These algorithms are beyond the scope of this text.

There are three additional issues to consider. The first is the choice of input signal. A step change in the manipulated variable will work, but measurement noise will tend to corrupt the model parameters. In order to be resistant to noise corruption, the signal input to the process should be persistently exciting (Gustavsson et al., 1977); that is, enough energy must be injected into the process to overcome the random noise and disturbances. Often a step manipulated variable move is not persistently exciting. A better choice is to use a random signal that is uncorrelated to the process noise and disturbances. A pseudorandom binary sequence (PRBS) is the best choice of a random signal. A typical

PRBS is shown as the lower trace in Figure 4.45. The switching time of the signal is the random part of the signal. The advantages of a PRBS input signal are:

1. It operates between two fixed limits.
2. It puts the most random energy into the process.
3. It can easily excite the slow and fast modes of the process.
4. The net effect of the PRBS is often unnoticed at the process output and is less than the PRBS magnitude.

In practice, the PRBS is added to the operating point values of the manipulated variable. A controller can be added to the loop provided it is not tuned too aggressively. In this case, the PRBS is added to the controller output.

The second issue to consider when using least-squares estimation is the pretreatment of the controlled variable samples. As for the reaction curve methods, filtering may be used to eliminate noise, though the least-squares algorithm does some filtering. A more serious problem is that any trends, or offset, in the data (both manipulated and controlled variable) must be handled. The offset may be handled in one of these ways:

1. Redefine the input and output data as deviations from the operating point. Given that the operating point is $(u_o, y_o)$, each data point is redefined as

$$\bar{u}(k) = u(k) - u_o \qquad \bar{y}(k) = y(k) - y_o$$

2. If the operating point is not known, then the sample mean of the data is subtracted from each data point. The sample means are calculated as

$$u_{\text{ave}} = \frac{1}{N}\sum_{k=1}^{N} u(k) \qquad y_{\text{ave}} = \frac{1}{N}\sum_{k=1}^{N} y(k)$$

   Each data point is redefined as

$$\bar{u}(k) = u(k) - u_{\text{ave}} \qquad \bar{y}(k) = y(k) - y_{\text{ave}}$$

3. Estimate the offset explicitly by including it in the model. The previous data pretreatment methods are not practical for on-line estimation. In the case of the ARMA model, the model including the offset is

$$\begin{aligned} y(k) &= a_1 y(k-1) + a_2 y(k-2) + \cdots + a_n y(k-n) \\ &\quad + b_1 u(k-1) + b_2 u(k-2) + \cdots + b_n u(k-n) + a + e(t) \end{aligned} \tag{4.116}$$

   where $\alpha$ represents the offset. The coefficient vector is

$$\theta = [a_1 \quad a_2 \quad \cdots \quad a_n \quad b_1 \quad \cdots \quad b_n \quad \alpha]^{\mathrm{T}}$$

and the regression vector is

$$\phi(k) = [y(k-1) \quad y(k-2) \quad \cdots \quad y(k-n) \quad u(k-1) \quad \cdots \quad u(k-n) \quad 1]^{\mathrm{T}}$$

4. Difference the data. The data points for the least squares are redefined as

$$\bar{u}(k) = u(k) - u(k-1) \qquad \bar{y}(k) = y(k) - y(k-1)$$

This method of data pretreatment is also useful for on-line identification.

The third issue is how to determine whether the model is good. Within a set of models, the best way to determine the particular model that has the best fit, is to compare these models using a different set of data. The model that has the smallest prediction error is then the best model. However, this approach may not be practical, especially when all the data points are used to generate each model. In this case, the model comparisons will have to be made using the one data set. For this case, the smallest model error is not generally given by the best model. Generally, the modeling error decreases as the number of parameters in the model increases, as shown in Figure 4.44. Even after a model contains enough parameters to correctly determine a model, additional parameters adjust themselves to features of the noise, decreasing $V_N$. Thus, the goal is to obtain a model whose modeling error is in the "knee" of the curve. An examination of the model parameters also validates a model. For example, if an ARMA model of an overdamped process is obtained, then all denominator roots should be positive and have a magnitude less than 1. The last coefficient of the MA model should be small, indicating that the model is long enough to reach steady state. If the MA model also includes an $h_0$ coefficient, which measures the correlation between the current input and output sample, a significant value indicates severe noise corruption and/or corruption due to closed-loop control (Otto, 1986).

All of these issues are addressed in the following example.

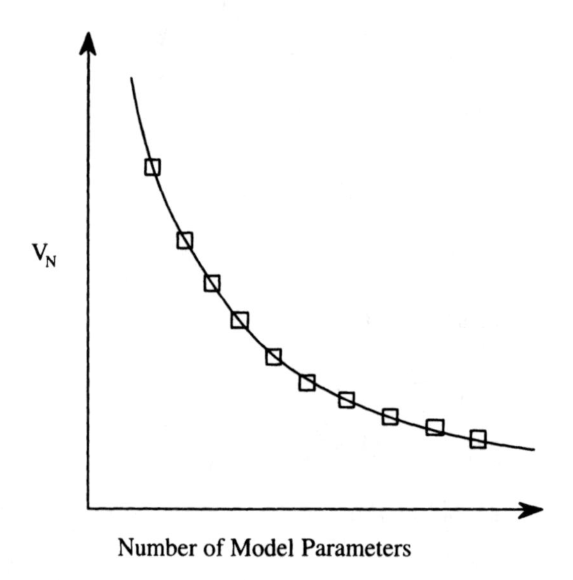


**FIGURE 4.44.** Modeling error as a function of the number of parameters.

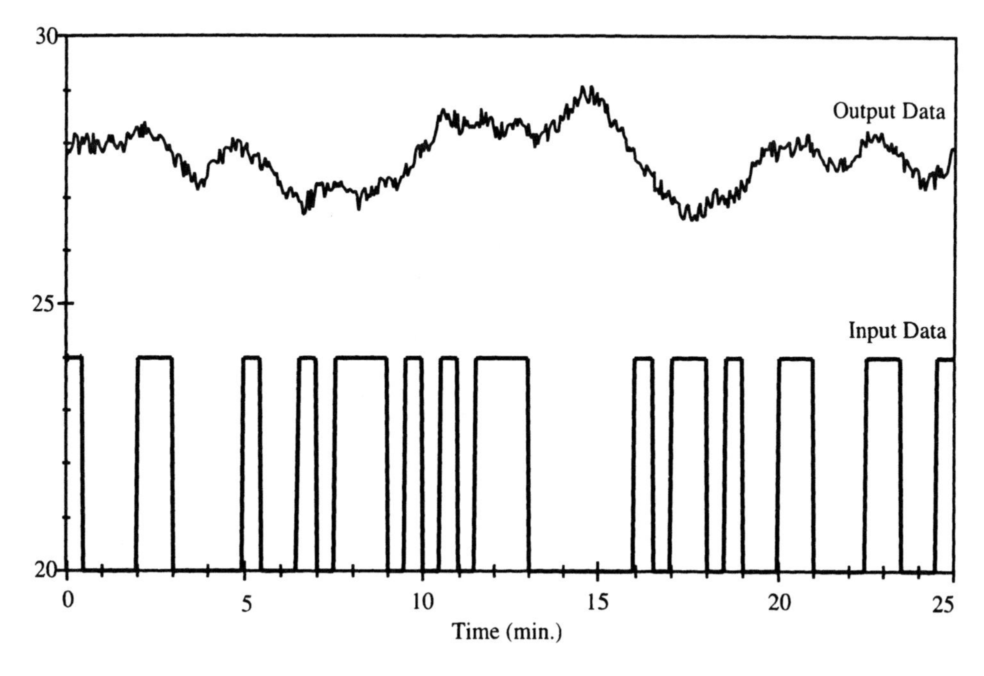


**FIGURE 4.45.** Input (lower trace) and output (upper trace) data for Example 4.15.

**Example 4.15.** Using least-squares estimation, find a model of a process whose input and output data are pictured in Figure 4.45. The process is not under control and is driven by a PRBS of magnitude ±2 units. The actual process is

$$G(s) = \frac{1.2e^{-1.5s}}{(0.5s + 1)(2s + 1)}$$

To generate a realistic problem, two sets of data are generated. The first set has low random noise uniformly distributed between ±0.05 units. The second set has moderate random noise uniformly distributed between ±0.20 units. The sample period is 0.5 min and 100 data samples are taken. Assume the process is overdamped and that the maximum deadtime is 2.5 min (five samples). Determine (a) the best ARMA model and (b) the best MA model.

***Solution.*** The least-squares algorithm outlined previously is used. The data points are pretreated by redefining them as deviations from the operating point,

$$\bar{u}(k) = u(k) - 22 \qquad \bar{y}(k) = y(k) - 28$$

(a) In order to determine a good ARMA model, the number of numerator and denominator coefficients are varied in order to determine the proper denominator order and deadtime of the model. The results are summarized in Table 4.3 for the low-noise process

**TABLE 4.3 Summary of ARMA Estimation Results for Low-Noise Data Set**

| $n_a$ | $n_b$ | $dt$ | $V_N$ | Comments |
|---|---|---|---|---|
| 1 | 6 | 0 | 0.0018 | |
| 2 | 7 | 0 | 0.0013 | |
| 3 | 8 | 0 | 0.0011 | |
| | | | | Looks like three samples of deadtime |
| 1 | 4 | 3 | 0.0309 | $A(z)$ root at 0.923 |
| 2 | 5 | 3 | 0.0021 | $A(z)$ roots at 0.794, 0.205 |
| 3 | 6 | 3 | 0.0013 | $A(z)$ roots at 0.781, 0.356, −0.635 |
| 2 | 4 | 2 | 0.0058 | $A(z)$ roots at $0.695 \pm j0.168$ |
| 2 | 6 | 4 | 0.0560 | $A(z)$ roots at 0.806, −0.190 |

**TABLE 4.4 Summary of ARMA Estimation Results for Moderate-Noise Data Set**

| $n_a$ | $n_b$ | $dt$ | $V_N$ | Comments |
|---|---|---|---|---|
| 1 | 6 | 0 | 0.0237 | |
| 2 | 7 | 0 | 0.0180 | |
| 3 | 8 | 0 | 0.0167 | |
| | | | | Looks like three samples of deadtime |
| 1 | 4 | 3 | 0.0533 | $A(z)$ root at 0.901 |
| 2 | 5 | 3 | 0.0259 | $A(z)$ roots at 0.842, −0.085 |
| 3 | 6 | 3 | 0.0195 | $A(z)$ roots at 0.806, −0.551, 0.161 |
| 1 | 3 | 2 | 0.1216 | $A(z)$ root at 0.895 |
| 1 | 5 | 4 | 0.0831 | $A(z)$ root at 0.806 |

data set and in Table 4.4 for the moderate-noise process data set. The symbols in the tables are defined as

$$n_a = \text{order of denominator polynomial, number of } a \text{ coefficients}$$
$$n_b = \text{order of numerator polynomial, number of } b \text{ coefficients}$$
$$dt = \text{number of } b \text{ coefficients assumed to be zero}$$
$$V_N = \text{modeling error, defined by (4.113)}$$

For both data sets, the order is determined by starting with a first-order model and a maximum of five samples of deadtime, increasing the order and examining the model error $V_N$. The model order is not clearly determined when the deadtime is assumed to be five samples since there is not a clear error decrease between model orders. For all three of these models, the first three numerator coefficients are close to zero and the fourth coefficient is significantly larger than zero, and so the deadtime is then assumed to be three samples (1.5 min) and the order is increased from first to third order. For the low-noise data set, the second-order model is clearly the best because of the large decrease in the modeling error when the order is increased from first to second order. The deadtime is confirmed by doing a least-squares fit when the deadtime is assumed to be two and four samples. Though the model error is small, the denominator roots fail the validity test since an overdamped model will have all roots real and positive. Therefore, the best deadtime is three samples. The discrete-time model is

$$G_{\text{SODT}}(z) = z^{-3} \frac{0.1183z^{-1} + 0.0552z^{-2}}{1 - 1.0993z^{-1} + 0.2425z^{-2}}$$

Using (4.83), on the discrete-time model coefficients, the continuous model is

$$G_{\text{SODT}}(s) = \frac{1.18e^{-1.5s}}{(2.17s + 1)(0.42s + 1)}$$

The process gain is obtained by averaging the gain obtained assuming each of the $b$ coefficients is correct. This model shows good comparison with the actual process when comparing the step response (Figure 4.46). The model could also be validated by comparing the response assuming a PRBS input with the actual data, but the comparison is very close.

For the moderate-noise data, the choice of the model order is not clear. For both the second- and third-order models, at least one denominator root is on the negative real axis. An overdamped process will have all roots positive and with a magnitude less than 1. The deadtime is confirmed by doing a least-squares fit when the deadtime is assumed to be two and four samples. The first-order model with three samples of deadtime has slightly less

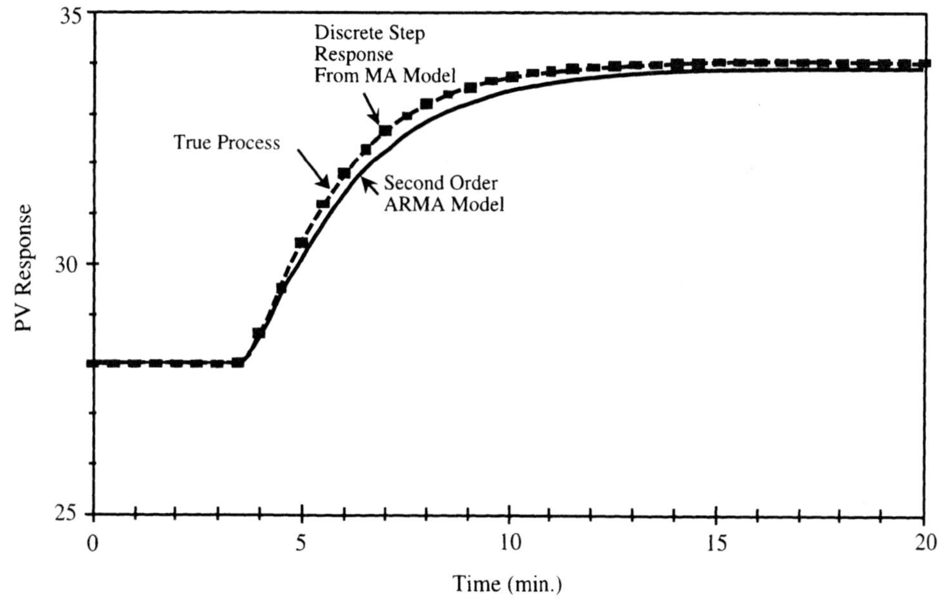


**FIGURE 4.46.** Model response for model obtained with low process data noise.

modeling error than the model with four samples of deadtime. Hence, the best model is the first-order model for the moderate-noise data set. The discrete-time model is

$$G_{\text{FODT}}(z) = z^{-3}\frac{0.1319z^{-1}}{1 - 0.9014z^{-1}} = z^{-3}\frac{0.1319}{z - 0.9014}$$

Using (4.82), on the discrete-time model coefficients, the continuous model is

$$G_{\text{FODT}}(s) = \frac{1.34e^{-1.5s}}{(5.0s + 1)}$$

As shown in Figure 4.47, this model does not compare favorably with the actual process model when comparing step responses.

(b) For the MA model, 32 coefficients are determined. The first coefficient $h_0$ is used as a measure of the amount of noise corruption in the model. The last coefficient $h_{31}$ is used primarily to determine if the model has reached steady state. If both of these parameters are small, then the model is assumed to be good. No iteration in model order or assumed deadtime is necessary. For the low-noise data set, $h_0 = 0.0017$ and $h_{31} = -0.009$, indicating a good model. The step response is obtained by a summation of the impulse response coefficients. The $i$th step response sample is

$$a_i = \sum_{j=1}^{i} h_j \quad \text{for } i \geq 1$$

As shown in Figure 4.46, the discrete-time step response from the MA model with the low-noise data set compares very favorably with the actual step response. Only the samples are

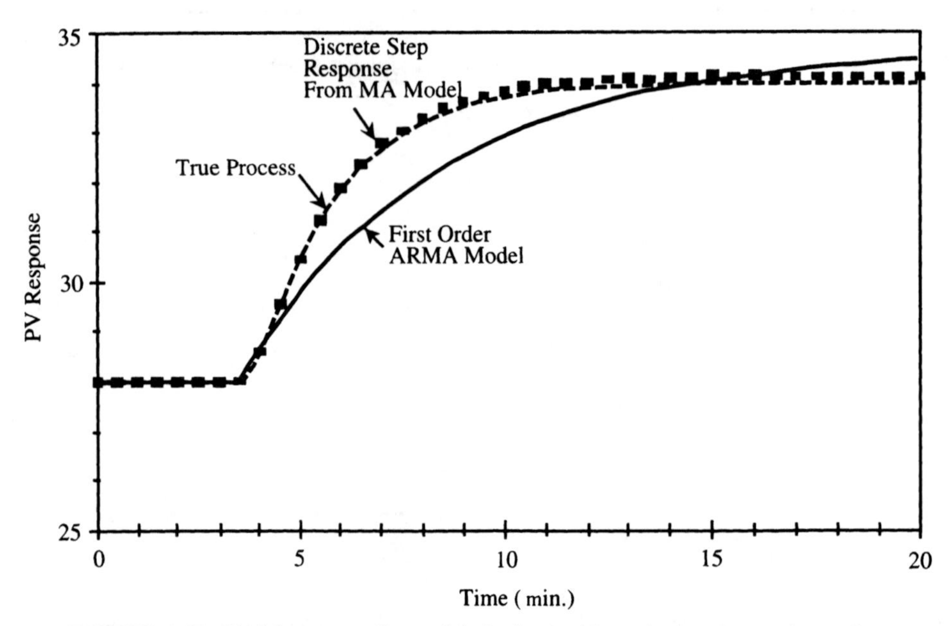


**FIGURE 4.47.** Model response for model obtained with moderate process data noise.

shown since it is a discrete-time model. For the moderate-noise data set, $h_0 = 0.0067$ and $h_{31} = -0.051$, indicating a reasonable model with some corruption. However, this MA model is still close to the actual process step response (Figure 4.47). Some model corruption is evident, though the MA model is clearly more accurate than the ARMA model.

## References

Carlson, G. E., *Signal and Linear Systems Analysis*, Wiley, New York, 1994.

Constantinides, A., *Applied Numerical Methods With Personal Computers*, McGraw-Hill, New York, 1987.

Denn, M., *Process Modeling*, Pitman, Marshfield, MA, 1986.

Dorf, R. C., and R. H. Bishop, *Modern Control Systems*, 7th ed., Addison-Wesley, Reading, MA, 1995.

Gibilaro, L. G., and S. P. Waldram, "The Evaluation of System Moments from Step and Ramp Response Experiments," *Chem. Eng. J.*, **4**, 197–198 (1972).

Goodwin, G. C., and K. S. Sin, *Adaptive Filtering Prediction and Control*, Prentice-Hall, Englewood Cliffs, NJ, 1984.

Gustavsson, I., L. Ljung, and T. Söderström, "Identification of Processes in Closed-Loop Identification and Accuracy Aspects," *Automatica*, **13**, 59–75 (1977).

Ljung, L., and T. Söderström, *Theory and Practice of Resursive Identification*, MIT Press, Cambridge, MA, 1983.

Luyben, W. L., *Process Modeling, Simulation, and Control for Chemical Engineers*, 2nd ed., McGraw-Hill, New York, 1990.

Marlin, T. E., *Process Control: Designing Processes and Control Systems for Dynamic Performance*, McGraw-Hill, New York, 1995.

Maron, M., and R. Lopez, *Numerical Methods, A Practical Approach*, 3rd ed., Wadsworth, Belmont, CA, 1991.

Otto, R. E., St. Louis, MO, personal communication, November 1986.

Stephanopoulos, G., *Chemical Process Control*, Prentice-Hall, Englewood Cliffs, NJ, 1984.

Sundaresan, K. R., C. Chandra Prasad, and C. Krishnaswamy, "Evaluating Parameters for Process Transients," *Ind. Eng. Chem. Process Des. Dev.*, **17**, 237–241 (1978).

Ziegler, J., and N. Nichols, "Optimum Settings for Automatic Controllers," *Trans. ASME*, **64**, 759–769 (1942).

Ziemer, R. E., W. H. Tranter, and D. R. Fannin, *Signals and Systems: Continuous and Discrete*, 2nd ed., Macmillan, New York, 1989.

# 5 Single-Loop Regulatory Control

**Chapter Topics**

- General features of feedback control
- PID controller
- PID controller tuning

*Scenario:* Controllers should be tuned at an appropriate operating point.

Based on his successes with the glass furnaces, the young engineer was once called to help in a situation where the operators were noticing that one of the controlled variables, a flow, never seemed to "line out," or assume a constant value. The control valve was not bouncing between the minimum and maximum values, and so they did not think it was a bad problem, just one that was nagging. The particular system had been in use for about two years. First, the engineer checked the tuning parameters. They were not the standard startup values for a flow loop, so he assumed that it had been properly tuned at least once. The upstream pressure was not perfect, but it was not oscillating like the flow he was observing. Not wanting to invest much more time in the solution, he tuned the loop using a closed-loop method and discovered that the previous settings were putting the loop close to the instability boundary. After the loop was properly tuned, he started to ask more questions. He discovered that the valve was an "equal-percentage" type of valve. Remembering that the gain of an equal-percentage valve is higher for higher flows, he checked the current flow value—around 80% of full scale. When he inquired about the current production rate versus the rate when the process was started up, he discovered that the throughput, and hence the flow rate of that loop, was considerably lower when the process was started up. The engineer made a mental note to tune loops at the point where the process had the maximum gain to avoid problems with instability when the process operating point changed.

## 5.1 INTRODUCTION

This chapter provides a foundation in the use and tuning of the loop process control object, which provides the regulatory control at the lower levels of the control hierarchy. For background, simple approaches to control are described and placed into the proper context. General feedback control concepts are then explained. The basic proportional-integral-derivative (PID) controller and all of its variants are described, and then the most popular tuning methods are described. Finally, the typical user interface to a PID controller is described.

In most systems, a simple PID controller is sufficient to handle the regulatory control functions. Enhancements to the PID controller and the situations in which they are applied are discussed in Chapter 6.

In order to place the material in this chapter in its proper context, the most common strategies for regulatory control are described and related to each other.

*Open-Loop Control.* In this control strategy, the manipulated variables are simply set to their design values and held there. Many people consider this as no control, but by setting the manipulated variables, the control system is influencing the process. For example, consider the control of a heat exchanger, depicted in Figure 5.1, which uses steam to heat cold water and produce hot water. The goal is to control the hot-water temperature to a desired value, ±2°C. An open-loop control strategy for this hot-water heater is shown in Figure 5.2. Given a desired temperature, the reference selector produces a steam valve position either by calculation or table look-up. Obviously, any variations in the cold-water temperature or steam temperature, among other external influences, will affect the hot-water temperature and the reference selector will not account for them.

This strategy is useful for those systems that have no or small disturbance effects. Obviously, if the system is subject to significant disturbances, then some more sophisticated control is needed.

*Feedforward Control.* If one knows the types of disturbances that can influence a process, then the reference selector can be modified to measure these disturbances and modify the calculation of the manipulated variable. For example, the heat exchanger of Figure 5.1 is subject to the following disturbances:

Cold-water temperature
Cold-water flow

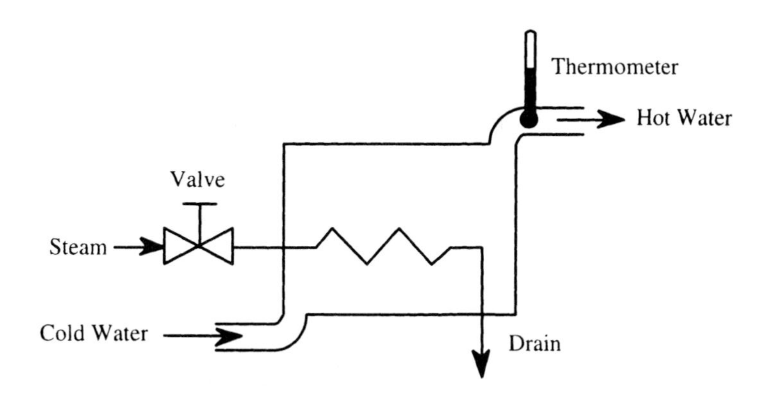


**FIGURE 5.1.** Example heat exchanger.

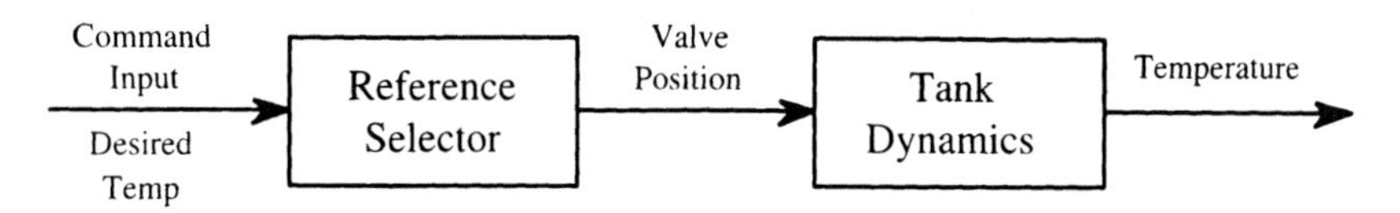


**FIGURE 5.2.** Open-loop control strategy for heat exchanger.

Steam temperature
Steam pressure
Ambient temperature

All of these influences are called *disturbances*, because they affect the output but are not manipulated. To construct a feedforward controller, the important disturbances are measured and used by the reference selector, along with the desired value of the output to calculate the manipulated variable value. For example, Figure 5.3 shows a possible feedforward controller for the heat exchanger. The important disturbances—cold-water temperature, cold-water flow, and steam temperature—are considered to be either the disturbances that have the biggest effect or the ones most likely to change. The other disturbances are unmeasured and still influence the process to some extent.

This strategy is useful for those cases where the process operating point does not change and the process is subject to significant disturbances whose effect on the controlled variable is known. The disturbances not used in this strategy must have small effects on the controlled variable.

*Manual Feedback Control.* The process of using knowledge of the output to take corrective action is called *feedback*. In manual feedback control, operating personnel monitor the controlled variable of interest and take corrective action in order to maintain the desired value. A manual feedback control scheme for the heat exchanger is shown in Figure 5.4. An operator periodically monitors the hot-water temperature and adjusts the steam valve position in order to maintain the hot-water temperature.

Manual control is useful in a large number of applications whose controlled variable is not critical, whose disturbances are few, or where the cost of automation is too high to justify automatic (open-loop or feedback) control. Manual control is complementary to the largely automatic control strategies presented in this text. Often, the dividing line between those parts of the process that are automated and those that remain under manual control is a crucial decision in a control project, as was shown in Chapter 2.

*Automatic Feedback Control.* When the operator in a manual feedback strategy is replaced by a controller device that continuously measures the controlled variable, then one has automatic feedback control. Automatic control is useful for every process and helps to eliminate disturbance effects. To maintain the controlled variable at the setpoint, the

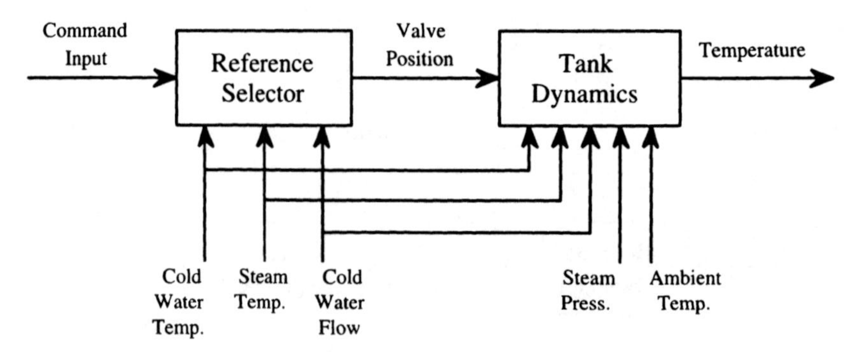


**FIGURE 5.3.** Feedforward control strategy for heat exchanger.

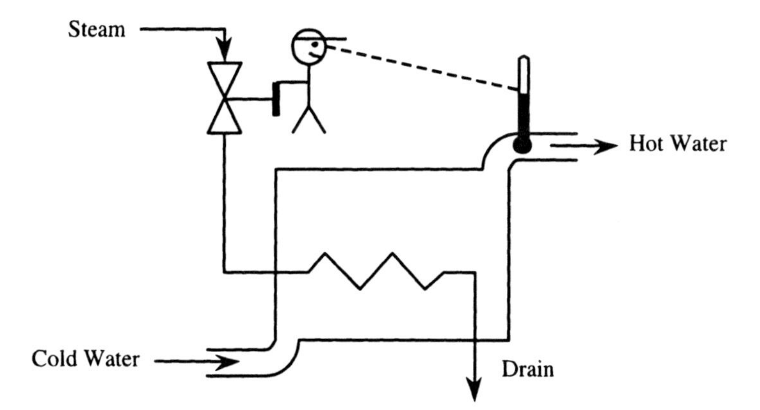


**FIGURE 5.4.** Manual feedback control of heat exchanger.

manipulated variable is adjusted each time there is a change in the load disturbance. For the heat exchanger example, the operator is replaced by an automatic feedback controller (Figure 5.5). This text emphasizes automatic feedback control, but the other types of regulatory controls will be used in the case study when appropriate. The combination of feedback and feedforward control is considered in the next chapter. The only drawback to automatic feedback control is its cost. Beyond the hardware cost, an installation, tuning, and maintenance cost is incurred.

Feedback control also applies to discrete processes. For example, a solenoid valve has only two valid states, open or closed. Often limit switches are attached to the valve to determine if the valve reaches the commanded state. If the controller commands the valve to open and the closed limit switch still indicates that the valve is still closed after 10 sec, then the valve has failed and the system must take corrective action, such as shut down the system. At the very least, an alarm is generated to inform the operator of the problem. In this example, the limit switches provide the feedback to the discrete controller.

*Emergency Controls.* The example of the discrete valve control in the previous paragraph also serves as an example of controls that handle abnormal or emergency conditions.

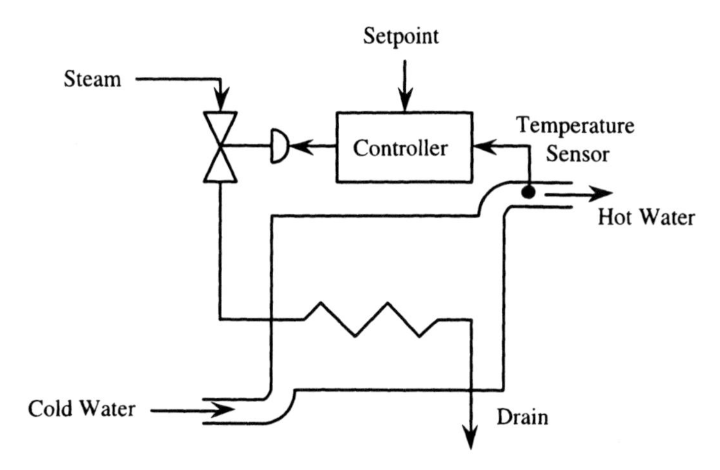


**FIGURE 5.5.** Automatic feedback control of heat exchanger.

Emergency controls are often used when a critical device fails. Even when continuous feedback control is employed, disturbances may force the controlled variable to violate safety limits. In these cases, emergency controls bring the process to a safe state. Emergency controls will be covered in Chapters 5, 6, and 8.

## 5.2 GENERAL FEEDBACK CONTROL

The block diagram of a typical automatic feedback control system is shown in Figure 5.6. Here, the disturbance effect is separated from the manipulated part of the process. A controller compares the process variable, the controlled variable measurement, with the setpoint and calculates a manipulated variable to maintain the setpoint. The manipulated variable influences the process through a final control element, for example, a valve or a motor controller. However, in many cases, the final control element, process, and measuring device are often lumped together and called the "process," as in Figure 5.7.

### 5.2.1 General Features of Feedback Control

In order to show the general features of feedback control, the transfer functions that relate the controlled variable to the setpoint (SP) and disturbance (D) are developed as follows.

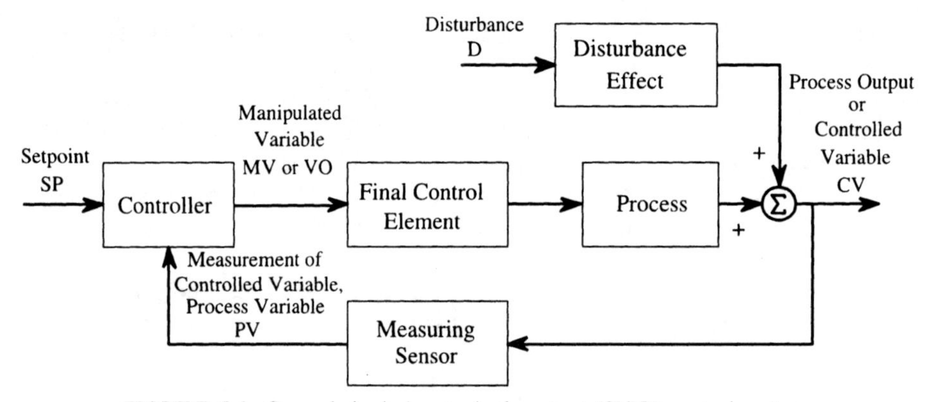


**FIGURE 5.6.** General single-input, single-output (SISO) control system.

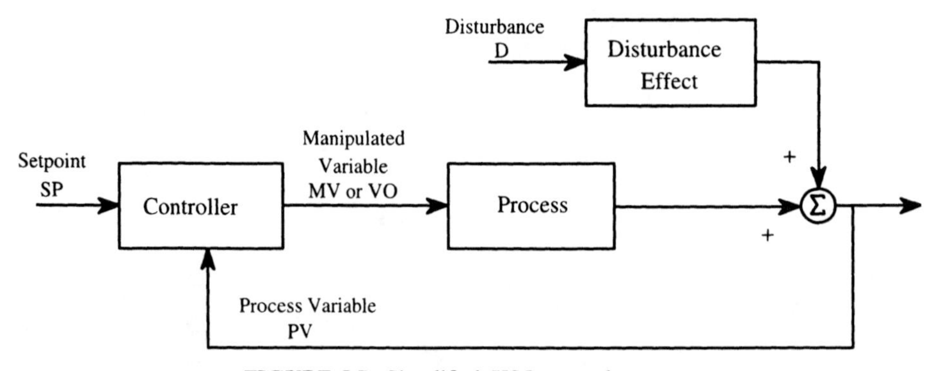


**FIGURE 5.7.** Simplified SISO control system.

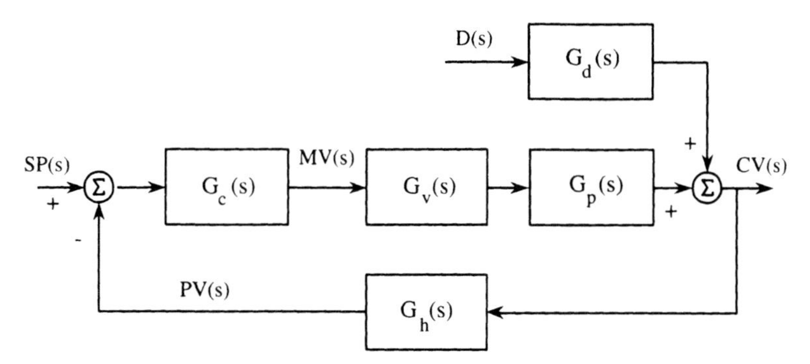


**FIGURE 5.8.** SISO control system with error as controller input.

Using the system of Figure 5.8, which is the same as Figure 5.6 but with a particular form to the controller, the expression for the controlled variable (CV) is

$$\begin{aligned} \mathrm{CV}(s) &= G_p(s)G_v(s)\mathrm{MV}(s) + G_d(s)\mathrm{D}(s) \\ &= G_p(s)G_v(s)G_c(s)[\mathrm{SP}(s) - \mathrm{PV}(s)] + G_d(s)\mathrm{D}(s) \end{aligned} \tag{5.1}$$

MV being the manipulated variable and PV the process variable. To obtain the transfer function from the setpoint to the controller variable output, set the disturbance input to zero and rearrange (5.1) to get

$$\frac{\mathrm{CV}(s)}{\mathrm{SP}(s)} = \frac{G_c(s)G_v(s)G_p(s)}{1 + G_c(s)G_v(s)G_p(s)G_h(s)} \tag{5.2}$$

To obtain the transfer function from the disturbance to the controller variable output, the setpoint is assumed to be zero and (5.1) is rearranged to get

$$\frac{\mathrm{CV}(s)}{\mathrm{D}(s)} = \frac{G_d(s)}{1 + G_c(s)G_v(s)G_p(s)G_h(s)} \tag{5.3}$$

Regardless of the particular controller algorithm, there are three general advantages of a feedback control system compared with an open-loop system:

1. Increased speed of response
2. Reduced error
3. Disturbance rejection
4. Reduced sensitivity to modeling errors

These features will be demonstrated with an example.

**Example 5.1.** Compare the open-loop and closed-loop (automatic feedback control) responses for the system whose transfer functions are

$$G_c(s) = 10 \qquad G_v(s) = 1 \qquad G_p(s) = \frac{0.5}{0.5s + 1}$$
$$G_h(s) = 1 \qquad G_d(s) = \frac{0.1}{s + 1}$$

Compare the responses for speed of response, steady-state error, and disturbance rejection. Also compare the responses when the gain of the process is doubled.

***Solution.*** The controller is the simplest type of controller (proportional only) but is sufficient to demonstrate the advantages of closed-loop feedback control. For the sake of this example, let the setpoint and disturbances be unit step functions $1/s$. The open-loop transfer function is just the process $G_p(s)$. The expression for the output $\mathrm{CV}(t)$ of the open-loop system for a unit step change in the process input is

$$\mathrm{CV_{OL}}(s) = \frac{0.5}{s(0.5s + 1)} = \frac{1}{s(s + 2)}$$
$$\mathrm{CV_{OL}}(t) = 0.5 - 0.5e^{-2t} \qquad t \geq 0 \tag{5.4}$$

The expression for the output of the closed-loop (automatic feedback) system for a unit step change in the setpoint is obtained by multiplying the input function and the transfer function (5.2):

$$\mathrm{CV_{CL}}(s) = \frac{1}{s}\frac{0.5/(0.5s + 1)}{1 + (10)(0.5)/(0.5s + 1)} = \frac{5}{s(0.5s + 6)} = \frac{10}{s(s + 12)}$$
$$\mathrm{CV_{CL}}(t) = 0.833 - 0.833e^{-0.0833t} \qquad t \geq 0 \tag{5.5}$$

Comparing the open- and closed-loop responses in Figure 5.9, the closed-loop response is clearly faster than the open-loop response. The process error is the difference between the input signal and the output signal. Clearly, Figure 5.9 shows that the final value of the closed-loop system is closer to the setpoint than for the open-loop system.

The disturbance rejection features of the closed-loop system is seen by considering the response of the system when the setpoint input is held at zero and considering a step input on the disturbance. In this case, the open-loop system does not mitigate the disturbance effect, so the system output is

$$\mathrm{CV_{DOL}}(s) = \frac{1}{s}G_d(s) = \frac{0.1}{s(s + 1)} = \frac{1}{s(s + 1)}$$
$$\mathrm{CV_{DOL}}(t) = 0.1 - 0.1e^{-t} \tag{5.6}$$

For the closed-loop system, the output response due to the disturbance is obtained by multiplying the input function times the transfer function from the disturbance to the output, (5.3)

$$\mathrm{CV_{CL}}(s) = \frac{1}{s}\frac{0.1\ (s + 1)}{1 +\ (10)(0.5)/(0.5s + 1)} = \frac{0.1(0.5s + 1)}{s(s + 1)(0.5s + 6)} = \frac{0.1(s + 2)}{s(s + 1)(s + 12)}$$
$$\mathrm{CV_{CL}}(t) = 0.0167 - 0.00909e^{-t} - 0.00758e^{-0.0833t} \qquad t \geq 0 \tag{5.7}$$

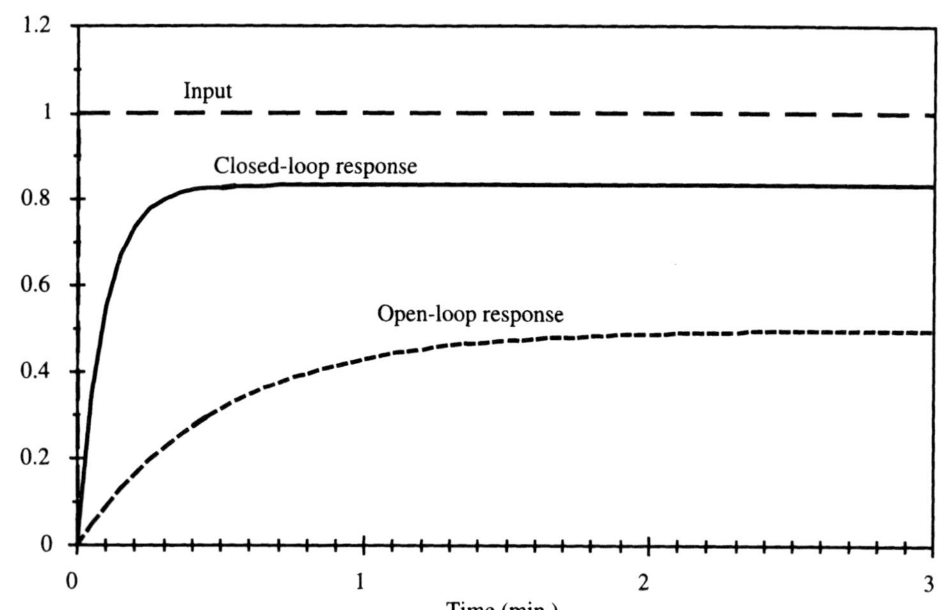


**FIGURE 5.9.** Comparison between open-loop and closed-loop responses for unit step setpoint change.

Comparing the open- and closed-loop responses in Figure 5.10, the effect of the disturbance for the closed-loop system is much smaller than for the open-loop system.

The sensitivity is shown only by example. Those interested in sensitivity analysis should refer to, for example, the text by Dorf and Bishop (1995). When the gain of $G_p(s)$ is changed to 1.0—a relative change of 100%—the effect of that change on the open- and closed-loop responses is as shown in Figure 5.11. The relative change in the magnitude of the closed-loop response is much less than the relative change in the gain of $G_p(s)$, while the relative change in the magnitude of the open-loop response is the same as the relative change in the gain of $G_p(s)$.

### 5.2.2 Control Loop Performance Measures

Ultimately, the purpose of the controller is to force the process variable to follow a setpoint value by adjusting the manipulated variable. Sometimes, the setpoint is not a constant but a trajectory. For most systems, it is impossible to make the process variable perfectly follow the setpoint. Hence, quantitative control performance measures are needed to judge the quality of the controller. These performance measures are based on the response to the two external inputs to the system: setpoint and disturbance changes.

**Setpoint Input Changes.** The most common control loop performance measures are based on changes to the setpoint. In many processes, the setpoint is held constant for long periods of time, as in the pressure of a steam header. In others, the setpoint is changed at frequent intervals, as in the temperature of a batch reactor. In still other cases, the setpoint assumes a profile. For example, in metal annealing, the furnace temperature is ramped from an initial temperature to a final temperature, held constant for a period of time, and then ramped back to the initial temperature. However, the common control performance

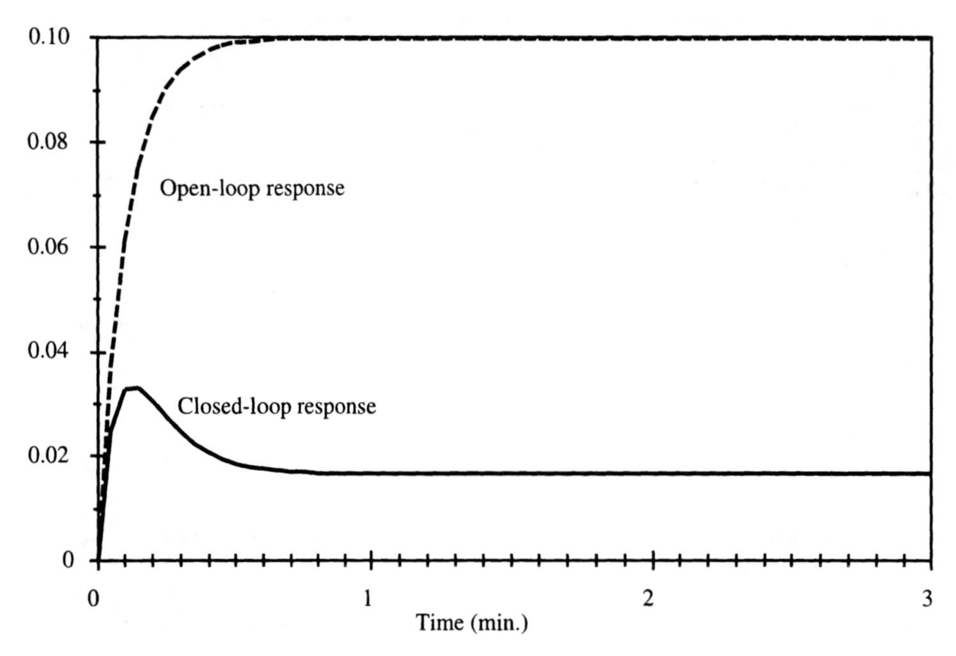


**FIGURE 5.10.** Comparison between open-loop and closed-loop responses for unit step disturbance change.

measures are based on a step change in the setpoint, which corresponds to the situation where an operator or supervisory controller makes an abrupt change in the setpoint and waits for the process variable to reach a steady-state value. A typical manipulated variable and process variable response are shown in Figure 5.12. The following aspects of these curves are used to evaluate the control system performance.

*Offset.* The difference between the setpoint and the process variable when the system has reached steady state is called the offset. For most systems, the desired offset is zero.

*Rise Time.* The time required for the process variable to go from 10 to 90% of the steady-state change is the rise time $T_r$. Generally, a short rise time is desired, although a short rise time often requires an oscillatory manipulated variable change. The short rise time is only relative and usually refers to a rise time shorter than the rise time of the system with no control. The rise time cannot be arbitrarily short, since given the constraints on the magnitudes of the manipulated variable, there is a shortest achievable rise time.

*Percent Overshoot.* The amount that the process variable proceeds beyond its final value is called the overshoot ($A$ in Figure 5.12). Obviously, if the process variable response is overdamped, it does not contain a peak whose value exceeds the final value and the overshoot is zero. The *percent overshoot* (PO), defined as

$$\text{PO} = \frac{A}{\Delta\text{PV}} \times 100\% \tag{5.8}$$

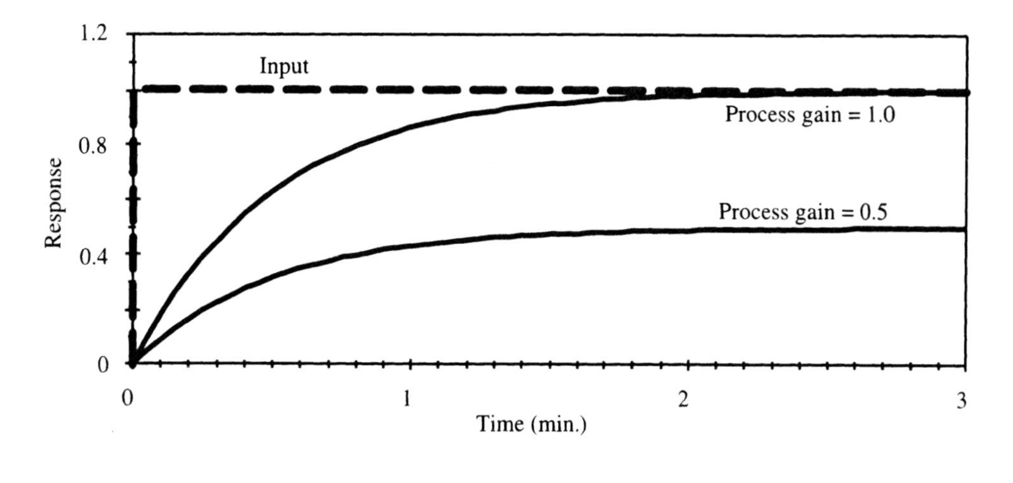


(a)

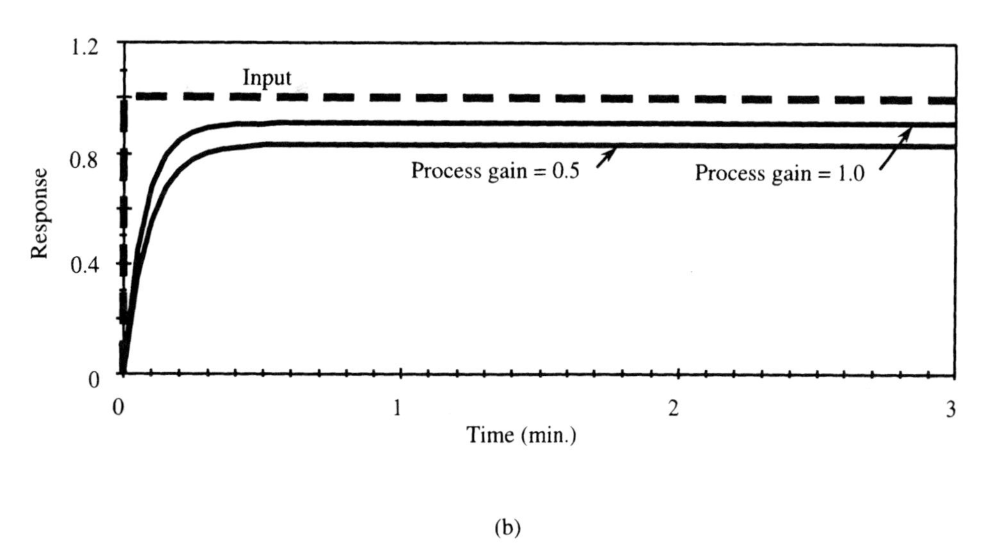


(b)

**FIGURE 5.11.** Open-loop and closed-loop responses for unit step setpoint change when process gain is doubled: (*a*) open loop; (*b*) closed loop.

is often used as the performance measure in place of the overshoot, since it is a relative measure that does not depend on the setpoint change magnitude. Generally, little or no overshoot is desirable in a process control system since overshoot has a detrimental effect on downstream processes.

*Peak Time.* The peak time $T_p$ is the time from the setpoint step change to the time of the first peak of an underdamped process variable response. The peak time and the rise time are closely related, and so usually only one of them is used.

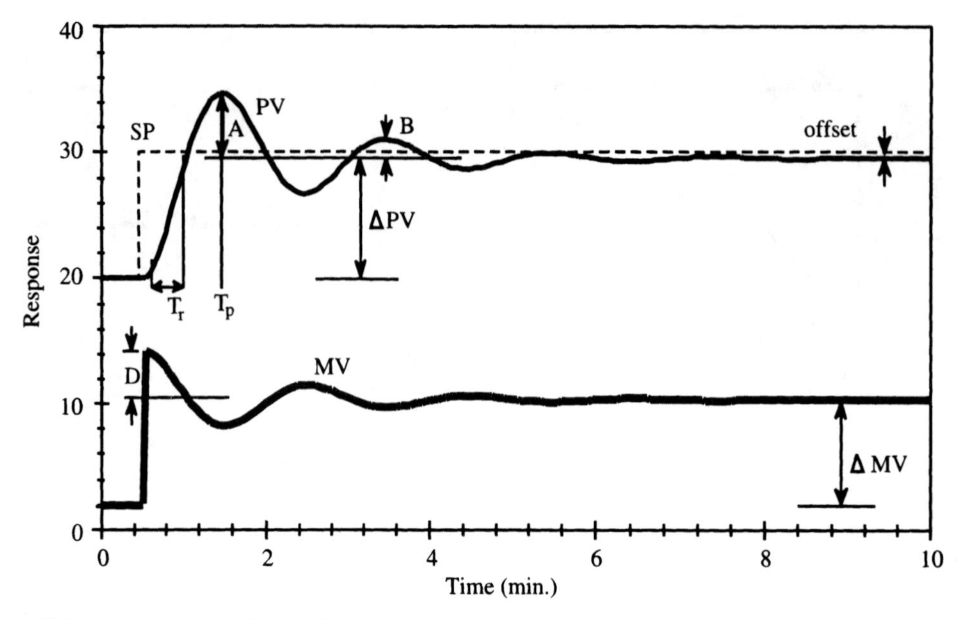


**FIGURE 5.12.** Typical controlled and manipulated variable response to a step setpoint change.

*Damping Ratio.* If the process variable response is underdamped, the damping ratio is the ratio of the second peak to the first peak in the response, ($B/A$ in Figure 5.12). Usually, a small ratio is desired.

*Settling Time.* The settling time $T_s$ is the time from the setpoint change to the time that the process variable response has settled within a certain percentage band of the final value, usually 2 or 5%. This measure is related to the rise time and peak time. Generally, a short rise time also means a short settling time. As for the rise time, a short settling time is desired.

*Integral Error.* The cumulative deviation of the process variable from the setpoint can be measured a number of ways:

Integral of the absolute error (IAE):

$$\text{IAE} = \int_0^\infty |\text{SP}(t) - \text{PV}(t)|\, dt \tag{5.9}$$

Integral of the squared error (ISE):

$$\text{ISE} = \int_0^\infty [\text{SP}(t) - \text{PV}(t)]^2\, dt \tag{5.10}$$

Integral of the product of time and the absolute error (ITAE):

$$\text{ITAE} = \int_0^\infty t|\text{SP}(t) - \text{PV}(t)|\, dt \tag{5.11}$$

Integral of the error (IE):

$$\text{IE} = \int_0^\infty [\text{SP}(t) - \text{PV}(t)]\, dt \tag{5.12}$$

The IAE is the most common performance measure in this group. When the process variable is a quality measurement, like concentration, the IAE basically measures the total amount of off-specification material produced by the system. The ISE is an appropriate measure when small deviations can be tolerated and large deviations are unacceptable. The ITAE penalizes deviations that endure for a long time. The IE, though simple, is not used since positive and negative deviations cancel each other.

*Manipulated Variable Percent Overshoot.* The manipulated variable overshoot is calculated in the same manner as the percent overshoot of the process variable. It is an indication of the amount of control effort needed to produce the closed-loop process variable response. A higher manipulated variable percent overshoot (MVPO) usually means more wear on the final control element, a valve in many process control systems:

$$\text{MVPO} = \frac{D}{\Delta\text{MV}} \times 100\% \tag{5.13}$$

**Disturbance Input Changes.** Disturbances are those uncontrolled inputs to the process that affect the controlled variable (and process variable). These inputs generally cause significant deviations from the setpoint if corrective action is not taken. The major types of disturbances encountered in process control are step disturbances, stochastic disturbances, and ramp disturbances.

*Step Disturbance.* A disturbance characterized by a large, abrupt change in its value is modeled as a step signal. A typical manipulated variable and process variable response to a step disturbance are shown in Figure 5.13. The performance measures for the process variable response are largely the same, except that the rise time and percent overshoot are replaced by the *maximum deviation*, which has meaning only in the context of a disturbance response. The maximum deviation measures the maximum effect of the disturbance on the process variable response, and obviously, a small value is desired.

*Stochastic Disturbance.* Stochastic disturbances are those seemingly random upsets that disturb the system from the steady state. Their average value is zero, so they do not have the major effects on the process variable that step disturbances have. Nevertheless, their effect can be dramatic on the expected performance, as will be demonstrated by the controller tuning examples. The only stochastic disturbance considered in this text is measurement noise, since a small amount of it is present in any real system and the

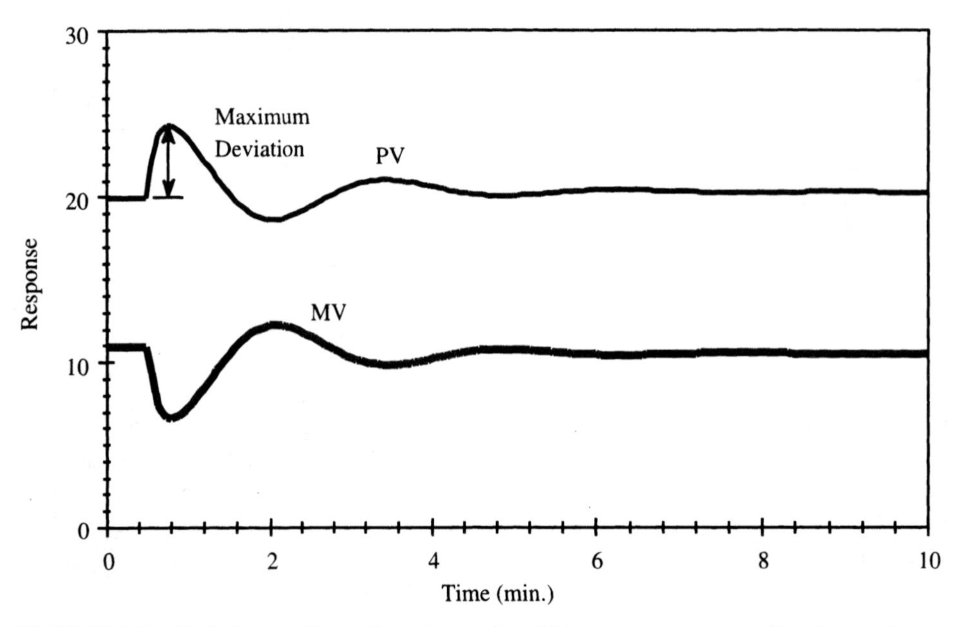


**FIGURE 5.13.** Typical controlled and manipulated variable response to a step disturbance change.

quantization process used to convert the measurement into a value to be used by the digital control algorithm introduces its own noise.

*Ramp Disturbance.* Ramplike disturbances generally correspond to environmental factors such as ambient outside temperature. They will be used occasionally to evaluate the performance of a control system but will not be directly used in the design procedure.

## 5.3 PID CONTROLLER

The most common industrial controller is based on the PID algorithm. After explaining each part of the basic algorithm, other parameters that must be set in a commercial PID controller are explained. The basic states of a PID (actually any industrial controller regardless of the algorithm) are described and then guidance concerning which terms should be selected is presented before the tuning procedures are presented.

### 5.3.1 PID Controller Terms

The three parts of the PID controller are (1) proportional, (2) integral, and (3) derivative. Each part of the algorithm is explained as follows:

1. *Proportional.* The manipulated variable is calculated as a constant (gain) multiplied by the error (setpoint minus process variable). The equation is commonly expressed as

$$\mathrm{MV}(t) = K_P e(t)$$

where $K_p$ is the proportional gain and $e$ is the error. On some controllers, the proportional gain is specified as proportional band (PB). Proportional gain and proportional band are related by

$$K_P = \frac{100}{\text{PB}}$$

The proportional action reduces the error between the setpoint and the process variable but does not eliminate it. Increasing the proportional gain also speeds up the response of the closed-loop system. However, there is usually a practical upper limit because increasing the proportional gain often causes increased oscillation. If the proportional gain is too high, the system may be unstable.

2. *Integral.* The manipulated variable calculation is based on the integrated error. Integral action drives the error to zero. Integral action is also called reset. The equation for integral action is commonly

$$\text{MV}(t) = \frac{1}{T_I} \int_0^t e(t)\, dt$$

where $T_I$ is called the integral time constant in units of minutes or seconds. On some controllers, the integral gain is specified, which is related to the integral time constant as

$$K_I = \frac{1}{T_I}$$

and whose units are repeats per minute or repeats per second. In digital controllers, the integration is approximated by a summation,

$$\text{MV}(n) = \frac{T}{T_I} \sum_{i=0}^{n} e(i) = \text{MV}(n-1) + \frac{T}{T_I}[e(n)]$$

where $T$ is the sample period, $n$ is the current sample number, and $e(n)$ is the error at the $n$th sampling instant.

The integral term is mainly used to force the offset to zero. However, it generally tends to slow the system down, in contrast to the proportional term. As the integral time constant is reduced, the system response becomes more oscillatory. A small enough value of the integral time constant will cause the system to become unstable.

When a controller has integral action, a persistent error will cause the integral term to increase (or decrease) to a value of large magnitude. The integral term does not become smaller until the sign of the error changes. This situation is called integral windup. Many commercial controllers limit the value of the integral term and/or modify the integral action when this situation is encountered to temporarily decrease the integral time constant as the magnitude of the error decreases but before the error changes sign.

3. *Derivative.* Also called rate action, the manipulated variable calculation is based on the rate of change of the process variable. The equation for derivative action is commonly

$$\text{MV}(t) = T_D \frac{d}{dt}[e(t)]$$

where $T_D$ is called the derivative time constant in minutes or seconds. On some controllers, the derivative gain is specified, which is equal to the derivative time constant. In reality, it is difficult to build an analog derivative circuit, so analog controllers actually approximate the derivative. As the Laplace transfer function, the derivative term can be approximated as

$$T_D s \approx \frac{T_D s}{\alpha s + 1} \tag{5.14}$$

or

$$T_D s \approx \frac{T_D s + 1}{\alpha T_D s + 1} \tag{5.15}$$

which is the easier of the two to implement, where $\alpha$ has a value of about 10. In digital controllers, the derivative may be approximated by a difference,

$$\text{MV}(n) = \frac{T_D}{T}[e(n) - e(n-1)] = \frac{T_D}{T}\Delta e(n)$$

where $T$ is the sample period and $n$ is the current sample number. The digital derivative may also be a discrete equivalent of (5.14) or (5.15) if the vendor wants the digital PID controller to closely approximate the response of an analog PID controller.

Major changes in load disturbances are anticipated by rate action. The derivative term also tends to cancel the effect of the integral action in slowing down the system response. However, the use of derivative action should be restricted to those processes where the process variable, the controlled variable measurement, is noise free. Derivative action tends to amplify any noise in the measurement.

Of course, there are many possible combinations of these three actions. For example, a P controller has only proportional action, a PI controller has proportional and integral action, and so on.

Often, the proportional, integral, and derivative controller actions are called the PID controller modes. However, the mode is also used to refer to whether the controller is in automatic or manual mode. To avoid confusion, the use of the term "mode" is omitted from this text. The proportional, integral, and derivative controller actions are called the PID controller terms and the state is used to refer to what is commonly called automatic/manual mode (Section 5.3.4). The $K_P$, $T_I$, and $T_D$ are called the PID tuning parameters.

### 5.3.2 Forms of the Equation

There are basically eight forms to the basic PID equation. Depending on the vendor, one may or may not have a chance to select the particular PID algorithm form. The various choices for algorithm type are:

Interacting (ISA standard) versus noninteracting (independent gains)
Derivative on error versus derivative on measurement
Positional versus velocity

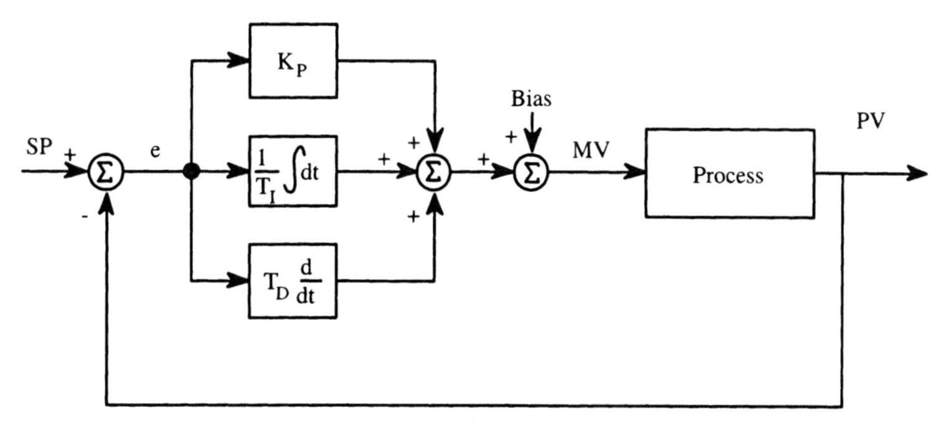


**FIGURE 5.14.** Noninteracting, derivative on error positional PID control.

In the noninteracting form of the PID equation, all three terms are calculated independently and summed to obtain the manipulated variable. In the interacting form, the $K_P$ modifies the integral and derivative terms. Derivative action on the measurement is actually preferred in most industrial situations, because then setpoint changes do not cause a large change in the manipulated variable. In the positional form of the PID algorithm, the value of the manipulated variable is calculated. In order to initialize the algorithm correctly, an initial bias term must be specified. In the velocity form of the PID, the change in the manipulated variable is calculated. Block diagrams of all combinations of interaction and derivative action are shown in Figures 5.14–5.17 for the positional form of the algorithm. For the continuous (analog) PID, the four possible equations for the positional forms of the algorithm are:

Noninteracting, derivative on error (Figure 5.14):

$$\mathrm{MV}(t) = K_P e(t) + \frac{1}{T_I}\int e(t) + T_D \frac{d}{dt} e(t) + \mathrm{Bias}$$

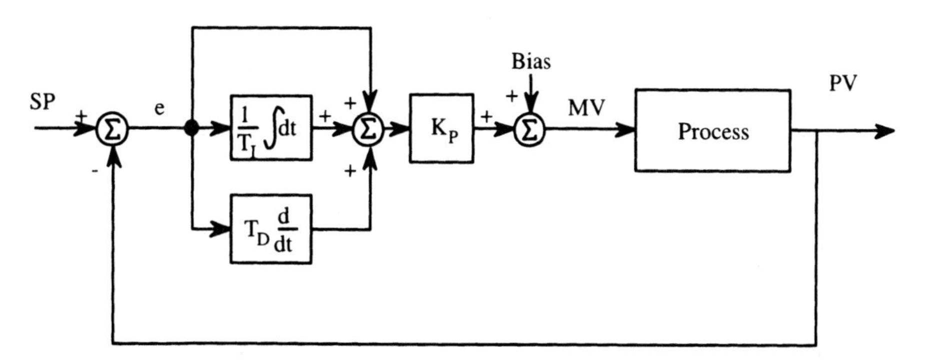


**FIGURE 5.15.** Interacting, derivative on error positional PID control.

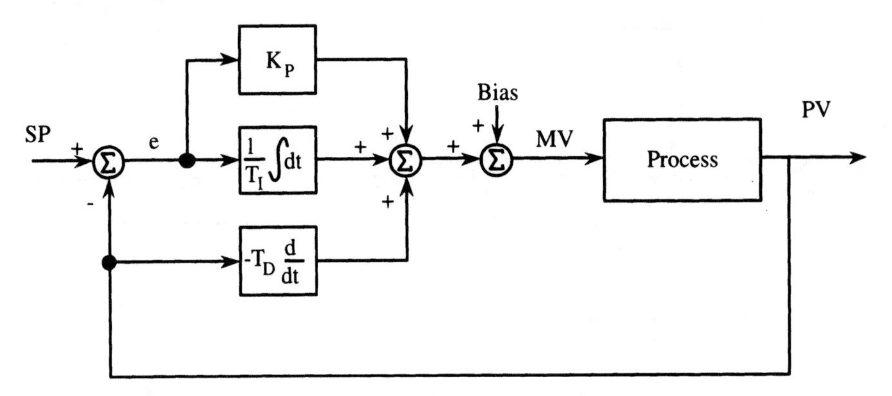


**FIGURE 5.16.** Noninteracting, derivative on measurement positional PID control.

Interacting, derivative on error (Figure 5.15):

$$\mathrm{MV}(t) = K_P\left(e(t) + \frac{1}{T_I}\int e(t) + T_D\frac{d}{dt}e(t)\right) + \mathrm{Bias}$$

Noninteracting, derivative on measurement (Figure 5.16):

$$\mathrm{MV}(t) = K_P e(t) + \frac{1}{T_I}\int e(t) - T_D\frac{d}{dt}\mathrm{PV}(t) + \mathrm{Bias}$$

Interacting, derivative on measurement (Figure 5.17):

$$\mathrm{MV}(t) = K_P\left(e(t) + \frac{1}{T_I}\int e(t) - T_D\frac{d}{dt}\mathrm{PV}(t)\right) + \mathrm{Bias}$$

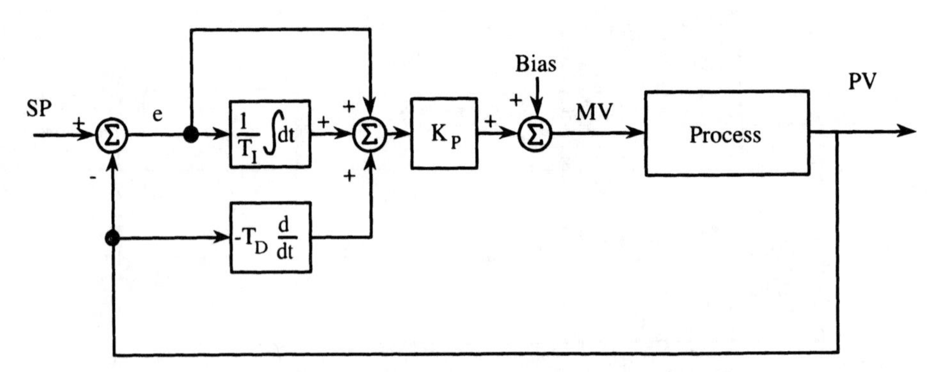


**FIGURE 5.17.** Interacting, derivative on measurement positional PID control.

For the positional form of the digital PID equation, the four possibilities are:

Noninteracting, derivative on error (Figure 5.14):

$$\mathrm{MV}(n) = K_P e(n) + \frac{T}{T_I}\sum_{i=0}^{n} e(i) + \frac{T_D}{T}\Delta e(n) + \mathrm{Bias}$$

Interacting, derivative on error (Figure 5.15):

$$\mathrm{MV}(n) = K_P\left(e(n) + \frac{T}{T_I}\sum_{i=0}^{n} e(i) + \frac{T_D}{T}\Delta e(n)\right) + \mathrm{Bias}$$

Noninteracting, derivative on measurement (Figure 5.16):

$$\mathrm{MV}(n) = K_P e(n) + \frac{T}{T_I}\sum_{i=0}^{n} e(i) - \frac{T_D}{T}\Delta \mathrm{PV}(n) + \mathrm{Bias}$$

Interacting, derivative on measurement (Figure 5.17):

$$\mathrm{MV}(n) = K_P\left(e(n) + \frac{T}{T_I}\sum_{i=0}^{n} e(i) - \frac{T_D}{T}\Delta \mathrm{PV}(n)\right) + \mathrm{Bias}$$

For the continuous PID controller, the velocity forms of the equations are not possible. So, the velocity form only applies to the digital PID algorithm. The four types of PID equations are obtained by subtracting $\mathrm{MV}(n-1)$ from $\mathrm{MV}(n)$. The block diagrams of these equations look similar to the diagrams for the positional PID (Figures 5.14–5.17) except that the Bias signal is replaced by $\mathrm{MV}(n-1)$:

Noninteracting, derivative on error:

$$\begin{aligned}\mathrm{MV}(n) - \mathrm{MV}(n-1) = {} & K_P[e(n) - e(n-1)] \\ & + \frac{T}{T_I}e(n) + \frac{T_D}{T}[e(n) - 2e(n-1) + e(n-2)]\end{aligned}$$

or,

$$\mathrm{MV}(n) = \mathrm{MV}(n-1) + K_P\Delta e(n) + \frac{T}{T_I}e(n) + \frac{T_D}{T}[e(n) - 2e(n-1) + e(n-2)]$$

Interacting, derivative on error:

$$\mathrm{MV}(n) = \mathrm{MV}(n-1) + K_P\left(\Delta e(n) + \frac{T}{T_I}e(n) + \frac{T_D}{T}[e(n) - 2e(n-1) + e(n-2)]\right)$$

Noninteracting, derivative on measurement:

$$\mathrm{MV}(n) = \mathrm{MV}(n-1) + K_P \Delta e(n) + \frac{T}{T_I} e(n) - \frac{T_D}{T} [\mathrm{PV}(n) - 2\mathrm{PV}(n-1) + \mathrm{PV}(n-2)]$$

Interacting, derivative on measurement:

$$\mathrm{MV}(n) = \mathrm{MV}(n-1) + K_p \left( \Delta e(n) + \frac{T}{T_I} e(n) - \frac{T_D}{T} [\mathrm{PV}(n) - 2\mathrm{PV}(n-1) + \mathrm{PV}(n-2)] \right)$$

Concerning these PID algorithm choices, only the derivative on error versus derivative on measurement may be critical to the control engineer. The choice of positional versus velocity forms of the equations has usually already been made by the vendor. The particular choice will have an impact on such issues as initialization, bumpless transfer, and constraint control, which is discussed in Chapter 6. The choice between interacting versus noninteracting affects the tuning rule equations. For example, if the tuning rules calculate the parameters for an interacting PID, divide the $T_I$ by $K_P$ and multiply the $T_D$ by $K_P$ to obtain the noninteracting parameters. If the user needs the derivative action, then derivative on measurement is the preferred choice. Of course, if derivative action is not needed, the choice of derivative on error versus derivative on measurement is a moot point. Derivative on measurement is preferred because a setpoint change does not induce a large change in the controller output, which happens if the derivative is on the error.

### 5.3.3 Other Controller Parameter Selections

Besides the main PID parameters $K_P$, $T_I$, and $T_D$ and the choices for the PID equation form, there are other parameter selections that must be made before the controller can be operational.

*Direct and Reverse Action.* The sign relationship of the manipulated variable to the process variable is specified through the direct/reverse selection. Direct and reverse action refers to the sign of the gain of either the process or the controller. When setting the direct/reverse parameter, the user should determine whether the vendor defines this parameter for the process or the controller. When referring to the process:

Specify the process as *direct acting* if an increase in MV causes an increase in PV.
Specify the process as *reverse acting* if an increase in MV causes a decrease in PV.

If referring to the controller:

Specify the control action as *direct acting* if the MV must increase to correct for an increasing PV.
Specify the control action as *reverse acting* if the MV must decrease to correct for an increasing PV.

Note that a reverse-acting process needs a direct-acting controller. For a direct-acting process, the error should be

$$\text{Error} = \text{SP} - \text{PV}$$

For a reverse-acting process, the error should be

$$\text{Error} = \text{PV} - \text{SP}$$

If one happens to set this parameter to the wrong value, the manipulated variable will often move to its maximum or minimum value and remain there, often causing the process variable to go to one extreme, generally an undesirable process variable response.

*Increase Open/Increase Close.* This selection is provided on controllers to compensate for valve actuator action (air-to-open or air-to-close) and current-to-pneumatic (I/P) action:

Specify *increase open* for an air-to-open valve since the valve opens with increasing current (or voltage) MV signal.

Specify *increase close* for an air-to-close valve since the valve closes with increasing current (or voltage) MV signal.

The type of valve is selected based on whether the valve should fail open or fail closed if the control system fails and/or system power is removed. This parameter is essentially another sign term on the manipulated variable. As for the direct/reverse parameter, if one sets this parameter to the wrong value, the closed-loop control system will be affected in the same manner as for an incorrect direct/reverse parameter. However, if one happens to set both the direct/reverse and the increase open/increase close to the incorrect value, the control loop will still function correctly, because the incorrect value of one will cancel the incorrect value of the other. However, the display device will show the incorrect value for the manipulated variable.

*Bumpless Transfer.* This important parameter is often overlooked when configuring a controller. Without bumpless transfer, when the controller state is switched from Manual to Automatic, the error between the process variable and setpoint generally causes an abrupt change in the manipulated variable. This abrupt change can damage equipment (e.g., "banging valves") and/or upset downstream processes with the process variable oscillations that often result. With bumpless transfer, when the controller state is switched from Manual to Automatic, the calculated manipulated variable change is either filtered or ramped to the calculated value. This gradual change in the manipulated variable is better tolerated by the equipment and the process. Depending on the controller vendor, bumpless transfer may be part of the algorithm and cannot be disabled. For others, bumpless transfer must be enabled by specifying a parameter.

*Filter Time Constant.* If the measurement of a process variable contains noise, then it may be removed by filtering the signal. Often the filtering may be selected as an option in the controller. Filtering the process variable increases the apparent process lag. A trade-off must be made between reduction in noise level and speed of response to process changes. Filtering also affects controller tuning and thus should be selected before the loop is tuned.

### 5.3.4 Basic Controller States

As stated earlier, the word "mode" is used by vendors to refer to whether the controller is in, for example, automatic or manual or cascade. To be consistent with the terminology in the control model (Chapter 3), the word *state* refers to how the controller executes its algorithm. The PID algorithm really only has two states: Manual and Automatic. Nevertheless, control vendors usually assign modes to the algorithm that indicate the source of the setpoint and/or the source of the controller output, the manipulated variable. The most common modes are manual, automatic, cascade, and computer. These modes indicate who or what has control of the setpoint/manipulated variable and generally provide some protection against inadvertent manipulation of the setpoint/manipulated variable by another device or operator. When the controller is in automatic mode, the controller calculates the manipulated variable according to the PID equation and the operator changes the setpoint. When the controller is in manual mode, the manipulated variable is set by the operator and the PID equation is not executed. When the controller is in cascade mode, the setpoint is supplied from another controller or device (not the operator) and the manipulated variable is calculated by the PID algorithm. When the controller is in computer mode, the manipulated variable is supplied by another device, usually a computer. The mode can be considered as a switch at the controller output and/or setpoint, as in Figure 5.18. Ordinarily, when the controller starts up, it is in the Manual state. When the controller state is switched from Manual to Automatic, most controllers implement some form of bumpless transfer, so that the manipulated variable does not make an abrupt change due to any error between the process variable and the setpoint. When the controller state is switched from Automatic to Manual, the manipulated variable is retained at its last calculated value or some default value (if the controller can be configured in this manner).

### 5.3.5 PID Term Selection

Depending on the controller manufacturer, different combinations of proportional, integral, and derivative terms may be selected. Also, by setting the derivative gain to

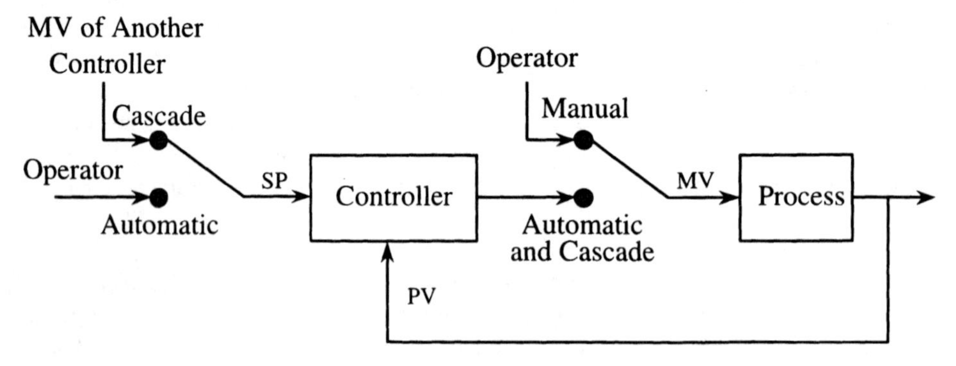


**FIGURE 5.18.** Controller manipulated variable and setpoint source determined by mode switch.

zero, the controller reduces to a PI controller. By setting the proportional gain and the derivative gain of a noninteracting PID controller to zero, the PID controller reduces to an I-only controller. The selection of controller terms should be based on the type of process, that is, how the process parameter reacts to a change in the manipulated parameter.

*Gain-Only Process: I Controller.* When the response of the process variable to a change in the manipulated parameter is instantaneous (Figure 4.12), the process is a gain only, and only the integral term is required. Example processes that are essentially gain only are:

Speed control of centrifugal compressors
Liquid flow control when a fast actuator is used

*First-Order Lag Process: PI Controller.* When the process can be adequately represented as a first-order lag (Figure 4.13), both proportional and integral terms should be used. The majority of industrial processes fall into this category. Example first-order processes are:

Gas pressure control
Liquid flow control with slow actuators
Concentration control

*Second-Order Low-Noise Process: PID Controller.* When the process is best represented as a second-order lag (Figure 4.14) and the process variable contains little noise, then the derivative term should be used with the proportional and integral terms. It is important to remember that the derivative term will amplify any noise present on the process variable. Therefore, the derivative term is generally used sparingly. An example low-noise second-order process is:

Temperature control loops, where the lag of the temperature bulb and thermowell is significant when compared to the process lag

*Integrating Process: P Controller.* If the process is best represented as an integrator (Figure 4.16), the process is not self-regulating, and only the proportional term is required. However, some integral action may be needed if the integration time is long (slope of process variable response curve is small). An example integrating process is:

Tank level control by regulation of the outflow or inflow.

## 5.4 PID CONTROLLER TUNING

There are many techniques of controller tuning. The ones presented here seem to be the most popular techniques. They will allow one to determine the parameters for reasonably good control. In many cases, the performance of the system may be adequate when tuned with these methods. Ultimately, the individual will generally use one or more of the following techniques to get the parameters "in the ballpark" and adjust the parameters until the desired performance is obtained, or until one runs out of time.

Most processes may be approximated as being linear over a small operating range. However, over a wide range of operation, most processes will exhibit some nonlinearity in the form of a process gain change. As a general rule, tuning should be done at the operating point where the maximum gain occurs. Then when the process moves to an operating point where the gain is lower, the control becomes more sluggish but does not become unstable.

The specification of the desired system response should include both the desired process variable and manipulated variable responses. Some general rules follow:

- There is a trade-off between speed of response and the amount of overshoot in the process response. Usually, a fast process variable response also causes the process variable to overshoot its desired value before settling to the desired value.
- The required manipulated variable changes to achieve the desired process variable response should be considered. An aggressive process variable response generally requires aggressive control. Excessive control action can wear out the final control element and also cause problems in those parts of the overall process that depend on the output as an input.
- For a given set of tuning parameters, the response to a setpoint change will not be the same as the response to a disturbance or load change.

There are two major categories of tuning methods: (1) closed-loop and (2) open-loop tuning. Closed-loop tuning refers to tuning the controller while it is operating in the Automatic state, and thus is operating in the closed loop. Open-loop tuning methods base the tuning of the PID parameters on the empirical first-order-plus-deadtime process model, obtained while the controller is in the Manual state. Both of these methods have their advantages and disadvantages.

Closed-loop tuning, whether done by the operator or autotuned by the controller, is probably the most popular technique of controller tuning. Autotuning techniques are generally vendor-specific and often proprietary and so are not discussed in this text. All tuning techniques, including autotuning, require that the process be "bumped" in order to evaluate the effect of the tuning parameters and/or calculate a new set of tuning parameters. Closed-loop tuning techniques are applicable to reasonably fast processes that can tolerate process variable oscillations. On the other hand, open-loop tuning techniques work well for slow processes but require a process model.

### 5.4.1 Settings for Plant Startup

Initial controller parameter values that may be used during a plant startup are given in Table 5.1. The values are conservative and will normally result in a sluggish process variable response. These parameter values are general guidelines and serve to get a plant operating, after which the tuning techniques outlined in the succeeding sections should be used to change the parameters to provide the desired response at normal operating conditions.

**TABLE 5.1 Initial PID Controller Parameter Values**

| | $K_P$ (gain) | $T_I$ (min) | $T_D$ (min) |
|---|---|---|---|
| Flow | 0.3 | 33.0 | 0 |
| Temperature | 1.3 | 0.2 | 0.05 |
| Level | 2.0 | Maximum | 0 |
| Gas pressure | 10.0 | Maximum | 0 |

### 5.4.2 Effects of Controller Terms

The following tuning techniques are based on the following observations about the process (closed-loop) response:

- As the controller proportional gain is increased, the response to setpoint changes becomes more oscillatory, commonly called underdamped (Figure 5.19*a*).

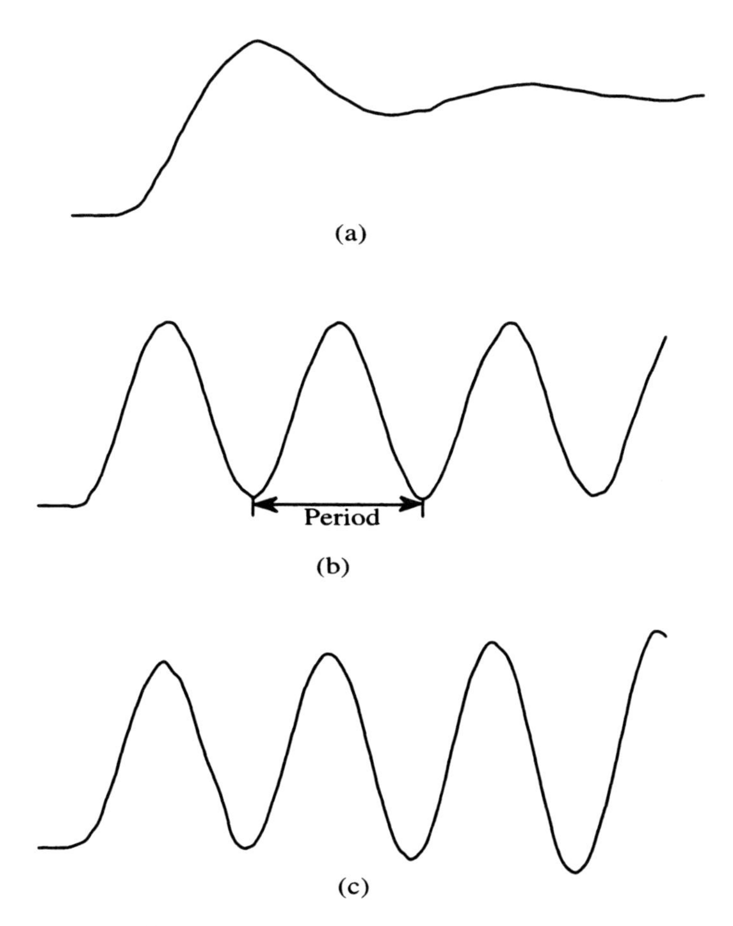


**FIGURE 5.19.** Types of oscillatory response: (*a*) underdamped; (*b*) sustained oscillation; (*c*) unstable.

- At some greater gain, the response of the control loop will become a steady-state oscillation (Figure 5.19*b*). The manipulated variable and process variable will be one-half of a cycle (180°) out of phase at this gain. The system is called "marginally stable."
- If the gain is increased past the point where steady-state oscillation is observed, the control loop will become unstable, and the oscillations will increase in amplitude (Figure 5.19*c*). The frequency of the oscillation is determined by the process dynamics.

The effect of the proportional, integral, and derivative gains on controller gains at different frequencies may be illustrated by examining the contribution of each controller term to a fast change in the error signal (high frequency) and a slow change in the error signal (low frequency), as shown in Figure 5.20:

The effect of the proportional gain is the same for all frequencies.
Integral gain is more responsive to low frequencies.
Derivative gain is more responsive to high frequencies.

Net controller gain may be seen as the sum of these three terms and will vary with frequency as shown in Figure 5.21. The process of determining the controller parameters may be seen as shaping the controller gain about the frequency at which oscillation would occur at a higher gain.

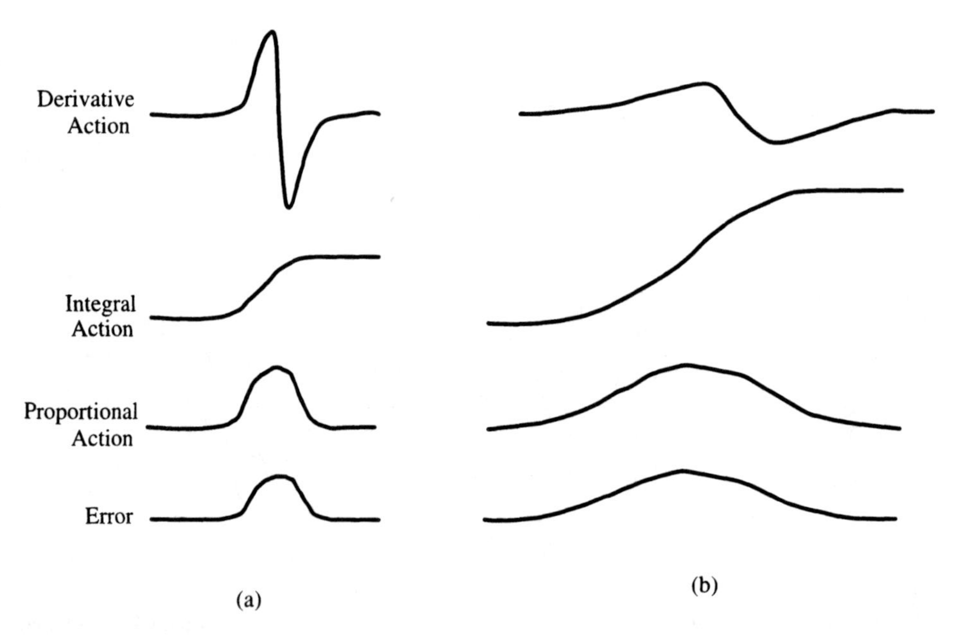


**FIGURE 5.20.** Contribution of each controller term to a change in error: (*a*) fast change; (*b*) slow change.

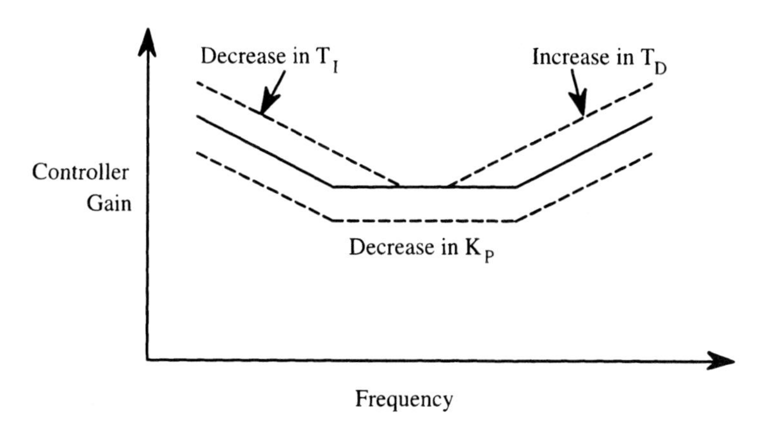


**FIGURE 5.21.** Controller gain as a function of frequency.

### 5.4.3 Closed-Loop Tuning Methods

Tuning may be accomplished while the controller is operating in the closed loop, that is, while it is in the Automatic state. The danger is that downstream processes may be affected by the oscillations that are produced during the tuning process. Depending on the particular process, these oscillations may or may not be acceptable. The one responsible for tuning will need to make that determination.

If the process responds fairly rapidly to changes in the manipulated variable, then tuning may be accomplished by observing the response to changes in the controller parameters. When a process responds slowly to changes in the manipulated variable, then the process should be characterized in order to speed up the tuning process. The Ziegler–Nichols and damped oscillation methods characterize the process. Based on the process characteristics, the parameters can be calculated.

***5.4.3.1 Tuning Based on Process Response.*** When a process responds rapidly to changes in the manipulated variable, tuning can be determined by observing the response to changes in controller parameters. In general, the parameters are tuned in the order proportional ($K_P$), integral ($T_I$), and derivative ($T_D$). Each parameter is increased until a sustained oscillation (stability boundary) is observed in the process variable response. Then the parameter is set to some percentage of the value that caused oscillation. The procedure is as follows:

1. Start any trending (plotting) of the process variable.
2. Set the $K_P$ and $T_D$ to their minimum values and the $T_I$ time constant to its maximum value. Now place the controller in the Automatic state.
3. If proportional action is not required, skip this step. Increase the proportional gain in small steps. After each adjustment, observe the process variable response to a setpoint change. When sustained oscillations (amplitude is not increasing or decreasing) are observed, decrease the proportional gain to 50% of the value that caused the sustained oscillations.
4. Decrease the integral time constant in small steps. After each adjustment, observe the process variable response to a setpoint change. When sustained oscillations are

observed, increase the integral time constant to 330% of the value that caused the sustained oscillations.

5. If derivative action is not required, skip this step. Increase the derivative time constant in small steps. After each adjustment, observe the process variable response to a setpoint change. When sustained oscillations are observed, decrease the derivative time constant to 30% of the value that caused the sustained oscillations.
6. Make any final adjustments in $K_P$, $T_I$, or $T_D$ to obtain the desired process variable response.

**Example 5.2.** Assume that the system has the general block diagram of Figure 5.8 with the following transfer functions:

$$G_p(s) = \frac{1.2e^{-0.5s}}{(0.5s+1)(2s+1)} \qquad G_d(s) = \frac{e^{-2s}}{5s+1} \qquad G_h(s) = G_v(s) = 1 \tag{5.16}$$

where time is measured in minutes. Tune the loop using the above method based on the process response. In reality, this process is too slow for this method but will be used as a comparison with the other tuning methods.

***Solution.*** The starting values of the controller parameters are

$$K_P = 0.1 \qquad T_I = 1000 \text{ min} \qquad T_D = 0 \text{ min}$$

Using proportional-only control, the gain $K_P$ is increased until the response exhibits a sustained oscillation when the setpoint is changed by a small amount. The value of $K_P$ for sustained oscillation is 5.0. The proportional gain is set to 2.5 and the integral time constant $T_I$ is decreased until the response exhibits a sustained oscillation. The value of the integral time constant for sustained oscillation is 1.05 min. The integral time constant is set to 3.5; a setpoint change response of the system with this PI controller is shown in Figure 5.22. If a PID controller is desired, then starting with the PI parameters, the derivative time constant $T_D$ is increased until a sustained oscillation is observed. The value of the derivative time constant for sustained oscillation is 1.4 min. The derivative time constant is set to 0.4; the response of the system with this PID controller to a setpoint change is observed in Figure 5.23. Note that the derivative action speeds up the system response and lowers the overshoot, but at the expense of slightly more aggressive manipulated variable action. The derivative also accentuates the small amount of noise on the measurement and causes the manipulated variable to "hunt" after the major manipulated variable move.

***5.4.3.2 Ziegler–Nichols Closed-Loop Tuning.*** This method was originally proposed by Ziegler and Nichols (1942). The tuning process is characterized by finding the gain at which the system is marginally stable and the frequency of oscillation at this point. From these two parameters, the controller parameters are calculated. The parameter calculations are intended to produce a closed-loop damping ratio of $\frac{1}{4}$ (Figure 5.24). The method is as follows:

1. Start any trending (plotting) of the process variable.

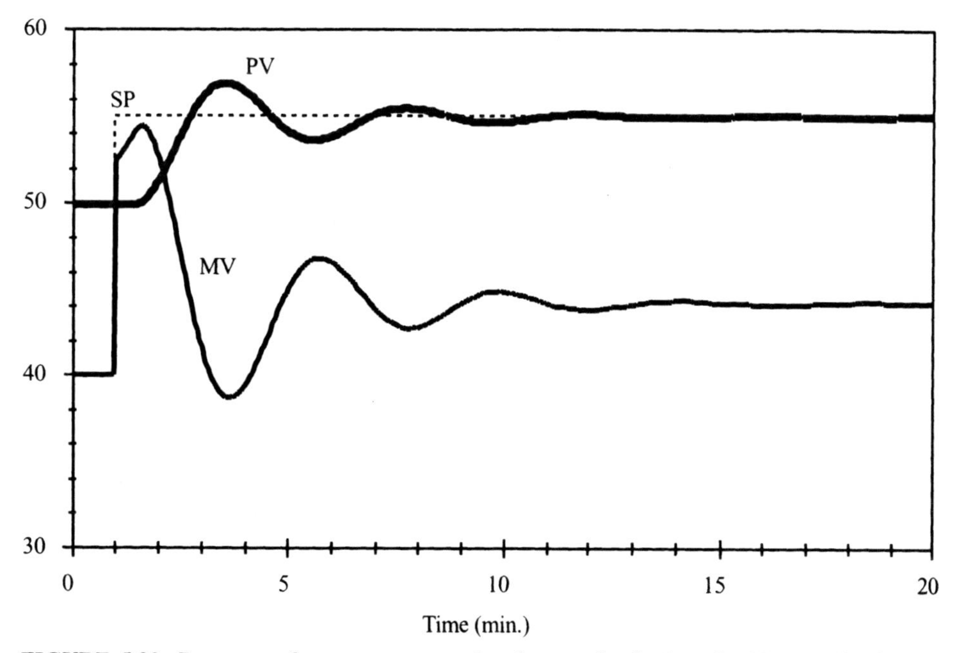


**FIGURE 5.22.** Response of system to a setpoint change of +5 when the PI controller is tuned according to the process response method.

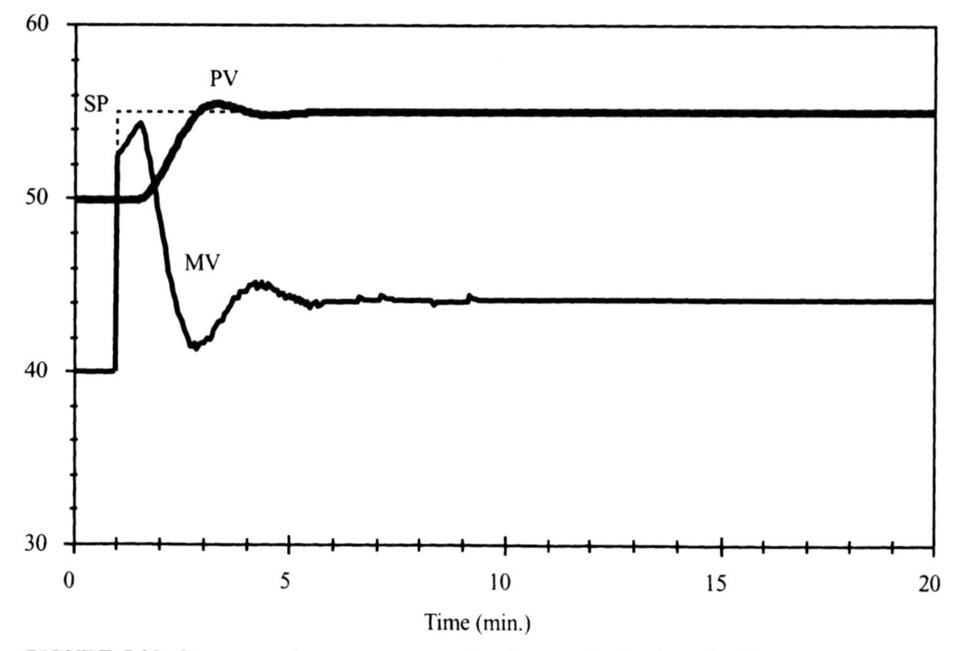


**FIGURE 5.23.** Response of system to a setpoint change of +5 when the PID controller is tuned according to the process response method.

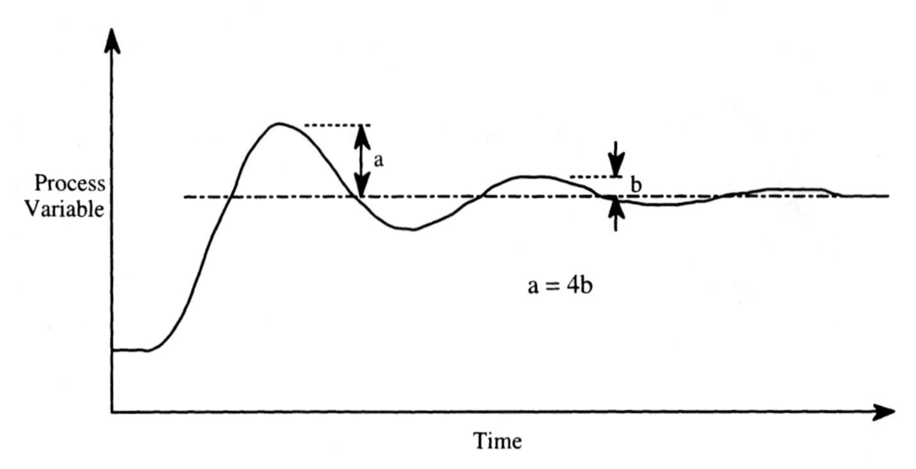


**FIGURE 5.24.** One-fourth damping ratio.

2. Set the $K_P$ and $T_D$ to their minimum values and the $T_I$ time constant to its maximum value. Now place the controller in the Automatic state.
3. Increase the proportional gain in small steps. After each adjustment, observe the process variable response to a setpoint change. When sustained oscillations (amplitude is not increasing or decreasing) are observed, note the value of the proportional gain and the period (in minutes) of the oscillations:

$$G_u = \text{proportional gain for sustained oscillations}$$
$$P_u = \text{period of oscillations (in minutes)}$$

4. Calculate the controller settings as shown in Table 5.2.
5. Make any final adjustments in $K_P$, $T_I$, or $T_D$ to obtain the desired process variable response.

**TABLE 5.2 Ziegler–Nichols Closed-Loop Tuning PID Parameters**

| | Interacting Equation | Noninteracting Equation |
|---|---|---|
| | *P controller* | |
| $K_P$ | $0.5G_u$ | $0.5G_u$ |
| | *PI controller* | |
| $K_P$ | $0.45G_u$ | $0.45G_u$ |
| $T_I$ (min) | $P_u/1.2$ | $P_u/1.2K_P$ |
| | *PID controller* | |
| $K_P$ | $0.6G_u$ | $0.6G_u$ |
| $T_I$ (min) | $P_u/2.0$ | $Pu/2.0K_P$ |
| $T_D$ (min) | $P_u/8.0$ | $K_PP_u/8.0$ |

**Example 5.3.** For the same process as in Example 5.2, tune a PI controller to control this process using the Ziegler–Nichols closed-loop tuning method.

***Solution.*** The starting values of the controller parameters are

$$K_P = 0.1 \qquad T_I = 1000 \text{ min} \qquad T_D = 0 \text{ min}$$

Using proportional-only control, the gain $K_P$ is increased until the response exhibits a sustained oscillation when the setpoint is changed by a small amount. The value of $K_P$ for sustained oscillation is 5.0 and the sustained oscillation appears as in Figure 5.25. Actually, the oscillation is growing slightly for this gain. However, this situation corresponds to tuning under actual process conditions, since one does not usually have the time to observe very many oscillations. Thus the values of the ultimate gain and ultimate period are

$$G_u = 5.0 \qquad P_u = 3.1 \text{ min}$$

If a PI controller is desired, the tuning parameters are

$$K_P = 0.45G_u = 0.45 \times 5.0 = 2.25 \qquad T_I = P_u/1.2 = 3.1/1.2 = 2.58 \text{ min}$$

The setpoint change response of the system with this PI controller is shown in Figure 5.26 and is similar to the response for the controller parameters obtained by the process response method.

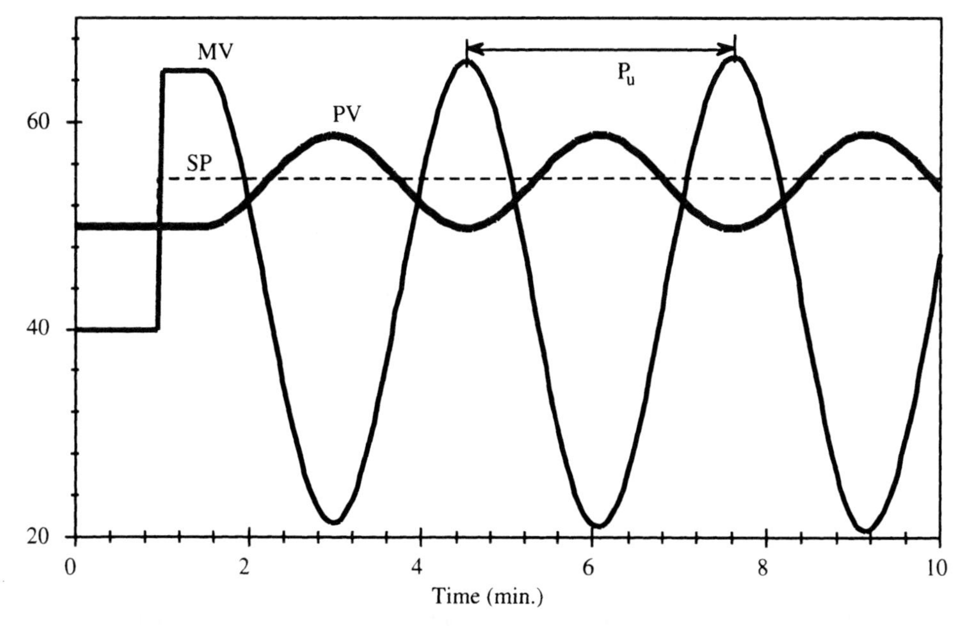


**FIGURE 5.25.** Sustained oscillation used in Ziegler–Nichols closed-loop tuning.

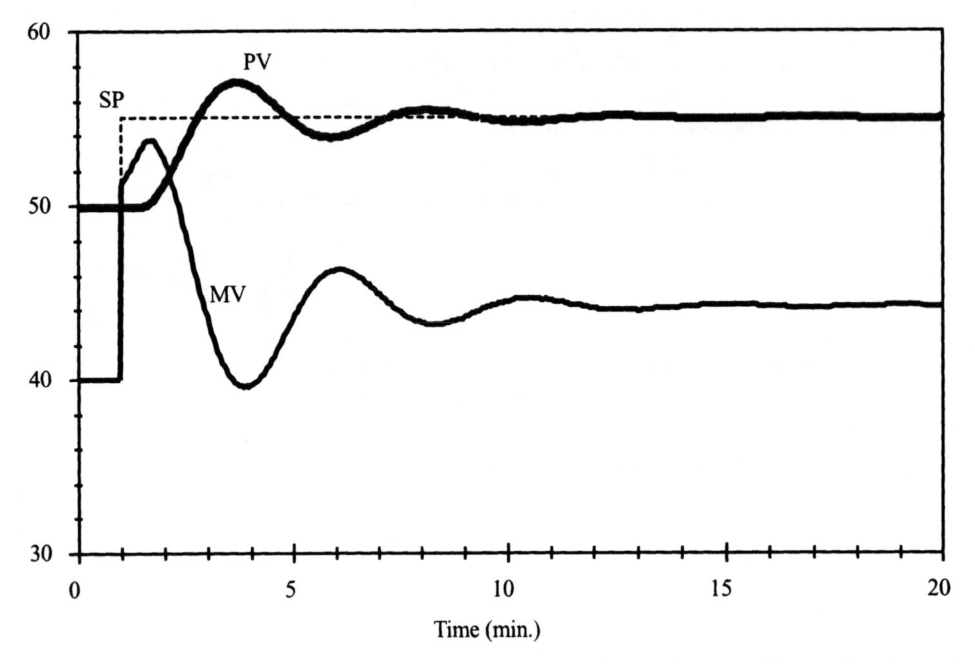


**FIGURE 5.26.** Response of system to a setpoint change of +5 when the PI controller is tuned according to the Ziegler–Nichols closed-loop method.

*5.4.3.3 Damped Oscillation Tuning Method.* When it is undesirable to allow sustained oscillations, the following method is used. The process is characterized by finding the gain at which the system has a damping ratio of $\frac{1}{4}$ (Figure 5.24), and the frequency of oscillation at this point. Similar to the Ziegler–Nichols method, the controller parameters are calculated from the gain and oscillation frequency. The method is as follows:

1. Start any trending (plotting) of the process variable.
2. Set the $K_P$ and $T_D$ to their minimum values and the $T_I$ time constant to its maximum value. Now place the controller in the Automatic state.
3. Increase the proportional gain in small steps. After each adjustment, observe the process variable response to a setpoint change. When a damping ratio of $\frac{1}{4}$ is observed, note the value of the proportional gain and the period (in minutes) of the oscillations:

$$G_d = \text{proportional gain for damping ratio of } \tfrac{1}{4}$$
$$P_d = \text{period of oscillations (in minutes)}$$

4. Calculate the controller settings as shown in Table 5.3.
5. Make any final adjustments in $K_P$, $T_I$, or $T_D$ to obtain the desired process variable response.

**Example 5.4.** For the same process as in Example 5.2, tune a PI controller to control this process using the damped oscillation tuning method.

**TABLE 5.3 Damped Oscillation Closed-Loop Tuning PID Parameters**

| | Interacting Equation | Noninteracting Equation |
|---|---|---|
| | *P controller* | |
| $K_P$ | $1.1G_d$ | $1.1G_d$ |
| | *PI controller* | |
| $K_P$ | $1.1G_d$ | $1.1G_d$ |
| $T_I$ (min) | $P_d/2.6$ | $P_d/2.6K_P$ |
| | *PID controller* | |
| $K_P$ | $1.1G_d$ | $1.1G_d$ |
| $T_I$ (min) | $P_d/3.6$ | $P_d/3.6K_P$ |
| $T_D$ (min) | $P_d/9.0$ | $K_P P_d/9.0$ |

***Solution.*** The starting values of the controller parameters are

$$K_P = 0.1 \qquad T_I = 1000 \text{ min} \qquad T_D = 0 \text{ min}$$

Using proportional-only control, the gain $K_P$ is increased until the response exhibits a response with quarter-wave damping when the setpoint is changed by a small amount. The

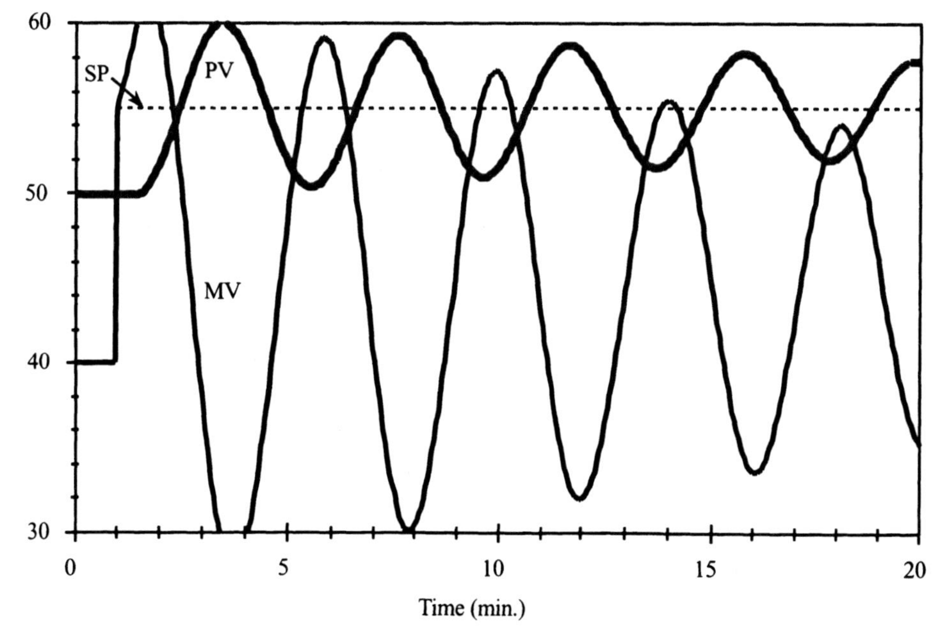


**FIGURE 5.27.** Response of system to a setpoint change of +5 when the PI controller is tuned according to the damped oscillation method.

value of $K_P$ for sustained oscillation is 2.75 and the period of the oscillation is 3.7 minutes. Thus,

$$G_d = 2.75 \qquad P_d = 3.7 \text{ min}$$

If a PI controller is desired, the tuning parameters are

$$K_P = 1.1G_d = 1.1 \times 2.75 = 3.0 \qquad T_I = P_d/2.6 = 3.7/2.6 = 1.42 \text{ min}$$

The setpoint change response of the system with this PI controller is shown in Figure 5.27. Obviously, in this case the method breaks down. Comparing the PI parameters to the previous two examples, the gain is too high and the integral time constant is too low. However, as indicated in the introductory remarks of this chapter, all of the tuning methods in this chapter are only designed to provide the engineer with a good starting point from which the parameters are further refined.

### 5.4.4 Open-Loop Tuning Methods

Tuning may be determined based on an empirical model of the process (Chapter 4). However, the process is not being controlled during such tests. This technique can be used to determine the tuning of slow processes. All of the open-loop tuning methods for self-regulating processes (reaches a steady-state value after a step change in the manipulated variable) use a first-order-plus-deadtime model of the process. An integrator-plus-deadtime model is used to tune non-self-regulating processes. Open-loop tuning methods are most amenable to slow processes or for those processes for which a model has been determined.

***5.4.4.1 Self-Regulating Processes.*** Given a first-order-plus-deadtime model that approximates the self-regulating process, there are a few methods one can use to calculate the tuning parameters. The first two self-regulating methods, Ziegler–Nichols open-loop (1942) and Cohen–Coon (1953), result in similar tuning parameters. The third method, proposed by Fertik (1975), yields a quite different set of tuning parameters. The first two methods yield a set of parameters that generate an oscillatory response to a setpoint change. The Fertik method generally yields a nonoscillatory response, which is usually desirable for most process control applications. The rationales behind these methods are:

| | |
|---|---|
| Ziegler–Nichols open-loop | Quarter-wave damping |
| Cohen-Coon | Quarter-wave damping |
| Fertik | Minimum ITAE, <8% overshoot |

Each method assumes a first-order-plus-deadtime process model of the form

$$G(s) = \frac{Ke^{-\theta_D s}}{\tau_1 s + 1}$$

and a process controllability $\alpha$ of the form

$$\alpha = \frac{\theta_D}{\tau_1}$$

Tuning using the Ziegler–Nichols open-loop and Cohen–Coon methods entails calculating $K_P$, $T_I$, or $T_D$ from the first-order-plus-deadtime model parameters. The calculations also depend on the type of controller (P only, PI, or PID). Table 5.4 summarizes these calculations for the interacting PID equation.

**Example 5.5.** For the same process as in Example 5.3, tune a PI and a PID controller to control this process using the Ziegler–Nichols open-loop tuning method. Show the response of the system to setpoint and disturbance changes.

***Solution.*** First, a first-order-plus-deadtime approximation to this process must be determined. Using the first-order-plus-deadtime method I from Chapter 4, the approximate model is obtained from a step response as

$$G_{\mathrm{FODT}}(s) = \frac{1.2e^{-0.7s}}{2.4s + 1}$$

**TABLE 5.4 Open-Loop Tuning PID Parameters for Self-Regulating Process**

| | Zeigler–Nichols Open Loop | Cohen–Coon |
|---|---|---|
| | *P controller* | |
| $K_P$ | $\frac{1}{K\alpha}$ | $\frac{1}{K}\left(\frac{1}{\alpha} + 0.333\right)$ |
| | *PI controller* | |
| $K_P$ | $\frac{0.9}{K\alpha}$ | $\frac{1}{K}\left(\frac{0.9}{\alpha} + 0.082\right)$ |
| $T_I$ (min) | $3.33\theta_D$ | $\theta_D\left[\frac{3.33 + 0.333\alpha}{1 + 2.2\alpha}\right]$ |
| | *PID controller* | |
| $K_P$ | $\frac{1.2}{K\alpha}$ | $\frac{1}{K}\left(\frac{1.35}{\alpha} + 0.270\right)$ |
| $T_I$ (min) | $2\theta_D$ | $\theta_D\left[\frac{2.5 + 0.5\alpha}{1 + 0.6\alpha}\right]$ |
| $T_D$ (min) | $0.5\theta_D$ | $\theta_D\left[\frac{0.37}{1 + 0.2\alpha}\right]$ |

Using the equations in Table 5.4, the PI controller parameters are determined to be

$$K_P = 2.57 \qquad T_I = 2.33 \text{ min} \qquad T_D = 0 \text{ min}$$

and the PID controller parameters are

$$K_P = 3.43 \qquad T_I = 1.4 \text{ min} \qquad T_D = 0.35 \text{ min}$$

The setpoint change response of the system with this PI controller is shown in Figure 5.28 and for the PID controller is shown in Figure 5.29. The responses to a disturbance of +5 are shown in Figures 5.30 and 5.31 for the PI and PID controllers, respectively. As seen in Example 5.2, the addition of the derivative term gives a better process variable response at the expense of more vigorous manipulated variable action.

Tuning using the Fertik method entails determining the type of controller (PI or PID), calculating the Fertik controllability $\alpha_F$ as

$$\alpha_F = \frac{\theta_D}{\theta_D + \tau_1} = \frac{T_d}{T_{ps}}$$

$$T_d = \theta_D \qquad T_{ps} = \theta_D + \tau_1$$

and then reading the normalized parameters from the set of graphs in Figure 5.32–5.36. The parameters may be optimized for setpoint response or for disturbance response. Note that the PID controller is not recommended for those processes whose Fertik controllability is greater than 0.5. These processes are dominated by deadtime.

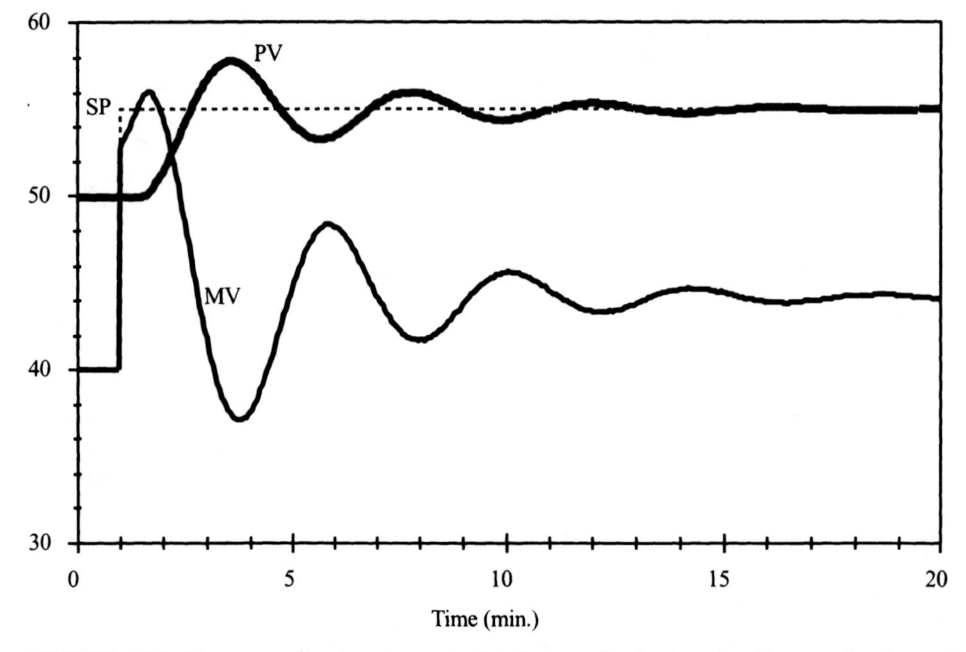


**FIGURE 5.28.** Response of system to a setpoint change of +5 when the PI controller is tuned according to the Ziegler–Nichols open-loop method.

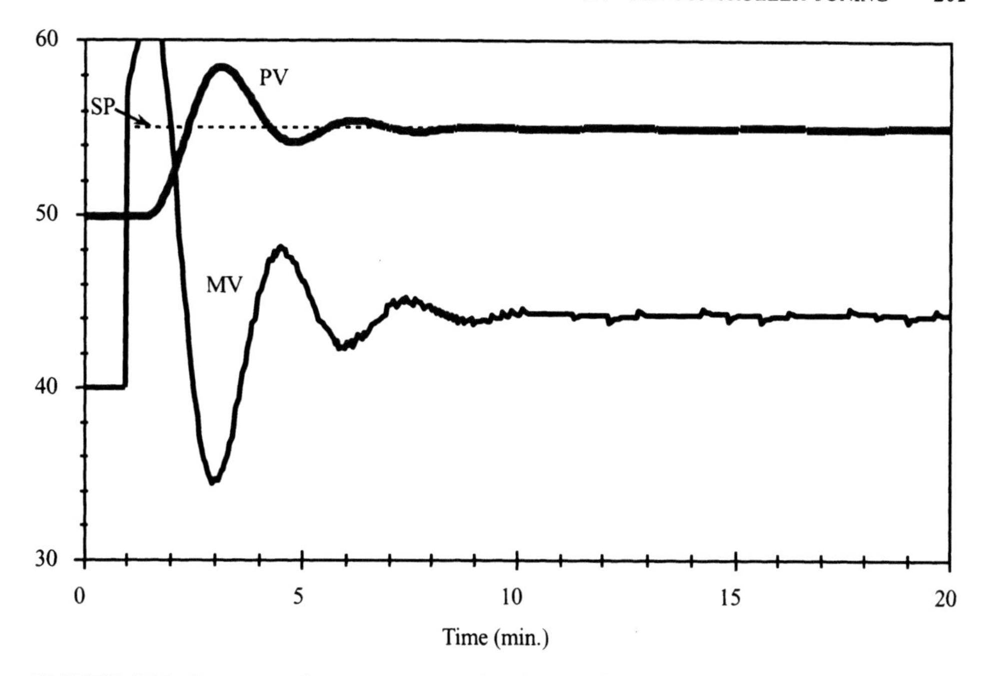


**FIGURE 5.29.** Response of system to a setpoint change of +5 when the PID controller is tuned according to the Ziegler–Nichols open-loop method.

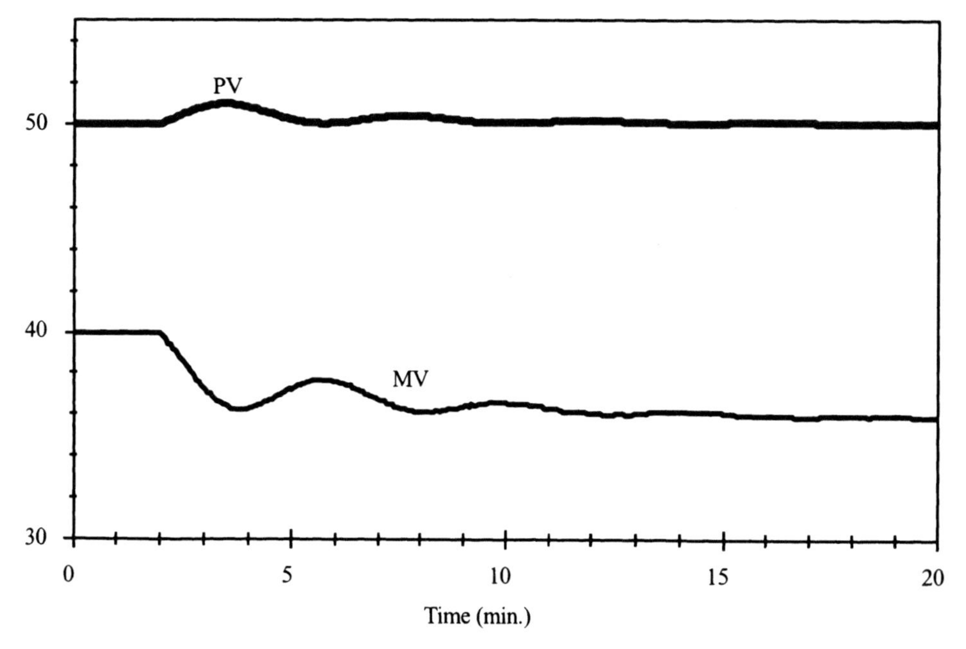


**FIGURE 5.30.** Response of system to a disturbance change of +5 when the PI controller is tuned according to the Ziegler–Nichols open-loop method.

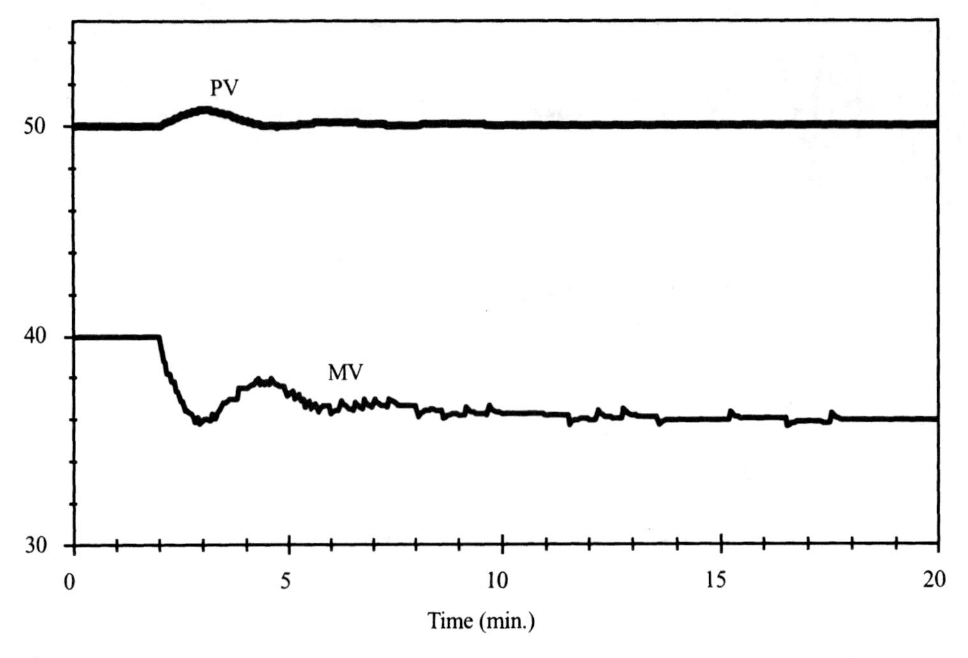


**FIGURE 5.31.** Response of system to a disturbance change of +5 when the PID controller is tuned according to the Ziegler–Nichols open-loop method.

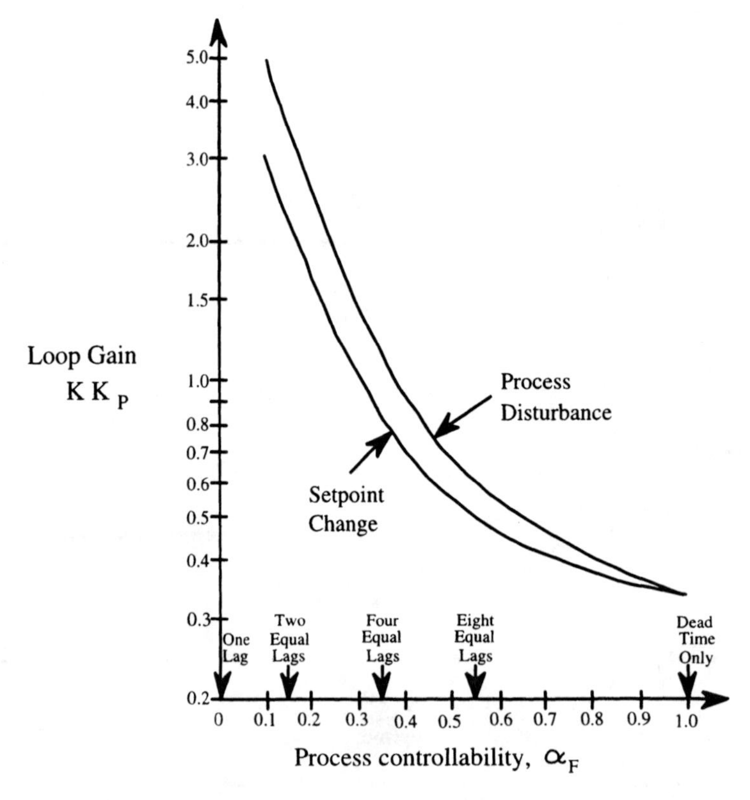


**FIGURE 5.32.** Fertik controller gain for PI control. (Copyright © 1975, ISA, all rights reserved, reprinted by permission, with changes.)

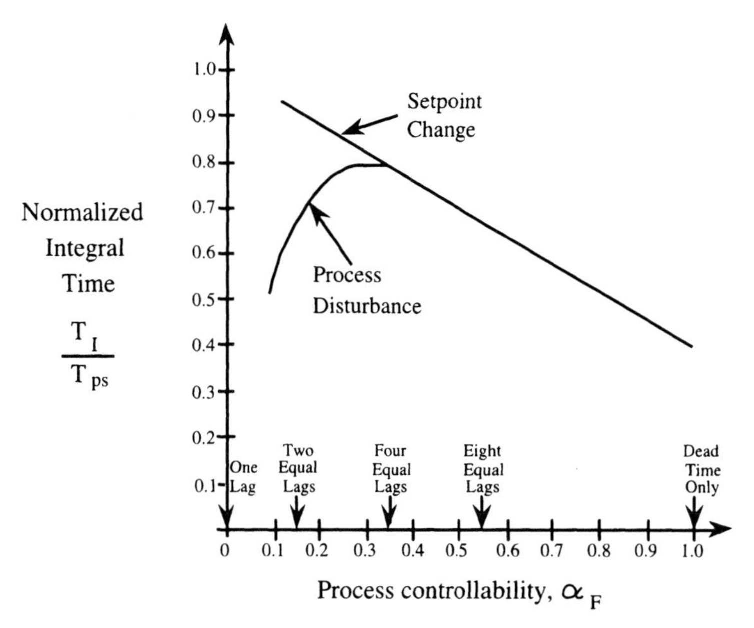


**FIGURE 5.33.** Fertik controller integral time for PI control. (Copyright © 1975, ISA, all rights reserved, reprinted by permission, with changes.)

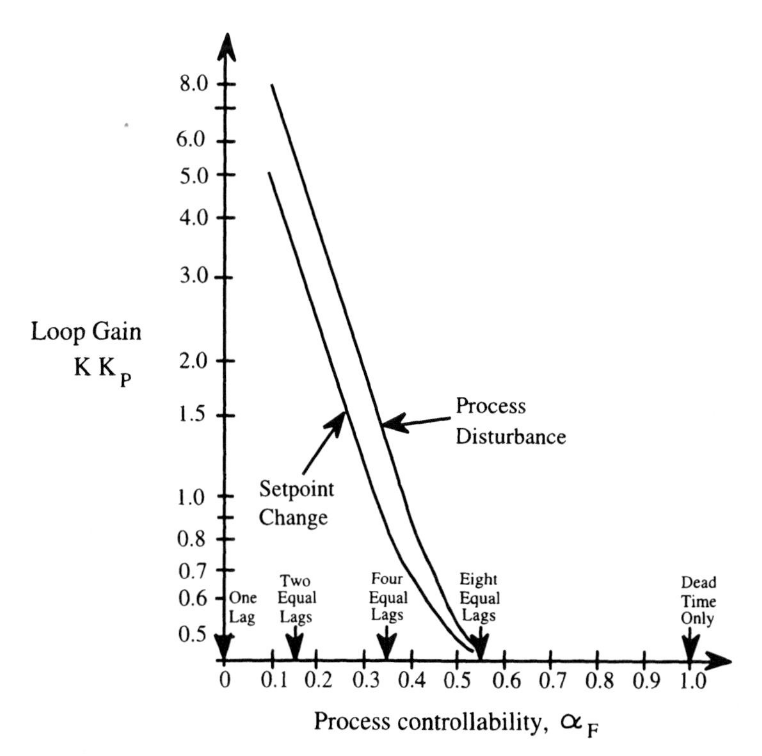


**FIGURE 5.34.** Fertik controller gain for PID control. (Copyright © 1975, ISA, all rights reserved, reprinted by permission, with changes.)

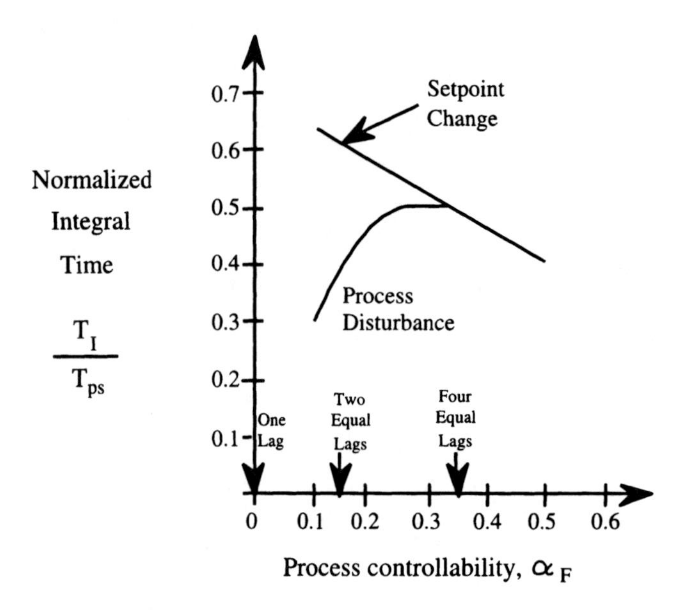


**FIGURE 5.35.** Fertik controller integral time for PID control. (Copyright © 1975, ISA, all rights reserved, reprinted by permission, with changes.)

**Example 5.6.** For the same process as in Example 5.2, tune a PI and a PID controller to control this process using the Fertik open-loop tuning method. Tune the controller for good setpoint response and show its setpoint response. Also tune the controller for good disturbance response and demonstrate its response to a step disturbance.

***Solution.*** The first-order-plus-deadtime approximation to this process is determined in the same manner as for Example 5.5:

$$G_{\text{FODT}}(s) = \frac{1.2e^{-0.7s}}{2.4s + 1}$$

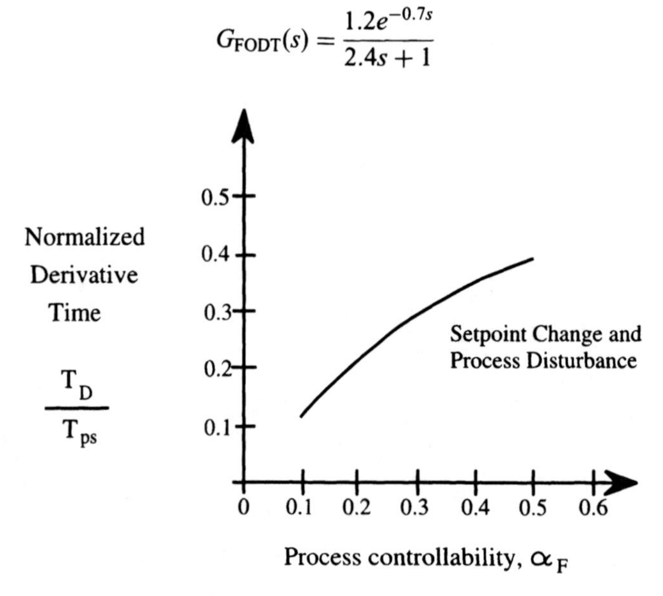


**FIGURE 5.36.** Fertik controller derivative time for PID control. (Copyright © 1975, ISA, all rights reserved, reprinted by permission, with changes.)

The Fertik controllability is determined as $\alpha_F = 0.23$. The value of $T_{ps} = 3.1$. Using the graphs in Figures 5.32 and 5.33, the PI controller parameters for a setpoint change are

$$KK_P \approx 1.4 \Rightarrow K_P = 1.4/1.2 = 1.17$$
$$T_I/T_{ps} \approx 0.86 \Rightarrow T_I = 0.86 \times 3.1 = 2.67 \text{ min}$$

and the PID controller parameters for a setpoint change are determined using the graphs in Figures 5.34–5.36:

$$KK_P \approx 1.9 \Rightarrow K_P = 1.9/1.2 = 1.58$$
$$T_I/T_{ps} \approx 0.57 \Rightarrow T_I = 0.57 \times 3.1 = 1.77 \text{ min}$$
$$T_D/T_{ps} \approx 0.24 \Rightarrow T_D = 0.24 \times 3.1 = 0.74 \text{ min}$$

The setpoint change response of the system with the PI controller is shown in Figure 5.37 and for the PID controller is shown in Figure 5.38. In contrast to the other methods presented in this chapter, the process variable exhibits a small overshoot to a setpoint change but a slower rise time. In a similar manner, the PI parameters for good disturbance response are

$$K_P = 1.67 \qquad T_I = 2.39 \text{ min}$$

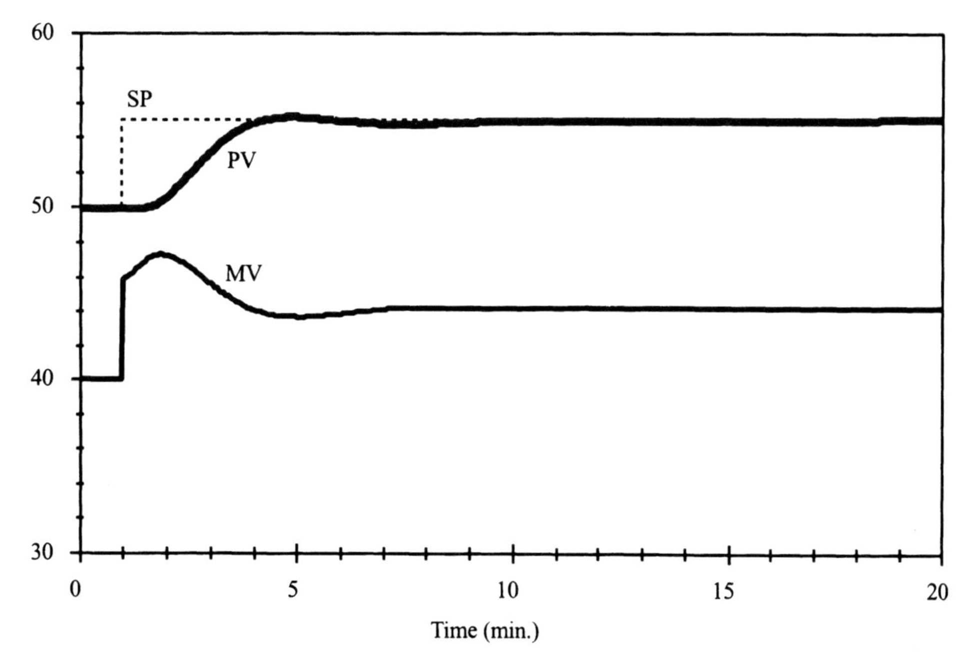


**FIGURE 5.37.** Response of system to a setpoint change of +5 when the PI controller is tuned according to the Fertik open-loop method.

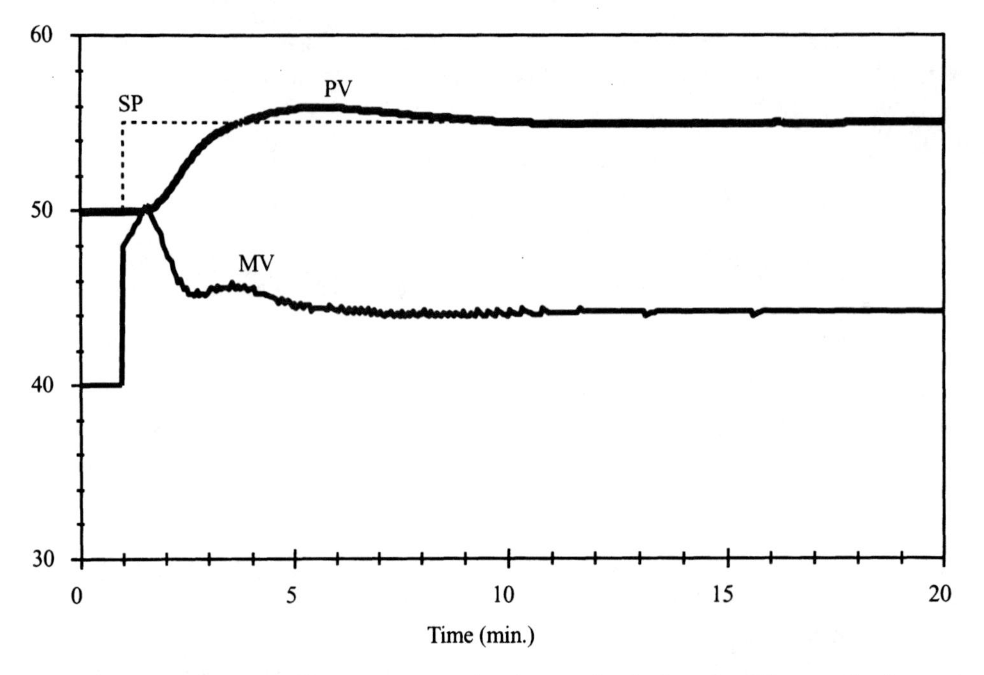


**FIGURE 5.38.** Response of system to a setpoint change of +5 when the PID controller is tuned according to the Fertik open-loop method.

and the PID parameters for good disturbance response are:

$$K_P = 2.5 \qquad T_I = 1.52 \text{ min} \qquad T_D = 0.74 \text{ min}$$

The response of the system to a step disturbance of +5 with the PI controller is shown in Figure 5.39 and for the PID controller is shown in Figure 5.40. Compared with the parameters determined with Ziegler–Nichols tuning, the parameters determined by the Fertik tuning method give a system response with smaller deviation but a slightly longer rise time.

*5.4.4.2 Integrating Processes.* This method was originally proposed by Ziegler and Nichols (1942) and may be used to determine the tuning parameters for a non-self-regulating process. After obtaining the parameters of the integrating process with deadtime,

$$G_{\text{ID}}(s) = \frac{K_s}{s} e^{-\theta_D s}$$

The controller settings are shown in Table 5.5. As usual, make any final adjustments in $K_P$, $T_I$, or $T_D$ to obtain the desired process variable response.

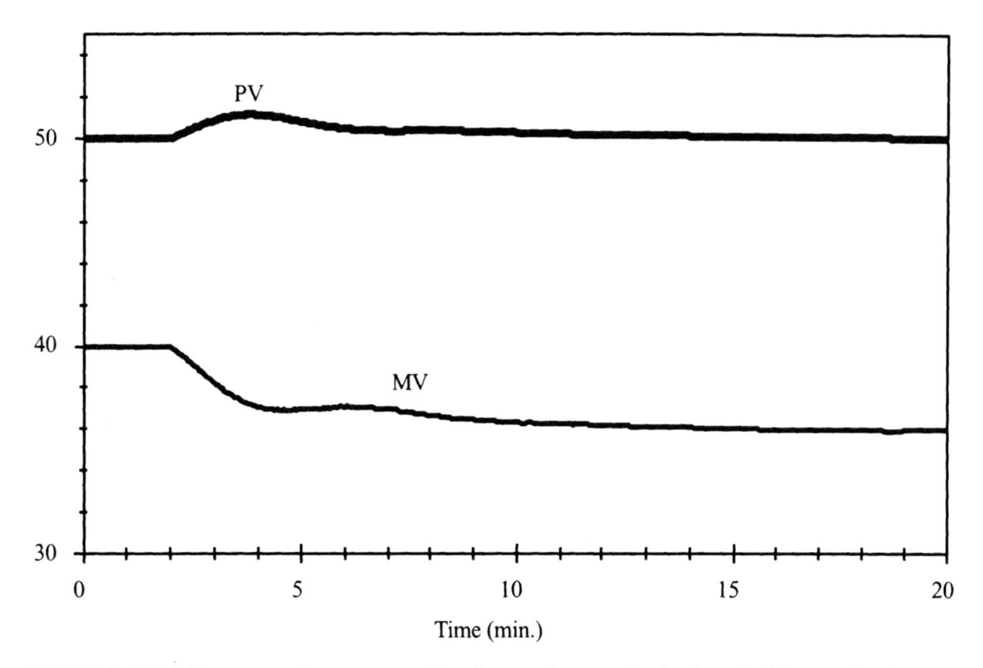


**FIGURE 5.39.** Response of system to a disturbance change of +5 when the PI controller is tuned according to the Fertik open-loop method.

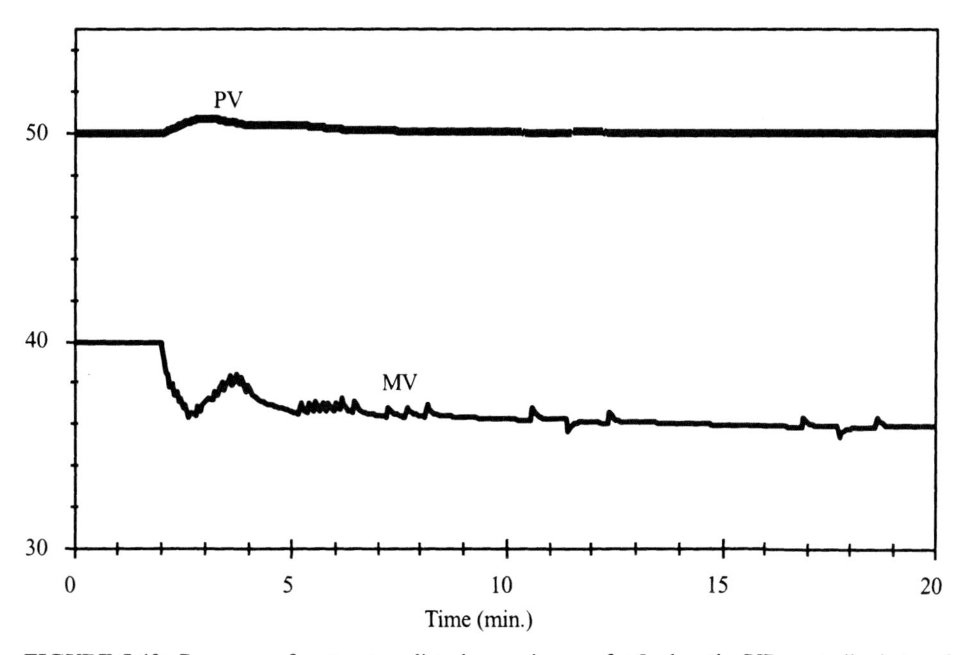


**FIGURE 5.40.** Response of system to a disturbance change of +5 when the PID controller is tuned according to the Fertik open-loop method.

**TABLE 5.5 Open-Loop Tuning PID Parameters for Integrating Process**

| | Interacting Equation | Noninteracting Equation |
|---|---|---|
| | *P controller* | |
| $K_P$ | $1/K_S\theta_D$ | $1/K_S\theta_D$ |
| | *PI controller* | |
| $K_P$ | $0.9/K_S\theta_D$ | $0.9/K_S\theta_D$ |
| $T_I$ (min) | $3.3\theta_D$ | $3.3\theta_D/K_P$ |
| | *PID controller* | |
| $K_P$ | $1.2/K_S\theta_D$ | $1.2/K_S\theta_D$ |
| $T_I$ (min) | $3.3\theta_D$ | $3.3\theta_D/K_P$ |
| $T_D$ (min) | $0.5\theta_D$ | $0.5\theta_D K_P$ |

**Example 5.7.** Assume that the system has the general block diagram of Figure 5.8 with the following transfer functions:

$$G_p(s) = \frac{0.3e^{-2s}}{s(5s+1)(s+1)} \qquad G_h(s) = G_v(s) = 1 \qquad G_d(s) = 0$$

where time is measured in minutes. Tune a PI and a PID controller to control this process using the Ziegler–Nichols open-loop tuning method for integrating processes.

***Solution.*** The integrator-plus-deadtime approximation to this process is determined in Example 4.13 as

$$G_{\mathrm{ID}}(s) = \frac{0.25e^{-6s}}{s}$$

The PI parameters are determined as

$$K_P = \frac{0.9}{K_S\theta_D} = \frac{0.9}{0.25 \times 6} = 0.6 \qquad T_I = 3.3\theta_D = 3.3 \times 6 = 19.8 \text{ min}$$

The PID parameters are determined as

$$K_P = \frac{1.2}{K_S\theta_D} = \frac{1.2}{0.25 \times 6} = 0.8$$

$$T_I = 3.3\theta_D = 3.3 \times 6 = 19.8 \text{ min} \qquad T_D = 0.5\theta_D = 0.5 \times 6 = 3 \text{ min}$$

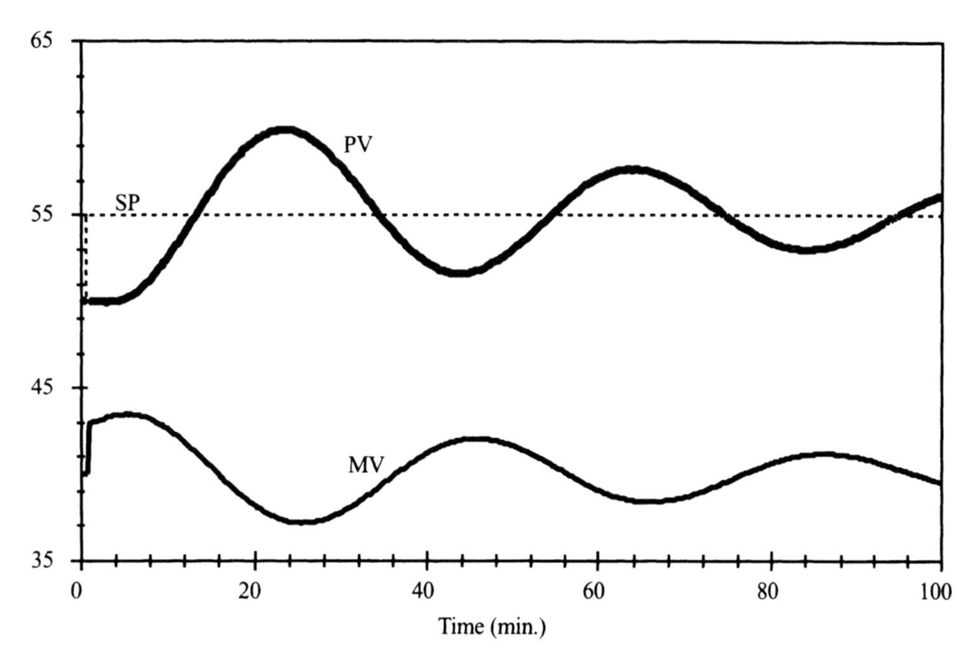


**FIGURE 5.41.** Response of non-self-regulating system to a setpoint change of +5 when the PI controller is tuned according to the Ziegler–Nichols open-loop method.

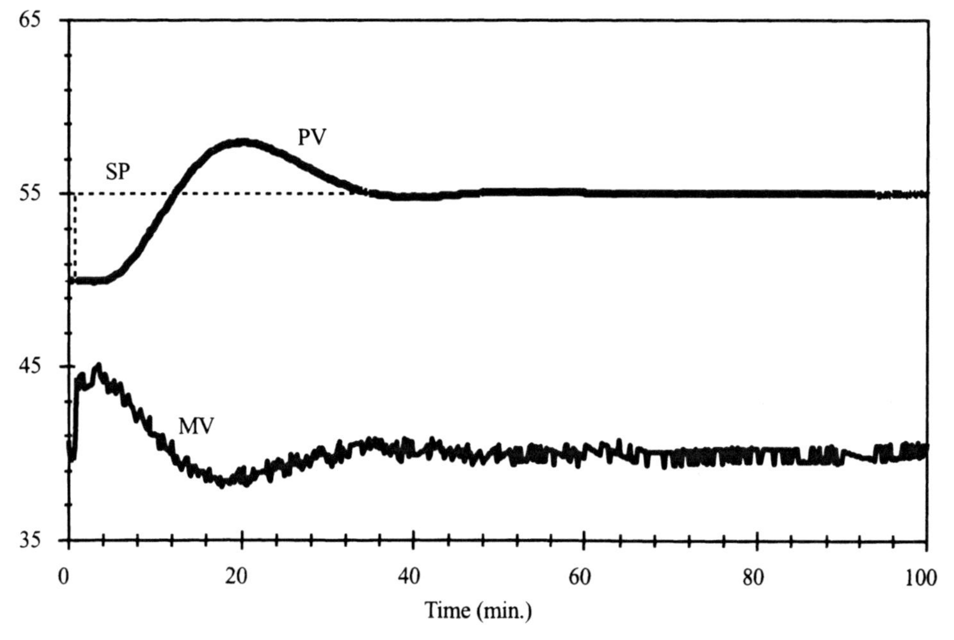


**FIGURE 5.42.** Response of non-self-regulating system to a setpoint change of +5 when the PID controller is tuned according to the Ziegler–Nichols open-loop method.

The responses of this system to a setpoint change are shown for the PI and PID controllers in Figures 5.41 and 5.42, respectively. As seen in the other examples, the derivative term speeds up the system response but accentuates the small amount of measurement noise.

### 5.4.5 Comparison of Tuning Methods

In this section all of the system responses, except for the controller tuned by the damped oscillation method, are evaluated for a setpoint change from 50 to 55 and for a +5 change in disturbance. Some of the responses have been shown with Examples 5.2–5.7. The setpoint responses are compared for the rise time, percent overshoot, settling time (2%), damping ratio, IAE, and manipulated variable percent overshoot. The disturbance responses are compared for the maximum deviation, settling time (to within ±0.1), IAE, and manipulated variable percent overshoot.

Comparing the responses to a setpoint change in Table 5.6, all but the Fertik method generate responses that are similar. Other comparisons follow:

*Rise Time.* The rise time is about the same for all of the tuning methods, except for the Fertik method, for which it is longer.

*Process Variable Overshoot.* The overshoot for the Fertik method is considerably smaller than the other methods. Surprisingly, the overshoot for the PID is higher than for the PI controller.

*Settling Time.* With the PID, the settling time is generally shorter than for the PI controller.

**TABLE 5.6 Comparison of System Responses for Setpoint Change**

| | Rise Time (min) | CV Overshoot (%) | Settling Time, 2% (min) | Damping Ratio | IAE | MV Overshoot (%) |
|---|---|---|---|---|---|---|
| | | | *Response* | | | |
| PI | 0.8 | 38.8 | 11.0 | 0.25 | 11.8 | 246 |
| PID | 0.9 | 10 | 4.0 | 0.19 | 6.8 | 245 |
| | | | *Ziegler–Nichols closed loop* | | | |
| PI | 0.9 | 41.7 | 12.0 | 0.23 | 11.9 | 230 |
| PID | 0.7 | 51.4 | 5.6 | 0.08 | 9.1 | 385 |
| | | | *Ziegler–Nichols open loop* | | | |
| PI | 0.8 | 56.4 | 17.7 | 0.35 | 15.2 | 284 |
| PID | 0.6 | 70.6 | 10.3 | 0.12 | 10.8 | 470 |
| | | | *Cohen–Coon* | | | |
| PI | 0.8 | 86.4 | 39.7 | 0.62 | 33.9 | 350 |
| PID | 0.6 | 93.2 | 22.5 | 0.49 | 20.8 | 573 |
| | | | *Fertik* | | | |
| PI | 2.0 | 3.6 | 10.2 | 0.04 | 9.9 | 75 |
| PID | 1.5 | 17.6 | 10.7 | 0.01 | 10.9 | 147 |

**TABLE 5.7 Comparison of Systems Responses for Disturbance Change**

| | Maximum Deviation | Settling Time, ±0.1 (min) | IAE | MV Overshoot (%) |
|---|---|---|---|---|
| | | *Response* | | |
| PI | 1.02 | 17.7 | 5.5 | 0 |
| PID | 0.83 | 17.8 | 5.5 | 0 |
| | | *Ziegler–Nichols closed loop* | | |
| PI | 1.04 | 14.5 | 4.6 | 0 |
| PID | 0.78 | 8.2 | 2.1 | 0 |
| | | *Ziegler–Nichols open loop* | | |
| PI | 1.00 | 13.3 | 3.6 | 0 |
| PID | 0.77 | 7.0 | 1.7 | 3 |
| | | *Cohen–Coon* | | |
| PI | 0.97 | 18.5 | 4.3 | 12 |
| PID | 0.77 | 12.4 | 2.1 | 10 |
| | | *Fertik* | | |
| PI | 1.14 | 16.0 | 5.7 | 0 |
| PID | 0.74 | 8.0 | 2.5 | 0 |

*Damping Ratio.* The Fertik tuning leads to a very small damping ratio. Also, the damping ratio with the PID controller is smaller than for the PI controller.

*Integrated Absolute Error.* The IAE is significantly higher for the Cohen–Coon method. Most, but not all, of the PID controllers produce a smaller IAE than the PI controller.

*Manipulated Variable Overshoot.* The overshoot for the Fertik method is considerably smaller than the other methods. In general, the overshoot for the PID is higher than for the PI controller.

Comparing the disturbance responses in Table 5.7, the performance for all of the tuning methods are very similar. In general, the PID controller gives a smaller deviation, shorter settling time, and smaller IAE.

## 5.5 USER INTERFACE CONSIDERATIONS

The user interface for a PID controller is typically simple and resembles the faceplate of the older analog controller. A typical faceplate as it would appear on a computer screen is shown in Figure 5.43. The process variable and setpoint values are shown as vertical bars whose height indicates the value. The units are typically shown along the side and the numeric value is shown somewhere close to the bars; here they are shown below. The controller output is displayed as a horizontal bar along the bottom of the faceplate. This bar may also be displayed vertically, usually as the rightmost bar. The state (Auto or Manual)

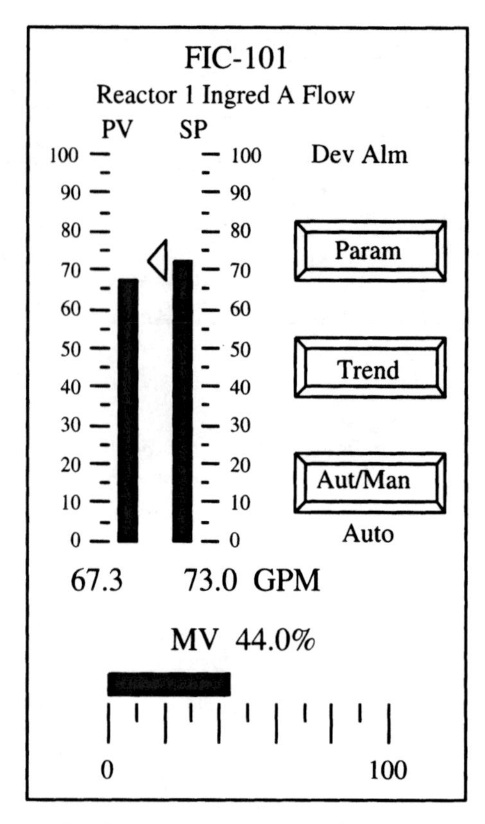


**FIGURE 5.43.** Typical PID faceplate.

is displayed and some means is provided to change the state. Since this faceplate indicates the state is Auto, the setpoint may be changed, which is indicated by the arrow between the PV and SP bars. For a touch-screen display, the operator changes the setpoint by pressing and sliding the arrow. The setpoint may also be changed by touching the setpoint value, which causes a keypad window to pop up, and keying in a new value. In the Manual state, the operator changes the controller output in a similar manner. Other tuning parameters are accessed by touching the "Param" button, which causes another window to pop up, displaying the other parameters (tuning parameters, alarm limits, output velocity limits, etc.). When the "trend" button is touched, a trend window pops up to the side of the faceplate and displays a plot of the process variable, setpoint, and controller, similar to a strip chart recorder. The trend display is typically a short-term trend of a few minutes.

## REFERENCES


Cohen, G. H., and G. A. Coon, "Theoretical Considerations of Retarded Control," *Trans. ASME*, **75** (7), 827–834 (1953).

Dorf, R. C., and R. H. Bishop, *Modern Control Systems*, 7th ed., Addison-Wesley, Reading, MA, 1995.

Fertik, H. A., "Tuning Controllers for Noisy Processes," *ISA Trans.*, **14** (4), 292–304 (1975).

Ziegler, J. G., and N. B. Nichols, "Optimum Settings for Automatic Controllers," *Trans. ASME*, **64**, 759–768 (1942).

# 6 Enhancements to Single-Loop Regulatory Control

**Chapter Topics**

- Cascade control
- Feedforward control
- Ratio control
- Split-range control
- Override control
- Single-loop model predictive control

*Scenario:* Simple enhancements to single-loop PID controllers can make dramatic process improvements.

At the glass bottle factory, our young engineer was involved in solving another problem with the glass bottles. The number of cracked bottles that were detected after they were cooled was unacceptably high. Various people looked at the problem and examined aspects of the process. Changing some of the machine parameters seemed to help, but only for a little while. After a day or so, the defect rate was just as high. In frustration, he plotted all of the available measurements around the bottle machines, including the hourly defect rate, for a period of a few days. He taped all of the plots together and arranged it around a conference room. He invited a few other engineers to examine the data with him and try to find the source of the problem. The revelation hit them all at once: The ambient outside temperature was the problem. The defect rate had a 24-hr cycle that was at a minimum during the day and a maximum at night. After the glass bottles came out of the forming machines, they were cooled. Most of the cooling air was drawn from the outside and the operators would adjust the cooling air flow rate, but not according to the outside air temperature. The glass bottles were cracking at night because of the cooler outside air. The cooling air flow was not being adjusted due to the outside air temperature. In order to solve the problem, feedforward control was added to the outside air flow controller, based on the outside air temperature. Previously, the air flow was set strictly by the machine demand in order to cool the bottles to a set temperature. Now a feedforward loop was added that measured the outside temperature and adjusted the air flow accordingly. With this change, the bottle defect dropped dramatically, and once again our young engineer was successful.

## 6.1 INTRODUCTION

While the PID single-loop controller works adequately in many process control situations, its performance can be improved if additional process information is known. This chapter presents these enhancements, from simplest to more complex. Each enhancement is

**TABLE 6.1 Situations for Single-Loop Control Enhancements**

| Enhancement | Situation |
|---|---|
| Cascade | Control of any nonflow process variable that uses a valve as its final control element. In this case, the controller for the nonflow variable is cascaded to a flow controller that manipulates the valve.<br>Other situations where the process can be described as a cascade combination of a fast process followed by a slow process and the secondary process disturbance effect needs to be mitigated. |
| Feedforward | Eliminate or reduce disturbance effect on the process variable. |
| Ratio | Blending or other applications where two streams need to be held in constant ratio to each other. |
| Split range | Process has one process variable and more than one manipulated variable. |
| Override | Protection of process equipment or personnel. |
| Model Predictive Control | PID alternative when the process has complex process dynamics. |

appropriate for certain situations, as shown in Table 6.1. Cascade control is used if the process can be best described as a cascade combination of two processes, especially if one of the processes is a flow process, but requires an additional measurement. Feedforward control attempts to eliminate all disturbance effects but requires an additional measurement for every disturbance. Ratio control is treated as a special example of feedforward control. In split range control, there are multiple manipulated variables for the same controlled variable. In the case where controlled variables must stay within allowable bounds control systems are built with signal selectors. Finally, unusual process dynamics are best handled by an alternative to the PID controller algorithm—model predictive control. Throughout the chapter, the typical user interface to each of these controllers is described.

## 6.2 CASCADE CONTROL

### 6.2.1 Cascade Control Objective

In many cases a controlled process may be considered as two processes in series, as shown in Figure 6.1. When a single controller is used in such a situation, as shown in Figure 6.2, poor control often results because disturbances in the secondary process are not measured until they have already affected the output of the primary process. Control performance can be improved by the addition of another feedback loop, as shown in Figure 6.3. This control scheme is called cascade control. The output of the outer loop becomes the setpoint of the inner loop. With cascade control, disturbances to the inner control loop are corrected by the inner controller and thus have little effect on the controlled output. For a cascade control loop to function properly, the dynamics of the inner loop must be at least as fast as the outer loop, and preferably faster. Usually, two loops are cascaded, although the strategy can be extended to more than two loops. In any case, only one loop directly manipulates the physical variable (usually a flow). The second loop output is the setpoint of the first loop, the third loop output is the setpoint of the second, and so on.

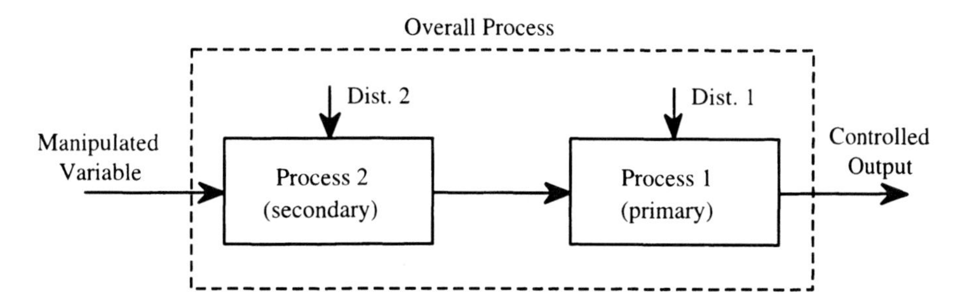


**FIGURE 6.1.** Process as cascade of two processes.

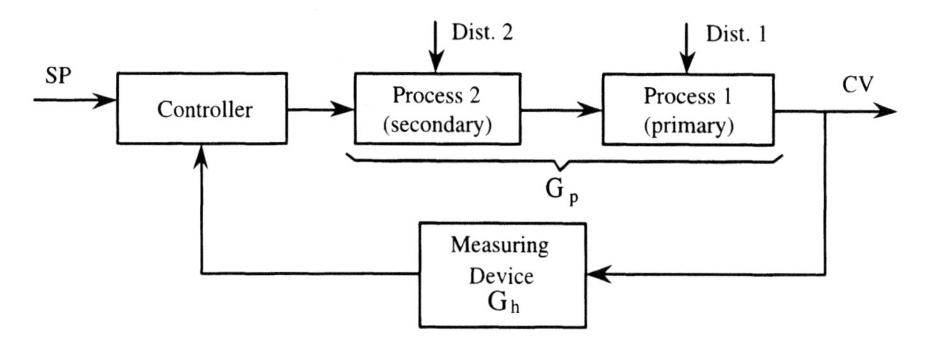


**FIGURE 6.2.** Single-loop control of two cascaded processes.

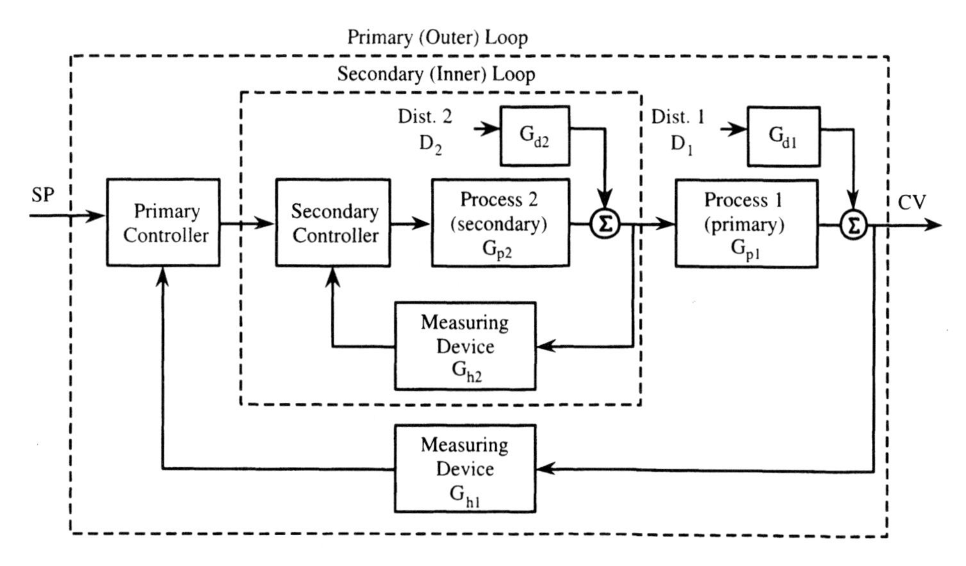


**FIGURE 6.3.** Cascade control.

As an example, consider the control of the heat exchanger in Figure 6.4. The objective is to control the outlet temperature, sensed by TT2501. With a single-loop controller, shown in Figure 6.5, the steam control valve is manipulated by the temperature controller to maintain a desired outlet temperature. However, disturbances in the steam header pressure, which cause the steam flow to change, are not detected by the temperature controller until their effect is seen as a change in the outlet temperature. This performance is demonstrated in Figure 6.6. In order to control the outlet temperature better, an inner loop is constructed that maintains the steam flow. The temperature controller then manipulates the flow controller setpoint, as shown in Figure 6.7. As shown in Figure 6.8, the outlet temperature response to a change in the steam header pressure is negligible.

The preceding example introduced the principles of cascade control. Marlin (1995) presented general design criteria for cascade control, summarized in Table 6.2. Cascade control should be considered when single-loop control does not provide acceptable performance and when an acceptable secondary variable measurement is available or can be added. Also, the secondary measurement must satisfy three criteria. First, the secondary variable must indicate the presence of a significant disturbance to the primary controlled

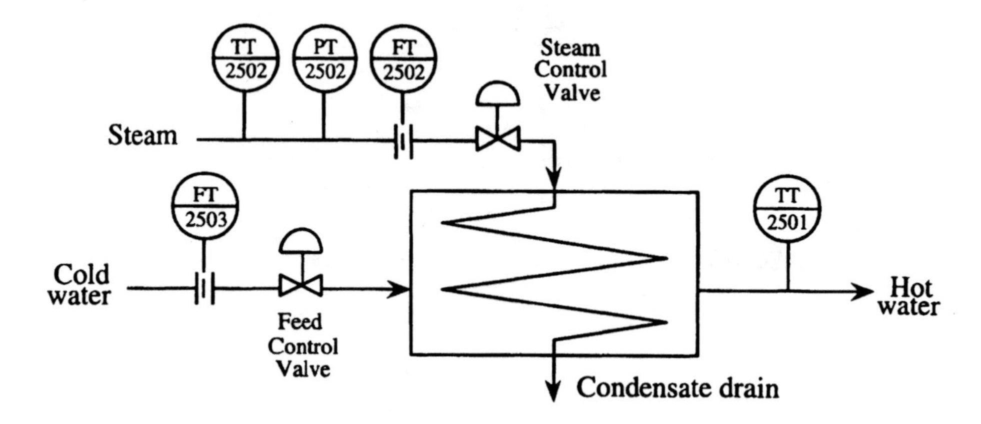


**FIGURE 6.4.** Heat exchanger process.

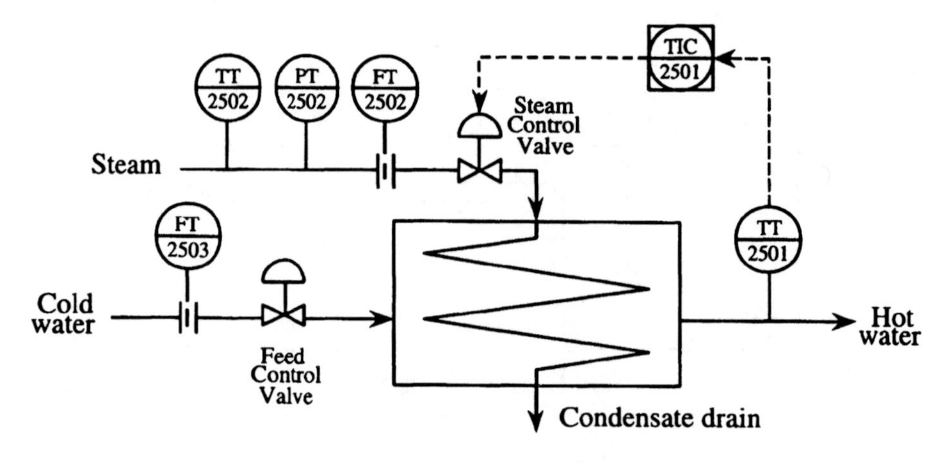


**FIGURE 6.5.** Single-loop control of heat exchanger.

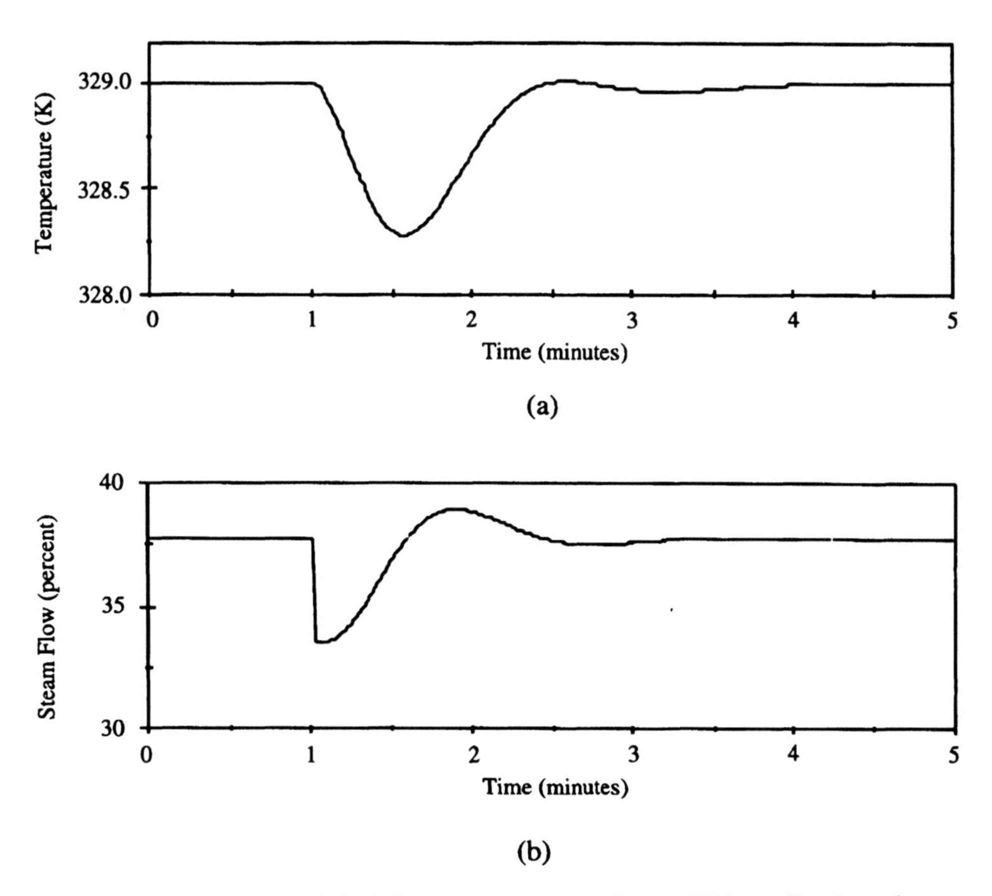


**FIGURE 6.6.** Performance of single-loop temperature control to a −10% step disturbance in steam pressure.

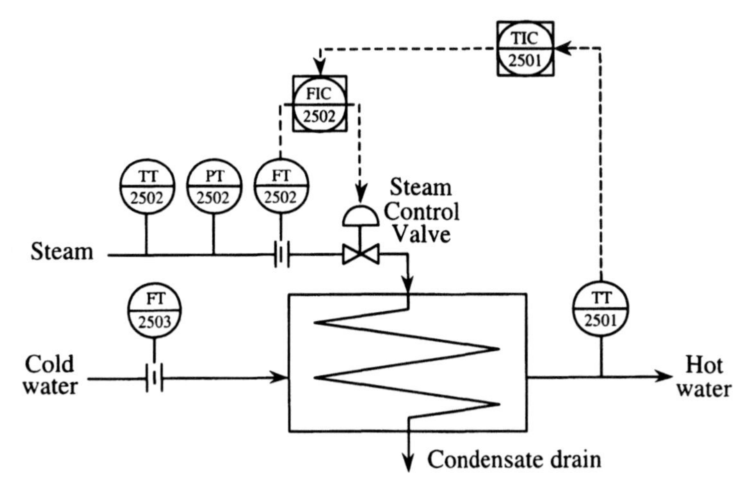


**FIGURE 6.7.** Cascade control of heat exchanger.

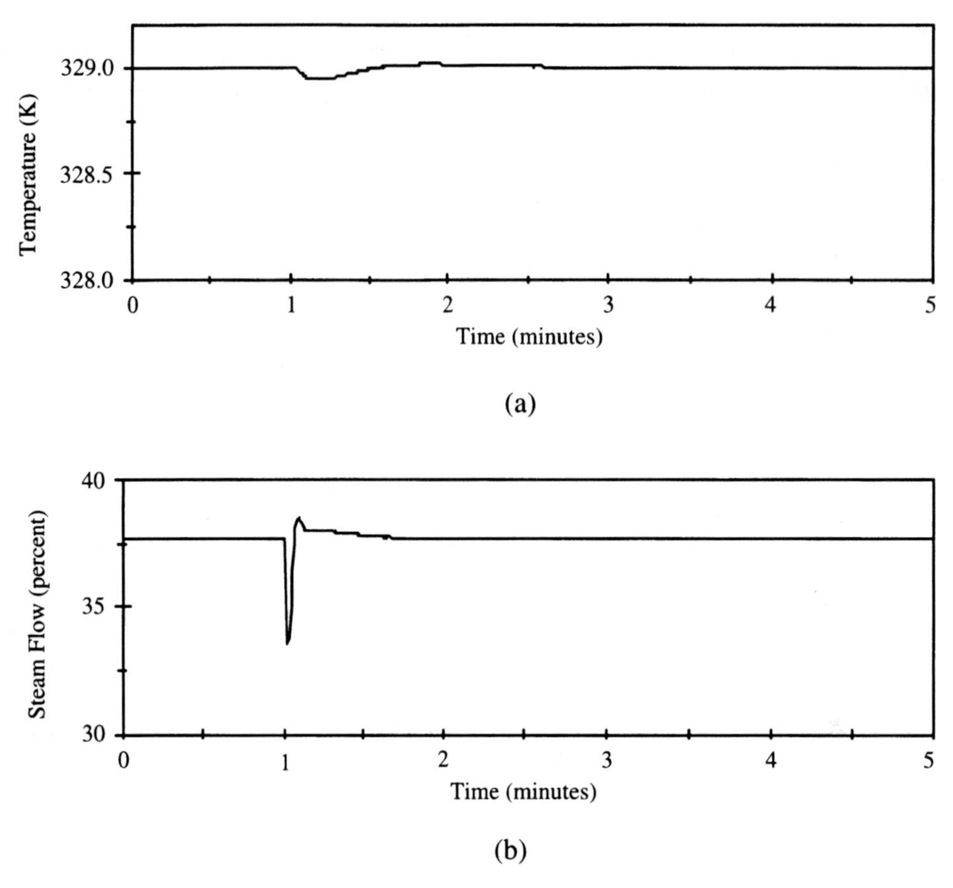


**FIGURE 6.8.** Performance of cascade temperature control to a −10% step disturbance in steam pressure.

**TABLE 6.2 Cascade Control Design Criteria**

| **Cascade control should be considered when:** |
|---|
| 1. Single-loop control does not provide satisfactory control performance. |
| 2. A measured secondary variable is available. |
| **A secondary variable must satisfy the following criteria:** |
| 3. The secondary variable must indicate the occurrence of an important disturbance. |
| 4. There must be a causal relationship between the manipulated and secondary variables. |
| 5. The secondary variable dynamics must be faster than the primary variable dynamics. |

variable. For the previous heat exchanger, the steam flow is the appropriate secondary variable, although the disturbance is a change in steam pressure. A change in steam pressure causes a change in the flow, and thus the flow is an indication of the pressure disturbance. Second, the secondary variable must be influenced by the manipulated variable; otherwise there can be no secondary feedback loop. For example, the steam control valve, the manipulated variable in the heat exchanger, influences the steam flow.

Finally, the dynamics of the secondary loop should be relatively faster than the primary loop so that it attenuates the disturbance before the primary controller variable is affected. A general guideline is that the secondary process should be at least three times faster than the primary process. The commonly encountered process control loops, in decreasing speed, are:

Gas pressure
Flow
Liquid level and pressure
Temperature
Composition

Flow is probably the most common secondary variable in cascade loops.

### 6.2.2 Performance of Cascade Control

Cascade control is primarily used to mitigate secondary process disturbances and it works best if the secondary dynamics are significantly faster than the primary dynamics. This last point is demonstrated by comparing the performance of a cascade control system and comparing its performance to conventional single-loop control. To keep the comparison simple, only step primary and secondary disturbances are considered. Marlin (1995) also considers stochastic and sinusoidal disturbances. To compare performance, the cascade system block diagram is as in Figure 6.3 with the following transfer functions:

$$G_{p1}(s) = \frac{1.0e^{-0.3s}}{0.7s + 1} \qquad G_{p2}(s) = \frac{1.0e^{-(0.3/\eta)s}}{(0.7/\eta)s + 1} \qquad G_{h1}(s) = G_{h2}(s) = 1.0$$

The single-loop block diagram is as in Figure 6.2 with the following transfer functions:

$$G_p(s) = G_{p1}(s)G_{p2}(s) = \frac{1.0e^{-(0.3+0.3/\eta)s}}{(0.7s + 1)[(0.7/\eta)s + 1)]} \qquad G_h(s) = 1.0$$

The disturbance transfer functions are

$$G_{d2}(s) = \frac{1.0}{(0.7/\eta)s + 1} \qquad G_{d1}(s) = \frac{1.0}{0.7s + 1}$$

The disturbances are unit steps. The relative dynamics between the secondary and the primary are defined by the variable $\eta$, which will be varied in the comparison. All controllers are PI controllers tuned according to the Fertik open-loop (disturbance) tuning method. The controllers are retuned for every value of $\eta$. The integrated absolute error, IAE, of the primary controlled variable is used as the performance measure. The relative performance is reported as the ratio of the cascade IAE to the single-loop IAE.

For a secondary unit step disturbance, the relative performance of the two controllers is shown as the solid line in Figure 6.9 as a function of the relative secondary/primary dynamics. The relative performance of the two controllers for a primary disturbance is shown as the dashed line. As expected, for secondary disturbances, the cascade control

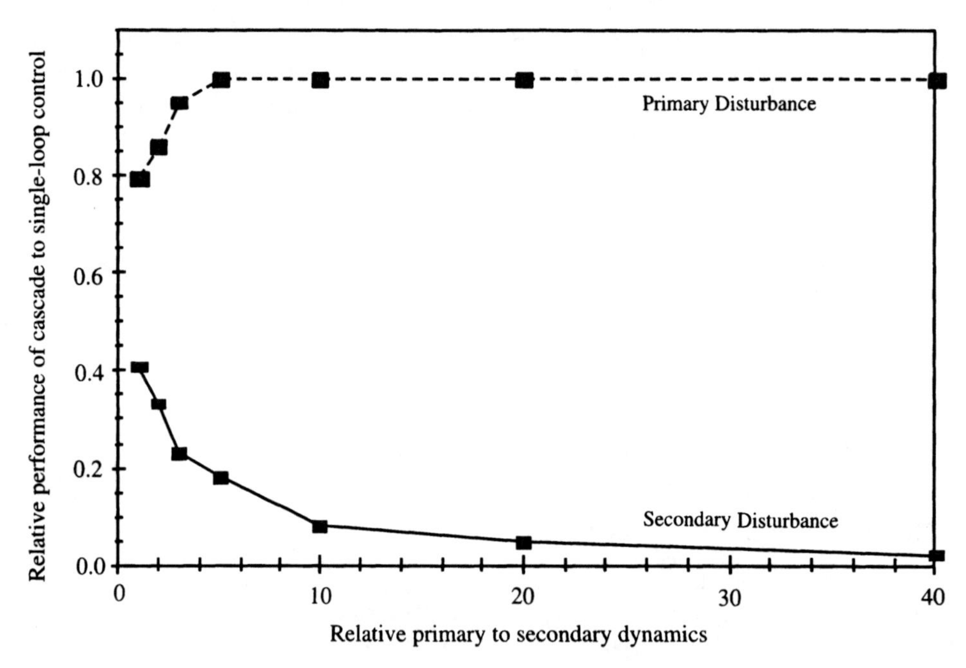


**FIGURE 6.9.** Relative performance ($IAE_{casc}/IAE_{sl}$) of cascade and single-loop control for a step disturbance in the secondary loop.

system performance improves as the secondary dynamics become faster. There is even some improvement when the secondary dynamics are the same as the primary dynamics. For primary disturbances, the cascade control system performance is about the same as for the single-loop system, illustrating that cascade control is most effective for secondary disturbances. Krishnaswamy et al. (1990) also compare cascade and feedback control with primary and secondary disturbances and reach comparable conclusions. Marlin (1995) extends this comparison to stochastic and sinusoidal disturbances with similar results.

### 6.2.3 Tuning Cascade Loops

The controllers that make up a cascade control strategy must be tuned individually. The tuning process should start with the controller that manipulates the final output:

1. Place all controllers involved in the cascade control strategy in the Manual state.
2. Tune the innermost controller, which manipulates the final output.
3. Leaving the tuned controller in the Automatic state, tune the controller that manipulates the setpoint of the previously tuned controller.

Though rare, it is possible to use more than two controllers in a cascade configuration. In this case, tuning starts with the innermost controller and proceeds outward, to the controller that manipulates its setpoint, and then to the controller that manipulates the setpoint of the second controller, and so on.

**Example 6.1.** For the heat exchanger control system of Figure 6.7, tune the two controllers. Use the Fertik open-loop tuning method and tune for a disturbance response. The model of the process, from Luyben (1990), is given by the equations

$$\rho_S V_S C_S \frac{dT_{SA}}{dt} = F_S \rho_S C_S (T_{So} - T_{S,\text{exit}}) - UA_H(T_{SA} - T_A)$$

$$\rho V C \frac{dT_A}{dt} = F \rho C (T_o - T_{\text{exit}}) + UA_H(T_{SA} - T_A)$$

$$T_{SA} = \tfrac{1}{2}(T_{So} + T_{S,\text{exit}})$$

$$T_A = \tfrac{1}{2}(T_o + T_{\text{exit}})$$

The steam flow is related to the steam valve position and steam pressure by

$$F_S = C_V v \sqrt{\frac{(P_0 - P_1)C_U}{\rho_S}}$$

where $v$ is the valve position, 0–100%, and $C_U$ converts psi to g/m min$^2$. The valve actuator dynamics are modeled by a lag with a 0.1-min time constant. The transport delay between the valve and the heat exchanger is modeled by a delay of 0.02 min. The model parameters are

$$\rho_s = \rho = 10^6 \text{ g/m}^3 \qquad C_S = C = 1 \text{ cal/g K}$$

$$V_s = 6 \text{ m}^3 \qquad V = 4 \text{ m}^3$$

$$U = 8 \times 10^5 \text{ cal/min m}^2 \text{ K} \qquad A_H = 20 \text{ m}^2$$

$$C_V = 4.58 \times 10^{-5} \qquad C_U = 2.482 \times 10^{10}$$

and the operating point is given by

$$F_S = 2 \text{ m}^3/\text{min} \qquad F = 5 \text{ m}^3/\text{min}$$

$$T_{So} = 370 \text{ K} \qquad T_{S,\text{exit}} = 274 \text{ K}$$

$$T_o = 293 \text{ K} \qquad T_{\text{exit}} = 329 \text{ K}$$

$$P_0 = 50 \text{ psi} \qquad P_1 = 2 \text{ psi}$$

First, the response of the steam flow to a change in the valve position must be obtained, shown in Figure 6.10. Also marked on the response plot is the information needed to obtain a first-order-plus-deadtime model:

$$G_p(s) = \frac{1.0}{0.10s + 1} e^{-0.02s} \tag{6.1}$$

Using the Fertik disturbance tuning method, the PI controller parameters for FIC2502 are

$$K_P = 3.5 \qquad T_I = 0.08 \text{ min}$$

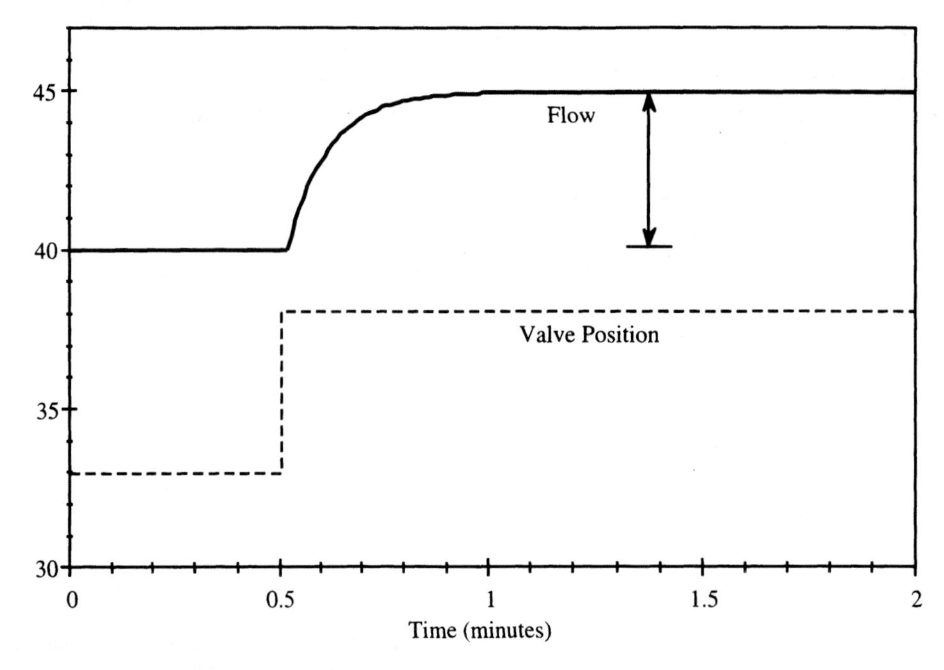


**FIGURE 6.10.** Response of steam flow to a +5% change in valve position.

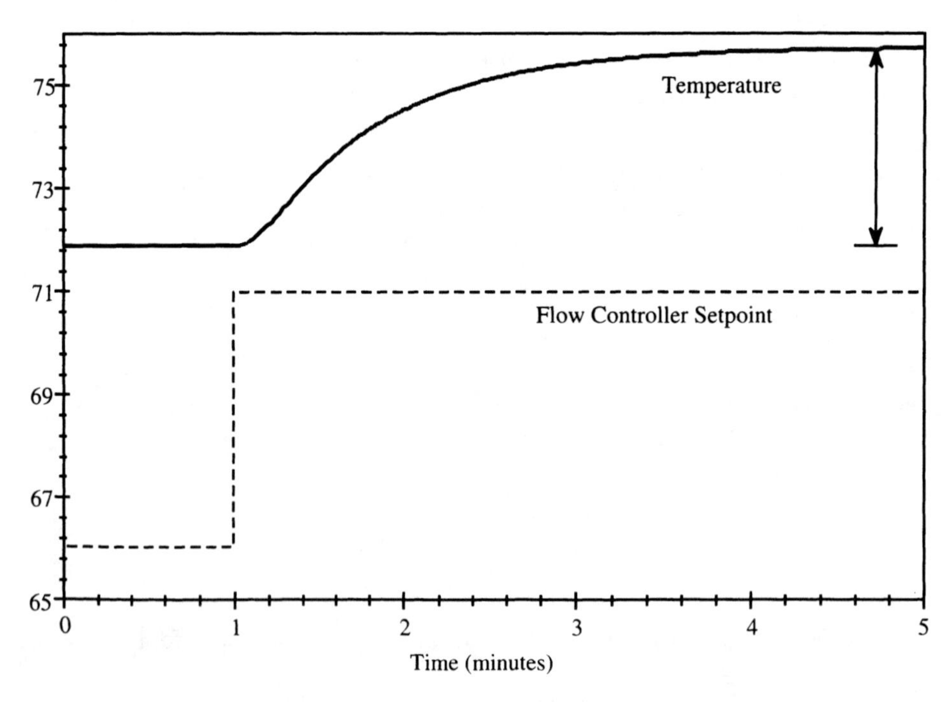


**FIGURE 6.11.** Response of hot-water temperature to a +5 change in the steam flow rate.

These parameters are entered into FIC2502, and then the response of the cold-water temperature to a change in the steam flow (the manipulated variable for TIC2501) is shown in Figure 6.11. The first-order-plus-deadtime approximation for the primary process is thus

$$G_p(s) = \frac{0.73}{0.65s + 1} e^{-0.19s} \tag{6.2}$$

Using the Fertik disturbance tuning method, the controller parameters for TIC2501, the primary loop, are

$$K_P = 3.0 \qquad T_I = 0.64 \text{ min}$$

The response of this control system is shown in Figure 6.8 for a $-10\%$ change in steam pressure and in Figure 6.12 for a $+10\%$ change in the steam pressure.

For noncascade feedback control, the entire process (from the steam valve to the hot-water temperature) can be approximated as

$$G_p(s) = \frac{0.73}{0.63s + 1} e^{-0.31s} \tag{6.3}$$

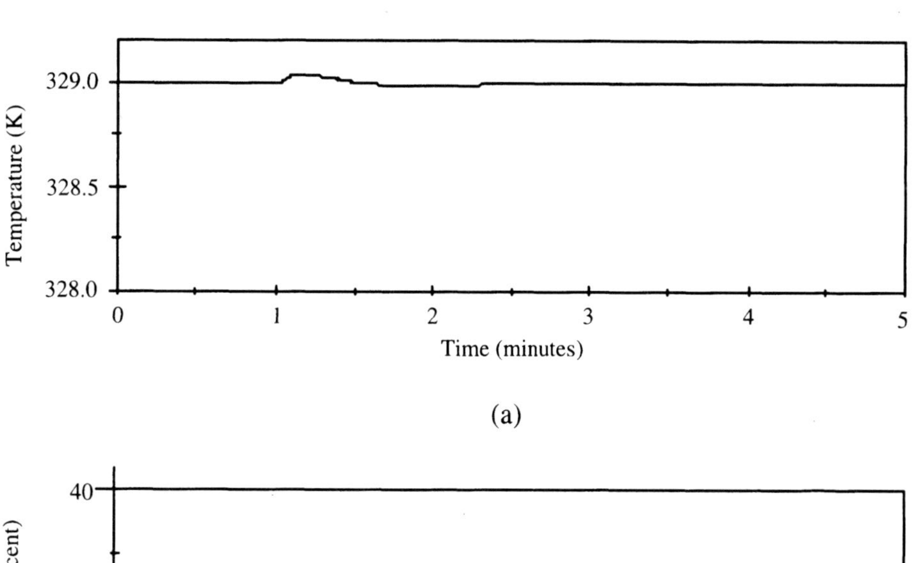


(a)

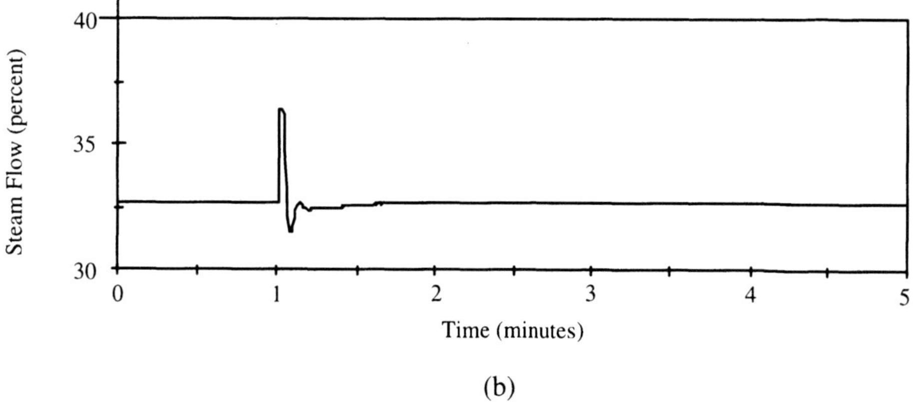


(b)

**FIGURE 6.12.** Response of cascade control system to a $+10\%$ change in steam pressure.

Using the Fertik disturbance tuning method, the controller parameters for TIC2501, where the controller output directly manipulates the steam valve, are

$$K_P = 1.71 \qquad T_I = 0.75 \text{ min}$$

The response of this control system is shown in Figure 6.6 for a -10% change in the steam pressure.

### 6.2.4 Other Examples of Cascade Control

*Valve Positioner* (Figure 6.13). Because of static and dynamic friction, valves often stick and do not achieve the position commanded by the controller output. When a small change in valve position is commanded, the actual valve position does not change until the commanded position is greater than a dead zone, as large as 3% (Buckley, 1970; Lloyd and Anderson, 1971). The result is poor control and cycling of the process variable. If the valve is being adjusted by a fast control loop and the process is not sensitive to the cycling, then no corrective action is necessary, though the performance is degraded to some extent. A *valve positioner* is a cascade controller that is physically included as part of the valve equipment. It receives the desired valve position from the primary controller and adjusts the air pressure to the valve in order to achieve the desired position. A typical valve positioner uses P-only control with a high gain that does not give perfect control but reduces the apparent dead zone to about one-tenth of its value without the valve positioner.

*Process Furnace* (Figure 6.14). Cascade control is used to regulate the temperature of a process stream that exits from a furnace. Because the dynamics between the valve position and the exit temperature are slow, direct adjustment of the valve by TIC1213 will lead to poor regulation to fuel pressure changes. The primary control loop adjusts the flow of the fuel in order to maintain the process stream exit temperature. The secondary controller adjusts the flow valve position and compensates for changes in the fuel pressure. Besides compensating for fuel pressure disturbances, the flow loop also compensates for the effects of a sticking valve.

*Distillation Column* (Figure 6.15). A common distillation loop adjusts the steam flow to the reboiler based on one of the tray temperatures. As for the heat exchanger example, the secondary flow controller, FIC1730, compensates for changes in the steam pressure before they are detected by the tray temperature.

*Batch Reactor* (Figure 6.16). A common example of cascade control is the temperature control of a jacketed batch reactor. The secondary controller regulates the cooling water

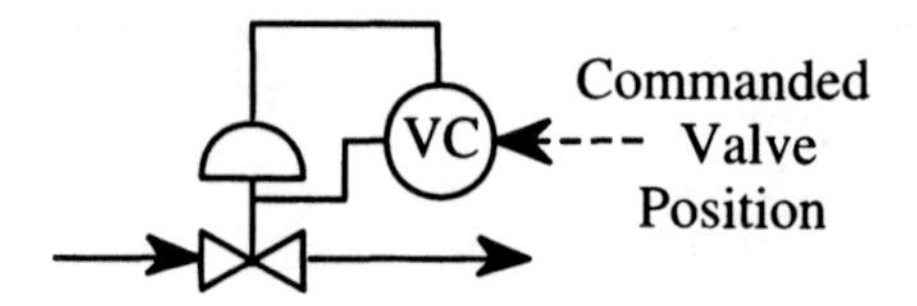


**FIGURE 6.13.** Valve positioner cascade control.

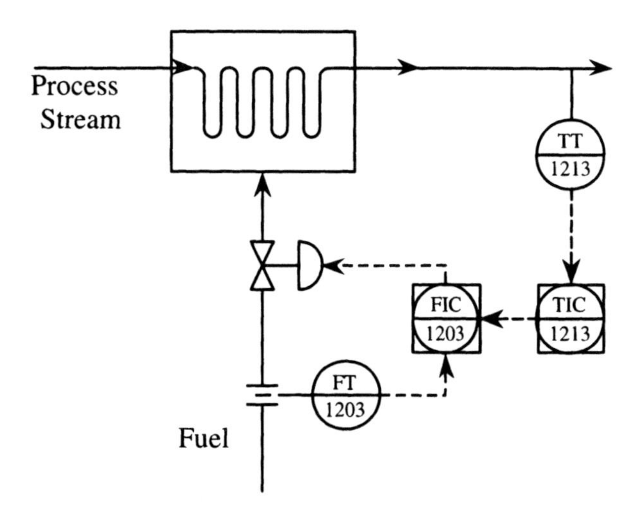


**FIGURE 6.14.** Process furnace cascade control.

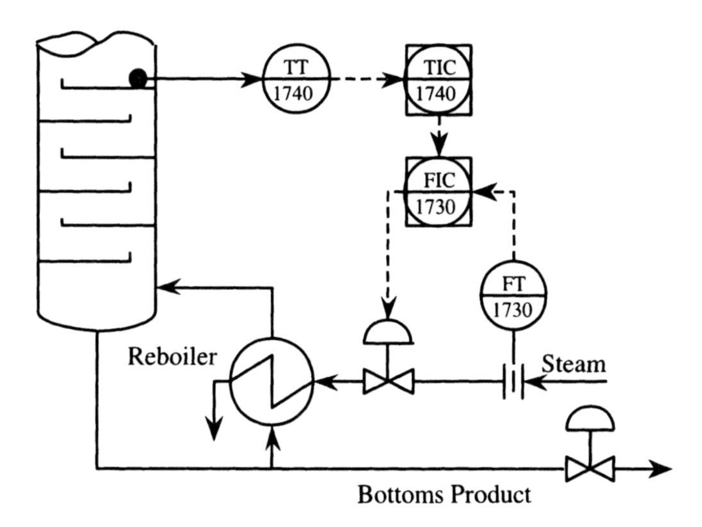


**FIGURE 6.15.** Distillation column reboiler cascade control.

flow and compensates for changes in the cooling water pressure while the primary controller regulates the reactor temperature.

### 6.2.5 Implementation and Operator Interface

Generally, cascade control is relatively easy to set up. Cascade control is common enough that the features necessary to connect the two single-loop controllers into a cascade configuration are always present. However, certain features that are ordinarily used may need to be disabled under cascade control. For example, if the output of a controller is always expressed as 0–100%, or as a raw analog output value (e.g., 0–4095), then this range must be known when configuring the setpoint scaling.

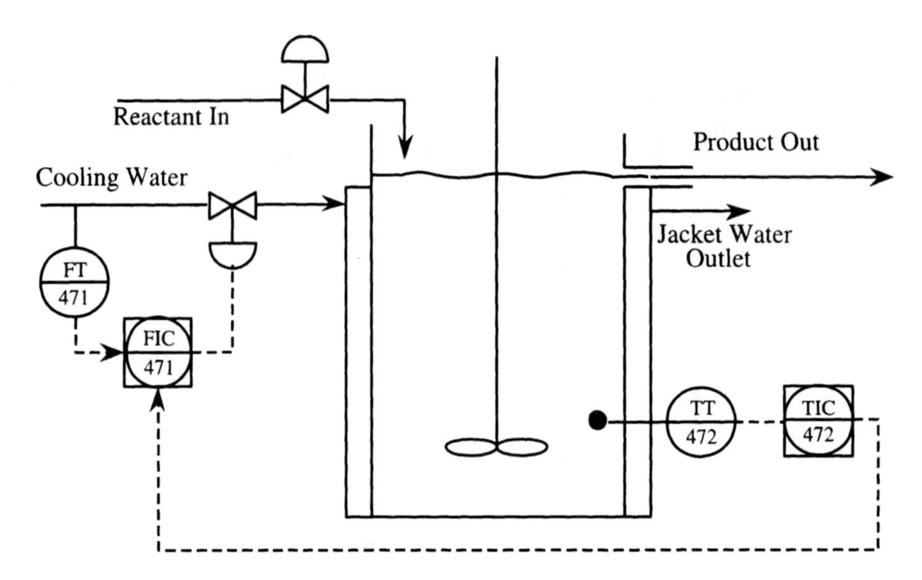


**FIGURE 6.16.** Exothermic batch reactor cascade control.

Integral windup is more of a problem in cascade loops, because the primary controller can easily calculate a secondary controller setpoint that is not achievable due to (1) configured secondary controller setpoint limits, (2) configured secondary manipulated variable limits, or (3) hard limits on the final control element (0–100%). In a typical situation, the secondary controller will force the manipulated variable to 0 or 100% and the primary PV will still be offset from the primary setpoint. Thus, the PID integrator term will continue to accumulate unless the controller has antireset windup.

Cascade control also requires more equipment. At a bare minimum, another sensor, and its wiring, to the controller, if it was not available earlier, is required. If stand-alone controllers are currently used, a more expensive controller may have to be purchased that handles cascade control. Increasingly, control systems are digital and the cost of cascade is generally the cost of additional sensor, wiring, and programming of the control system.

In a conventional control system, each controller has its own faceplate, which is displayed by selecting the loop by pressing a button, pressing a touch-screen graphic, clicking on it with a mouse, or keyboard entry. The faceplate is similar to that of a conventional PID controller (Figure 5.43). The only real difference is shown in the faceplate for the secondary controller. When the primary controller is in the Automatic state, the secondary controller displays a cascade status to inform the operator that the setpoint of the secondary controller cannot be changed by the operator. For example, the faceplate displays for the loops of the heat exchanger controller of Figure 6.7 could appear as shown in Figure 6.17. However, as described in Chapter 3, a better approach is to combine the two controllers into one controller and display it as one faceplate, shown in Figure 6.18. With this approach, the operator immediately knows the state of the controller combination and does not need to know the underlying controller states. The traditional three-bar display can change with the controller state. When controlling the steam flow, the PV, SP, and MV are flow, desired flow, and valve position, respectively. When controlling hot-water temperature, the PV, SP, and MV are temperature, desired temperature, and flow setpoint, respectively.

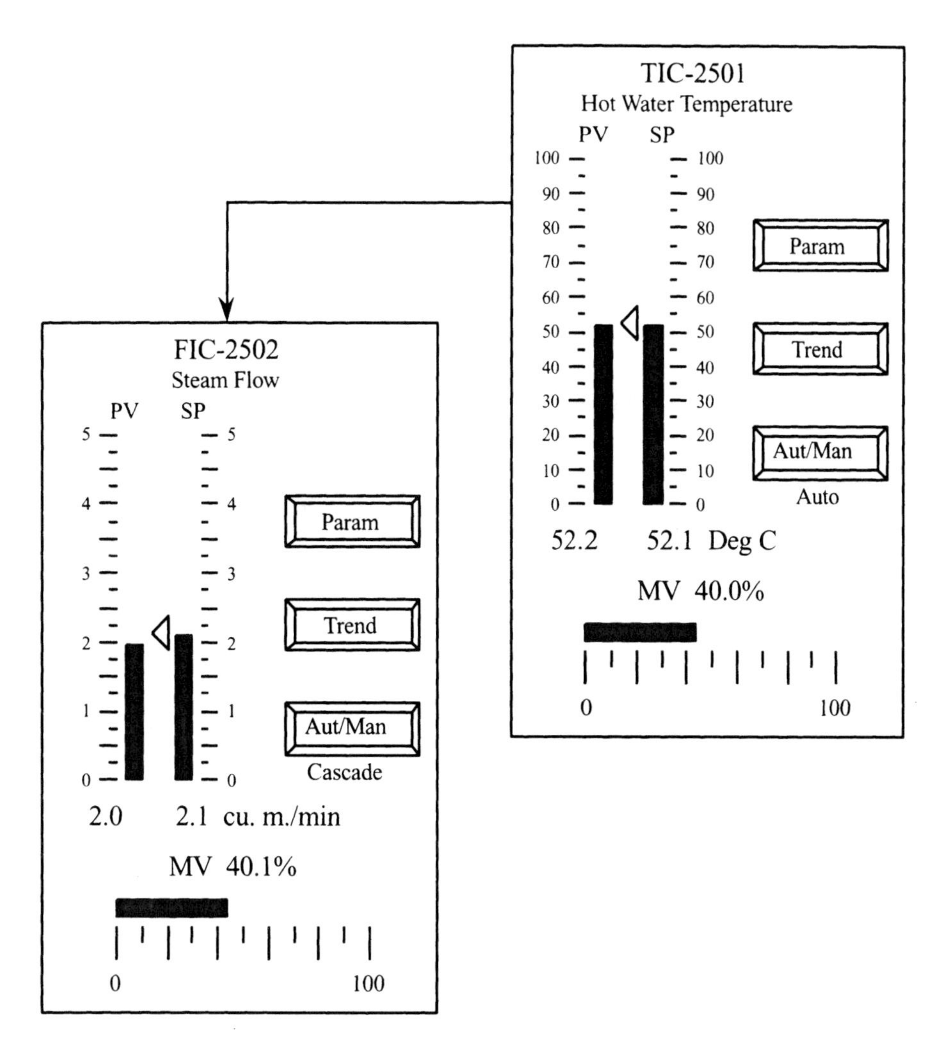


**FIGURE 6.17.** Conventional faceplate display for heat exchanger cascade control.

## 6.3 FEEDFORWARD CONTROL

Feedback control mitigates the effect of a disturbance by measuring its effect on the process variable and then adjusting the manipulated variable accordingly. In contrast, feedforward control measures the disturbance and changes the manipulated variable in order to counteract the disturbance before it affects the process variable. Feedforward control was introduced in Chapter 5. An example to motivate feedforward control is presented first, followed by design procedures.

### 6.3.1 Motivation for Feedforward Control

Consider the heat exchanger of Figure 6.4. The object of a control system is to manipulate the steam flow in order to maintain a desired hot-water temperature. If a feedback control system is constructed, the resulting system is subject to the following disturbances:

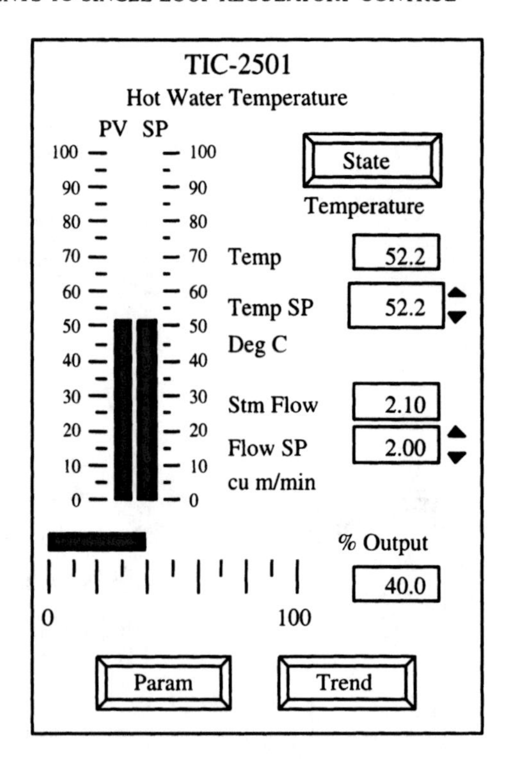


**FIGURE 6.18.** State-based heat exchanger faceplate display.

Cold-water temperature
Cold-water flow
Steam temperature
Steam pressure
Ambient temperature

A feedforward controller could be constructed for each of these disturbances. However, it may not be appropriate to construct a feedforward controller for each disturbance. For example, steam pressure disturbances are more appropriately handled with a cascade controller. The ambient temperature does not vary rapidly and so can be easily handled by the feedback controller. Of the remaining disturbances, the major one is the hot outlet water flow, since it can change rapidly depending on the demand. A block diagram of the combination feedforward and cascade control system is shown in Figure 6.19. Response to a +2-gpm change in hot-water flow with and without dynamic feedforward control is shown in Figure 6.20. Note that the feedforward control does not completely cancel the disturbance effect due to imperfect process knowledge.

### 6.3.2 Feedforward Control Design

A general block diagram of a feedforward-only control system is shown in Figure 6.21. The process variable can be expressed as

$$\mathrm{PV}(s) = \mathrm{D}(s)[G_d(s) + G_{\mathrm{ff}}(s)G_p(s)]$$

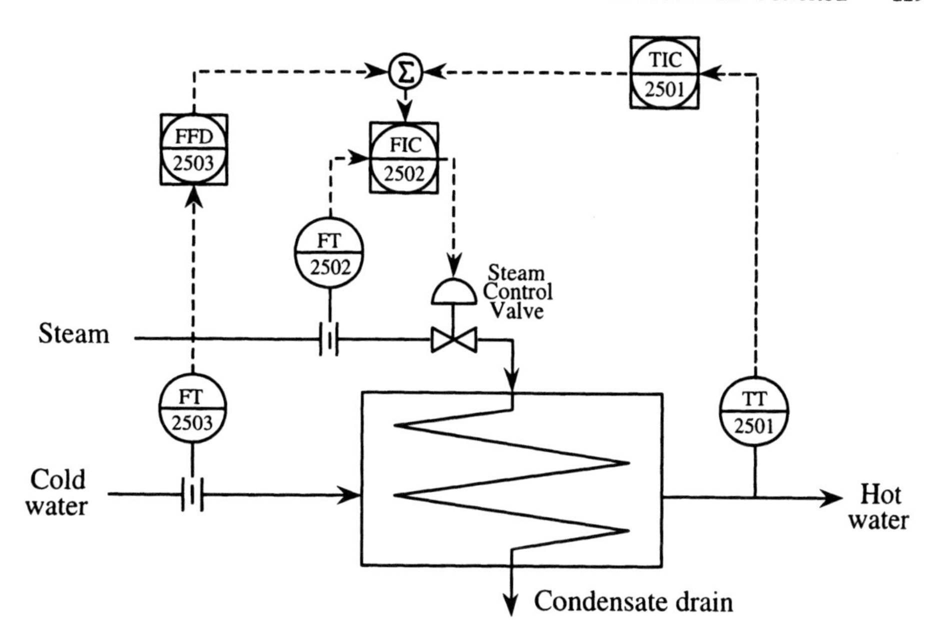


**FIGURE 6.19.** Heat exchanger with feedforward control.

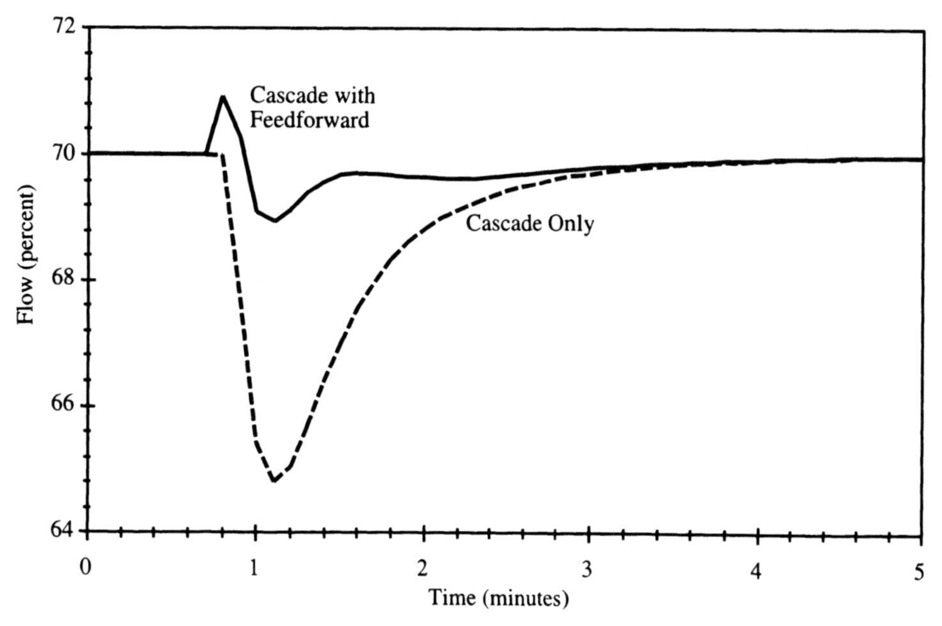


**FIGURE 6.20.** Response to a +2 change in cold-water flow without and with dynamic feedforward control.

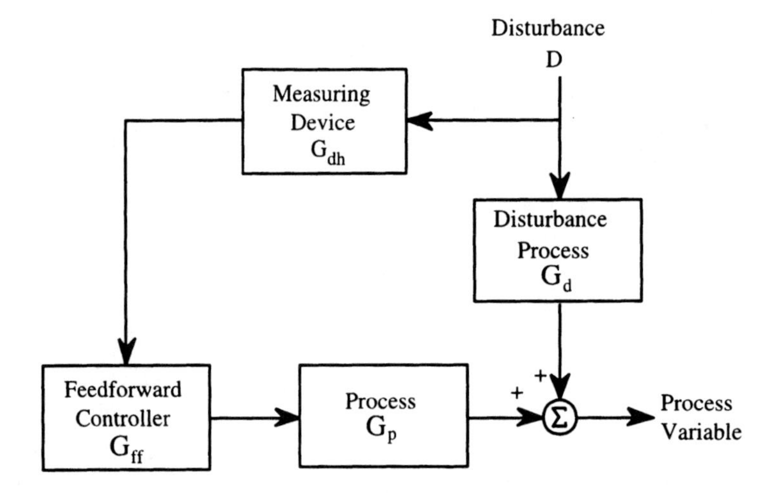


**FIGURE 6.21.** General block diagram of a feedforward control system.

In order for the disturbance to be completely rejected,

$$G_d(s) + G_{ff}(s)G_p(s) = 0$$

Therefore, the feedforward controller $G_{ff}(s)$ is given as

$$G_{ff}(s) = -\frac{G_d(s)}{G_p(s)} \tag{6.4}$$

Practically, there are two types of feedforward controllers. The $G_{ff}(s)$ of (6.4) is generally called a *dynamic feedforward controller.* A simpler feedforward controller is a steady-state, or *static, feedforward controller,*

$$G_{ff,ss}(s) = \lim_{s \to 0} G_{ff}(s) = \frac{\lim_{s \to 0} G_d(s)}{\lim_{s \to 0} G_p(s)} \tag{6.5}$$

The static feedforward controller has the advantage that it requires only limited information about the disturbance and the process. A common form of the feedforward controller is developed by assuming first-order-plus-deadtime models of the disturbance and the process as follows:

$$G_d(s) = \frac{K_d e^{-\theta_d s}}{\tau_d s + 1} \qquad G_p(s) = \frac{K_p e^{-\theta_p s}}{\tau_p s + 1}$$

Using (6.4), the dynamic feedforward controller transfer function is

$$G_{ff}(s) = -\frac{K_d\,(\tau_p s + 1)}{K_p\,(\tau_d s + 1)} e^{-(\theta_d - \theta_p)s} \tag{6.6}$$

The controller consists of three parts: (1) gain, (2) lead/lag, and (3) deadtime. Thus in a control system these three blocks are combined to implement a dynamic feedforward controller. Depending on the disturbance measurement, a bias may be subtracted from the feedforward measurement so the net effect of the feedforward controller is zero at the operating point. Also, the feedforward controller output is added to the normal operating point of the manipulated variable, which may be a feedback controller output. Thus, the general block diagram of a feedforward controller implementation is shown in Figure 6.22.

When designing a dynamic feedforward controller, the *realizability* of the resulting controller becomes an issue. In order for the feedforward transfer function (6.6) to be realizable, the deadtime must be nonnegative,

$$\theta_d - \theta_p \geq 0 \quad \text{or} \quad \theta_d \geq \theta_p$$

Physically, this inequality makes sense. In order to completely cancel the effect of the disturbance, the combination of the feedforward controller and the process must have a smaller deadtime than the deadtime from the disturbance measurement and the process variable. What does one do if the designed dynamic feedforward controller has a negative deadtime (positive exponent of $e$)? The best one can do is to use a feedforward controller with a zero deadtime. It will not completely cancel the disturbance effect but may be the best one can do. The only other alternative is to physically move the manipulated variable in order to decrease the apparent process deadtime. However, depending on the system, this may not be possible. A system with this problem will be considered as one of the examples.

Another issue concerns the practical implementation of (6.6). If $\tau_d = 0$, then one must implement a transfer function of the form

$$G_{\text{ff}}(s) = -\frac{K_d}{K_p}(\tau_p s + 1)e^{-(\theta_d - \theta_p)s}$$

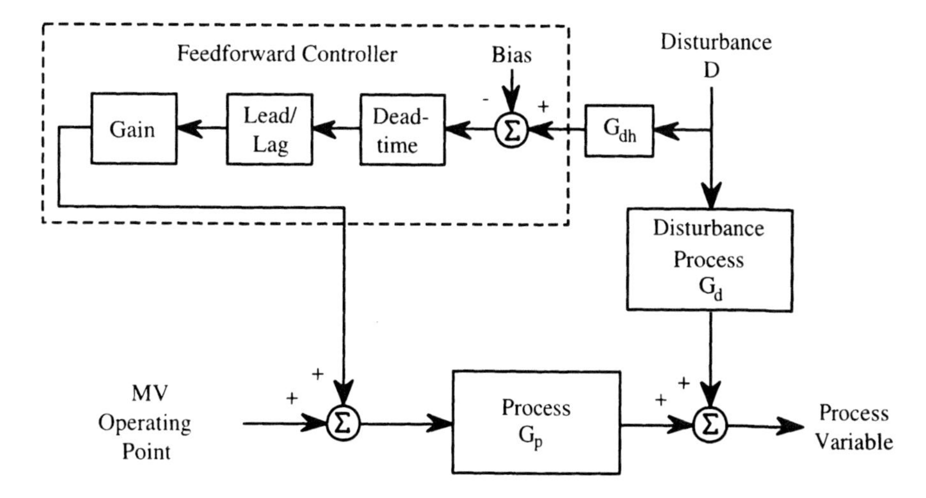


**FIGURE 6.22.** Feedforward controller implementation.

The term $\tau_p s + 1$ is difficult to implement, as either an analog circuit or a digital algorithm. If this situation arises, then the term is approximated as

$$\tau_p s + 1 \approx \frac{\tau_p s + 1}{\alpha \tau_p s + 1} \quad \text{where } \alpha \geq 10$$

In order to design a feedforward controller using (6.4) or (6.5), a model of the disturbance effect and a process model must be known. When the system contains a feedback controller, there is a simple way to design a static feedforward controller. Monitor the disturbance and the manipulated variable. The static feedforward gain is the positive of the ratio of the manipulated variable change to the disturbance change. For example, a typical disturbance and corresponding manipulated variable response is shown in Figure 6.23. The static feedforward gain is

$$G_{\text{ff,ss}}(s) = +\frac{\Delta\text{MV}}{\Delta\text{DV}} \tag{6.7}$$

**Example 6.2.** For the heat exchanger of Figure 6.4, design a dynamic feedforward controller that compensates for changes in the hot-water flow. Evaluate the performance of this feedforward controller for a +2- and −2-gpm change in the hot-water flow, assuming the same operating point as in Example 6.1.

***Solution.*** The system does not have a cascade controller. The feedforward controller manipulates the steam valve position. So, the first-order-plus-deadtime approximation of the process, from Example 6.1, is

$$G_p(s) = \frac{0.73}{0.63s + 1} e^{-0.31s} \tag{6.8}$$

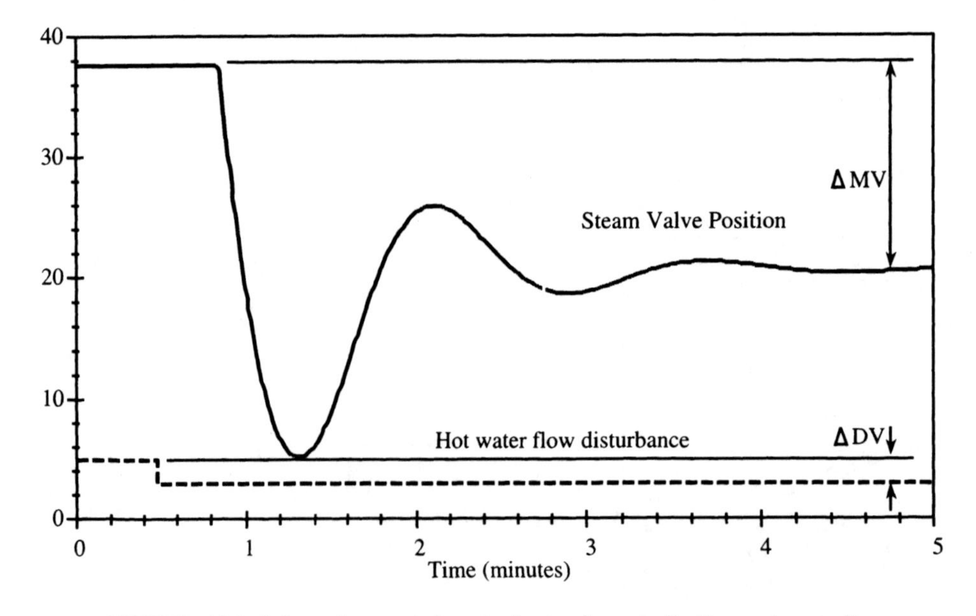


**FIGURE 6.23.** Information needed to obtain simple static feedforward controller.

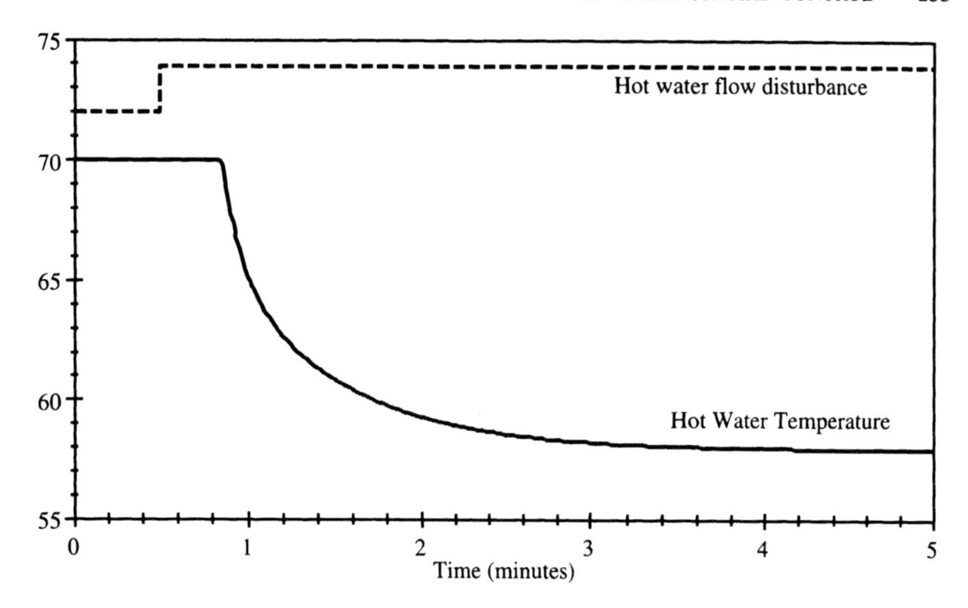


**FIGURE 6.24.** Hot-water temperature response to a +2-gpm change in hot-water flow.

The disturbance model must now be obtained. For a +2-gpm change in the hot-water flow at 0.5 min (and no feedback control on the temperature), the resulting hot-water temperature response is as shown in Figure 6.24. Using this information, the first-order-plus-deadtime approximation of the disturbance model is,

$$G_d(s) = \frac{-6.025}{0.38s + 1} e^{-0.35s} \tag{6.9}$$

Using (6.4), the feedforward controller is designed as

$$G_{ff}(s) = -\frac{G_d(s)}{G_p(s)} = 8.25 \frac{0.63s + 1}{0.38s + 1} e^{-0.04s}$$

The response of a feedforward-only control for a +2- and −2-gpm change in the hot-water flow is shown in Figure 6.25. Note that the feedforward control does not completely compensate for the disturbance. The transient deviations in the early part of the temperature response are due to modeling errors that arise because the heat exchanger is modeled by a first-order transfer function when the actual process is second order. The steady-state offset is due to the valve nonlinearity. The gain of the system is actually a little smaller. Thus a larger feedforward gain will be needed to completely cancel this disturbance. However, a different disturbance will require a different gain. Because of this problem, feedforward control is most often used with a feedback controller that will compensate for feedforward modeling errors.

**Example 6.3.** For the heat exchanger of Figure 6.4, design a static feedforward controller that compensates for changes in the hot-water flow. Rather than use the results from Example 6.2, obtain the static feedforward gain with a feedback controller and (6.7).

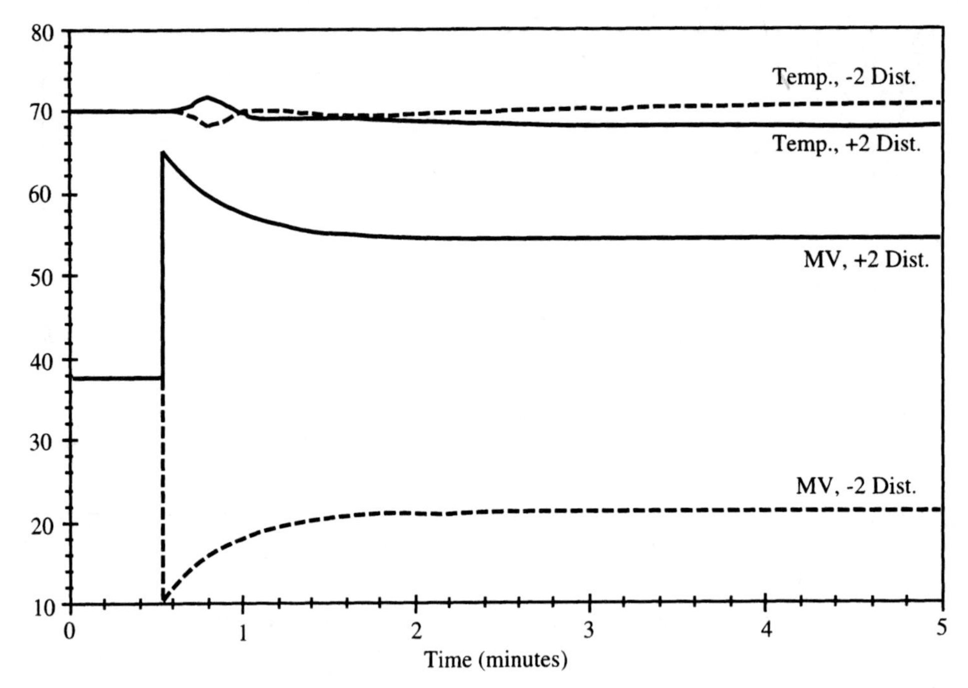


**FIGURE 6.25.** Response of dynamic feedforward-only control of heat exchanger to +2- and −2-gpm change in hot-water flow.

Evaluate the performance of this feedforward controller for a +2- and −2-gpm change in the hot-water flow, assuming the same operating point as in Example 6.1.

***Solution.*** Using the noncascade feedback controller of Example 6.1, a +2-gpm hot-water flow and the response of the controller manipulated output are shown in Figure 6.26. The static feedforward controller is thus

$$G_{\mathrm{ff,ss}}(s) = \frac{\Delta \mathrm{MV}}{\Delta \mathrm{DV}} = \frac{58.27 - 37.70}{7 - 5} = 10.29 \tag{6.10}$$

The response of this static feedforward-only control for a +2- and −2-gpm change in the hot-water flow is shown in Figure 6.27.

Changing the gain of $G_{ff}(s)$ to 10.215 will completely cancel the +2-gpm disturbance.

General design criteria for feedforward control (Marlin, 1995) are summarized in Table 6.3. The first two items lead one to consider feedforward control. In order for feedforward control to even be considered, feedback control performance must be unsatisfactory and a feedforward measurement must be available. In addition, the feedforward measurement must satisfy three criteria. First, the feedforward variable must measure an important disturbance. That is, the feedforward variable must have a significant effect on the process variable and occur frequently. Second, the manipulated variable *must not* influence the feedforward variable, since this is not a feedback controller. If this rule is not met, then the variable should be used for cascade control. The steam pressure disturbance in the heat exchanger example violates this rule; hence cascade control should be used to compensate

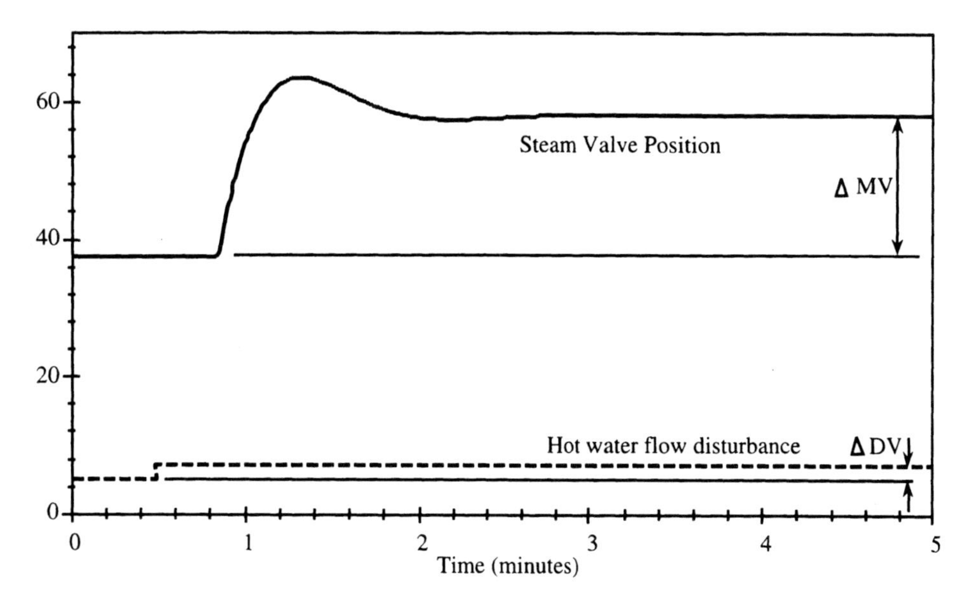


**FIGURE 6.26.** Information from feedback control to obtain static feedforward controller gain.

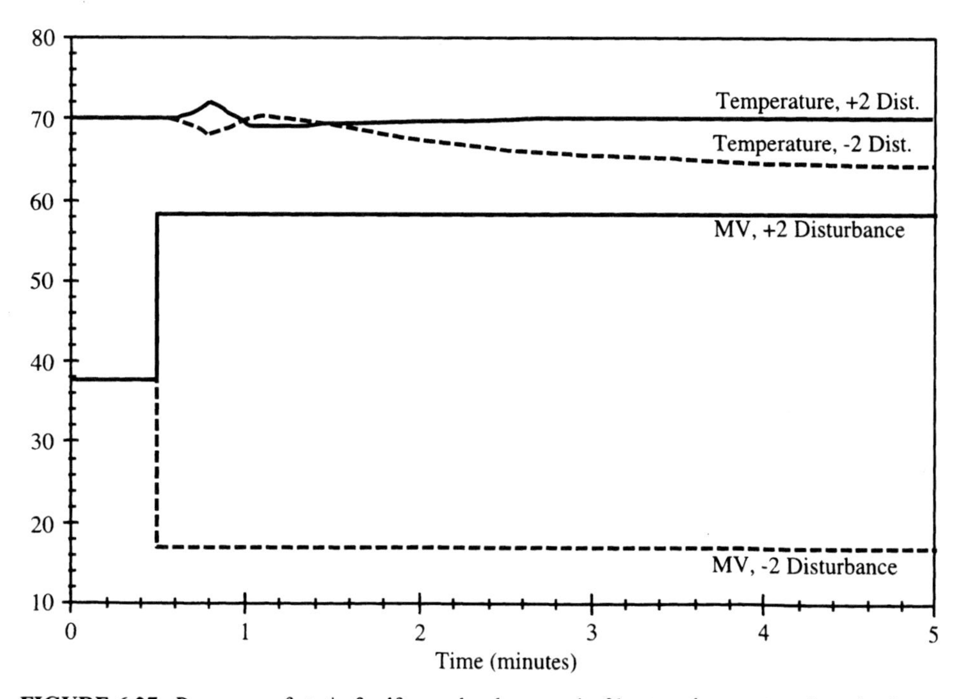


**FIGURE 6.27.** Response of static feedforward-only control of heat exchanger to +2- and −2-gpm change in hot-water flow.

**TABLE 6.3 Feedforward Control Design Criteria**

| **Feedforward control should be considered when:** |
| --- |
| 1. Feedback control does not provide satisfactory control performance. |
| 2. A measured disturbance (feedforward) variable is available. |
| **A feedforward variable must satisfy the following criteria:** |
| 3. The variable must indicate the occurrence of an important disturbance. |
| 4. There must *not* be a causal relationship between the manipulated and feedforward variables. |
| 5. The feedforward variable dynamics must not be significantly faster than the manipulated variable dynamics (when feedback control is also present). |

for it. The final requirement is needed when feedforward control is used in conjunction with feedback control. The intent of feedforward control is to compensate for the disturbance before the feedback controller can measure the disturbance effect. If the disturbance dynamics are faster than the dynamics of the variable that the feedforward controller manipulates, then it cannot completely cancel the disturbance effect and the feedback controller also generates a correction. These two corrections tend to cause a much larger overshoot in the process variable.

### 6.3.3 Feedforward Combined with Feedback

The more important relative advantages and disadvantages of feedforward and feedback control are summarized in Table 6.4. The feedforward controller does not introduce instability into the closed-loop system since there is no feedback loop from the process variable back to the disturbance. The feedforward controller offset and its sensitivity to modeling error was demonstrated in Example 6.2. As can be seen from Table 6.4, the disadvantages of feedforward control are countered by feedback control. Therefore, by combining feedback with feedforward, a control system is obtained that combines the strengths of each with few drawbacks. When feedforward control is combined with

**TABLE 6.4 Advantages and Disadvantages of Feedforward and Feedback Control**

| Feedforward | Feedback |
| --- | --- |
| *Advantages* | |
| 1. Compensates for disturbance before effect is seen at controlled output | 1. Gives zero steady-state offset (if controller has integral action) |
| 2. Does not introduce instability into the closed-loop response | 2. Insensitive to modeling errors |
| | 3. Does not require identification and measurement of any disturbance |
| *Disadvantages* | |
| 1. Offset generally nonzero | 1. Compensates for disturbance after effect is seen at controlled output. |
| 2. Sensitive to modeling errors | 2. Introduces instability when tuned improperly |
| 3. Requires an extra sensor and disturbance model | |

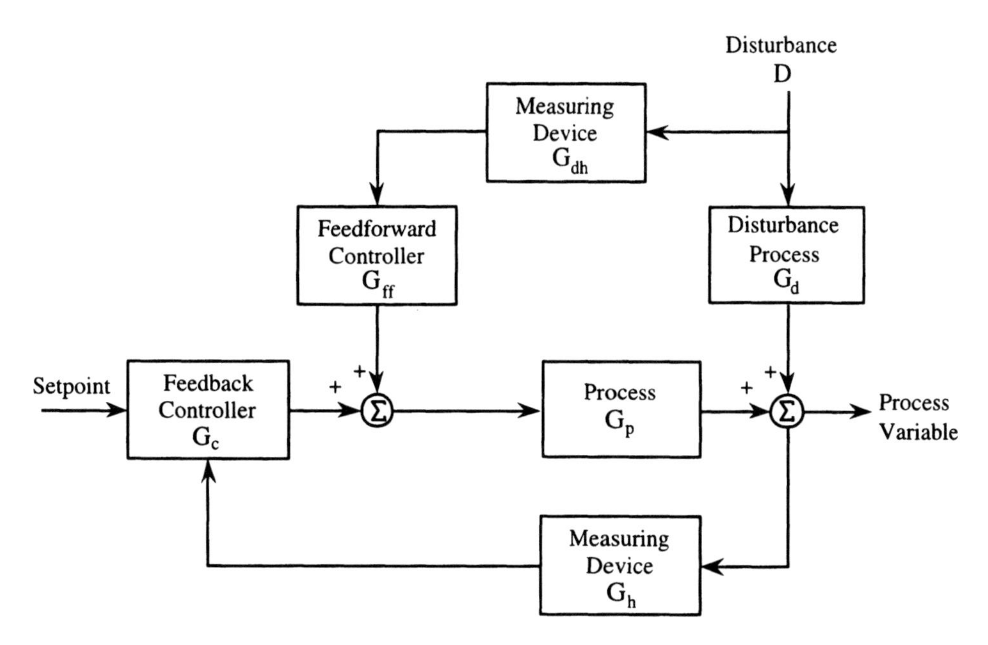


**FIGURE 6.28.** Block diagram of combined feedback and feedforward control.

feedback control, the block diagram of this system appears as in Figure 6.28. The characteristics of feedforward–feedback control are thus:

1. If the disturbance transfer function and the process transfer function are known exactly and the deadtime of the process is less than the deadtime of the disturbance transfer function, the feedforward controller generally completely compensates for the disturbance.
2. If the disturbance transfer function and the process transfer function are known only approximately and the deadtime of the process is less than the deadtime of the disturbance transfer function, the feedforward controller does not completely compensate for the disturbance. The advantages of feedback–feedforward control are illustrated by the following examples.

**Example 6.4.** For the heat exchanger of Figure 6.4, design a feedback and feedforward controller that compensates for changes in the hot-water flow. Evaluate the performance of this feedforward controller for a +2- and −2-gpm change in the hot-water flow, assuming the same operating point as in Example 6.1.

***Solution.*** There are two solutions to this problem, since the feedback controller can be a cascade or noncascade controller. The noncascade control system with feedforward control is shown in Figure 6.29. The parameters for TIC2501 are the same as for the noncascade controller of Example 6.1. The feedforward controller is the one designed in Example 6.2. The response of this combination feedback–feedforward control system is shown in Figure 6.30. Clearly, the performance of this system is better than either the feedback or the feedforward system alone.

The cascade control system with feedforward control is shown in Figure 6.19. The parameters for TIC2501 and FIC2502 are the same as for Example 6.1. Since the process

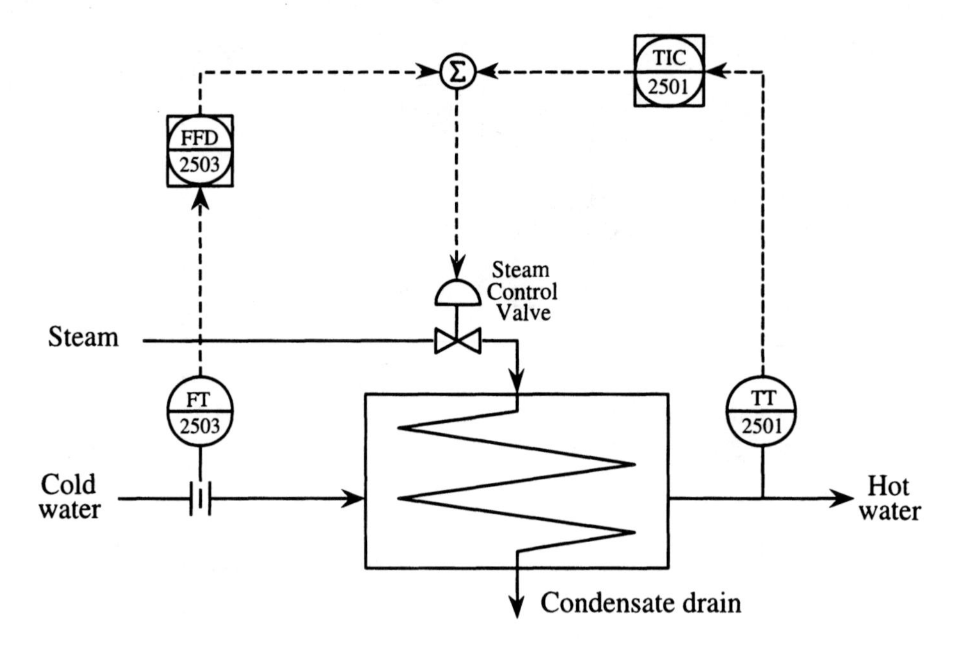


**FIGURE 6.29.** Feedforward with feedback control for heat exchanger.

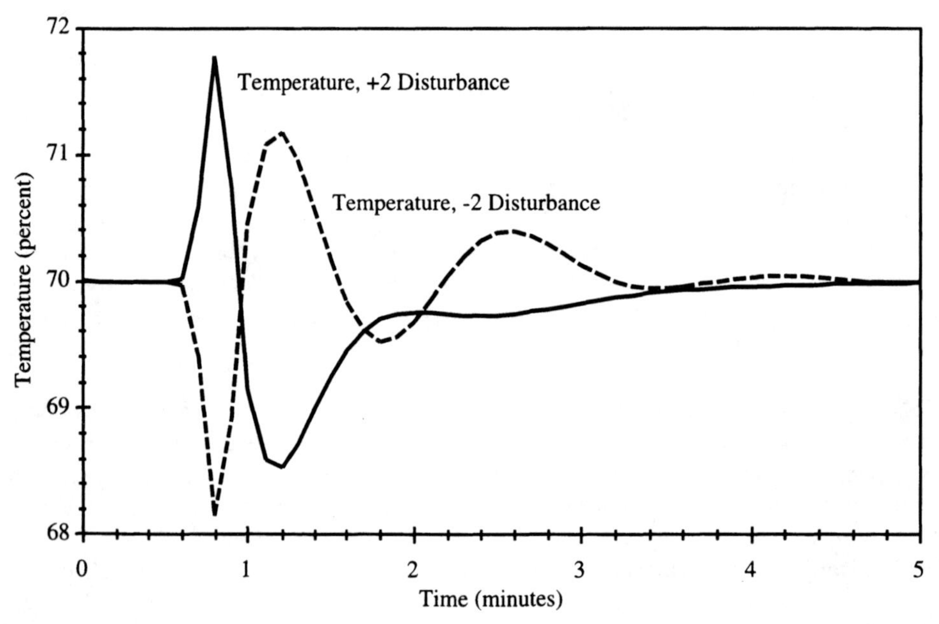


**FIGURE 6.30.** Response of dynamic feedforward and feedback control of heat exchanger to +2- and −2-gpm change in hot-water flow.

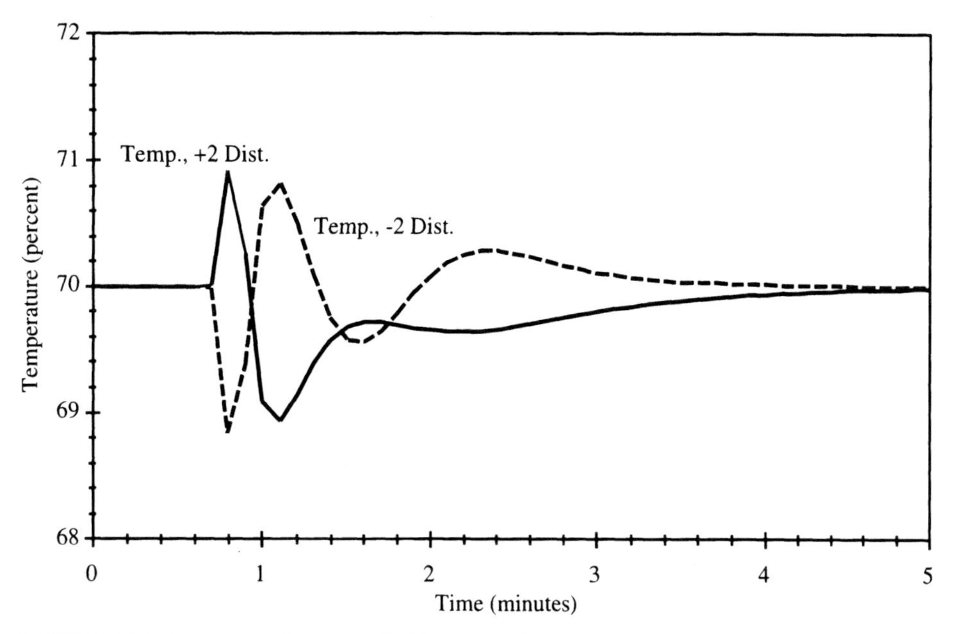


**FIGURE 6.31.** Response of dynamic feedforward and cascade control of heat exchanger to +2- and −2-gpm change in hot-water flow.

now includes the flow loop, the feedforward controller must be redesigned. Using (6.2) and (6.9), the dynamic feedforward controller is

$$G_{\mathrm{ff}}(s) = -\frac{G_d(s)}{G_p(s)} = 8.25\frac{0.65s + 1}{0.38s + 1}e^{-0.16s}$$

The response of this combination cascade–feedforward control system is shown in Figure 6.31. When the static feedforward controller of Example 6.3 is combined with the cascade control system, the responses to hot-water flow disturbances are shown in Figure 6.32.

The performance of these feedforward–feedback control systems are summarized in Table 6.5. There is not really much sacrifice to using a static feedforward controller versus a dynamic feedforward controller. Feedback control does eliminate the offset in the disturbance responses from Example 6.2. However, it does not completely eliminate all of the problems due to disturbance and process modeling errors.

### 6.3.4 Other Feedforward Examples

*Boiler Level* (Figure 6.33). The objective is to keep the boiler level constant. The principal disturbances are the feedwater pressure and the steam flow from the boiler, dictated by varying demand in the plant. The former is handled by cascade control and the latter is handled with a feedforward controller.

*Continuous-Flow Stirred-Tank Reactor* (Figure 6.34). This process has the objective to maintain the product composition. The inlet flow and the inlet temperature are the main

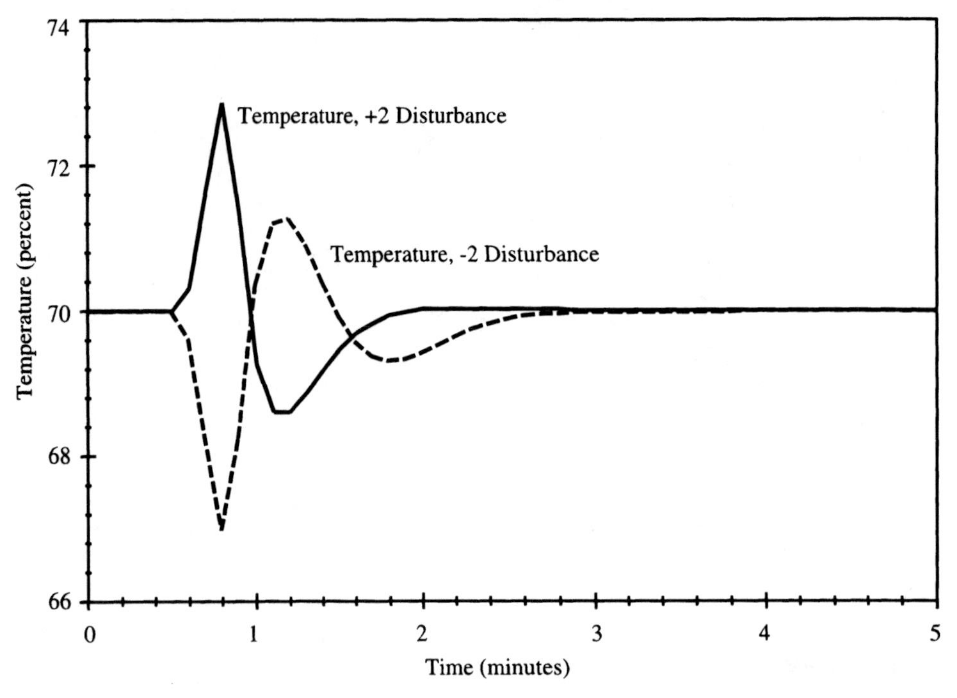


**FIGURE 6.32.** Response of static feedforward and cascade control of heat exchanger to +2- and −2-gpm change in hot-water flow.

disturbances. Both of these measurements are used in the feedforward control of the composition. Note that cascade control is used for the coolant flow, and so the feedforward controllers modify the setpoint to the flow controller.

*Distillation Column* (Figure 6.35). The control objective is to control the overhead or bottom composition. Since distillation columns often have slow dynamics, they are good candidates for feedforward control. The system shown in Figure 6.35 shows a feedforward control system constructed to compensate for feed rate changes. The same measurement is used for two feedforward controllers. A similar system could be constructed for changes in feed composition.

**TABLE 6.5 Summary Performance of Feedback–Feedforward Control of Example 6.4**

| | IAE, +2 gpm Disturbance | IAE, −2 gpm Disturbance |
|---|---|---|
| Dynamic feedforward, noncascade feedback control | 1.53 | 1.29 |
| Dynamic feedforward, cascade feedback control | 1.15 | 0.85 |
| Static feedforward, cascade feedback control | 1.30 | 1.58 |

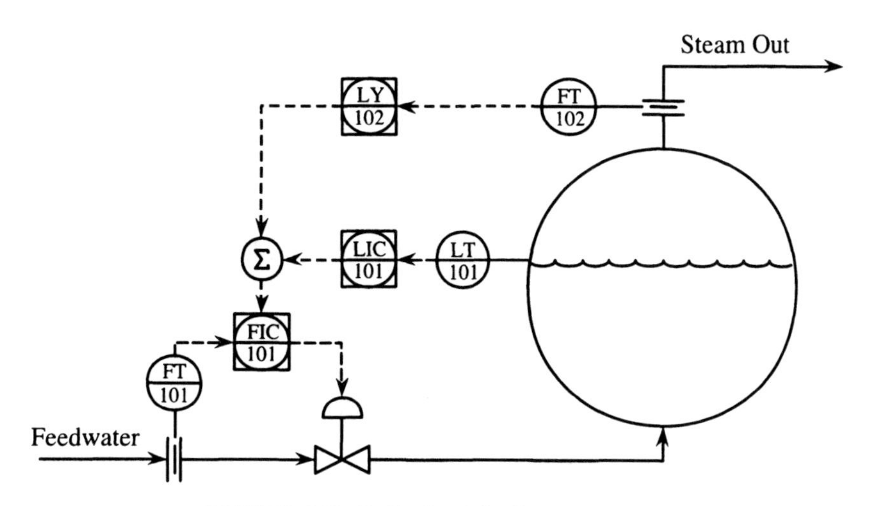


**FIGURE 6.33.** Boiler level feedforward control.

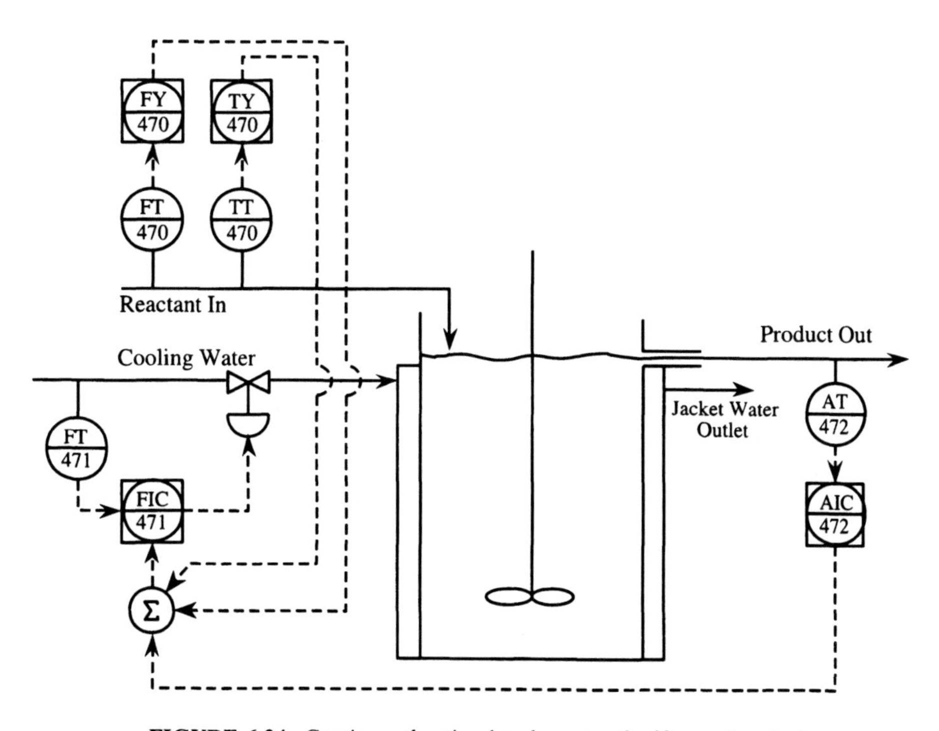


**FIGURE 6.34.** Continuously stirred tank reactor feedforward control.

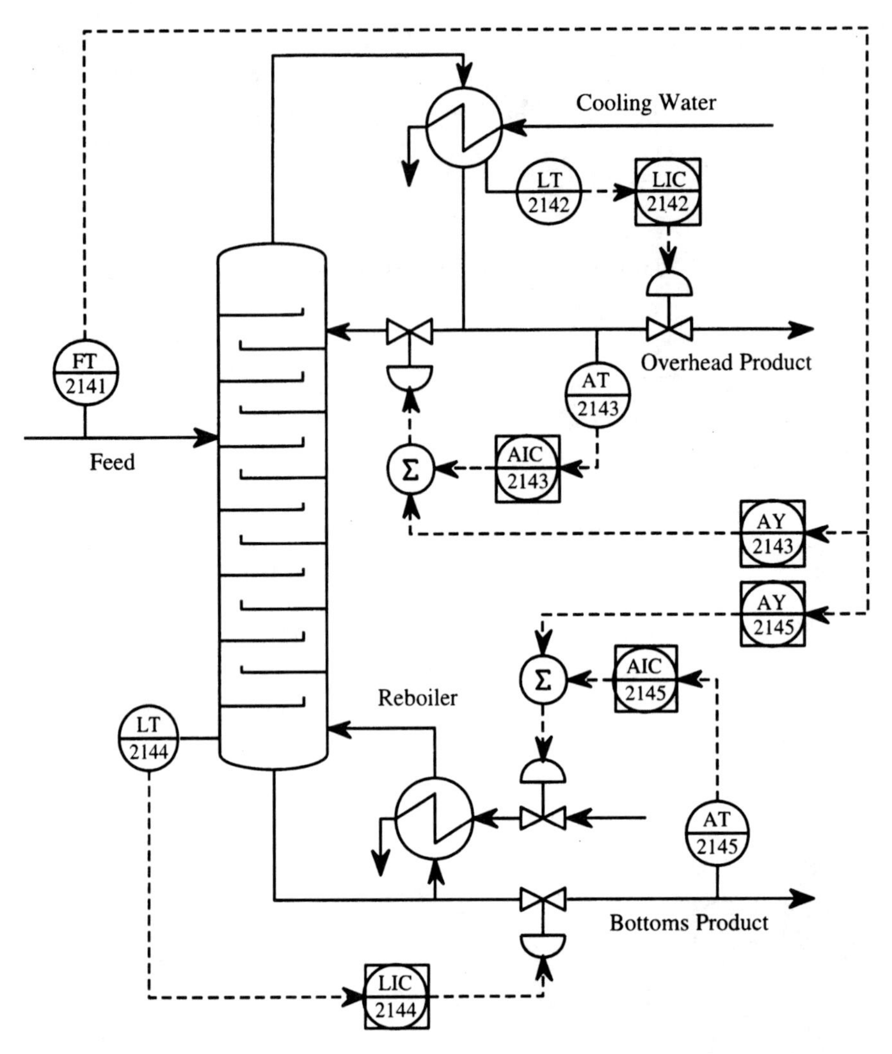


**FIGURE 6.35.** Distillation column feedforward control.

## 6.4 RATIO CONTROL

Ratio control is used to control the ratio of the flow rates of two or more streams. It is commonly used in such situations as adding two ingredients to a batch reactor where a constant ratio between the two ingredients must be maintained as they are added. Another common situation is the blending of additives to a base stock chemical. Ratio control may also be considered as a special form of feedforward control.

In a ratio control scheme, all stream flow rates are measured and all but one stream is controlled. The uncontrolled stream is usually called the *wild stream*. The flows of the other streams are set based on a ratio of the wild stream. There are two configurations of a ratio control system. In Figure 6.36*a*, both flow rates are measured and their ratio is calculated. This ratio is the measurement to a PID controller whose setpoint is the desired ratio. The controller then adjusts a valve to control the flow of the manipulated stream. The alternative scheme, in Figure 6.36*b*, multiplies the wild stream flow by the desired ratio to generate the setpoint to a PID controller for the manipulated stream. Both of these schemes are feedforward schemes, because the downstream ratio is not measured and used in a feedback controller. Thus, in a situation where two or more streams are blended, both of these schemes do not guarantee zero offset between the actual and desired composition of the blended stream. In order to accurately control the blended stream composition, a ratio control system similar to Figure 6.37 must be used. This system is a modification of the ratio control system of Figure 6.36*a*. The output of the blended composition controller, AIC1503, is the ratio setpoint to the feedback flow controller.

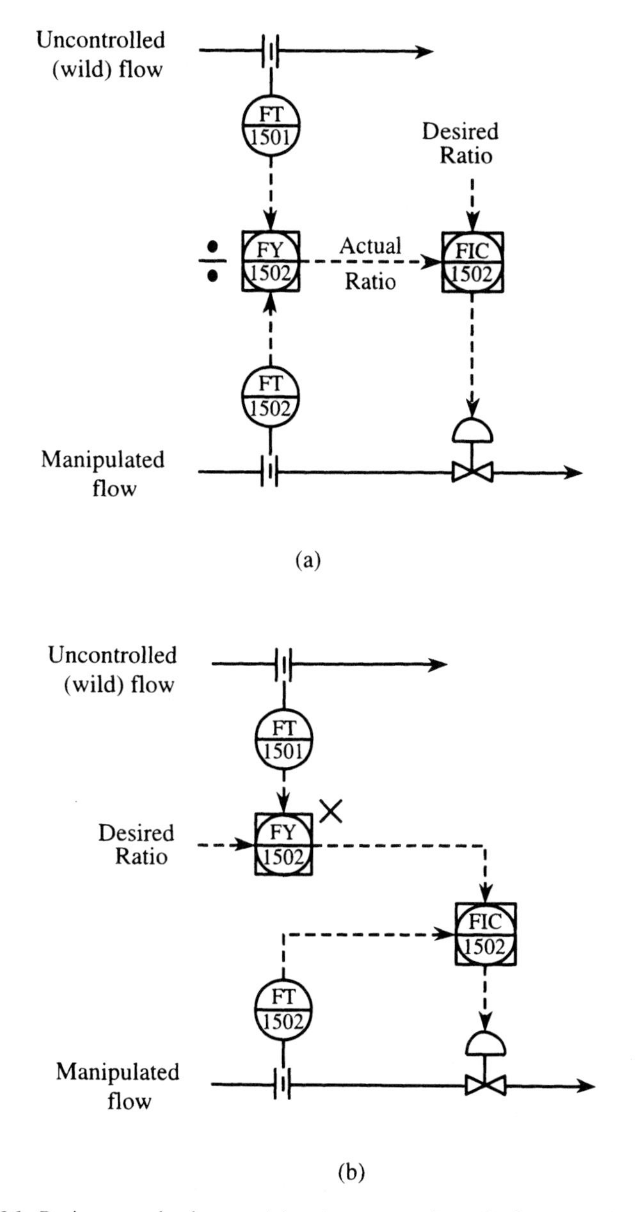


**FIGURE 6.36.** Ratio control schemes: (*a*) ratio computation; (*b*) flow setpoint computation.

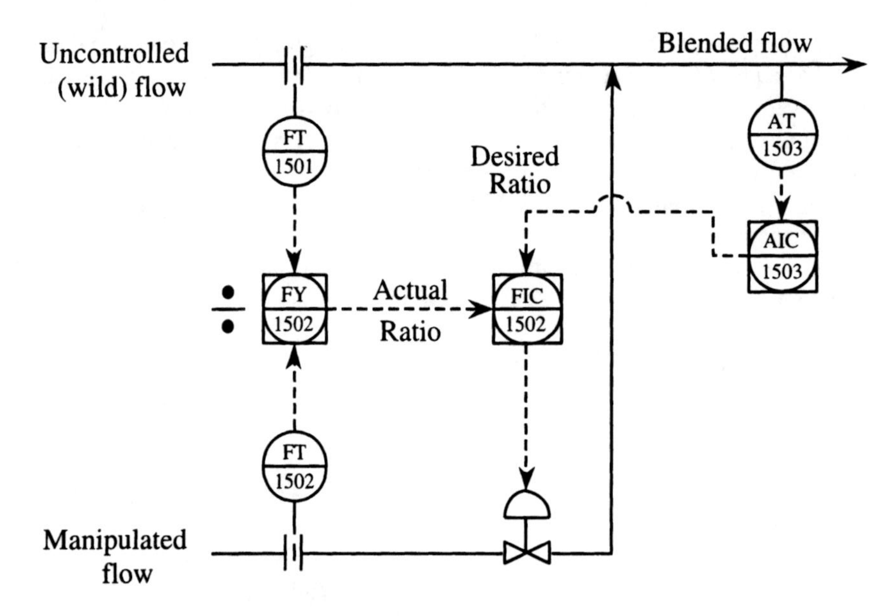


**FIGURE 6.37.** Ratio control for zero offset in the blended composition.

Ratio control is used in the following process control situations:

1. Control the ratio of two reactants into a continuous or batch reactor at a desired value.
2. Hold the ratio of two or more blended streams constant in order to maintain the blend composition at the desired value.
3. Hold the ratio of a recycle stream to the forward stream constant.
4. Maintain the fuel–air ratio of a burner for most efficient combustion.
5. Keep a constant ratio between the feed flow rate and steam flow rate for a distillation column reboiler.
6. Maintain a constant reflux ratio in a distillation column.

## 6.5 SPLIT-RANGE CONTROL

The cascade, feedforward, and ratio control schemes considered in the previous sections had more than one measurement and one manipulated variable. In contrast, the split-range control configuration has *one measurement and more than one manipulated variable.* Since the PID controller has only one output, it must be split into several parts, each affecting one manipulated variable. The control of the single controlled variable is accomplished by coordinating the actions of several manipulated variables. Split-range control systems are used in certain situations in order to improve control quality or to optimize the process, as demonstrated by the following examples.

Consider the batch reactor system of Figure 6.38. The vessel is charged with reactant and then heated with steam to bring the reaction mass up to a desired temperature. Cooling

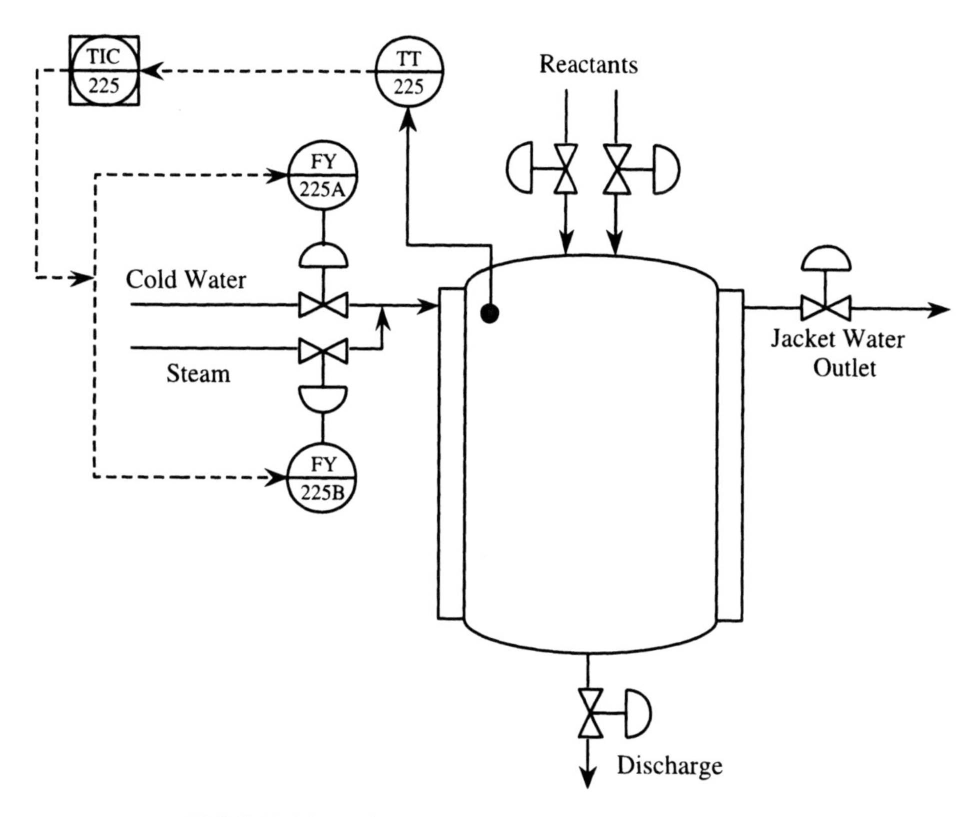


**FIGURE 6.38.** Split-range control for batch reactor temperature.

water must then be added to the jacket to remove the exothermic heat of reaction and control the reactor temperature to a certain profile. There is one measurement, the reactor temperature, TT225, and two manipulated variables, FY225A and FY225B. The one controller output must be split between the two valves, which can be accomplished in one of two ways. The valves can be configured as *split-range valves*. Normally, the 4–20 mA output of the controller is converted to a 3–15 psig signal that drives the valve. When the valves are set up as split-ranged valves, they are set up to respond to either the upper half of the range of this pressure signal or the lower half of the range. The valve position for the two valves corresponding to the controller output are shown in Figure 6.39. In operation, when the controller is switched to Automatic in order to heat up the reactor, the controller output is 100%, meaning the steam valve is fully open. As the temperature approaches the desired initial value, the steam valve is incrementally closed, and it is fully closed when the temperature is at the desired initial value. At this point, the reaction is exothermic and the cooling valve must be opened in order to maintain the reactor temperature. The cooling valve position is manipulated in order to follow a desired temperature profile during the remainder of the batch process. The controller bias parameters must be set up so that the controller output is 50% when the error between the actual and desired temperature is zero. The cold-water valve, FY225A, is set up to be reverse acting (air-to-close) on the low half of the actuating signal range. The steam valve, FY225B, is set up to be direct acting (air-to-open) on the upper half of the actuating signal range. For safety, the cold-water valve is

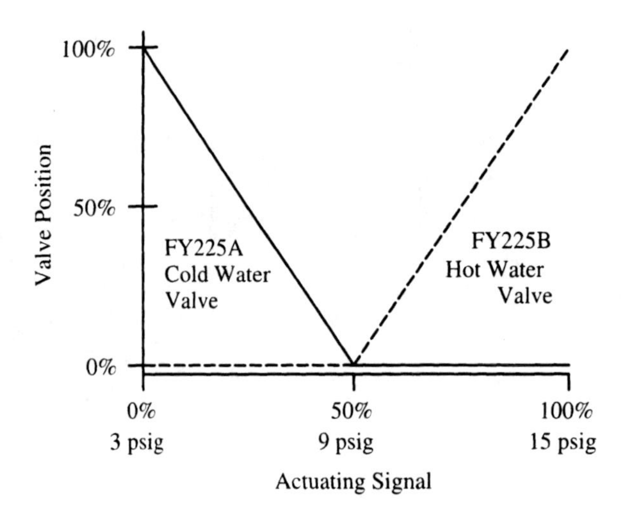


**FIGURE 6.39.** Split-ranging valve positions as function of actuating signal for batch reactor temperature control.

fully open and the steam valve is fully closed at zero pressure. Upon loss of air pressure or controller failure, the system always cools down the reactor. As an alternative to split ranging the valves, many commercial controllers will do the split ranging internal to the controller and present two actuating signals, each for one valve. The cost is higher since two current-to-pressure converters are required. But, the split point, the point at which both

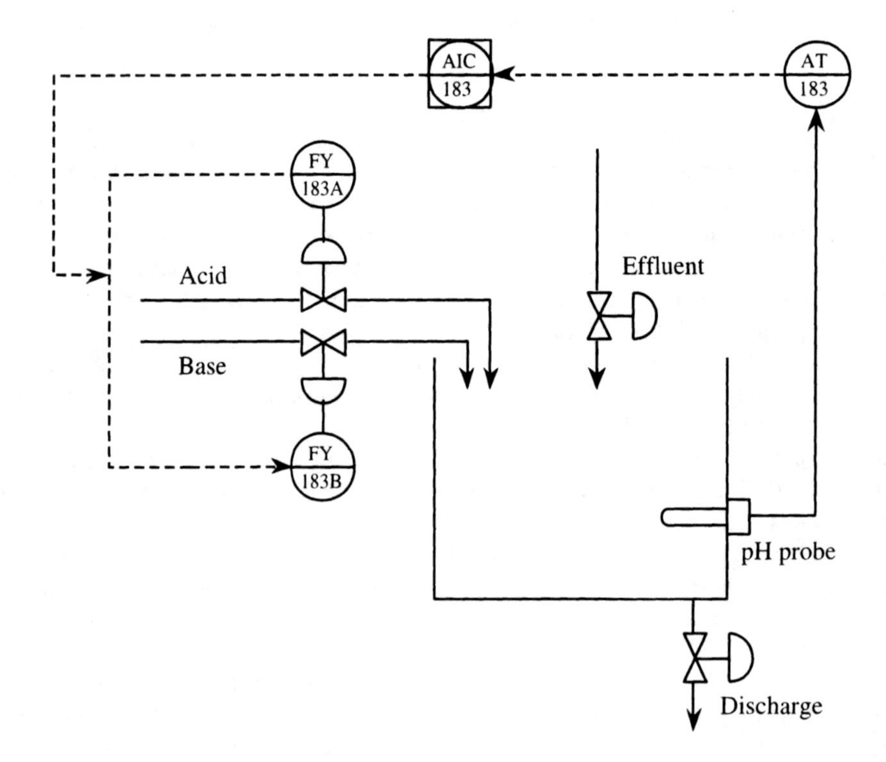


**FIGURE 6.40.** Split-range control of effluent.

valves are closed, can easily be adjusted by changing one parameter instead of adjusting two valve actuators.

In most systems of this nature, the heating and cooling dynamics are different. There are three solutions to this dilemma. The controller could be tuned strictly for the cooling phase if control during the heating phase is not critical. The controller tuning constants could represent some compromise between the cooling and heating tuning. In order for the control to be most effective, the controller tuning constants are changed when it switches from heating to cooling.

As another example of split-range control, consider the waste neutralization process in Figure 6.40. The goal is to neutralize the plant effluent, which may be acidic or basic. The acid and base addition valves and controller are set up so that when the effluent is basic, the acid valve, FY183A, is some percent open and the base valve, FY183B, is fully closed. The situation is reversed when the effluent is acidic. In this case, the split ranging is more appropriately set up as part of the controller, since, for safety, both valves should be closed when air pressure is zero.

A third example of split-range control is the control of pressure in a steam header (Figure 6.41). Four parallel boilers discharge steam into a common steam header and then

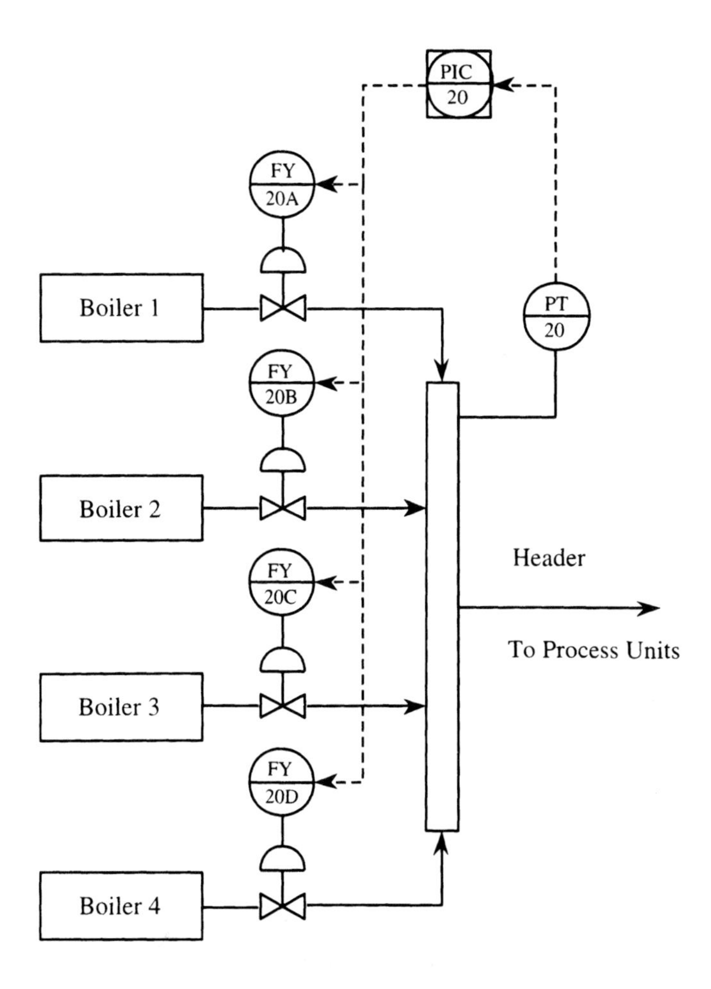


**FIGURE 6.41.** Split-range control of steam header pressure.

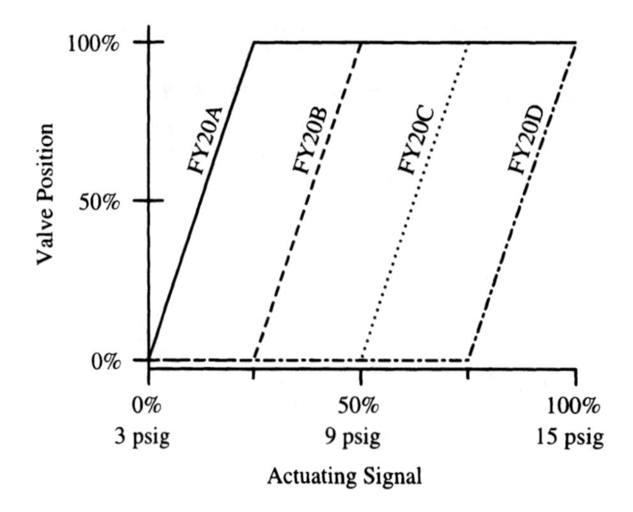


**FIGURE 6.42.** Split-ranging valve positions for steam header control.

on to various plant processing units. The control objective is to maintain constant steam pressure when the steam demand changes. One could manipulate the boiler outlet valves and set up their actuation ranges as in Figure 6.42. This type of scheme will work for fast changes in steam pressure. However, there is usually another split-range control of the furnace firing rate for each boiler based on the steam flow rate.

## 6.6 OVERRIDE CONTROL

During startup, normal operation, or shutdown of a process, it is possible that conditions can arise that lead to damage to equipment and plant personnel. In such situations, the normal control action must be changed and a different control action must prevent the control variable from exceeding an allowable upper or lower limit. Special types of control switches are used to accomplish this function. The *high-selector switch* (HSS) is used to select the highest of two or more signals, and the *low-selector switch* (LSS) is used to select the lowest of two or more signals. Generally, a HSS is used to prevent a variable from exceeding an upper limit and a LSS is used to prevent a variable from exceeding a lower limit. However, in order to apply them correctly, each situation must be analyzed since the type of actuator, dictated by safety concerns, determines the type of selector switch.

Consider the steam pressure header control used as an example in Chapter 3 (Stephanopoulos, 1984), repeated in Figure 6.43. The high-pressure steam is "let down" to the low-pressure steam line. The pressure in the low-pressure steam line is controlled by PIC101. To prevent the high-pressure line from excessive pressures, a HSS transfers control action from PIC101 to PIC102 when the pressure in the high-pressure steam line exceeds an upper limit. The actuating valve, PY102, is air-to-open so that it closes on loss of air or controller failure. Since the goal is protection of the high-pressure line, the correct type of selector switch is a HSS, since it will select the signal that opens the valve the most. PIC101 is a reverse-acting controller whose setpoint is the desired pressure of the low-pressure steam line. PIC102 is a direct-acting PI controller whose setpoint is the value

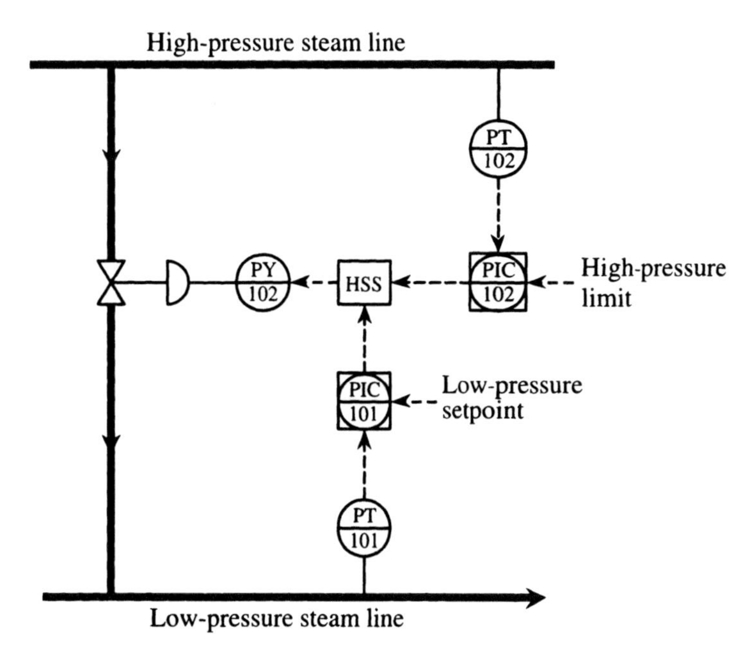


**FIGURE 6.43.** Override control for steam header.

of the high-pressure limit. When the pressure in the high-pressure steam line exceeds the setpoint, the output of PIC102 increases until it is larger than the output of PIC101, causing it to control the position of valve PY101. In reality, the setpoint of PIC102 is set to a value that is below the real high-pressure limit. The PIC102 controller is generally not tuned tightly, so the value of PT102 will generally be significantly greater than the setpoint before the output of PIC102 is high enough to assume control of PY102.

As another example, consider the chemical reactor of Figure 6.44. The control objective is (1) to maximize the conversion while (2) maintaining the effluent concentration above a minimum value, $C_{A,\min}$, and (3) preventing the reactor temperature from exceeding a maximum value $T_{\max}$. The cooling-water flow valve is air-to-close (fail-open). Since a cascade controller is used, the type of valve is of no consequence to the temperature and composition controllers. The type of signal selector is chosen based on whether the minimum or maximum cooling-water flow should be selected. In this case, the maximum cooling-water flow is desired, so a HSS is appropriate. To both controllers, the output actuation is increase-open since an increase in the manipulated variable increases the cooling-water flow. To the temperature controller, TIC472, the process is reverse acting since an increase in the cooling-water flow causes a decrease in the reactor temperature. In contrast, for the composition controller, AIC472, the process is direct acting since an increase in the cooling-water flow causes an increase in the effluent composition (decrease in the conversion rate).

The first two control objectives are met by maintaining the effluent composition at its minimum value. So, normally the composition controller output determines the setpoint to FIC471. An increase in the feed concentration disturbance will cause AIC472 to decrease the cooling-water flow in order to increase the conversion rate. If this change causes the reactor temperature to approach the maximum $T_{\max}$, the output of TIC472 increases until it

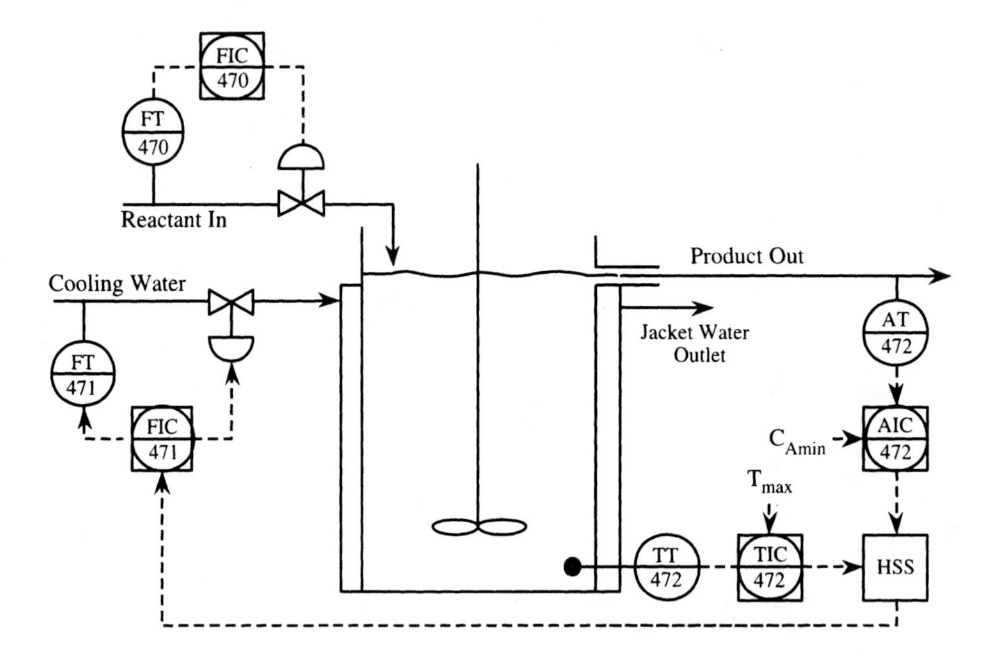


**FIGURE 6.44.** Override control for reactor (cascade cooling water flow control).

is higher than the output of AIC472 in order to decrease the reactor temperature. The resulting concentration is higher than desired, but the reactor is protected from damage.

Though generally not desirable, if the cooling-water control valve is controlled not by a cascade flow controller but directly from the composition and temperature controllers, the control system is as in Figure 6.45. Since the control valve is air-to-close (fail-open), the minimum signal must be selected in order to produce the maximum cooling-water flow. In addition, for both controllers output actuation is increase-close.

Sometimes a limit is the result of equipment performance. For example, in the reactor example, the feed flow rate may be such that the cooling-water flow valve is at its maximum and the reactor temperature is increasing toward its maximum. Clearly, the feed flow should be decreased. An automatic way to set the feed flow in order to get maximum throughput and not violate the other constraints is to monitor the cooling-water flow valve position and decrease the feed flow when the cooling-water valve nears its maximum value of 100%. Such a controller is called a *valve position controller* (VC) and is shown as VC470 in Figure 6.46. The feed flow setpoint is adjusted by the output of VC470.

In practice, the setpoint of the valve position controller is set to a value well away from the actual equipment limitation. This approach assures that the composition/temperature control responds to faster disturbances and maintains the composition/temperature. In this example, a typical value of the VC470 setpoint is 90%, though smaller values might be appropriate if the composition/temperature disturbances are large. In addition, the valve position controller is generally a PI controller that is tuned loosely in order to not upset the composition controller.

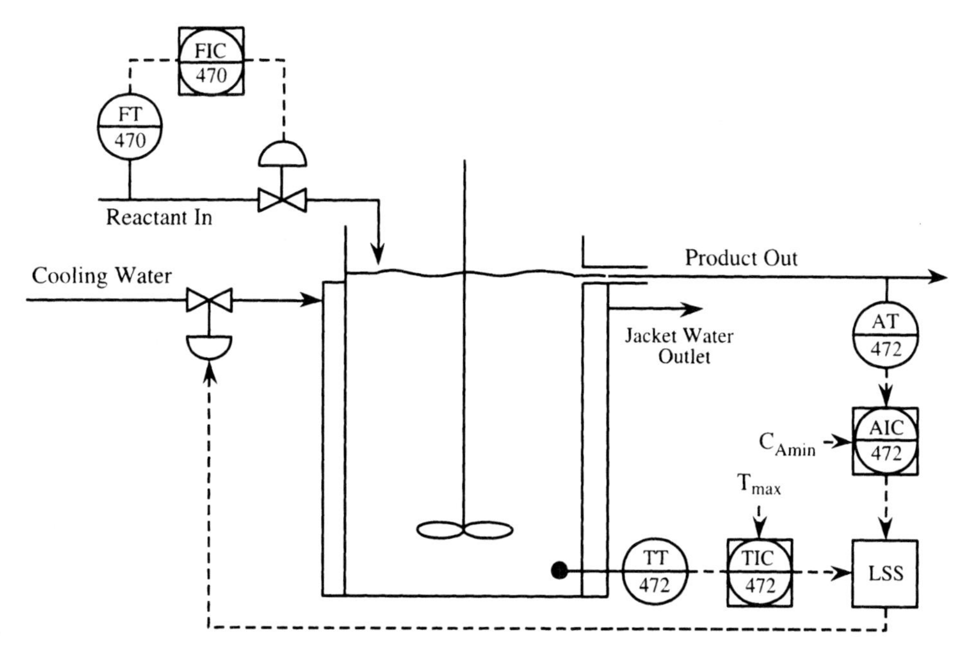


**FIGURE 6.45.** Override control for reactor (not cascade cooling water flow control).

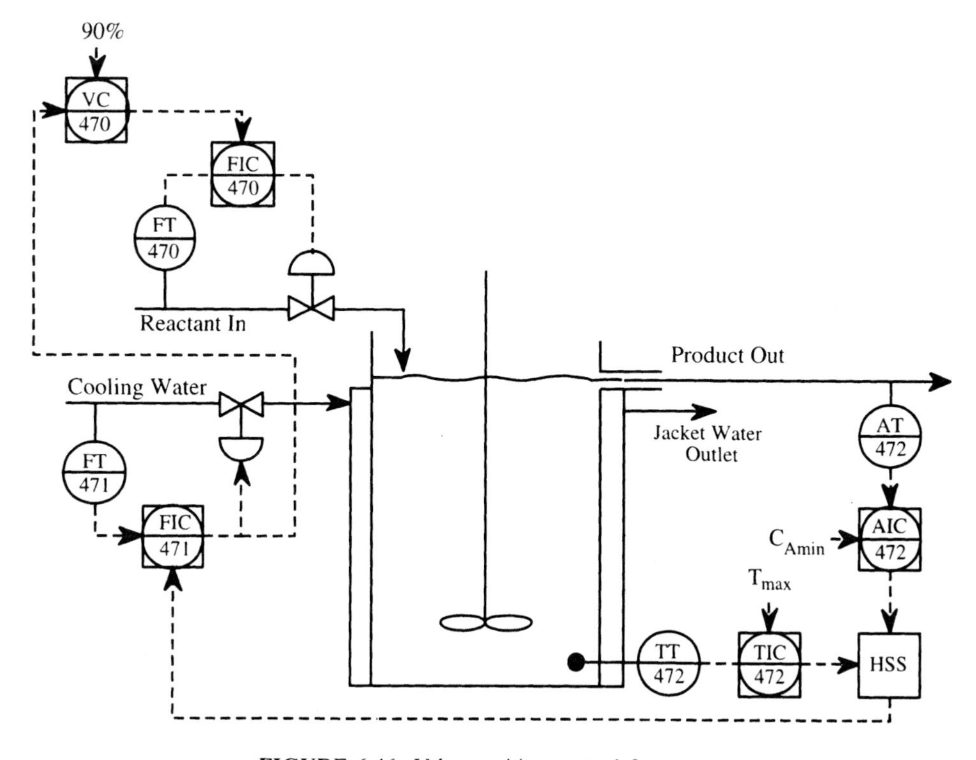


**FIGURE 6.46.** Valve position control for reactor.

## 6.7 SINGLE-LOOP MODEL PREDICTIVE CONTROL

The limited successful application of modern control theory to chemical process control motivated the development of digital model-based predictive control algorithms such as Dynamic Matrix Control (DMC) (Cutler and Ramaker, 1980), Model Algorithmic Control (MAC) (Richalet et al., 1978; Rouhani and Mehra, 1982), Internal Model Control (IMC) (Garcia and Morari, 1982, 1985a,b), and Forward Modeling Control (FMC) (Erickson and Otto, 1991). These digital control algorithms use an impulse response or step response model of the process to predict the trend of the process outputs and to compute the required change in the process inputs to bring the outputs to their desired values. These model-based predictive control schemes were formulated to deal with the deadtime and unusual behavior of complex multi-input, multi-output (MIMO) chemical industrial processes. Consequently, these control algorithms have been applied to many chemical processes with favorable results (e.g., Ricker et al., 1989; Van Hoof et al., 1989; Wassick and Camp, 1988).

In this section, two single-input, single-output (SISO) model predictive control (MPC) algorithms are reviewed. The first is Forward Modeling Control (FMC), a single-point MPC algorithm independently developed by Otto (1986) and Marchetti and co-workers (1983). This particular MPC algorithm is conceptually similar to other model-based predictive digital controllers but is a computationally simpler algorithm. The transfer function is developed and used to examine the system for stability. Robustness issues, feedforward control, and controller tuning are also discussed. Examples of controller performance are given. The operator interface to this controller is also discussed. Second, the SISO version of the DMC is developed and compared with the FMC. In Chapter 7, both of these controller algorithms are extended to processes with multiple inputs and outputs.

### 6.7.1 Single-Loop Forward Modeling Control

Unlike other model predictive controllers, the FMC algorithm calculates the controller action to minimize the error at only one future sample (the output horizon) of the predicted outputs. Other model-based predictive controllers minimize the error at multiple samples of the predicted outputs. Therefore, only a scalar inversion is needed, in contrast to the matrix inversion used in DMC (Cutler and Ramaker, 1980). In addition, the FMC algorithm does not require the derivation of an approximate process inverse as in IMC (Garcia and Morari, 1985a). The response models are used directly in the calculation of the controller outputs. In operation FMC and DMC perform similarly.

***6.7.1.1 Algorithm Development.*** The process is modeled by a discrete moving average (MA) model with $n$ samples:

$$y(k) = h_1 m(k-1) + h_2 m(k-2) + \cdots + h_n m(k-n) + \hat{d}(k) \tag{6.11}$$

where $y(k)$ is the process variable measurement at the $k$th sample, $m(k)$ is the value of the manipulated variable at the $k$th sample measured as close to the process as practical, $h_j$ are the model coefficients, and $\hat{d}(k)$ is the discrepancy between the model and the measurement at the $k$th sample. A MA model of the process has the following significant advantages:

1. No a priori assumptions about model order and time delay are necessary.
2. Unusual process dynamics are handled naturally and do not require the specification of model structure.
3. The coefficients of the model can be obtained from simple step response data.
4. The step response model, obtained by integrating the impulse response model, has high intuitive appeal to process operators.
5. With prediction error identification methods, multivariable moving average (impulse response) models will, in the limit, converge to the true parameter values (Stoica and Söderström, 1982).

The chief disadvantages of a MA model are:

1. There is nonminimality of the representation, that is, a relatively larger number of parameters are used in the model, compared with the smaller number of parameters associated with a low-order transfer function model.
2. There is an a priori assumption of the time to steady state, that is, $n$, the number of coefficients in the model, must be chosen sufficiently large.

However, the advantages of a MA model far outweigh the disadvantages.

It is important to note that the value of $\hat{d}(k)$ is determined by two separate factors: unmeasured process disturbances and modeling errors. It is impossible to separate the two without making further assumptions.

The single-loop FMC computes a prediction of the process output $T(k)$ into the future with the following assumptions:

1. The manipulated variable is held constant into the future:

$$m(k) = m(k+1) = m(k+2) = \cdots$$

2. The future discrepancies remain at their current value:

$$\hat{d}(k) = \hat{d}(k+1) = \hat{d}(k+2) = \cdots$$

The process output predictions are given by the following equations [where $T_j(k)$ is the predicted process variable at the $j$th sample in the future, given the information of the current, $k$th, sample]:

$$T_1(k) = h_1 m(k) + h_2 m(k-1) + \cdots + h_n m(k-n+1) + \hat{d}(k)$$

$$T_2(k) = h_1 m(k) + h_2 m(k) + h_3 m(k-1) + \cdots + h_n m(k-n+2) + \hat{d}(k)$$

$$\vdots$$

$$\begin{aligned} T_j(k) &= (h_1 + h_2 + \ldots + h_j) m(k) + h_{j+1} m(k-1) + \cdots + h_n m(k-n+j) + \hat{d}(k) \\ &= a_j m(k) + h_{j+1} m(k-1) + \cdots + h_n m(k-n+j) + \hat{d}(k) \end{aligned} \tag{6.12}$$

$$\vdots$$

$$T_n(k) = a_n m(k) + \hat{d}(k) \tag{6.13}$$

where $a_j = \sum_{i=1}^{j} h_i$ is the $j$th coefficient of the process discrete-time step response. Note that

$$\begin{aligned} T_j(k) &= T_{j+1}(k-1) + a_j[m(k) - m(k-1)] + \hat{d}(k) - \hat{d}(k-1) \\ &= T_{j+1}(l-1) + a_j \Delta m(k) + \Delta\hat{d}(k) \end{aligned} \tag{6.14}$$

where $\Delta m(k) = m(k) - m(k-1)$ and $\Delta\hat{d}(k) = \hat{d}(k) - \hat{d}(k-1)$. Expressed in words, the prediction of the $j$th sample in the future is updated from the previous prediction of the $(j+1)$st sample by the change in the unmeasured disturbances and the changes in the manipulated variables.

The controller is designed that yields at sample $k$ a controller output $m(k)$ that, if it were kept constant from here on, would drive the prediction after $P$ samples in the future to the desired setpoint $s(k)$:

$$\begin{aligned} s(k) &= T_P(k) \\ &= a_P m(k) + h_{P+1} m(k-1) + \cdots + h_n m(k-n+P) + \hat{d}(k) \end{aligned} \tag{6.15}$$

Substituting for $\hat{d}(k)$ from (6.11) into (6.15),

$$\begin{aligned} s(k) - y(k) = {} & a_P m(k) + h_{P+1} m(k-1) + \cdots + h_n m(k-n+P) \\ & - h_1 m(k-1) - \cdots - h_n m(k-n) \end{aligned} \tag{6.16}$$

Converting to $Z$-transforms,

$$S(z) - Y(z) = [a_P + h_{P+1} z^{-1} + \cdots + h_n z^{-n+P} - (h_1 z^{-1} + \cdots + h_n z^{-n})] M(z) \tag{6.17}$$

In conventional controller terms (Figure 6.47), the controller transfer function $G_c(z)$ is given as

$$\begin{aligned} G_c(z) &= \frac{M(z)}{S(z) - Y(z)} \\ &= \frac{1}{a_P + h_{P+1} z^{-1} + \cdots + h_n z^{-n+P} - (h_1 z^{-1} + \cdots + h_n z^{-n})} \end{aligned} \tag{6.18}$$

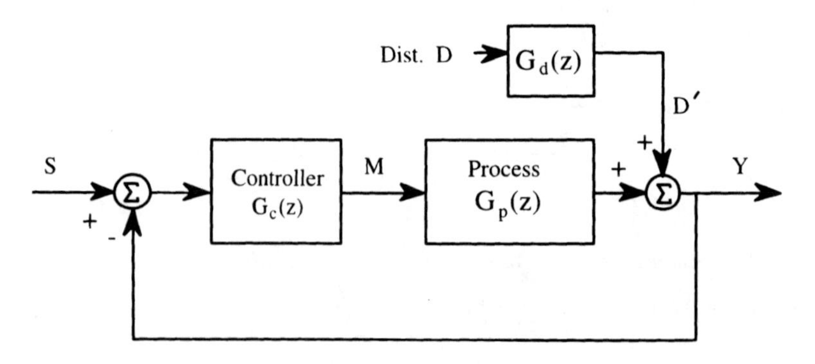


**FIGURE 6.47.** Conventional control system configuration.

One way to examine stability of the closed-loop system is to examine the closed-loop system poles. However, by converting the controller to the internal model configuration (Figure 6.48), stability analysis is simpler. Garcia and Morari (1985a) show that the output of the system in Figure 6.48 is given by

$$Y(z) = \frac{G_{cp}(z)G_p(z)}{1 + G_{cp}(z)(G_p(z) - G_m(z))}[S(z) - D'(z)] + D'(z) \tag{6.19}$$

If exact process modeling is assumed, then $G_m(z)$ is set to $G_p(z)$ and (6.19) becomes

$$Y(z) = G_{cp}(z)G_p(z)[S(z) - D'(z)] + D'(z) \tag{6.20}$$

Therefore, the closed-loop system is stable if $G_{cp}(z)$ and $G_p(z)$ are stable. Assuming the process $G_p(z)$ is stable, stability of the controller $G_{cp}(z)$ is sufficient for closed-loop stability. For the FMC, $G_{cp}(z)$ is (Otto, 1986)

$$G_{cp}(z) = \frac{1}{a_P + h_{P+1}z^{-1} + \cdots + h_n z^{-n+P}} \tag{6.21}$$

For system stability, the controller poles are given by the roots of the equation

$$a_P z^{n-P} + h_{P+1}z^{n-P-1} + h_{P+2}z^{n-P-2} + \cdots + h_n = 0 \tag{6.22}$$

Thus the loop is stable if and only if the roots of (6.22) are strictly inside the unit circle. The order of (6.22) can be high since typically $n \geq 20$ and the roots tend to cluster close to $z = 1$, making the task of accurately determining the roots difficult. As a way around this difficulty, Jury (1964) presented two sufficient conditions for the stability of (6.22) that does not involve determining the roots. The roots of (6.22) lie strictly within the unit circle if either of the following conditions is met:

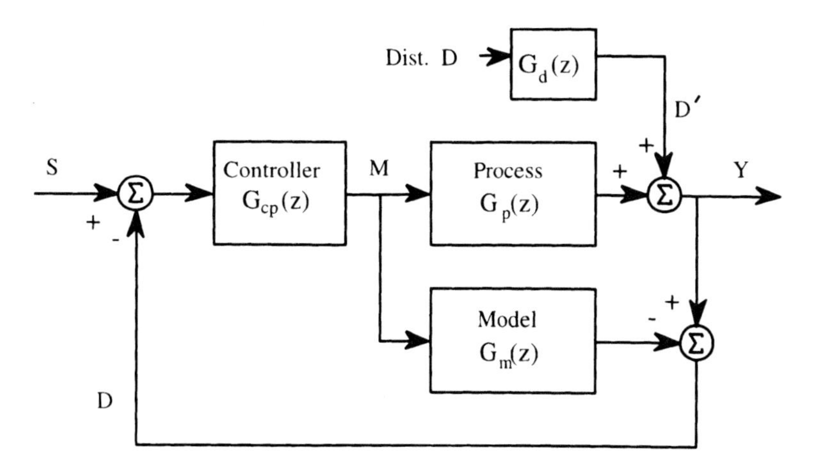


**FIGURE 6.48.** Internal model control configuration.

Theorem 6.1:

$$a_P > h_{P+1} > h_{P+2} > \cdots > h_n > 0 \tag{6.23}$$

Theorem 6.2:

$$|a_P| > |h_{P+1}| + |h_{P+2}| + |h_{P+3}| + \cdots + |h_n| \tag{6.24}$$

In practice, the smallest value of $P$ that satisfies at least one of the two theorems is the minimum value of $P$ that guarantees stability. Note that if $P$ is set equal to $n$ (steady-state control), both theorems show that the loop is stable; hence stability is guaranteed if the modeling is exact. The value of $P$ is called the *output horizon* and is the main tuning parameter.

The two stability theorems may be interpreted in terms of the graphical step response. Theorem 6.2 will be satisfied for any *monotonic* step response if point $P$ is picked beyond the 50% point to steady state. For a monotonic step response, Theorem 6.1 can be interpreted as follows: All inequalities except the leftmost one are satisfied if $P$ is picked *beyond* the last inflection point on the step response. The leftmost inequality can always be satisfied if one considers increasing the number of points in the model while decreasing the control interval to maintain a constant settling time. The step response coefficient $a_P$ will remain constant during this process while the MA model coefficients $h_{P+1}, \ldots, h_n$ will become indefinitely small. Monotonic step responses and their stability theorem interpretation are depicted in Figure 6.49. Nonmonotonic step responses do not admit to graphical interpretation.

*6.7.1.2* ***Filtering for Robustness.*** If modeling errors exist, stability cannot be guaranteed even for $P = n$. Furthermore, even with exact modeling, a policy of setting $P = n$ may produce a controller that moves the manipulated variable too vigorously when large amounts of noise are present in the measurement. The controller needs to be modified to produce robustness (tolerance to modeling errors) and noise rejection. Garcia and Morari (1985a) have shown that if a filter $F(z)$ is added to the controller input, the closed-loop system can be made stable for arbitrarily large modeling errors (other than the wrong sign on the model gains) by filtering heavily enough. The filter $F(z)$ is of the form

$$F(z) = \frac{1}{1 + f - fz^{-1}}$$

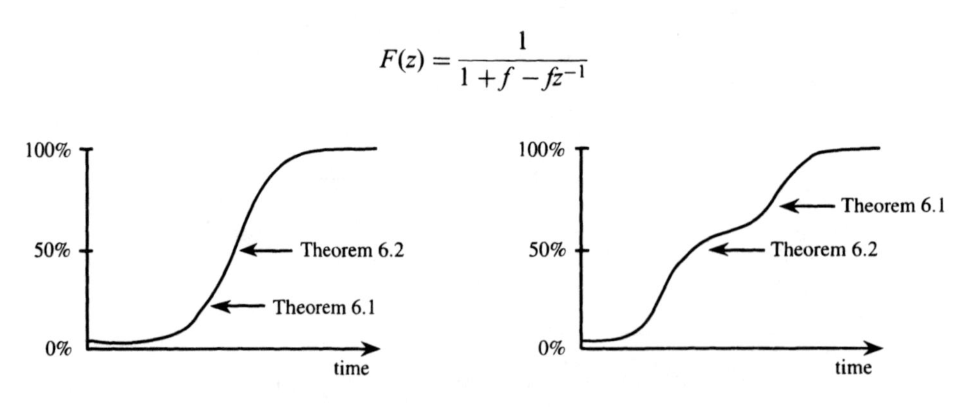


**FIGURE 6.49.** Stability theorems for monotonic step responses.

where $f$ is picked as

$$f = \begin{cases} 0 & \text{for } P \leq n \\ \frac{1}{3}(P-n) & \text{for } P > n \end{cases}$$

When the output horizon is larger than the length of the step response model, the controller input is filtered and provides robustness to modeling error. Therefore, adjustment of the output horizon $P$ can move the controller from high-performance control to noise-rejecting, sluggish, robust control.

*6.7.1.3* ***Feedforward Control.*** Feedforward control is easily accommodated by FMC. Given a model of the effect of a *measured* disturbance $d$ on the process outputs of the same form as the discrete-time MA model used for the process:

$$y(k) = h'_1 d(k-1) + h'_2 d(k-2) + \cdots + h'_n d(k-n) \tag{6.25}$$

Under the additional assumption that the measured disturbance remains at its present value, the prediction vector at the $j$th sample in the future is

$$\begin{aligned} T_j(k) = {} & a_j m(k) + h_{j+1} m(k-1) + \cdots + h_n m(k-n+j) + \hat{d}(k) \\ & + a'_j d(k) + h'_{j+1} d(k-1) + \cdots + h'_n d(k-n+j) \end{aligned}$$

where $a'_j = \sum_{i=1}^{j} h'_i$ is the $j$th coefficient of the measured disturbance step response. With this change, Equation (6.14) becomes

$$T_j(k) = T_{j+1}(k-1) + a_j\,\Delta m(k) + a'_j\,\Delta d(k) + \Delta\hat{d}(k) \tag{6.26}$$

Therefore, the only change to the control algorithm is to update the process variable predictions with the expected effect of the measured disturbances. No other change to the control algorithm is necessary.

*6.7.1.4* ***Controller Tuning.*** The single-loop FMC has only two types of adjustments. The above development focused on the output horizon $P$, which smoothly takes the control action from

1. extremely aggressive control by setting $P$ to its minimum value to
2. steady-state control where the controller moves the manipulated variable only to statically compensate the process to
3. extremely sluggish, noise-rejecting, robust control that should be stable in all practical situations.

In addition to the output horizon, the controller has only one other tuning parameter, the control interval $\Delta t$. The number of points in the model $n$ is usually fixed. For ease of communication with the engineer commissioning the loop and with operating personnel, the two types of parameters can be renamed:

1. the *closed-loop settling time* (CLST) of the controlled variable, $P\,\Delta t$, and
2. the *open-loop settling time* (OLST) of the process, $n\,\Delta t$.

In practice, the OLST of the process is rarely changed. The CLST is approximate (except for $P = n$ with exact modeling) but is useful for explaining the trade-off between loop performance and robustness. Decreasing the CLST tends to make the system less tolerant to modeling errors. This feature has strong intuitive appeal and is well received by operating personnel.

***6.7.1.5 Algorithm Implementation.*** If the controller were implemented in the form of (6.21), the controller output would be based on the current error signal, $s - \tilde{d}$, and the past values of $m$. By themselves, the past values of $m$ are not useful to process operators. An indication of where the process is going is far more useful to process operators. Therefore, the controller is implemented to recursively process a prediction of the future trend of the process variable. No past values of $m$ are needed. When a predictor is used within the control loop, the structure of the controller with feedforward changes to the configuration shown in Figure 6.50.

**Example 6.5.** For the heat exchanger of Figure 6.4, design a single-loop forward modeling controller with feedforward that compensates for changes in the hot-water flow. Evaluate the performance of this feedforward controller for a +2- and −2-gpm change in the hot-water flow, assuming the same operating point as in Example 6.1.

***Solution.*** In order to appropriately handle changes in the steam pressure, the forward modeling controller is cascaded with the steam flow controller, as shown in Figure 6.51. The feedforward measurement is an input to the controller, rather than a modification of the controller output, as is done for PID control. Examining the step response of the hot-water temperature to a step change in the steam flow setpoint (Figure 6.11) and the response to a hot-water flow disturbance (Figure 6.24), the settling time of the process is no larger than 4 min. So the OLST is assumed to be 4 min. The step response models have 100 samples, giving a controller sampling period of 0.04 min. The step response model of the process is obtained from response data similar to Figure 6.11, except that the starting flow setpoint is 37.7%, the operating point of the process. Using, the stability theorems (6.23) and (6.24), the minimum CLST is 0.28 min. The disturbance step response is assumed identical to Figure 6.24. The response of this model predictive controller with feedforward to disturbances in the hot-water flow is shown in Figure 6.52. The response for the minimum CLST and a larger value of the CLST are also shown. As expected, the

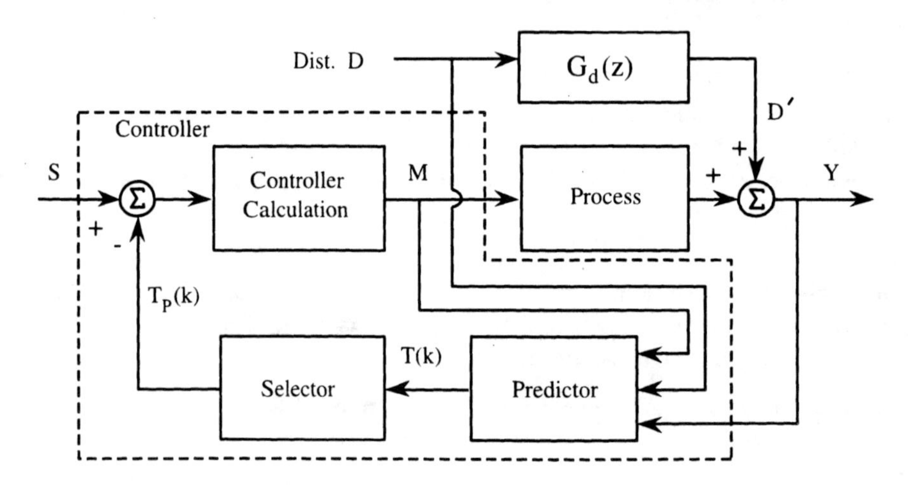


**FIGURE 6.50.** Model predictive control form of system.

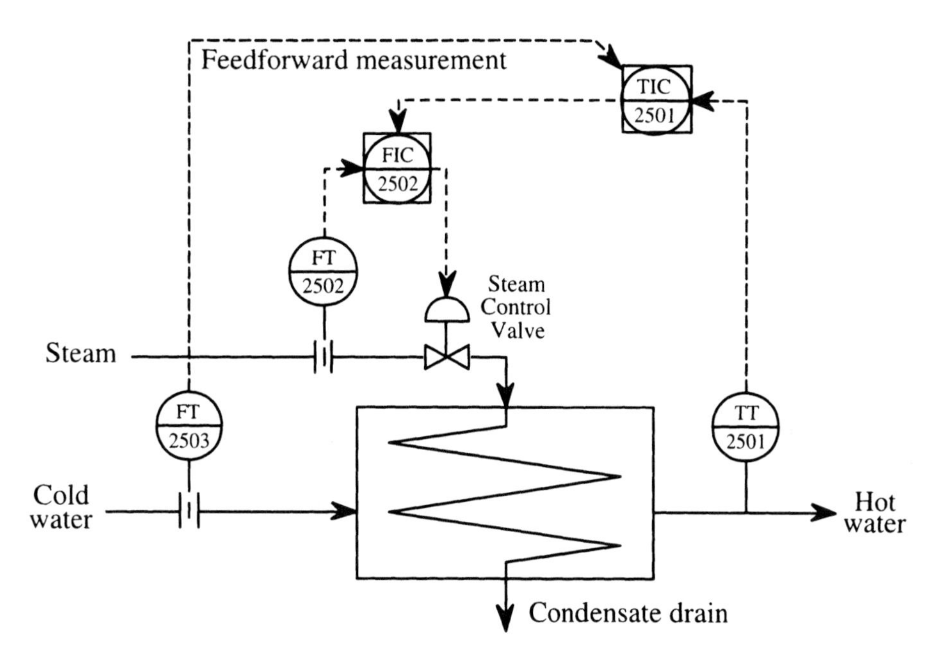


**FIGURE 6.51.** Model predictive control of heat exchanger with cascade and feedforward.

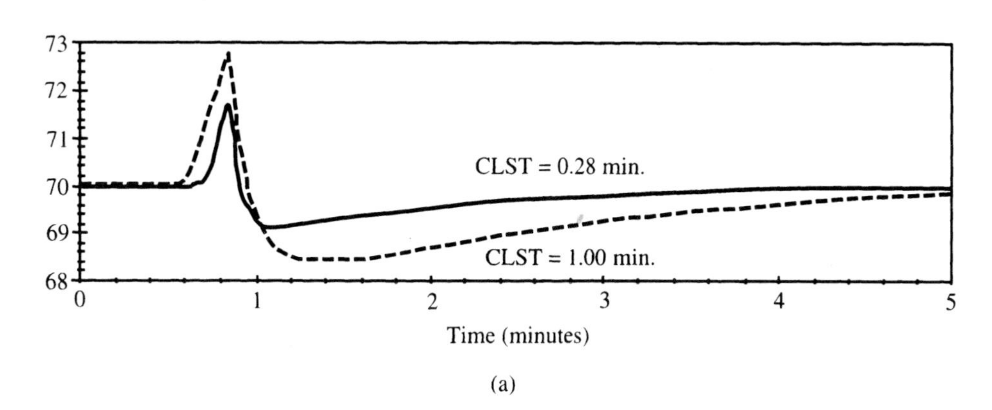


(a)

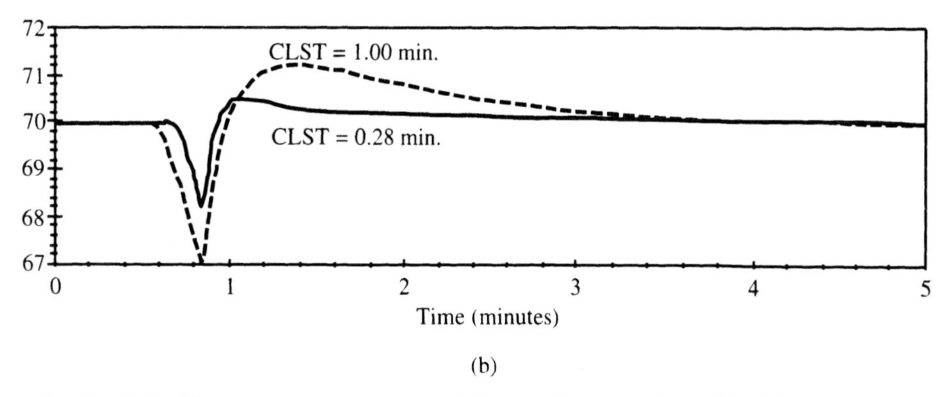


(b)

**FIGURE 6.52.** Disturbance response of model predictive controller with different closed-loop settling times: (*a*) +2-gpm change in hot-water flow; (*b*) −2-gpm change in hot-water flow.

**TABLE 6.6 Summary Performance of FMC Model Predictive Control of Example 6.5**

| CLST (min) | Disturbance (gpm) | IAE |
|---|---|---|
| 0.28 | +2 | 1.36 |
| 0.28 | −2 | 0.66 |
| 1.00 | +2 | 3.68 |
| 1.00 | −2 | 2.16 |

control is tighter for the smaller CLST. As for the PID feedback–feedforward control system (Example 6.4), the disturbance modeling error is evident, and not surprising since this system has some nonlinearity. The performance is summarized in Table 6.6 and is comparable to that of the PID feedback–feedforward control.

### 6.7.2 Single-Loop Dynamic Matrix Control

The DMC algorithm is developed in a similar manner as for the FMC algorithm. This development follows the presentation in Balhoff and Lau (1985). The process is modeled by a discrete MA model (6.11). The prediction assumes more than one manipulated variable move in order to bring the process variable back to its setpoint. The prediction is developed assuming $M$ future manipulated variable moves and $N$ predicted process variable measurements ($N < n$, the step response or MA model length). The samples of the prediction are expressed as

$$\begin{aligned}
\Delta T_1(k) &= a_1\,\Delta m(k) \\
\Delta T_2(k) &= a_2\,\Delta m(k) + a_1\,\Delta m(k+1) \\
&\;\;\vdots \\
\Delta T_M(k) &= a_M\,\Delta m(k) + a_{M-1}\,\Delta m(k+1) + \cdots + a_1\,\Delta m(k+M-1) \\
&\;\;\vdots \\
\Delta T_N(k) &= a_N\,\Delta m(k) + a_{N-1}\,\Delta m(k+1) + \cdots + a_{N-M+1}\,\Delta m(k+M-1)
\end{aligned}$$

where the notation is the same as for the prediction equations developed for FMC. These prediction equations can be expressed in matrix form as

$$\begin{bmatrix} \Delta T_1(k) \\ \Delta T_2(k) \\ \vdots \\ \Delta T_M(k) \\ \vdots \\ \Delta T_N(k) \end{bmatrix} = \begin{bmatrix} a_1 & 0 & \cdots & 0 \\ a_2 & a_1 & \cdots & 0 \\ \vdots & \vdots & & \vdots \\ a_M & a_{M-1} & \cdots & a_1 \\ \vdots & \vdots & & \vdots \\ a_N & a_{N-1} & \cdots & a_{N-M+1} \end{bmatrix} \begin{bmatrix} \Delta m(k) \\ \Delta m(k+1) \\ \vdots \\ \Delta m(k+M-1) \end{bmatrix} \tag{6.27}$$

or concisely as

$$\Delta \mathbf{T}(k) = \mathbf{A}\,\Delta \mathbf{m} \tag{6.28}$$

where $\mathbf{A}$ is commonly called the *dynamic matrix* and has $N$ rows and $M$ columns. The set of moves $\Delta \mathbf{m}(k)$ is calculated to produce changes in the predicted output $\Delta \mathbf{T}(k)$ that force the prediction of the error to zero at each future sample. This is accomplished by setting the prediction vector equal to the vector of predicted errors $\hat{\mathbf{e}}$,

$$\hat{\mathbf{e}} = \mathbf{A}\,\Delta \mathbf{m} \tag{6.29}$$

Note that $\hat{\mathbf{e}}$ only contains the effects of past moves and not the prediction of future moves. The system of equations defined by (6.29) is called overdetermined since there are more equations than unknowns and hence ordinary matrix inversion cannot be used to determine $\Delta \mathbf{m}$. In such a situation, the error of the solution is minimized,

$$\mathbf{e}^* = \hat{\mathbf{e}} - \mathbf{A}\,\Delta \mathbf{m} \tag{6.30}$$

Actually, $\Delta \mathbf{m}$ is calculated to minimize the sum of the squares of the solution error vector components,

$$\min_{\Delta \mathbf{m}} \sum_{i=1}^{N} (e_i^*)^2 = \min_{\Delta \mathbf{m}(k)} (\mathbf{e}^*)^{\mathrm{T}} \mathbf{e}^* \tag{6.31}$$

where $(\cdot)^{\mathrm{T}}$denotes a matrix transpose. The solution is further complicated by the fact that the moves calculated by (6.31) are usually too violent. Thus a penalty term is added to (6.31) to mitigate the control action. The manipulated variable moves are calculated to minimize

$$\min_{\Delta \mathbf{m}} \left[ \sum_{i=1}^{N} (e_i^*)^2 + \lambda \sum_{i=1}^{M} [\Delta m(k+i-1)]^2 \right] \tag{6.32}$$

where $\lambda$ is called the *move suppression factor*. In matrix form, (6.32) is written as

$$\min_{\Delta \mathbf{m}} [(\hat{\mathbf{e}} - \mathbf{A}\,\Delta \mathbf{m})^{\mathrm{T}} (\hat{\mathbf{e}} - \mathbf{A}\,\Delta \mathbf{m}) + \lambda (\Delta \mathbf{m})^{\mathrm{T}} (\Delta \mathbf{m})] \tag{6.33}$$

The unconstrained solution to (6.33) is

$$\begin{aligned} \Delta \mathbf{m} &= (\mathbf{A}^{\mathrm{T}}\mathbf{A} + \lambda \mathbf{I})^{-1} \mathbf{A}^{\mathrm{T}} \hat{\mathbf{e}} \\ &= \mathbf{S}\hat{\mathbf{e}} \end{aligned} \tag{6.34}$$

where $\mathbf{I}$ is the identity matrix and $(\cdot)^{-1}$ is ordinary matrix inversion. In practice, only the first current move, $\Delta m(k)$ is calculated by multiplying the first row of $\mathbf{S}$ by the error prediction vector $\hat{\mathbf{e}}$.

To complete the controller, the prediction must be updated with the actual process variable measurement. The update is accomplished in the same manner as for the single-loop FMC (6.14).

Balhoff and Lau (1985) derive the transfer function of the SISO DMC as

$$G_c(z) = \frac{K_0}{1 + \sum_{i=1}^{\infty} \xi_i z^{-1}}$$

where

$$K_0 = \sum_{i=1}^{N} s_{1i} \ (s_{1i} \text{ is an element of the first row of } \mathbf{S})$$

$$\xi_i = \sum_{j=1}^{N} s_{1j}(a_{i+j} - a_i)$$

However, since $a_i = a_n$ for $i > n$ (step response model constant after $n$th sample), $\xi_i = 0$ for $i \geq n$ and the denominator polynomial of the transfer function has finite order,

$$G_c(Z) = \frac{K_0}{1 + \sum_{i=1}^{n-1} \xi_i z^{-i}}$$

Feedforward control is accomplished in the same manner as for the FMC algorithm. The prediction is updated with the expected effect of the disturbance before the manipulated variable move is calculated.

In addition to the sample period and the step response model length, the DMC algorithm has three adjustments:

$N$—*Output horizon*, the future time over which the control performance is evaluated. Typically, this parameter is 20–50 samples. The closed-loop system should approach steady state in $N$ samples.

$M$—*Manipulated variable* (input) *horizon*, the number of future manipulated variable moves considered in the calculation. Typically, this parameter is $\frac{1}{4}$ to $\frac{1}{3}$ of $N$.

$\lambda$—*Move suppression factor*. Increasing this value tends to slow the manipulated variable moves, which degrades control loop performance. However, increasing the value also improves the robustness of the system to model mismatch.

**Example 6.6.** For the heat exchanger of Figure 6.4, design a dynamic matrix controller, with feedforward, that compensates for changes in the hot-water flow. Evaluate the performance of this feedforward controller for a +2- and −2-gpm change in the hot-water flow, assuming the same operating point as in Example 6.1.

***Solution.*** As for the forward modeling controller, the dynamic matrix controller is cascaded with the steam flow controller, as shown in Figure 6.51. As for single-loop FMC, the process settling time is 4 min. The step response models are assumed to have 50 samples, giving a controller sampling period of 0.08 min. The step response models of the process and disturbance are obtained in the same manner as for single-loop FMC. The output horizon $N$ is 25 samples and the input horizon $M$ is 5 samples. The response of this

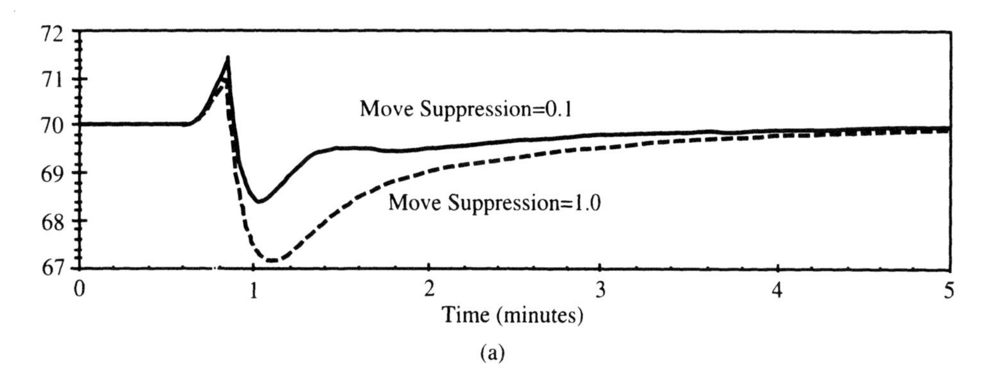


(a)

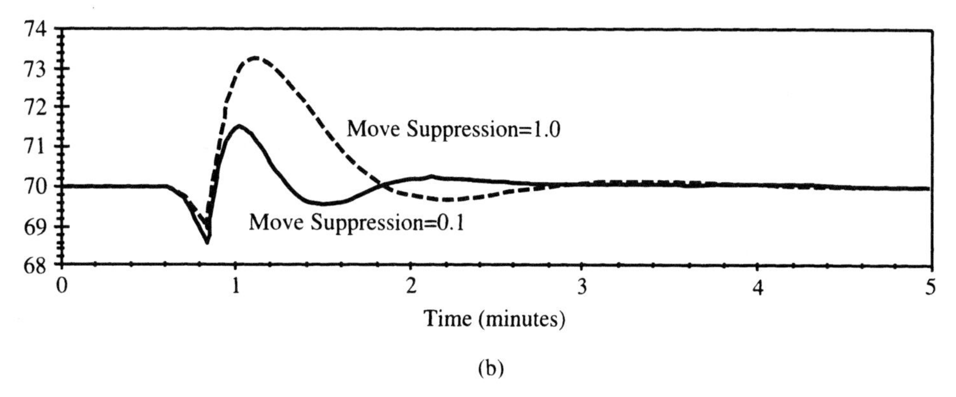


(b)

**FIGURE 6.53.** Disturbance response of DMC with different move suppression factors: (*a*) +2-gpm change in hot-water flow; (*b*) −2-gpm change in hot-water flow.

dynamic matrix controller with feedforward to disturbances in the hot-water flow is shown in Figure 6.53. The responses for a smaller and a larger value of the move suppression factor are also shown. As expected, the control is tighter for the smaller move suppression factor. As for the PID feedback–feedforward control system (Example 6.4), a disturbance modeling error is evident, and not surprising since this system has some nonlinearity. The performance is summarized in Table 6.7 and is comparable to that of both FMC and PID feedback–feedforward control.

**TABLE 6.7 Summary Performance of DMC Control of Example 6.6**

| Move Suppression | Suppression Horizon | Disturbance (gpm) | IAE |
|---|---|---|---|
| 0.1 | 5 | +2 | 1.54 |
| 0.1 | 5 | −2 | 0.92 |
| 1.0 | 5 | +2 | 3.34 |
| 1.0 | 5 | −2 | 2.20 |

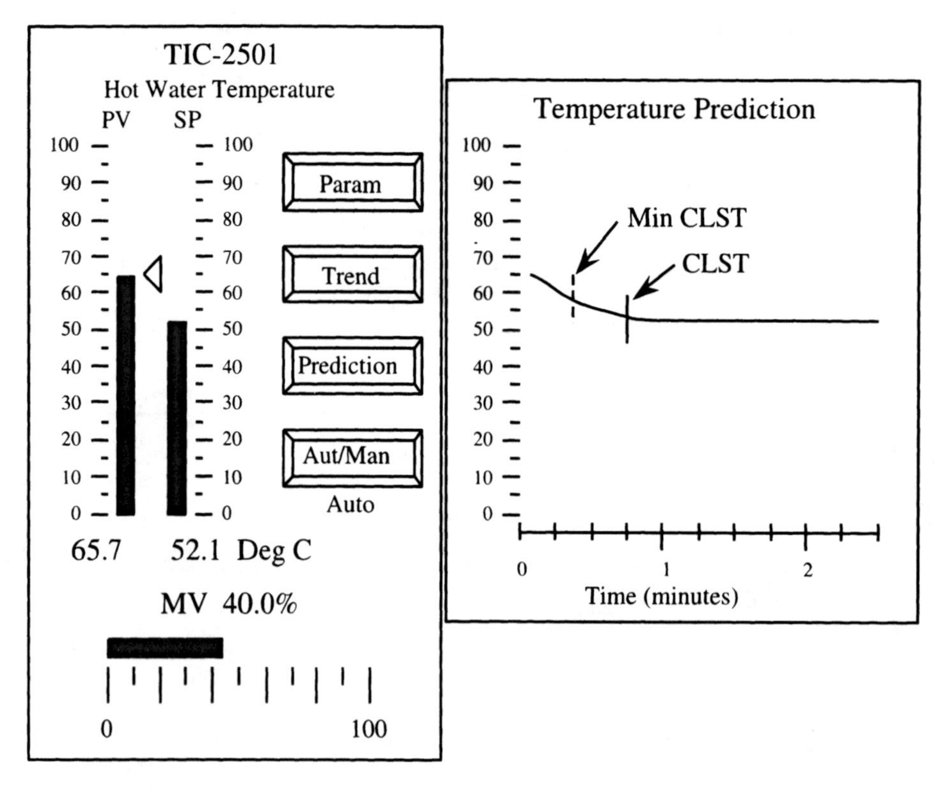


**FIGURE 6.54.** Typical operator interface for SISO FMC model predictive controller.

### 6.7.3 User Interface

When either the FMC or DMC algorithm is formulated to process a process variable prediction (Otto, 1986), this information can be made available to the operator. Commonly, this information is presented in a window separate from the conventional faceplate (Figure 6.54). This prediction window is commonly invoked by the operator and may or may not be a permanent window. For FMC, the current CLST and the minimum CLST is also displayed on the prediction display. For DMC, the input and output horizons are displayed in a similar manner. The step response model may also be displayed below the prediction window but is generally done this way only if the controller also contains an on-line step response identifier. Though not discussed previously, when FMC or DMC is in a Manual state, the algorithm continues to calculate a manipulated variable move but does not implement it. This *suggested MV* can be shown to the operator as an arrow pointing to the appropriate value on the MV bar display. The suggested MV is the next move that the controller will implement when switched to the Automatic state.

## 6.8 BATCH CONTROL CONSIDERATIONS

In batch control systems, PID controllers are often used to control process variables such as temperature, flow, and pressure. However, there are important differences between most continuous processes and batch processes. Namely, regulatory control in batch processes must deal with:

*Non–Steady State.* Batch processes are rarely in steady state, that is, holding a variable at a constant value. For example, vessels are being charged, vessels are being heated, and vessel temperatures are following a profile. It is more critical that the process follow the profile, rather than follow a constant value.

*Zero Load.* Material removal from a vessel may not be possible at certain times, and thus no overshoot in the process variable can be tolerated. For example, when charging a vessel with a certain ingredient amount, a larger amount of the ingredient is often not tolerated. So, the control of the ingredient charging must not allow any overshoot of the final amount. A ruined batch will occur if there is too much overshoot.

*Zero Deviation Blending.* Related to the zero-load situation, when blending ingredients as they are loaded into a vessel, the ratio of the two or more ingredients must be maintained as the vessel is being charged. Again, there generally is no opportunity to compensate for loading too much of one ingredient.

*Process Changes May Be Frequent.* The varying amounts and materials in a multiproduct batch process may cause a significant variation in process dynamics. In such situations, the controllers must be tuned in order to handle all possible situations.

*Endpoint Control.* In many batch process units regulatory control is required to achieve an endpoint to the batch operating cycle. The objective is to converge on the desired control objective, such as final temperature, pH, and composition. The control objective is to achieve the endpoint without violating operating constraints, some of which could be hazardous.

The control of batch reactor temperature can be a problem, especially if the reaction is exothermic. In a typical batch sequence, the reaction vessel is heated from an initial temperature to an optimal reaction temperature. However, as the reactor is heated, the reaction becomes exothermic and thus heat must be removed by cooling. The typical PID controller performance is less than satisfactory since it will initially apply full steam until the temperature approaches the desired value and overshoots the desired temperature for a period of time before lowering the steam flow and maybe applying cooling water to bring the temperature to the desired value. This problem may be dangerous, especially if the overshoot is larger than the cooling water can handle, causing the process to proceed out of control. This particular problem is due to reset windup, which forces the process to overshoot in order to decrease the integral term of the PID controller. Antireset windup schemes will mitigate the problem but not totally eliminate it. The most common way to solve this problem is to linearly ramp the reactor temperature from the initial value to the reaction temperature. The temperature is maintained by adjusting the steam or cooling water as necessary, often in a split-range scheme, as shown in Figure 6.38. In critical temperature control applications, there will be alarms for a measured temperature ramp slope that is too low or too high, indicating loss of control.

While using a setpoint ramp and/or antireset windup is a possible solution to prevent the overshoot of the process variable in a batch system, another possible solution is to use model predictive controllers, such as the forward modeling controller or the dynamic matrix controller. Since they use a process model to produce a prediction of the process variable, little or no overshoot in the process variable is achieved since the control action will be reduced in anticipation of reaching the target value. The drawback to model predictive controllers is their need for an accurate process model.

## REFERENCES

Balhoff, R. A., and H. K. Lau, "A Transfer Function Form of the Dynamic Matrix Control and Its Relationship with Some Classical Controllers," paper TP7-2:30, presented at the American Control Conference, Boston, MA, 1985.

Buckley, P., "A Control Engineer Looks at Control Valves," in Lovett, O. (Ed.), Paper 2.1, *Final Control Elements, First ISA Final Control Elements Symposium*, May 14–16, Instrument Society of America, Research Triangle Park, NC, 1970.

Cutler, C. R., and B. L. Ramaker, "Dynamic Matrix Control—A Computer Control Algorithm," paper WP5-B, *Proc. JACC*, vol. 1., San Francisco, CA, American Automatic Control Council, Green Valley, AZ, 1980.

Erickson, K. T., and R. E. Otto, "Development of A Multivariable Forward Modeling Controller," *Ind. Eng. Chem. Res.*, **30**, 482–490 (1991).

Garcia, C. E., and M. Morari, "Internal Model Control. 1. A Unifying Review and Some New Results," *Ind. Eng. Chem. Process Des. Dev.*, **21**, 308–323 (1982).

Garcia, C. E., and M. Morari, "Internal Model Control. 2. Design Procedures for Multivariable Systems," *Ind. Eng. Chem. Process Des. Dev.*, **24**, 472–484 (1985a).

Garcia, C. E., and M. Morari, "Internal Model Control. 3. Multivariable Control Law Computation and Tuning Guidelines," *Ind. Eng. Chem. Process Des. Dev.*, **24**, 484–494 (1985b).

Jury, E. I., *Theory and Application of the Z-Transform Method*, Huntington, New York, 1964.

Krishnaswamy, P., G. Rangaiah, R. Jha, and P. Deshpande, "When to Use Cascade Control," *Ind. Eng. Chem. Res.*, **29**, 2163–2166 (1990).

Lloyd, S. G., and G. D. Anderson, *Industrial Process Control*, Fisher Controls, Marshalltown, IA, 1971.

Luyben, W. L., *Process Modeling, Simulation, and Control for Chemical Engineers*, 2nd ed., McGraw-Hill, New York, 1990.

Marchetti, J. L., D. A. Mellichamp, and D. E. Seborg, "Predictive Control Based on Discrete Convolution Models," *Ind. Eng. Chem. Process Des. Dev.*, **22**, 488–495 (1983).

Marlin, T. E., *Process Control: Designing Processes and Control Systems for Dynamic Performance*, McGraw-Hill, New York, 1995.

Otto, R. E., St. Louis, MO, personal communication, November 1986.

Richalet, J., A. Rault, J. L. Testud, and J. Papon, "Model Predictive Heuristic Control: Applications to Industrial Processes," *Automatica*, **14**, 413–428 (1978).

Ricker, N. L., T. Subrahmanian and T. Sim, "Case Studies of Model Predictive Control in Pulp and Paper Production," in McAvoy, T. J., Y. Arkun, and E. Zafiriou (eds.), *Model Based Process Control*, Pergamon, Oxford, 1989.

Rouhani, R., and R. K. Mehra, "Model Algorithmic Control (MAC): Basic Theoretical Properties," *Automatica*, **18**, 401–414 (1982).

Stephanopoulos, G., *Chemical Process Control*, Prentice-Hall, Englewood Cliffs, NJ, 1984.

Stoica, P., and T. Söderström, "Uniqueness of Prediction Error Estimates of Multivariable Moving Average Models," *Automatica*, **18**, 617–620 (1982).

Van Hoof, A. H., C. R. Cutler, and S. G. Finlayson,"Application of a Constrained Multi-variable Controller to a Hydrogen Plant," *Proc. Am. Control Conf.*, Pittsburgh, PA, vol. 2, 1555–1560 (1989).

Wassick, J. M., and D. T. Camp, "Internal Model Control of an Industrial Extruder," *Proc. Am. Control Conf.*, Atlanta, GA, vol. 3, 2347–2352 (1988).

# 7 Multivariable Regulatory Control

**Chapter Topics**

- Multiloop control
- Decoupling control
- Singular-value decomposition
- Multivariable model predictive control

*Scenario:* Multivariable control can effectively control a process with moderate interaction and is not necessarily complex.

Our engineer was called in to help in the control of an ammonia reform furnace. Because of the construction of the particular furnace, any disturbance to the process caused the operator to spend an excessive amount of time adjusting the controllers in order to bring the furnace temperatures back to the normal operating point. The primary reformer in an ammonia plant produces hydrogen by reforming a mixture of desulfurized natural gas and steam. An isometric view of the reform furnace is shown in Figure 7.1. The steam–gas mixture enters the furnace and is distributed to five rows of catalyst tubes suspended in the furnace. The catalyst tubes are interspersed between six rows of burners, resulting in a highly coupled system. Each burner row has a fuel pressure regulator that controls the fuel to that burner row. Any attempt to change the temperature of any one of the catalyst tubes requires many adjustments of the fuel pressure regulators. Because of the highly interactive nature of the process, the furnace temperatures cycled ±30 to ±50°F, which shortens the catalyst tube life. Because of his success at solving other process control problems, our engineer was asked to help. After examining the problem, it became obvious that a multivariable control scheme was needed. A simple decoupling control scheme was designed and is documented later in this chapter. The first time the algorithm was placed on-line, the five tube temperatures converged to their setpoint within 100 min. With further tweaking of the controller, the algorithm could bring the temperatures to their setpoint within 30 min. In addition, the normal band of temperatures was held to within ±5°F. This particular control design was successful and was also implemented in other plants (Otto and Nieman, 1979).

## 7.1 INTRODUCTION

The vast majority of control loops in a plant are single loop in nature and are handled quite adequately by PID controllers. However, as illustrated by the above scenario, many unit processes are inherently multivariable; that is, they have more than one controlled variable and more than one manipulated variable. While in many situations one can adequately determine the "correct" way to pair measurements and manipulated variables and

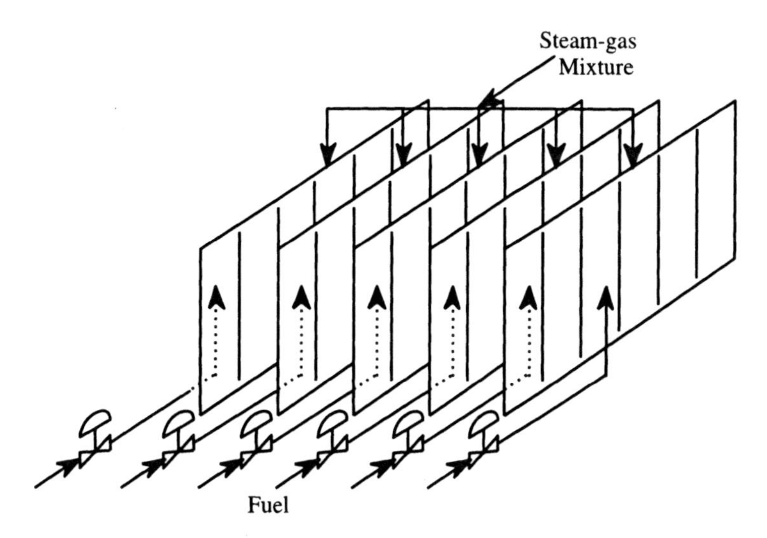


FIGURE 7.1. Isometric view of ammonia reform furnace. (Reprinted from *Industrial Process Design*, with changes. Copyright © 1979, American Institute of Chemical Engineers.)

construct some arrangement of single-loop controllers, there are units for which this approach will not give acceptable results. For systems with interaction, it is often better to analyze the entire system and, with some analytical guidance, determine the best control approach.

The objective of multivariable control is to simultaneously control more than one variable in a process unit or cell. Thus, in the control hierarchy, a multivariable controller is part of unit supervision or cell supervision, as illustrated in Figure 7.2. In general, the multivariable controller object takes measurements from indicator and status objects and manipulates the setpoints of loop objects. In some cases, the controller directly manipulates the device objects. Inherent in any multivariable control strategy is the need for procedural elements, such as startup, steady-state, and shutdown procedures. The multivariable control strategy generally deals with the steady-state operation and that is the assumption in this chapter. Procedural elements are treated in Chapters 8 and 9.

What sets multivariable control apart from single-loop control or single-loop control with enhancements is the *interaction* in the process. Interaction is evident when a change in one manipulated variable affects more than one process variable. For example, in the ammonia reform furnace example, a change in one of the fuel pressure regulators influenced two tube row temperatures. Process interaction results from coupled process relationships. For example, in a batch reactor, a change in the flow rate of the cooling jacket water influences not only the reactor temperature but also the final concentration of the product, through the reaction rate constant. In a distillation column, a change in the steam flow to the reboiler influences the composition of the product not only at the column bottom but also at the top of the column.

There are two basic approaches to multivariable control. The first, *multiloop* control, is basically an extension of single-loop control. In this approach, the single-loop controllers are connected to the process in such a way (called *loop pairing*) as to reduce the process interaction. An extension of multiloop control is *decoupling* control where a decoupler is

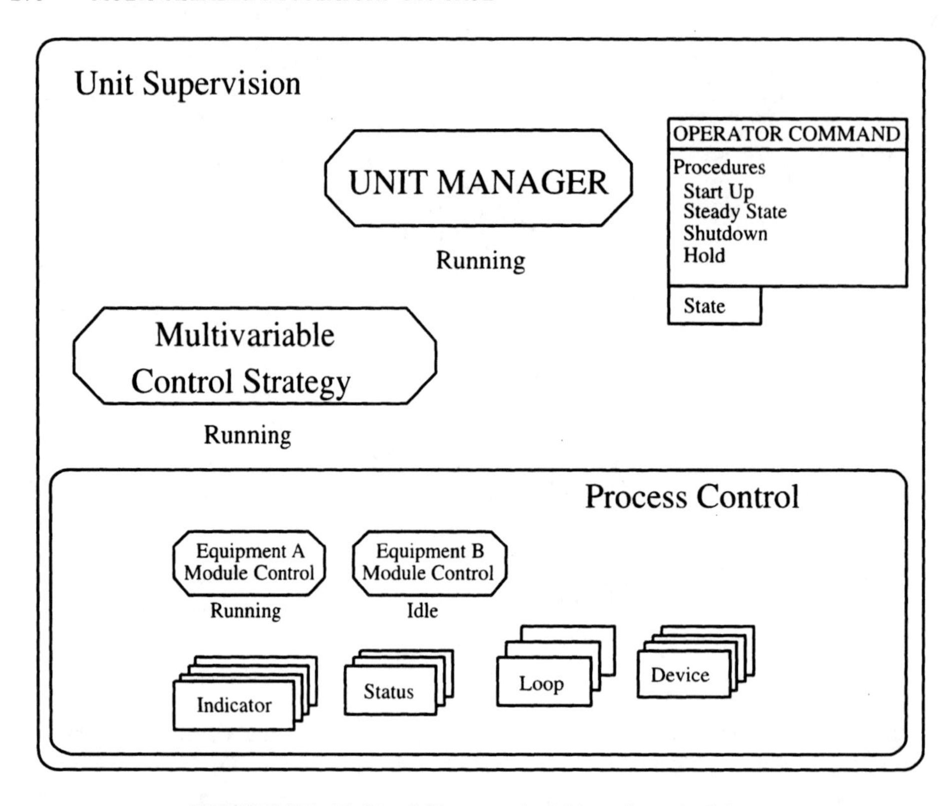


**FIGURE 7.2.** Multivariable control within unit supervision.

constructed so that the process interaction is "factored out" and the single-loop controllers perceive little process interaction. The second basic approach, called *full multivariable* control, uses all measurements to simultaneously calculate the manipulated variables.

As an example, a distillation column is shown in Figure 7.3 with its single-loop controllers that are the unit loop objects. The objective is to simultaneously control the product concentration in the distillate (>95%) and the product concentration in the bottoms (<5%). A possible multiloop control scheme for the column is shown in Figure 7.4. The bottom product concentration measurement is paired with the reboiler steam flow setpoint to form one control loop. The distillate concentration measurement is paired with the reflux flow to form the second loop. Whether this control scheme leads to good control depends on the constituents being separated. A decoupled control scheme is shown in Figure 7.5. In this design, a decoupler is inserted between the single-loop controller outputs and the manipulated variables. The decoupler is designed so the single-loop controllers perceive little process interaction. With a full multivariable controller, connected as shown in Figure 7.6, there is no concept of "loops" since all measurements are considered simultaneously in order to calculate the manipulated variables.

Note that the outputs of the multivariable controllers for the distillation column (Figures 7.4–7.6) manipulate the setpoint of a single-loop controller rather than directly manipulate the valve position. This cascade arrangement is used mainly to mitigate the effects of pressure disturbances in the flow streams. For full multivariable controllers, this arrangement is mandatory since they may only execute once every minute and therefore a lower level single-loop controller must maintain the flow.

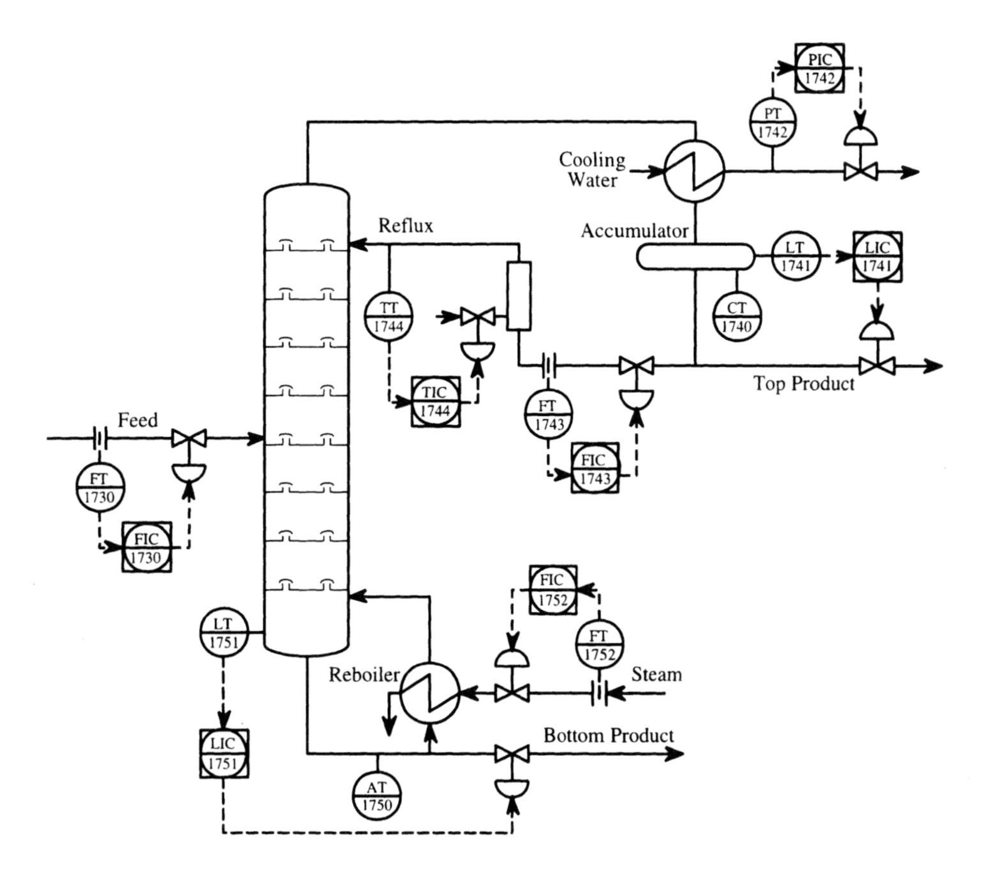


**FIGURE 7.3.** Distillation column.

This section is only intended to summarize the most important techniques of multivariable control. Multiloop control is presented first, followed by decoupling control. The singular-value decomposition is introduced and its use in designing multiloop control and in choosing the manipulated and process variables is illustrated. Finally, model predictive control, the most frequently used true multivariable process control technique, is presented. Illustrative examples are used throughout this chapter. More information may be found in Marlin (1995), Stephanopoulos (1984), Seborg et al. (1989), and the other references cited throughout this chapter.

## 7.2 MULTILOOP CONTROL

The *multiloop* approach, which uses multiple single-loop controllers, was the first approach used for control of multivariable processes. Over the past decades, many multiloop strategies have been developed and continue to be used. For example, Figure 7.4 depicts a multiloop control scheme for a distillation column (Wood and Berry, 1973). The control for most of the variables in the process are adequately handled by single-loop controllers. The feed flow, cooling-water pressure, accumulator level, reflux temperature, and bottom level are all easily controlled by conventional PID controllers. However, the

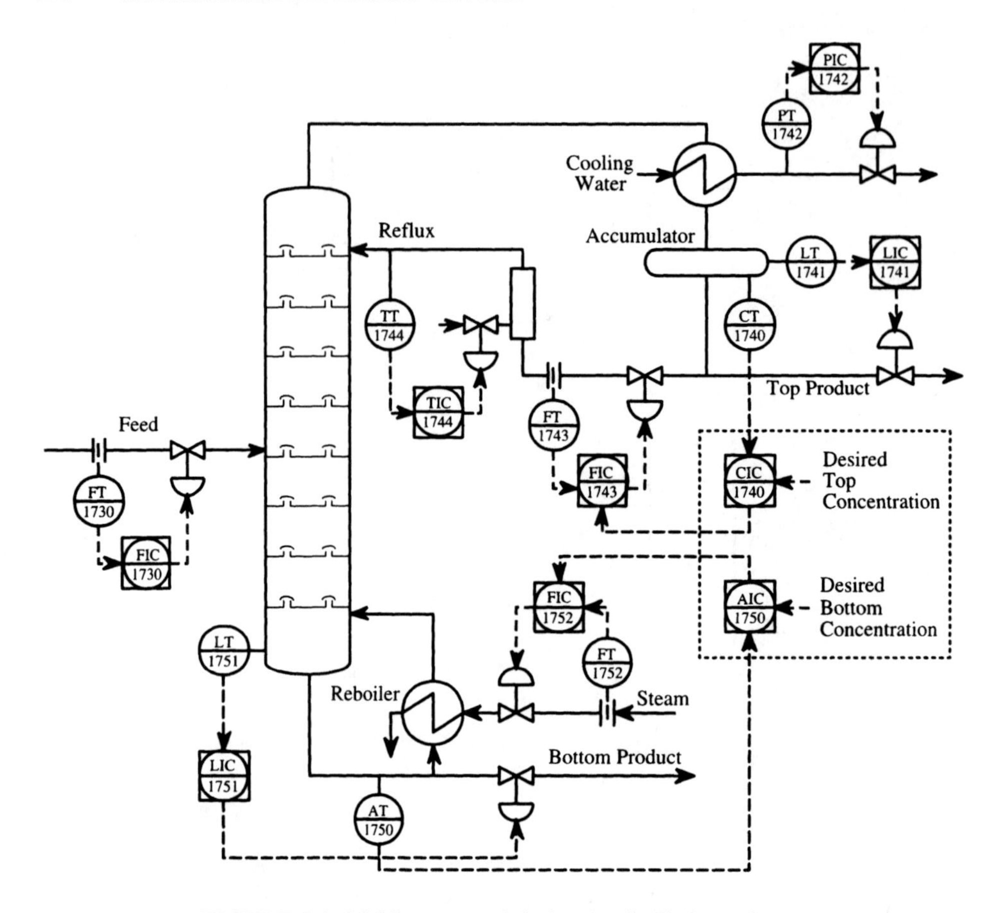


**FIGURE 7.4.** Multiloop control design for distillation column.

top and bottom product compositions are both affected by the reflux flow and steam flow. This part of the process is interactive, and some guidance is needed to pair each measurement with the appropriate manipulated variable.

A multivariable process has *interaction* if a given process manipulated variable (input) affects more than one controlled variable (output). This interaction is the key problem in multiloop control systems. After some examples illustrating the problem of interaction, the relative-gain array (RGA) is defined and then used to guide in the problem of pairing loops. Finally, guidelines for tuning the paired loops are given.

### 7.2.1 Interaction and Its Effect on Multivariable System Behavior

Interaction in the process is the basic problem encountered when designing a multiloop control scheme. Generally, when connecting the SISO PID controllers, the manipulated variable should be paired with the process variable that is affected most by a change in the manipulated variable. However, this choice is not always obvious for processes that have more than two inputs and two outputs. In addition, the presence of interaction in the process will influence the best achievable control performance. The following examples

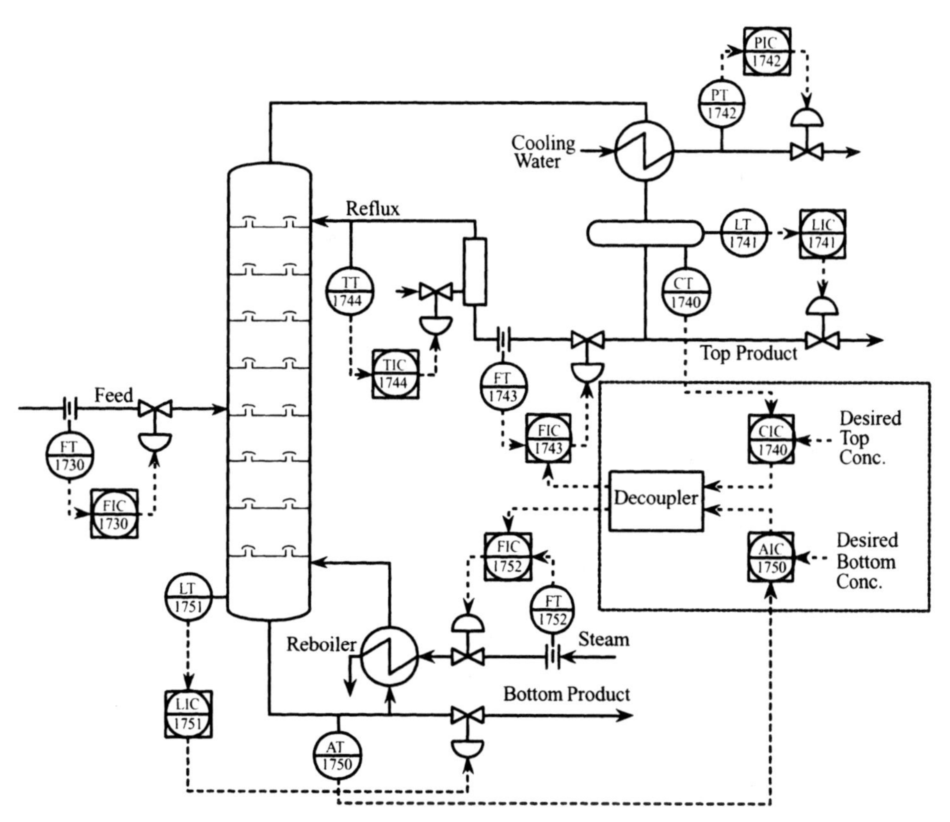


**FIGURE 7.5.** Decoupling control design for distillation column.

will illustrate this pairing problem when the process has interaction. In addition, the effect of interaction on the controllability and process operating window is demonstrated.

**Example 7.1.** Develop a model of the blending process depicted in Figure 7.7 where pure A is mixed with a solvent. The controlled variables are the production rate $F_m$ and the blended product composition $X_m$ the mass fraction of A in the blended mixture. The two manipulated variables are the flows in each inlet stream. The model should also include the effects of a disturbance in the solvent pressure.

***Solution.*** The blending is modeled with the following assumptions:

1. Constant inlet concentrations
2. Constant inlet pressure of A
3. Perfect mixing where flows meet
4. Solvent and A have equal densities and temperatures

The overall material balance at the mixing point is

$$F_m = F_A + F_S \tag{7.1}$$

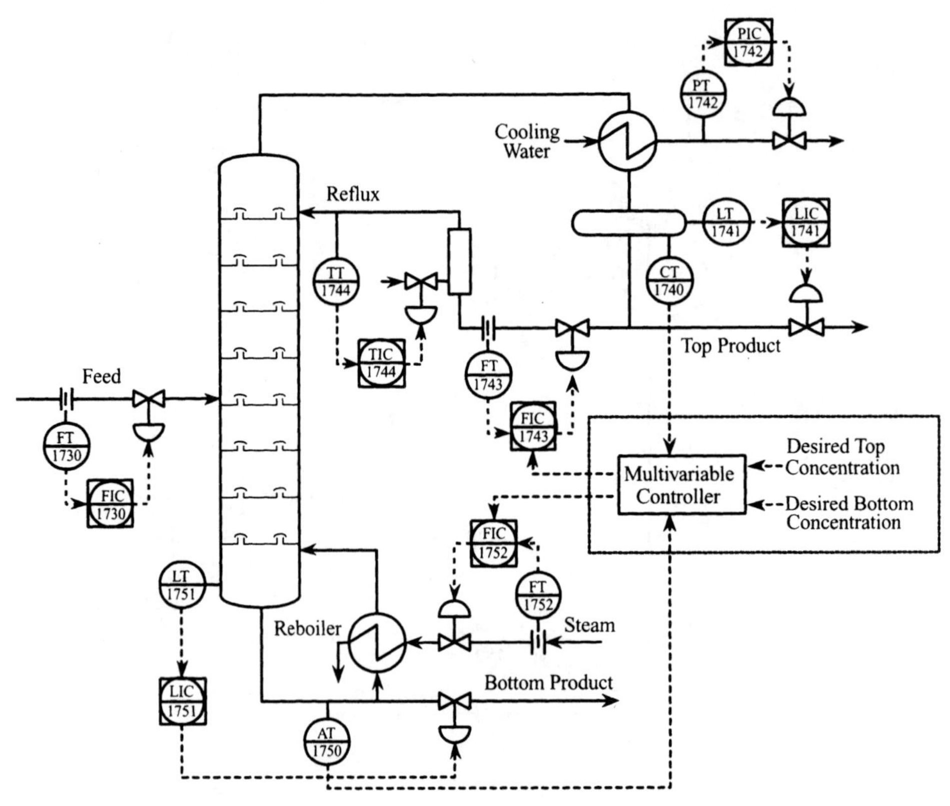


**FIGURE 7.6.** Multivariable control design for distillation column.

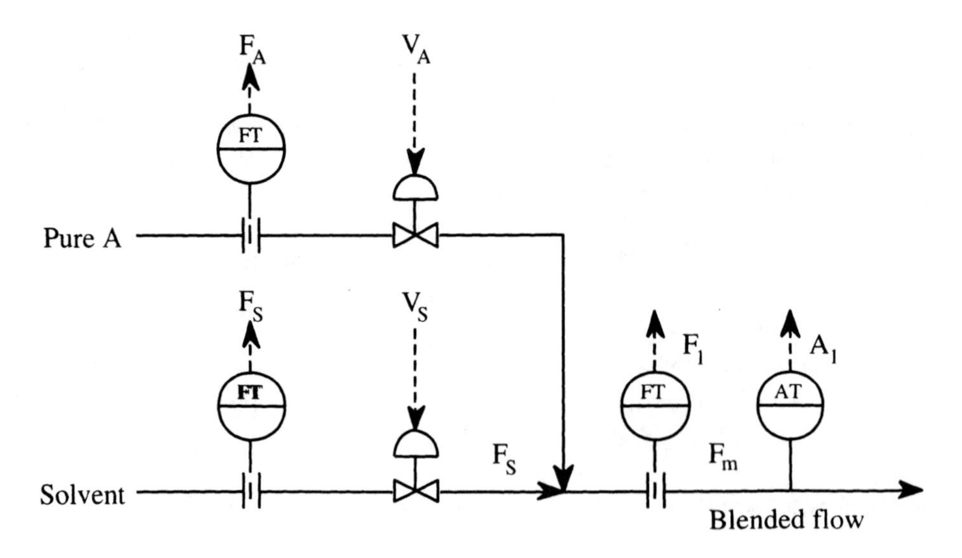


**FIGURE 7.7.** Blending process.

The material balance on component A is

$$F_m X_m = F_A$$

or

$$X_m = \frac{F_A}{F_S + F_A} \tag{7.2}$$

Equation (7.2) can be linearized about an operating point $(F_{A0}, F_{S0})$ to produce

$$\bar{X}_m(t) = \left(\frac{F_S}{(F_S + F_A)^2}\right)_{\text{op.pt.}} \bar{F}_A(t) + \left(\frac{-F_A}{(F_S + F_A)^2}\right)_{\text{op.pt.}} \bar{F}_S(t) \tag{7.3}$$

where the bar over the variables indicates deviation from the operating point. Since the system is liquid filled, any change in the valve positions will instantly change the component flow rate and the blended flow rate. It is assumed that the valve dynamics are negligible and the valves have equal-percentage flow characteristics, that is, the flow is linear with respect to the valve position ($F_A = V_A$, $F_S = V_S$). The flow and concentration sensors do have some dynamics associated with them. The flow sensor is assumed to have a first-order response with a gain of 1.0, time constant of 0.2 sec, and dead time of 0.1 sec. The concentration sensor has a slower response and is assumed to have a first-order response with a gain of 1.0, time constant of 2.0 sec and dead time of 0.2 sec. The solvent pressure disturbance $D$ is assumed to have the same effect on the solvent flow as the valve position.

Equations (7.1) and (7.3) are combined with the sensor dynamics and disturbance to give the following linearized dynamic model:

$$\begin{aligned} A_1(s) = {} & \frac{(F_S/(F_S + F_A)^2)_{\text{op.pt.}} e^{-0.2s}}{2.0s + 1} V_A(s) + \frac{(-F_A/(F_S + F_A)^2)_{\text{op.pt.}} e^{-0.2s}}{2.0s + 1} V_S(s) \\ & + \frac{(-F_A/(F_S + F_A)^2)_{\text{op.pt.}} e^{-0.2s}}{2.0s + 1} D(s) \end{aligned} \tag{7.4}$$

$$F_1(s) = \frac{1.0e^{-0.1s}}{0.2s + 1} V_A(s) + \frac{1.0e^{-0.1s}}{0.2s + 1} V_S(s) + \frac{1.0e^{-0.1s}}{0.2s + 1} D(s) \tag{7.5}$$

Shown as matrices, the process is

$$\begin{bmatrix} A_1(s) \\ F_1(s) \end{bmatrix} = \begin{bmatrix} \dfrac{(F_S/(F_S + F_A)^2)_{\text{op.pt}} e^{-0.2s}}{2.0s + 1} & \dfrac{(-F_A/(F_S + F_A)^2)_{\text{op.pt}} e^{-0.2s}}{2.0 + 1} \\ \dfrac{1.0e^{-0.1s}}{0.2s + 1} & \dfrac{1.0e^{-0.1s}}{0.2s + 1} \end{bmatrix} \begin{bmatrix} V_A(s) \\ V_S(s) \end{bmatrix} + \begin{bmatrix} \dfrac{(-F_A/(F_S + F_A)^2)_{\text{op.pt}} e^{-0.2s}}{2.0s + 1} \\ \dfrac{1.0e^{-0.1s}}{0.2s + 1} \end{bmatrix} D(s)$$

Interaction is evident in the model since each measurement is influenced by both manipulated variables.

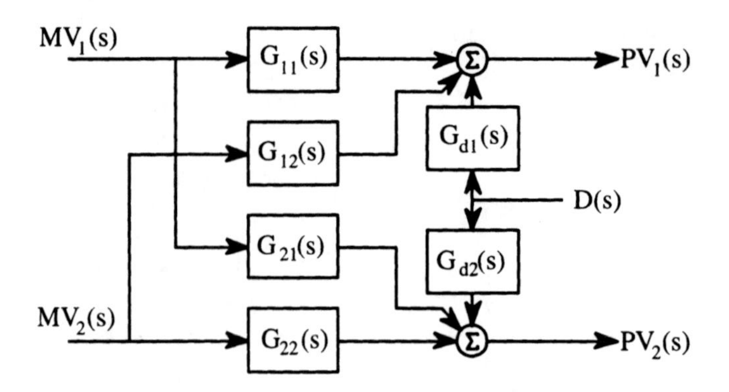


**FIGURE 7.8.** Block diagram of a 2 × 2 process.

The typical block diagram of a two-input, two-output process is shown in Figure 7.8. The relationship between inputs and outputs is often presented in matrix form as

$$\begin{bmatrix} PV_1(s) \\ PV_2(s) \end{bmatrix} = \begin{bmatrix} G_{11}(s) & G_{12}(s) \\ G_{21}(s) & G_{22}(s) \end{bmatrix} \begin{bmatrix} MV_1(s) \\ MV_2(s) \end{bmatrix} + \begin{bmatrix} G_{d1}(s) \\ G_{d2}(s) \end{bmatrix} D(s) \tag{7.6}$$

Each element in the second and fourth matrices is a transfer function relating one input to one output. The interaction in the process comes from the off-diagonal terms $G_{12}(s)$ and $G_{21}(s)$.

*Controllability.* One important issue in multivariable control is the independence of the relationships between the manipulated and process variables. A *controllable* process is one in which the relationships are independent. Formal definitions exist for the *controllability* of a process (e.g., Brogan, 1991; Rugh, 1996). For the purposes of this text, a simpler definition is used (Marlin, 1995): A system is controllable if the determinant of the gain matrix is nonzero. Note that this definition applies only to square systems. In order for nonsquare systems to be candidates for multiloop control, measurements and/or manipulated variables need to be added or deleted in order to make the number of manipulated variables equal to the number of measurements.

**Example 7.2.** Evaluate the controllability of the blending process.

***Solution.*** The gain matrix is

$$\begin{bmatrix} \dfrac{F_s}{(F_S+F_A)^2} & \dfrac{-F_A}{(F_S+F_A)^2} \\ 1 & 1 \end{bmatrix}$$

and its determinant is

$$\frac{F_S+F_A}{(F_S+F_A)^2} = \frac{1}{F_S+F_A} \neq 0$$

Therefore, this process is controllable with the selected measurements and manipulated variables.

*Operating Window.* Another important consideration is the possible range of the system variable values. The term *operating window* is used to describe this range. The operating window can be shown in various ways (Marlin, 1995). In one approach, the controlled variables are used as coordinates to indicate the range of possible setpoints, with disturbances set to zero. Another common approach uses the disturbance variables as coordinates and indicates the range of disturbance values that can be handled by the control system.

**Example 7.3.** Assuming that the maximum component flow rates for the blending example are $F_{A,\max}$ and $F_{S,\max}$, draw the operating window of attainable flow rate and composition.

***Solution.*** The attainable total flow rate $F_1$ and composition $A_1$, are shown in Figure 7.9. Obviously, the total flow is limited by the sum of the two component flows. The diagonal line on the left showing the limit of the composition is determined by setting the solvent flow to its maximum value and then solving (7.2) for the composition. The right diagonal line is obtained in a similar manner, except the flow of component A is fixed at its maximum value and the solvent flow is varied.

*System Behavior.* When single-loop controllers are connected to the $2 \times 2$ system of Figure 7.8, one possible block diagram of the resulting system appears as in Figure 7.10. The alternative configuration is to swap the two controller outputs. Clearly, a change in the setpoint of one controller also influences the other controller because of the off-diagonal terms $G_{12}(s)$ and $G_{21}(s)$. However, rather than do a rigorous treatment of the system

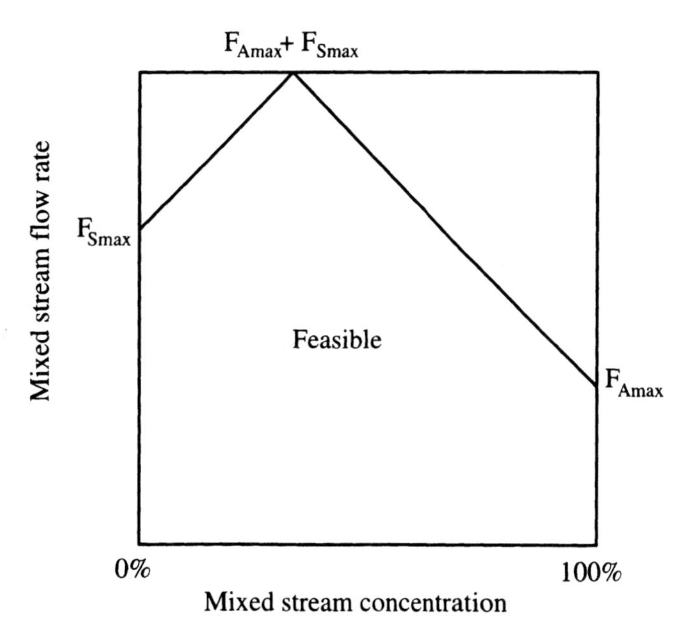


**FIGURE 7.9.** Operating window for blending process. [From Marlin (1995). Copyright © 1995, McGraw-Hill. Reproduced by permission.]

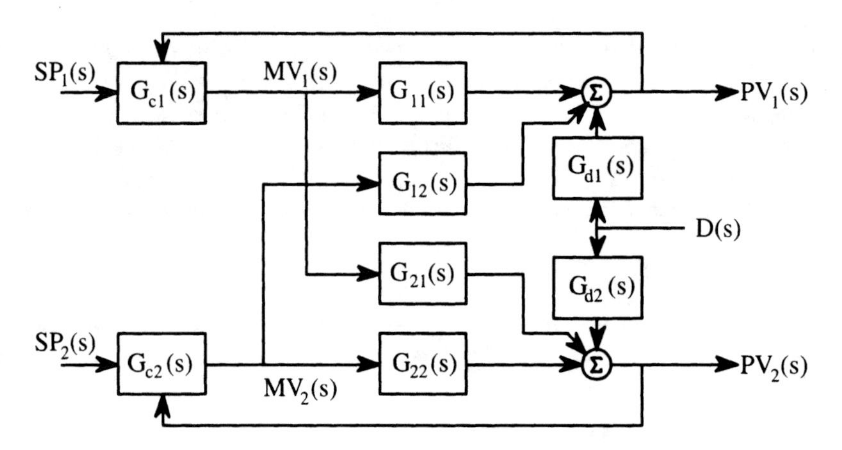


**FIGURE 7.10.** Block diagram of a 2 × 2 process with single-loop controllers.

behavior, the system behavior to differing magnitudes of the off-diagonal terms is demonstrated by a simple two-input, two-output example.

**Example 7.4.** Evaluate the performance of a control system for the blending system of Example 7.1. The outlet flow $F_1$ is 1.0. Evaluate the performance for $A_1 = 0.05, 0.20, 0.45$ when the flow setpoint is changed from 1.0 to 1.2.

***Solution.*** For the operating point $A_1 = 0.05$, the component flows are $F_A = 0.05$ and $F_S = 0.95$. The model of the process at this operating point is

$$\begin{bmatrix} A_1(s) \\ F_1(s) \end{bmatrix} = \begin{bmatrix} \dfrac{0.95e^{-0.2s}}{2.0s+1} & \dfrac{-0.05e^{-0.2s}}{2.0s+1} \\ \dfrac{1.0e^{-0.1s}}{0.2s+1} & \dfrac{1.0e^{-0.1s}}{0.2s+1} \end{bmatrix} \begin{bmatrix} V_A(s) \\ V_S(s) \end{bmatrix} + \begin{bmatrix} \dfrac{-0.05e^{-0.2s}}{2.0s+1} \\ \dfrac{1.0e^{-0.1s}}{0.2s+1} \end{bmatrix} D(s) \tag{7.7}$$

The controllers are connected as shown in Figure 7.11. The concentration controller, AIC1503, manipulates $V_A$, and the flow controller, FIC1503, manipulates $V_B$. The controllers are tuned assuming no interaction; that is, AIC1503 is tuned assuming the process is $G_{11}(s)$ and FIC1503 is tuned assuming the process is $G_{22}(s)$. The controllers are tuned using the Fertik tuning for a disturbance response. For the process at the operating point $A_1 = 0.20$, the model of the process at this operating point is

$$\begin{bmatrix} A_1(s) \\ F_1(s) \end{bmatrix} = \begin{bmatrix} \dfrac{-0.80e^{-0.2s}}{2.0s+1} & \dfrac{-0.20e^{-0.2s}}{2.0s+1} \\ \dfrac{1.0e^{-0.1s}}{0.2s+1} & \dfrac{1.0e^{-0.1s}}{0.2s+1} \end{bmatrix} \begin{bmatrix} V_A(s) \\ V_S(s) \end{bmatrix} + \begin{bmatrix} \dfrac{-0.20e^{-0.2s}}{2.0s+1} \\ \dfrac{1.0e^{-0.1s}}{0.2s+1} \end{bmatrix} D(s)$$

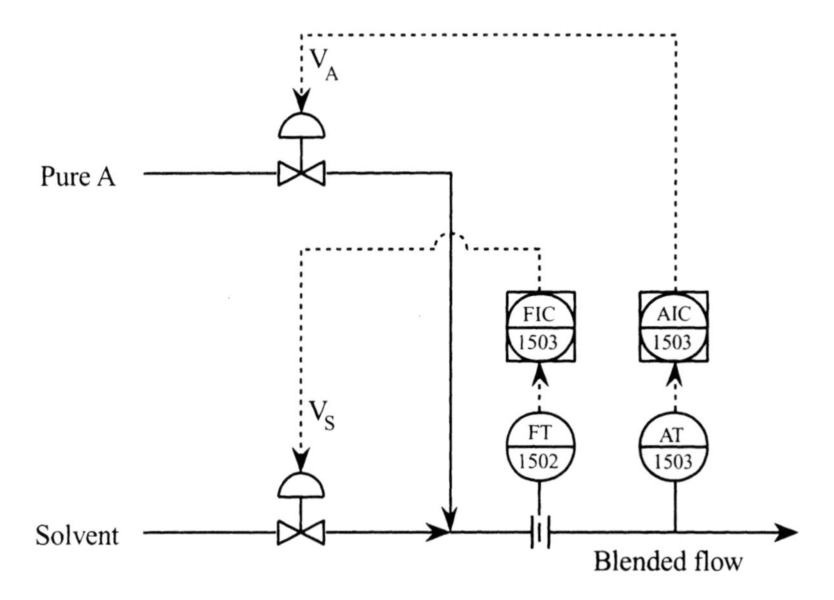


**FIGURE 7.11.** Multiloop control system for blending process.

At the operating point $A_1 = 0.45$, the model of the process is

$$\begin{bmatrix} A_1(s) \\ F_1(s) \end{bmatrix} = \begin{bmatrix} \dfrac{-0.55e^{-0.2s}}{2.0s+1} & \dfrac{-0.45e^{-0.2s}}{2.0s+1} \\ \dfrac{1.0e^{-0.1s}}{0.2s+1} & \dfrac{1.0e^{-0.1s}}{0.2s+1} \end{bmatrix} \begin{bmatrix} V_A(s) \\ V_S(s) \end{bmatrix} + \begin{bmatrix} \dfrac{-0.45e^{-0.2s}}{2.0s+1} \\ \dfrac{1.0e^{-0.1s}}{0.2s+1} \end{bmatrix} D(s)$$

The flow and concentration responses for a flow setpoint change of 0.2 at time 0.5 sec is shown in Figure 7.12. The responses are shown as deviations from the operating point. As shown by the responses, the larger the gain of the off-diagonal term, the more interaction, indicated by the disturbance in the concentration loop whose setpoint is not being changed. Since the relative magnitudes of the gains in the first row of the process when $A_1 = 0.45$ are close, one may be tempted to try interchanging the manipulated variables for the two controllers. After this interchange, the two controllers are retuned according to the appropriate process model. The flow setpoint change responses are shown in Figure 7.13. Clearly, the system is getting closer to instability. Although not shown, the response to a disturbance is marginally unstable—clearly a situation one would want to avoid.

In the previous example, the correct pairing of the manipulated and measured variables was not too difficult. However, the situation becomes more difficult for a system with more than two inputs and two outputs. One certainly wants to avoid loop pairings that are unstable. In addition, the control system designer needs to know those loops for which good control will be difficult. The next section presents such a quantitative measure.

### 7.2.2 Relative-Gain Array

As was shown in the previous section, interaction in the process is an important consideration when designing a multiloop control system. The relative gain array (RGA) was developed in order to measure the process interaction and provide a tool in the design

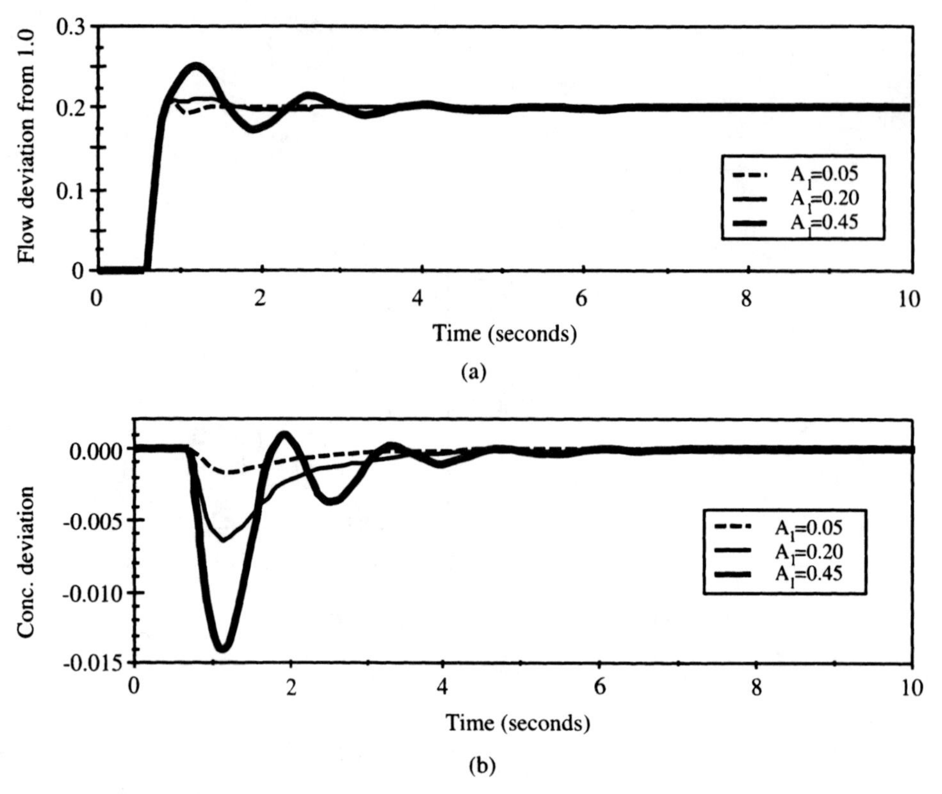


**FIGURE 7.12.** Response of blending control system at various operating points, shown as deviation: (*a*) flow response as deviation from operating point; (*b*) concentration response.

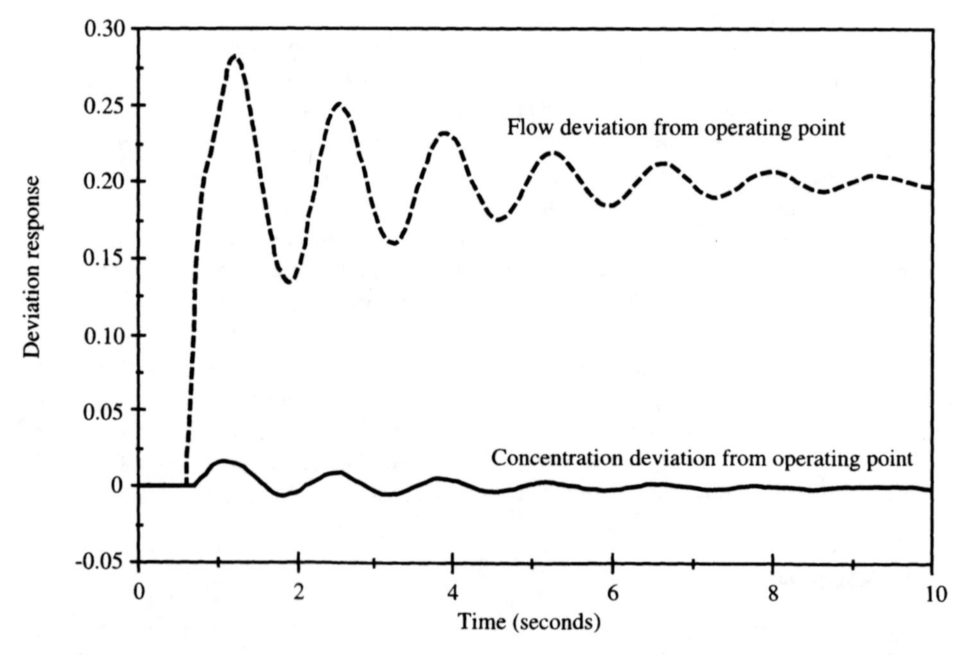


**FIGURE 7.13.** Response of blending control system when $A_1$ is 0.40 and the controller manipulated variables are interchanged.

of multiloop control systems. The RGA was developed by Bristol (1966) and extended by others (McAvoy, 1983; Shinskey, 1988). In this section, the RGA is defined and some of its important properties are summarized. Control-related interpretations of the RGA are presented and will be used in the next section to choose the loop pairings when designing a multiloop control system.

The RGA is a matrix composed of elements, called $\lambda_{ij}$. The element in the $i$th row and $j$th column $\lambda_{ij}$ is the ratio of the steady-state gain between the $i$th process variable and the $j$th manipulated variable when all other manipulated variables are constant divided by the steady-state gain between the same two variables when all other process variables are constant.

$$\lambda_{ij} = \frac{[\partial \mathrm{PV}_i/\partial \mathrm{MV}_j]_{\mathrm{MV}_k=\mathrm{const},k\neq j}}{[\partial \mathrm{PV}_i/\partial \mathrm{MV}_j]_{\mathrm{CV}_k=\mathrm{const},k\neq i}} = \frac{[\partial \mathrm{PV}_i/\partial \mathrm{MV}_j]_{\text{other loops in manual}}}{[\partial \mathrm{PV}_i/\partial \mathrm{MV}_j]_{\text{other loops in automatic}}} \tag{7.8}$$

In the definition, it is assumed that all controllers have integral mode so that when in Automatic the steady-state values of the process variables are maintained constant (i.e., at zero). If the relative gain is 1.0, then the process gain for that loop is unaffected by the other control loops and no interaction exists. Therefore, the amount that the relative gain deviates from 1.0 indicates the amount of interaction present in the process.

Rather than apply the definition of (7.8), the RGA can be computed directly from the steady-state gain matrix **K** in the following manner:

$$\mathrm{RGA} = [\mathbf{K}^{-1}]^{\mathrm{T}} \otimes \mathbf{K} \tag{7.9}$$

where $\otimes$ is called the *Hadamard product* (Johnson, 1974) and is an *element-by-element* multiplication. If using a matrix calculation package like MATLAB, Equation (7.9) is rendered as

```
rga = k.*(inv(k)')
```

where "k" is the matrix of process steady-state gains.

Some important properties of the RGA are:

1. The relative gain is scale independent. That is, it will not change if the transmitter or manipulated variable ranges or units are changed.
2. The rows and columns sum to 1.0. For a 2 × 2 system, the remainder of the RGA may be determined when only $\lambda_{11}$ is given,

| | $\mathrm{MV}_1$ | $\mathrm{MV}_2$ |
|---|---|---|
| $\mathrm{PV}_1$ | $\lambda_{11}$ | $1-\lambda_{11}$ |
| $\mathrm{PV}_2$ | $1-\lambda_{11}$ | $\lambda_{11}$ |

3. The relative gain can be sensitive to errors in the gain matrix. As an example, consider a 2 × 2 system whose gain matrix is

$$\mathbf{K} = \begin{bmatrix} K_{11} & K_{12} \\ K_{21} & K_{22} \end{bmatrix}$$

It can be shown (McAvoy, 1983) that the (1, 1) element of the RGA is

$$\lambda_{11} = \frac{K_{11}K_{22}}{K_{11}K_{22} - K_{12}K_{21}} = \frac{1}{1 - K_{12}K_{21}/K_{11}K_{22}} \tag{7.10}$$

Obviously, when the products $K_{11}K_{22}$ and $K_{12}K_{21}$ are close in value, a small change in any of them may cause the sign of $\lambda_{11}$ to change. Since the sign of the relative gain is important in determining loop pairings, the values of the gain matrix should be accurately determined. Generally, one should determine the model gains from an analytic model. Gains calculated from empirical models may be used, but with caution.

The control-related interpretations of the RGA are as follows and will be used in succeeding sections of this chapter.

$\lambda_{ij} < 0$ — If this loop is opened (controller placed in Manual) the remainder of the multiloop system may become unstable. The negative relative gain arises because the open- and closed-loop process gains are of opposite signs.

$\lambda_{ij} = 0$ — Generally, this value indicates that either (1) the manipulated variable has no effect on the measurement, a situation to be avoided when determining pairings, or (2) the manipulated variable has an effect only when other loops are in Automatic. Pairing variables in the latter situation is possible in certain situations (Marlin, 1995), but generally, pairing on zero elements is avoided.

$0 < \lambda_{ij} < 1.0$ — The steady-state loop gain with the other loops in Automatic is larger than the same gain when the other loops are in Manual.

$\lambda_{ij} = 1.0$ — The process variable is only affected by a change in the manipulated variable. However, the manipulated variable may also affect other process variables, called *one-way interaction*.

$1.0 < \lambda_{ij}$ — The steady-state loop gain with the other loops in Automatic is smaller than the same gain when the other loops are in Manual.

$\lambda_{ij} = \infty$ — The process gain is zero when the other loops are in Automatic, and thus it is not possible to control this variable in a multiloop system.

**Example 7.5.** Determine the RGA for the different operating points of Example 7.4. Relate the RGA values to the control performance observed in Example 7.4.

***Solution.*** The steady-state gain matrix and the resulting RGA for each operating point is shown below:

$$A_1 = 0.05 \qquad \mathbf{K} = \begin{bmatrix} 0.95 & -0.05 \\ 1 & 1 \end{bmatrix} \qquad \text{RGA} = \begin{bmatrix} 0.95 & 0.05 \\ 0.05 & 0.95 \end{bmatrix}$$

$$A_1 = 0.20 \qquad \mathbf{K} = \begin{bmatrix} 0.80 & -0.20 \\ 1 & 1 \end{bmatrix} \qquad \text{RGA} = \begin{bmatrix} 0.80 & 0.20 \\ 0.20 & 0.80 \end{bmatrix}$$

$$A_1 = 0.45 \qquad \mathbf{K} = \begin{bmatrix} 0.55 & -0.45 \\ 1 & 1 \end{bmatrix} \qquad \text{RGA} = \begin{bmatrix} 0.55 & 0.45 \\ 0.45 & 0.55 \end{bmatrix}$$

The RGAs confirm the results from Example 7.4. The control when $A_1 = 0.05$ and $A_1 = 0.20$ is reasonably good, while the RGA for $A_1 = 0.45$ indicates that interaction is a problem, borne out by the simulation results of Example 7.4. Also, the RGA indicates that interchanging the manipulated variables is not recommended for any of the operating points.

With minimal process information, the RGA is one measure of the process interaction. Now this information needs to be used to determine the correct way to pair each measurement with a manipulated variable.

### 7.2.3 Loop Pairings

The following guidelines are suggested by McAvoy (1983) when using the RGA to determine the loop pairings for multiloop control:

| | |
|---|---|
| $\lambda_{ij} \leq 0$ | Avoid pairing these loops. |
| $0 < \lambda_{ij} \leq 0.67$ | Multiloop control will probably not function acceptably. Decoupling control should be considered. |
| $\lambda_{ij} = 1.0$ | The best possible pairing. |
| $0.67 < \lambda_{ij} \leq 1.5$ | Interaction with other loops is relatively small. Multiloop control will be reasonably good. |
| $2 < \lambda_{ij} \leq 10$ | A dynamic decoupler may need to be constructed in order to provide acceptable control. |
| $25 < \lambda_{ij}$ | Control system will function poorly. Control strategies other than multiloop or decoupling control should be considered. It may be necessary to consider other measurements and/or manipulated variables for the controllers. |

While the RGA is an important tool in designing multiloop systems, three additional considerations should be given to those loop pairings suggested by the RGA. The first is a test for instability. Niederlinski (1971) devised a stability test, later corrected by Grosdidier et al. (1985), that can be used to eliminate unworkable loop pairings. The controller settings do not have to be known, just that each controller includes integral action.

*Niederlinski Stability Theorem.* Given a multivariable system with $\mathbf{K}$ as the matrix of steady gains where $\mathbf{K}$ is arranged so that the controller pairings are diagonal, that is, $PV_1$ is paired with $MV_1$, $PV_2$ is paired with $MV_2$, and so on. Furthermore, each of the $n$ feedback controllers in the multiloop configuration contains integral action. The closed-loop system is unstable if

$$\frac{\det \mathbf{K}}{\prod_{i=1}^{n} K_{ii}} < 0 \tag{7.11}$$

This condition is a sufficient condition; that is, when the left-hand side of (7.11) is negative, the system *will* be unstable. However, if the left-hand side of (7.11) is positive, the system may or may not be stable. Further system analysis is needed. The stability of the system generally depends on the controller tuning parameters.

The second consideration on loop pairing is the dynamics of the various loops. The RGA is calculated strictly from process gain information and no dynamic information is used. Therefore, the dynamics of the suggested pairings should be verified. When possible, the dynamics between the suggested manipulated–process variable pair should have fast dynamics and small deadtimes. Example 7.8 demonstrates this consideration.

The third consideration is a rough evaluation of the expected disturbance response. The relative disturbance gain (RDG) was proposed by Stanley et al. (1985) as a complement to the RGA in evaluating multiloop control pairings. The RDG for the $i$th loop $\beta_i$ is defined as the ratio of the change in controller output required to counteract changes in the disturbance $d$ when all other process variables are constant divided by the change in controller output required to counteract the disturbance when all other manipulated variables are constant (no control over other process variables):

$$\beta_i = \frac{[\partial \mathrm{MV}_i/\partial d]_{\mathrm{PV}_k=\mathrm{const},k\neq i}}{[\partial \mathrm{MV}_i/\partial d]_{\mathrm{MV}_k=\mathrm{const},k\neq i}} = \frac{[\partial \mathrm{MV}_i/\partial d]_{\text{other loops in automatic}}}{[\partial \mathrm{MV}_i/\partial d]_{\text{other loops in manual}}} \tag{7.12}$$

As for the RGA, it is assumed that all controllers have integral mode. For a $2 \times 2$ process matrix of steady-state gains of the form

$$\begin{bmatrix} \mathrm{PV}_1 \\ \mathrm{PV}_2 \end{bmatrix} = \begin{bmatrix} K_{11} & K_{12} \\ K_{21} & K_{22} \end{bmatrix} \begin{bmatrix} \mathrm{MV}_1 \\ \mathrm{MV}_2 \end{bmatrix} + \begin{bmatrix} K_{F1} \\ K_{F2} \end{bmatrix} D$$

the two relative disturbance gains are

$$\beta_1 = \lambda_{11}\left(1 - \frac{K_{F2}K_{12}}{K_{F1}K_{22}}\right) \qquad \beta_2 = \lambda_{11}\left(1 - \frac{K_{F1}K_{21}}{K_{F2}K_{11}}\right)$$

where $\lambda_{11}$ is calculated as in Equation (7.10). The RDG indicates whether the interaction resulting from a particular disturbance is favorable or unfavorable. If $\beta_i > 1$, then the interaction is unfavorable since the controller in the multiloop configuration has a larger manipulated variable change than when in a SISO configuration. If $\beta_i < 1$, then the interaction is favorable. The RDG is meant to supplement, rather than replace, the RGA. As shown by various examples in Stanley et al. (1985), pairing on a large $\lambda_{ij}$ is acceptable if disturbance rejection is more important than other control system responses, that is, responses to setpoint changes.

In the remainder of this section, various examples are used to illustrate the use of the RGA in determining the pairing of manipulated and process variables.

**Example 7.6.** The following transfer function models for the methanol–water distillation column in Figure 7.3 are given by Wood and Berry (1973). The distillate and bottom products ($X_D$ and $X_B$) are the measurements, the reflux flow $F_R$ and the reboiler steam flow $F_S$ are the manipulated variables, and the feed flow $F$ is the disturbance:

$$\begin{bmatrix} X_D(s) \\ X_B(s) \end{bmatrix} = \begin{bmatrix} \dfrac{12.8e^{-s}}{16.7s+1} & \dfrac{-18.9e^{-3s}}{21s+1} \\ \dfrac{6.6e^{-7s}}{10.9s+1} & \dfrac{-19.4e^{-3s}}{14.4s+1} \end{bmatrix} \begin{bmatrix} F_R(s) \\ F_S(s) \end{bmatrix} + \begin{bmatrix} \dfrac{3.8e^{-8s}}{14.9s+1} \\ \dfrac{4.9e^{-7s}}{13.2s+1} \end{bmatrix} F(s)$$

Find the loop pairings for a multiloop control system and comment on its expected performance.

***Solution.*** The gain matrix is

$$\mathbf{K} = \begin{bmatrix} 12.8 & -18.9 \\ 6.6 & -19.4 \end{bmatrix}$$

Using (7.9), the RGA, with the suggested loop pairings circled, is

| | $F_R$ | $F_S$ |
|---|---|---|
| $X_D$ | (2.01) | −1.01 |
| $X_B$ | −1.01 | (2.01) |

The RGA indicates that $X_D$ should be controlled with $F_R$ and $X_B$ should be controlled with $F_S$. In addition, these two loops will interact with each other. The relative gains for these two loops indicate that decoupling control should probably be considered; decoupling is examined by Wood and Berry (1973). The control of this process is considered in later chapter examples.

Application of the Niederlinski stability theorem (7.11) shows that this loop pairing is not unstable. The process dynamics for the loop pairings are good, each having the minimum process deadtime for the row or column of the process matrix. The RDGs are $\beta_1 = -0.52$ and $\beta_2 = 1.21$, indicating that the interaction is beneficial for distillate composition control and slightly unfavorable for bottoms concentration control.

**Example 7.7.** The following gain matrix is derived for a heat exchanger system (Figure 7.14) in McAvoy (1983):

$$\begin{bmatrix} T_1 \\ \theta_2 \\ \theta_4 \end{bmatrix} = \begin{bmatrix} -0.3289 & 0.01725 & 0.01357 \\ -0.1566 & 0.00820 & -0.04847 \\ -1.007 & -0.03065 & 0.02715 \end{bmatrix} \begin{bmatrix} \mathrm{MV}_1 \\ \mathrm{MV}_2 \\ \mathrm{MV}_3 \end{bmatrix}$$

where the manipulated variables are $F_3$, $A/B$, and $C/D$. Find the loop pairings for a multiloop control system and comment on its expected performance.

***Solution.*** Using (7.9), the RGA, with the suggested loop pairings circled, is

| | $\mathrm{MV}_1$ | $\mathrm{MV}_2$ | $\mathrm{MV}_3$ |
|---|---|---|---|
| $T_1$ | 0.2755 | (0.6070) | 0.1175 |
| $\theta_2$ | 0.0918 | 0.0258 | (0.8824) |
| $\theta_4$ | (0.6327) | 0.3672 | 0.0001 |

The RGA indicates that $T_1$ should be controlled with $\mathrm{MV}_2$, $\theta_2$ should be controlled with $\mathrm{MV}_3$, and $\theta_4$ should be controlled with $MV_1$. In addition, the $\theta_2$–$\mathrm{MV}_3$ loop is relatively unaffected by the other two loops, while the $T_1$–$\mathrm{MV}_2$ and $\theta_4$–$\mathrm{MV}_1$ loops will interact

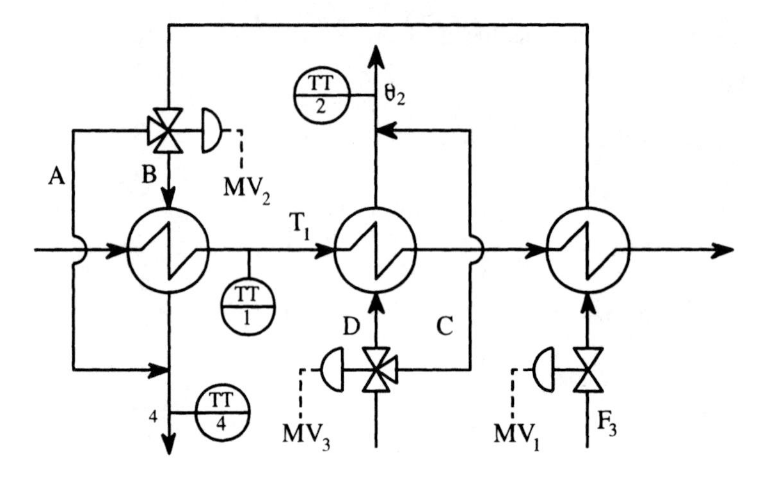


**FIGURE 7.14.** Heat exchanger system. (Copyright © 1983, ISA, all rights reserved. Reprinted by permission, with changes.)

each other. The relative gains for these two loops indicate that decoupling control should probably be considered; decoupling is examined by McAvoy (1983).

To verify the pairings using the Niederlinski stability theorem, the steady-state gain matrix must be rearranged so the variable pairings are on the diagonal:

$$\begin{bmatrix} T_1 \\ \theta_2 \\ \theta_4 \end{bmatrix} = \begin{bmatrix} 0.01725 & 0.01357 & -0.3289 \\ 0.00820 & -0.04847 & -0.1566 \\ -0.03065 & 0.02715 & -1.007 \end{bmatrix} \begin{bmatrix} \mathrm{MV}_2 \\ \mathrm{MV}_3 \\ \mathrm{MV}_1 \end{bmatrix}$$

Application of the Niederlinski stability theorem (7.11) shows that this loop pairing is not unstable.

**Example 7.8.** McAvoy (1983) derived the following gain matrix for an atmospheric crude tower system, depicted in Figure 7.15 using data from DiBiano (1981):

$$\begin{bmatrix} T_1 \\ T_2 \\ T_3 \\ T_4 \\ T_5 \end{bmatrix} = \begin{bmatrix} 0.1271 & 1.549\times 10^{-3} & -6.197\times 10^{-3} & -4.086\times 10^{-3} & 5.132\times 10^{-4} \\ 0.01986 & -0.03623 & -0.08309 & -0.07475 & -0.4149 \\ 0.02835 & 0.04301 & -0.02991 & -0.06642 & -0.2872 \\ 0.1590 & 0.1501 & 0.1423 & 0.05604 & -0.03342 \\ 0.1288 & 0.1214 & 0.1178 & 0.1022 & -0.02733 \end{bmatrix} \times \begin{bmatrix} \mathrm{MV}_1 \\ \mathrm{MV}_2 \\ \mathrm{MV}_3 \\ \mathrm{MV}_4 \\ \mathrm{MV}_5 \end{bmatrix}$$

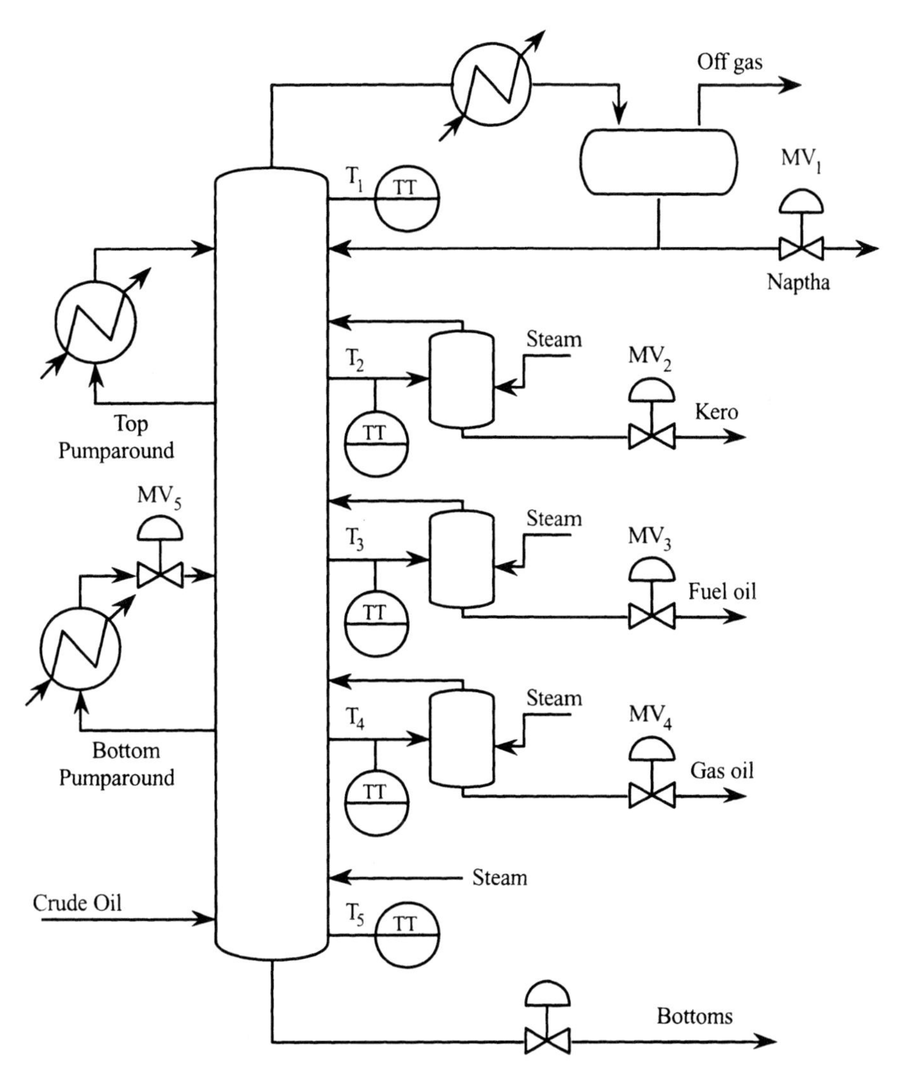


**FIGURE 7.15.** Atmospheric crude tower system.

Find the loop pairings for a multiloop control system and comment on its expected performance.

***Solution.*** Using (7.9), the RGA, with the suggested loop pairings circled, is:

| | $MV_1$ | $MV_2$ | $MV_3$ | $MV_4$ | $MV_5$ |
|---|---|---|---|---|---|
| $T_1$ | (0.928) | $4.65 \times 10^{-3}$ | 0.0682 | $-2.04 \times 10^{-3}$ | $1.17 \times 10^{-3}$ |
| $T_2$ | 0.0199 | 0.623 | −1.36 | 0.0598 | (1.66) |
| $T_3$ | −0.0425 | (1.09) | 0.748 | −0.0797 | −0.718 |
| $T_4$ | 0.159 | −2.18 | (3.84) | −0.869 | 0.0435 |
| $T_5$ | −0.0644 | 1.46 | −2.30 | (1.89) | 0.0137 |

The pairing is explained in the following manner. Considering the first row; the only reasonable pairing is $MV_1$–$T_1$. The fourth and fifth columns indicate that the only reasonable pairings are $MV_4$–$T_5$ and $MV_5$–$T_2$. These choices only leave $MV_2$ and $MV_3$ to be paired with $T_3$ and $T_4$. Because of the negative relative gain for $MV_2$–$T_4$, the only reasonable pairings are $MV_2$–$T_3$ and $MV_3$–$T_4$. Examining the pairings, the $MV_3$–$T_4$ loop looks particularly troublesome. The pairings $MV_4$–$T_5$ and $MV_5$–$T_2$ have interaction to a lesser extent. Decoupling control should be considered. In addition, one of the loop pairings, $MV_5$–$T_2$, would probably not be made in practice. With this pairing, the bottom pumparound controls the kero temperature near the top of the tower, meaning a sluggish loop response. In summary, an alternative control scheme should be considered and is presented by McAvoy (1983).

To verify the pairings using the Niederlinski stability theorem, the steady-state gain matrix must be rearranged (second through the fourth columns are shifted right one column and the fifth column is placed in the second column) so the variable pairings are on the diagonal:

$$\begin{bmatrix} T_1 \\ T_2 \\ T_3 \\ T_4 \\ T_5 \end{bmatrix} = \begin{bmatrix} 0.1271 & 5.132 \times 10^{-4} & 1.549 \times 10^{-3} & -6.197 \times 10^{-3} & -4.086 \times 10^{-3} \\ 0.01986 & -0.4149 & -0.03623 & -0.08309 & -0.07475 \\ 0.02835 & -0.2872 & 0.04301 & -0.02991 & -0.06642 \\ 0.1590 & -0.03342 & 0.1501 & 0.1423 & 0.05604 \\ 0.1288 & -0.02733 & 0.1214 & 0.1178 & 0.1022 \end{bmatrix} \times \begin{bmatrix} MV_1 \\ MV_5 \\ MV_2 \\ MV_3 \\ MV_4 \end{bmatrix}$$

Application of the Niederlinski stability theorem (7.11) shows that these loop pairing are not unstable.

### 7.2.4 Tuning of Multiloop Control

The interaction in a multivariable process does affect the allowable range of parameters for each of the SISO controllers. Generally, the range of parameters for which the closed-loop system is stable is smaller than if the system had no interaction. This statement will be demonstrated by the next example. Other examples with mathematical analysis of the allowable range of tuning parameters are given by Marlin (1995) and Seborg et al. (1989).

**Example 7.9.** For the distillation column of Example 7.6, demonstrate the effect of the interaction on the controller tuning. In particular, examine the response to a feed flow disturbance.

**TABLE 7.1 Wood and Berry Column Variable Summary**

| Variable | Description | Operating Point | 0–100% Range |
|---|---|---|---|
| $X_D$ | Overhead composition | 96.25 mol% methanol | 90–100 |
| $X_B$ | Bottoms composition | 0.50 mol% methanol | 0–10 |
| $F_R$ | Reflux flow rate | 1.95 lb/min | 1–3 |
| $F_S$ | Steam flow rate | 1.71 lb/min | 1–4 |

***Solution.*** The control loop pairings are as determined in Example 7.6. In order to make the simulation reasonable, the process gains are adjusted to reflect sensor and manipulated variable ranges. The ranges of the variables and the operating conditions are shown in Table 7.1. Since the RGA and RDG are scale independent, changes in the manipulated and process variable scaling do not change their results. The process model now is

$$\begin{bmatrix} X_D(s) \\ X_B(s) \end{bmatrix} = \begin{bmatrix} \dfrac{2.56e^{-s}}{16.7s+1} & \dfrac{-5.67e^{-3s}}{21s+1} \\ \dfrac{1.32e^{-7s}}{10.9s+1} & \dfrac{-5.82e^{-3s}}{14.4s+1} \end{bmatrix} \begin{bmatrix} F_R(s) \\ F_S(s) \end{bmatrix} + \begin{bmatrix} \dfrac{38e^{-8s}}{14.9s+1} \\ \dfrac{49e^{-7s}}{13.2s+1} \end{bmatrix} F(s) \tag{7.13}$$

where time is measured in minutes. The $F_R$–$X_D$ PI controller is tuned with Fertik's method (Chapter 5) for a disturbance assuming the process transfer function is the one in the (1, 1) position of the transfer function matrix. The controller parameters are $K_P = 2.34$ and $T_I = 7.08$ min. In a similar manner, the tuning parameters for the $F_S$–$X_B$ PI controller are $K_P = -0.54$ and $T_I = 12.18$ min. The responses to a +15% (0.34 magnitude) feed flow disturbance are shown in Figure 7.16 as deviations from the operating point. The response is too oscillatory to be acceptable. The controllers need to be detuned by some factor. When the controller gains are halved, the responses appear as in Figure 7.17. Since an interacting form of the controller is used, the integral time is also effectively doubled. Further detuning causes the response to be more sluggish, not improving the control. McAvoy (1981) also investigates the setpoint response of this multiloop system and comes

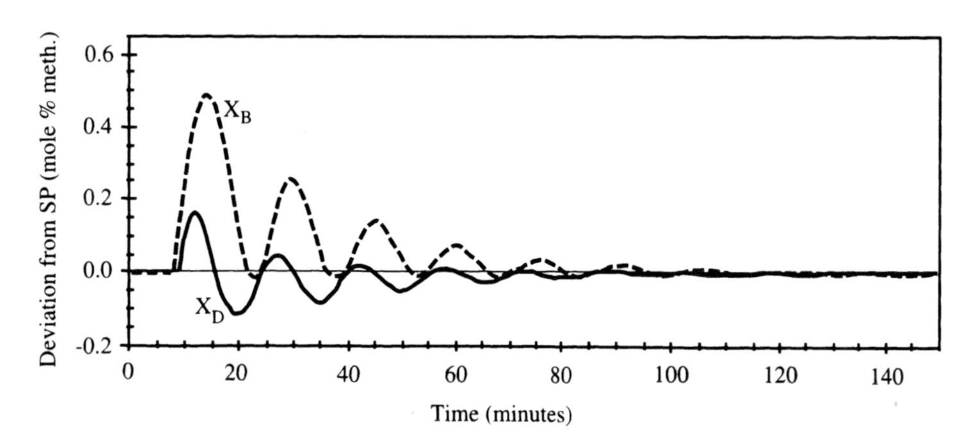


**FIGURE 7.16.** Response of Wood–Berry distillation column using multiloop control with controllers tuned by Fertik's method.

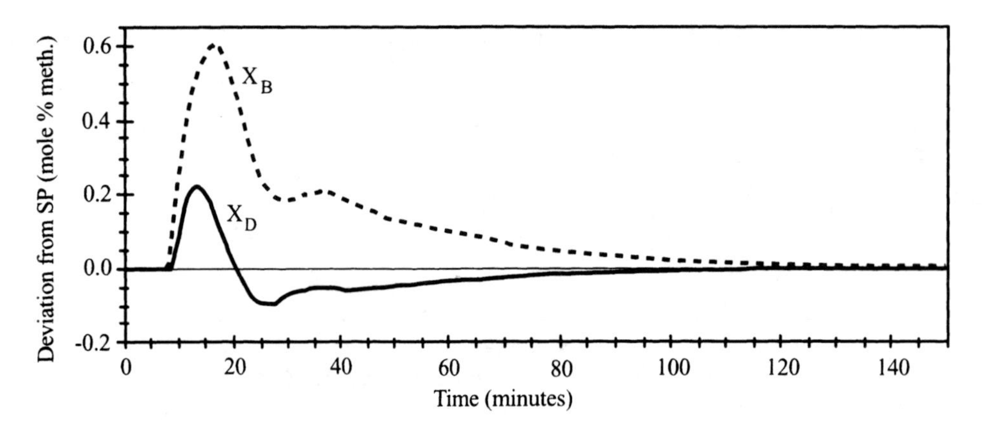


**FIGURE 7.17.** Response of Wood–Berry distillation column with detuned multiloop controllers.

to a similar conclusion. However, the controller tuning for reasonable setpoint response with multiloop control leads to a response that is more sluggish than shown in Figure 7.17.

*Trial-and-Error Tuning.* In practice, most people tune multiloop controllers using trial and error even though it is often tedious. Typically, the initial tuning is done assuming no interaction and then the gains are decreased, typically by a factor of 2 or more, as was done in Example 7.9. The tuning is tested with a simulation or on-line. If tested on-line, the initial gains may be reduced even further to ensure the system is stable during the initial test. The tuning parameters are generally conservative so that if the process operating conditions change, retuning is not required. The success of this approach depends on the expertise of the one doing the tuning, but the results are reasonable if the interaction is not too severe.

*Optimization.* Optimization algorithms may be used with a process simulation to calculate the controller tuning parameters to optimize some aspect of the transient response (Edgar and Himmelblau, 1988). Optimization is justified when the interaction is strong, which may result in a lengthy trial-and-error approach or cause process upsets if tuned on-line.

**Example 7.10.** For the distillation column of Example 7.9, find the optimum PI parameters for the two controllers to minimize the integrated absolute error (IAE) of the response to a feed flow disturbance.

***Solution.*** The function to be minimized is the sum of the IAE for each loop. The optimization features of VisSim are used to determine the four controller parameters. For the $F_R$–$X_D$ PI controller, $K_P = 8.21$ and $T_I = 6.46\,\text{min}$. The tuning parameters for the $F_S$–$X_B$ PI controller are $K_P = -0.57$ and $T_I = 10.0\,\text{min}$. The responses to a +15% (0.34 magnitude) feed flow disturbance are shown in Figure 7.18 as deviations from the operating point. The responses of the manipulated variables, though not shown, are highly oscillatory and would not be tolerated in an actual distillation column. So, the function to be minimized is modified to include velocity limits on the manipulated variables. In particular, the integrated absolute value of the manipulated variable velocity is penalized

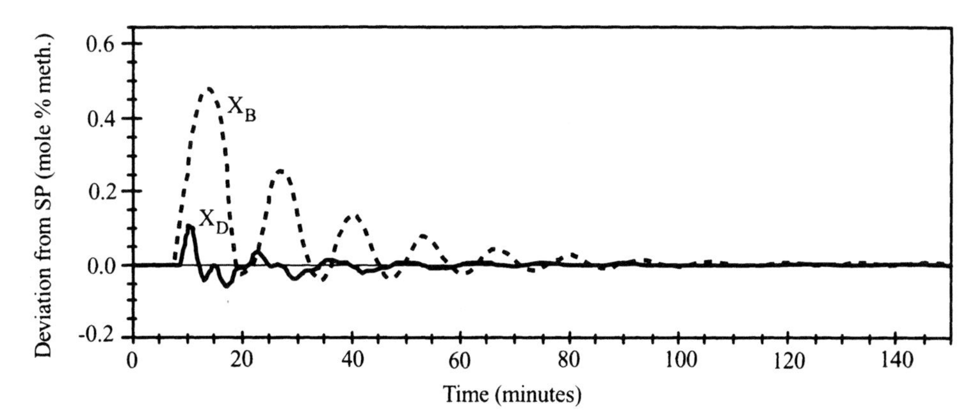


**FIGURE 7.18.** Response of Wood–Berry distillation column with controllers tuned to minimize the total IAE.

10 times higher than the IAE when the velocity is greater than 0.1% per minute. For this optimization, the tuning parameters for the $F_R$–$X_D$ PI controller are $K_P = 0.554$ and $T_I = 7.17$ min and for the $F_S$–$X_B$ PI controller are $K_P = -0.503$ and $T_I = 9.31$ min. The responses to a +15% (0.34 magnitude) feed flow disturbance are shown in Figure 7.19. Compared with Example 7.9 (Figure 7.17), the bottom composition response is improved but the distillate (top) composition response is worse.

*Approximate, Noniterative Approach.* There is no generally accepted method for quickly estimating initial tuning constants for multiloop controllers, though many methods have been proposed. The following method, proposed by Marlin (1995), is reasonably simple and provides some insight into the key process-related issues. It is developed for 2 × 2 systems but does not easily extend to higher order systems. The key points of the method are shown here, but the method is not developed.

The relative importance of both process variables is assumed to be equal. In addition, the method is based on the closed-loop stability of the system and hence the same tuning is

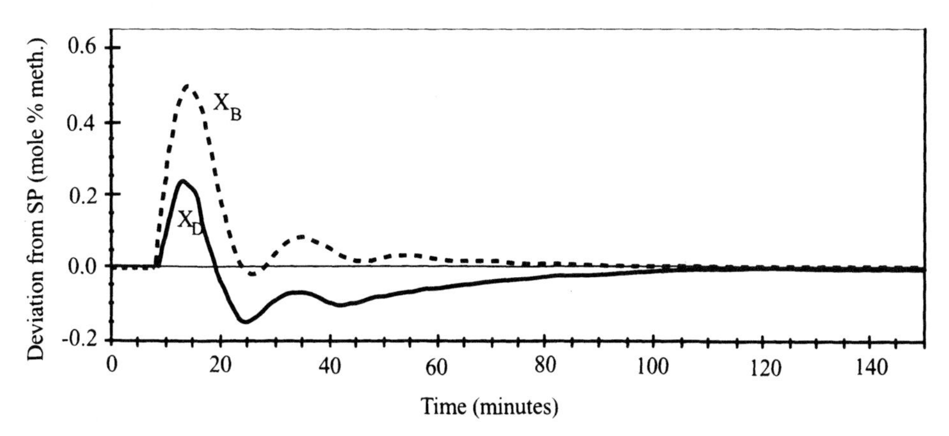


**FIGURE 7.19.** Response of Wood–Berry distillation column with controllers tuned to minimize the total IAE and manipulated variable velocities > 0.1.

applied for both setpoint changes and disturbances. Three cases for tuning loop 1 are considered: (1) loop 1 much faster than loop 2, (2) loop 1 much slower than loop 2, and (3) both loops with the same dynamics.

1. Loop 1 is much faster than loop 2: Loop 1 stability is not strongly affected by the other loop, so loop 1 is tuned in the same manner as for no interaction.
2. Loop 1 is much slower than loop 2: Loop 1 stability is affected by the change in the closed-loop process gain, so the single-loop controller gain is multiplied by $\lambda_{11}$ if $\lambda_{11} < 1$. If $\lambda_{11} \geq 1$, then the single-loop controller gain is not changed.
3. Both loops have about the same dynamics: Loop 1 stability is affected by changes in both phase and gain. The exact relationship between the detuning constants for the controller gain and integral time constant is shown graphically in Marlin (1995). The approximate relationship is

$$K_{P,\text{ml}} \approx \begin{cases} (1.2\lambda_{11} - 0.2)K_{P,\text{sl}} & 0.5 \leq \lambda_{11} \leq 1.0 \\ 0.5K_{P,\text{sl}} & \lambda_{11} > 1.0 \end{cases}$$

$$T_{I,\text{ml}} \approx \begin{cases} T_{I,\text{sl}}/\lambda_{11} & 0.5 \leq \lambda_{11} \leq 1.0 \\ T_{I,\text{sl}} & \lambda_{11} > 1.0 \end{cases}$$

where the ml and sl subscripts denote multiloop and single-loop tuning constants, respectively.

## 7.3 DECOUPLING CONTROL

The previous tuning methods for multiloop controllers did not eliminate the interaction in the process, but merely detuned the controllers in order to accommodate the interaction. Through the process of *decoupling*, an attempt is made to eliminate the interaction, at least as perceived by the controllers. A simple method of decoupling is simply to decide that one (or more) of the process variables is less important and loosely tune that loop. With this approach, the interaction of the detuned loop with the tightly tuned loop is reduced. This point is illustrated by the following example.

**Example 7.11.** For the distillation column of Example 7.9, assume the distillate composition is the important controlled variable and tune the two PI controllers to primarily minimize the IAE of the distillate response to a feed flow disturbance.

***Solution.*** The function to be minimized is the IAE for the distillate plus 0.10 times the IAE for the bottoms plus a penalty for manipulated variable velocities greater than 0.1% per minute, as in Example 7.10. The optimization features of VisSim are used to determine the four controller parameters. For the $F_R$–$X_D$ PI controller, $K_P = 0.56$ and $T_I = 5.55$ min. The tuning parameters for the $F_S$–$X_B$ PI controller are $K_P = -0.33$ and $T_I = 14.0$ min. The responses to a +15% (0.34 magnitude) feed flow disturbance are shown in Figure 7.20 as deviations from the operating point. Compared with the responses of the optimization of Example 7.10 (Figure 7.19), the distillate response is actually slightly worse with this minimization.

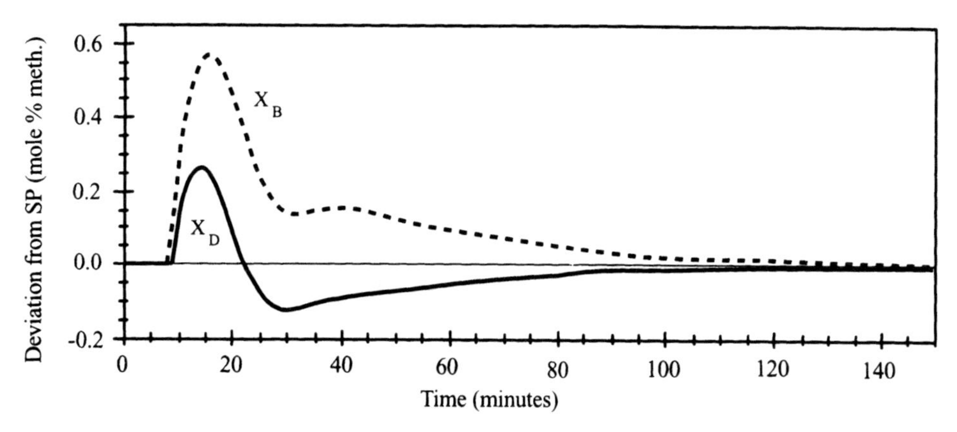


**FIGURE 7.20.** Response of Wood–Berry distillation column with controllers tuned to minimize the distillate IAE and manipulated variable velocities $> 0.1$.

Clearly, whether detuning one or more less important loops to achieve some measure of decoupling depends on the economics of the process. Often, it is not desirable to keep only a few variables tightly controlled while loosening others. Two other decoupling approaches are used in such a situation. In the first approach, knowledge about the process is used to redefine the manipulated variables and/or sensors in order to produce a noninteracting system. In the second approach, the original sensors and manipulated variables are retained and the control calculation is altered, commonly called *explicit decoupling*.

### 7.3.1 Decoupling Based on Process Knowledge

Often, a particular process can be decoupled simply by redefining the manipulated variables and/or the sensors. This method is introduced by reconsidering the blending example (Example 7.1). Actually any blending process is better controlled using a ratio control scheme (Chapter 6). For this process, redefine the two manipulated variables as $F_m$ and $X_m$,

$$F_m = F_A + F_S \tag{7.14}$$

$$X_m = \frac{F_A}{F_A + F_S} \tag{7.15}$$

With this approach, the flow controller measures the total flow and adjusts the sum of the two flows accordingly and the composition controller manipulates the ratio in order to obtain the desired composition. The setpoints to the two component flow controllers are calculated by rearranging (7.14) and (7.15) to obtain

$$F_A = X_m F_m \qquad F_S = F_m - F_A$$

The blending system now appears as in Figure 7.21. The cascade flow controllers are not mandatory but provide better control when the pressure of A or the solvent changes.

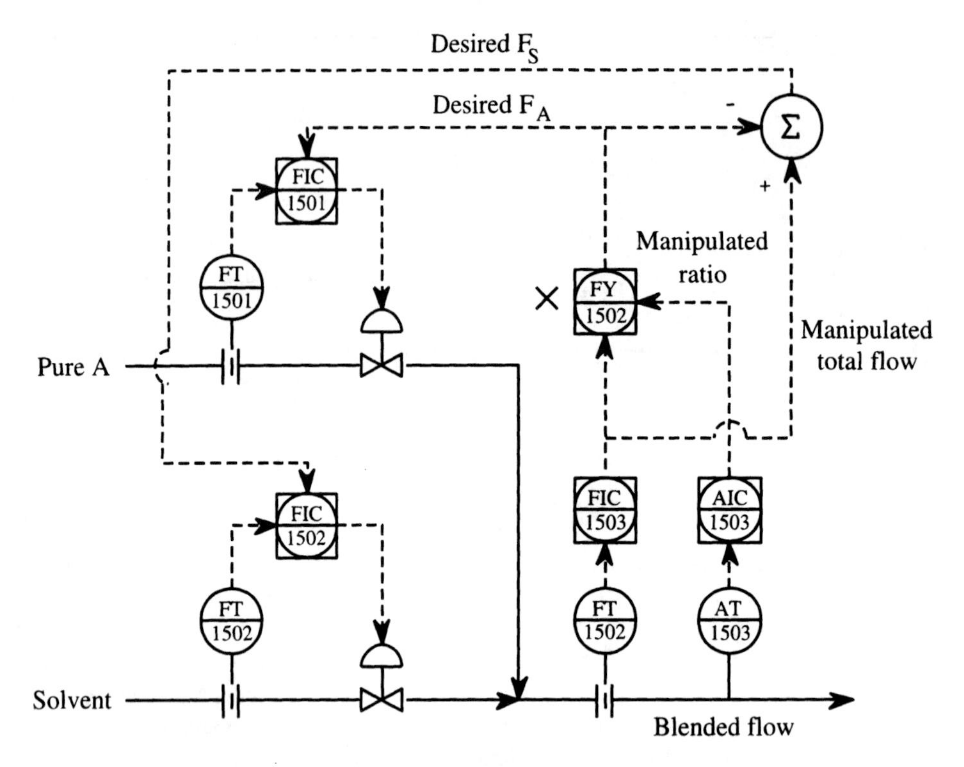


**FIGURE 7.21.** Decoupled blending system.

Whether this approach will work is dependent on the process. Other example processes that can be decoupled by redefining the manipulated variables or sensors are presented by McAvoy (1983) and Marlin (1995).

### 7.3.2 Explicit Decoupling

In the third approach to decoupling, the original manipulated variables and sensors are kept and the control calculation is modified while retaining the multiloop configuration. With an explicit decoupler, the process, as perceived by the controllers, is approximately diagonal and thus is noninteracting. A general explicit decoupling control system for a two-input, two-output process is shown in Figure 7.22. The decoupler is inserted between the multiloop controller outputs and the process manipulated variables and can be either static or dynamic. One specific form of the decoupler, called a *simplified decoupler,* has $D_{11}(s) = D_{22}(s) = 1.0$. It can be shown (Marlin, 1995; Shinskey, 1981) that the off-diagonal terms of the decoupler are

$$D_{12}(s) = -\frac{G_{12}(s)}{G_{11}(s)} \qquad D_{21}(s) = -\frac{G_{21}(s)}{G_{22}(s)} \tag{7.16}$$

Through block diagram manipulations and with the decoupler in the form of (7.16), the entire system block diagram is equivalent to the system in Figure 7.23. The decoupler transfer functions in (7.16) can be considered feedforward controllers, which compensate for disturbances. Here, the disturbance is the manipulated variable change from the interacting feedback controller. As for the feedforward controller, the decoupler transfer

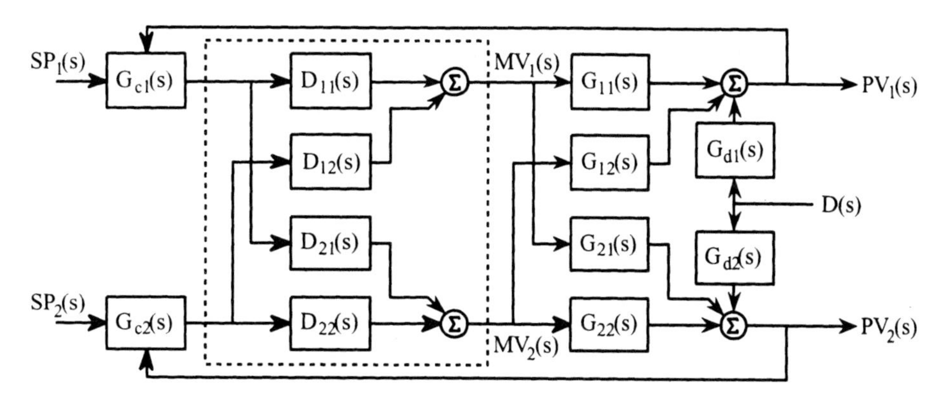


**FIGURE 7.22.** System with explicit decoupler.

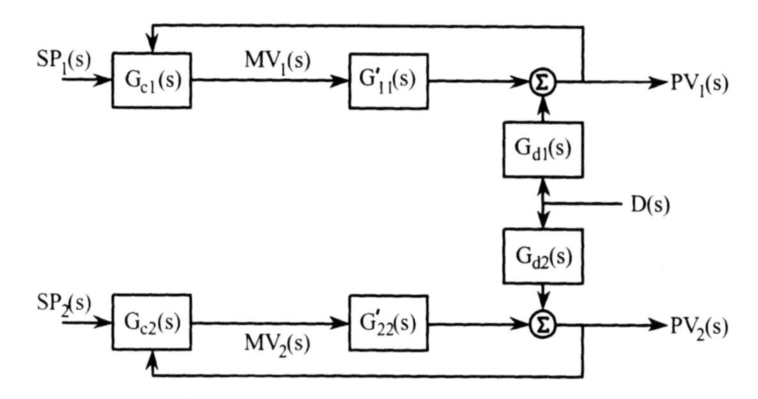


**FIGURE 7.23.** Equivalent block diagram of system with explicit decoupler.

function is not realizable if the deadtime of the denominator transfer function is greater than the deadtime of the numerator transfer function. In that case, perfect decoupling is not possible and the controllers will see some process interaction.

A *static simplified decoupler* is designed when only the gains of the decoupler transfer functions in (7.16) are implemented. Obviously, the decoupler is not perfect and the two controllers will interact. As for the RGA, this decoupler only requires gain information about the process and therefore does not require accurate modeling in order to determine the dynamics.

**Example 7.12.** For the distillation column of Example 7.9, design a dynamic and static simplified decoupler and compare their response to a feed flow disturbance.

**Solution.** Using (7.16), the process transfer functions from (7.13) are used to calculate the decoupler off-diagonal transfer functions,

$$D_{12}(s) = -\frac{G_{12}(s)}{G_{11}(s)} = -\frac{-5.67e^{-3s}/(21s+1)}{2.56e^{-s}/(16.7s+1)} = 2.21\frac{16.7s+1}{21s+1}e^{-2s}$$

$$D_{21}(s) = -\frac{G_{21}(s)}{G_{22}(s)} = -\frac{1.32e^{-7s}/(10.9s+1)}{-5.82e^{-3s}/(14.4s+1)} = 0.227\frac{14.4s+1}{10.9s+1}e^{-4s} \qquad (7.17)$$

and the dynamic decoupler is written in matrix form as

$$\mathbf{D}(s) = \begin{bmatrix} 1 & 2.21\dfrac{16.7s+1}{21s+1}e^{-2s} \\ 0.227\dfrac{14.4s+1}{10.9s+1}e^{-4s} & 1 \end{bmatrix} \tag{7.18}$$

In this case all of the transfer functions are realizable. The static decoupler is

$$\mathbf{D} = \begin{bmatrix} 1 & 2.21 \\ 0.227 & 1 \end{bmatrix} \tag{7.19}$$

If an approximate first-order-plus-deadtime (FODT) model is found for each process, as perceived by the controller, and the Fertik disturbance tuning is used, the $F_R$–$X_D$ PI controller parameters are $K_P = 3.20$ and $T_I = 5.27$ min and the $F_S$–$X_B$ tuning parameters are $K_P = -0.36$ and $T_I = 6.24$ min when the dynamic decoupler (7.18) is used. However, these parameters lead to a fairly oscillatory response to a change in the $X_D$ setpoint, even though the bottoms composition is not affected. Thus, the $K_P$ for the $F_R$–$X_D$ controller is decreased to 1.20. A similar phenomenon is noted for the static decoupler. In this case both controller gains are reduced to mitigate the oscillations in the setpoint since some interaction is present. The final tuning parameters for the $F_R$–$X_D$ PI controller are $K_P = 1.20$ and $T_I = 6.25$ min and for the $F_S$–$X_B$ controller the tuning parameters are $K_P = -0.50$ and $T_I = 11.5$ min. The responses to a +15% (0.34 magnitude) feed flow disturbance are shown in Figure 7.24 for the dynamic decoupler and in Figure 7.25 for the static decoupler. Compared with each other, the dynamic decoupler has a response with less maximum deviation from the setpoint and a smaller settling time. Compared with the multiloop tuning of Example 7.9 (Figure 7.17) or Example 7.10 (Figure 7.19), the response of both decouplers is worse than the multiloop examples.

Shinskey (1981) points out some difficulties with the decoupler shown in Figure 7.22. First, the controllers cannot be initialized easily, that is, have their outputs adjusted to an initial value prior to a change from the Manual to the Automatic state. Assume the control valves are positioned individually. Before changing the controllers to Automatic, both

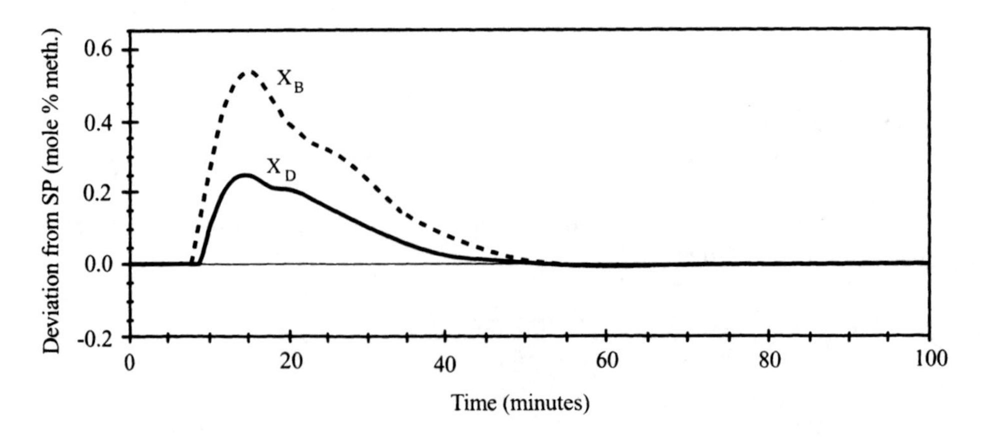


**FIGURE 7.24.** Response of Wood–Berry distillation column with dynamic simplified decoupler.

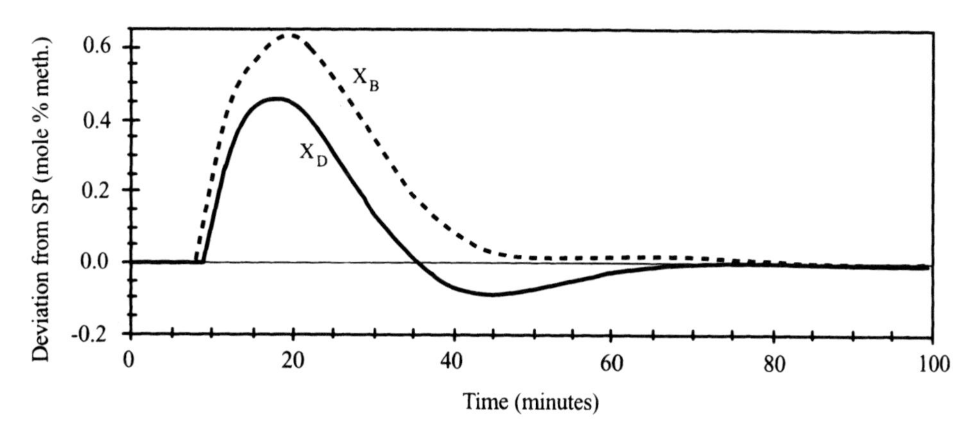


**FIGURE 7.25.** Response of Wood–Berry distillation column with static simplified decoupler.

controller outputs must be matched to produce the initial valve positions. Suppose only one controller output is adjusted to produce the current valve position and then placed in Automatic. When the other controller output is adjusted in preparation for its switch to Automatic, it upsets the controller already in Automatic. Second, a problem appears if one of the manipulated variables encounters a constraint. In this situation, the controlled variable cannot be controlled. Nevertheless, the controller output continues to change, thereby affecting the other controller. Since both controllers are competing for the single unconstrained manipulated variable, control of both loops is lost. Shinskey rearranges the decouplers in the configuration shown in Figure 7.26, wherere HIC is a manual station (or equivalent). By using the actual valve position for the feedforward to the other controller, the initial controller outputs are easily back-calculated from the two valve positions as

$$MV_1' = MV_1 - D_{12}MV_2 \qquad MV_2' = MV_2 - D_{21}MV_1$$

Constraints may be imposed at the input or output of each manual station as long as the signal to the decoupler is the true valve position. When a valve becomes constrained, the unconstrained controller follows it as in a feedforward system.

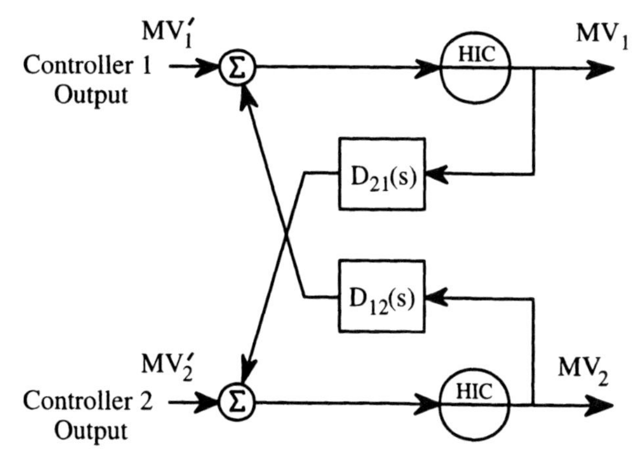


**FIGURE 7.26.** Alternate arrangement of explicit decoupler. (Copyright © 1981, ISA, all rights reserved. Reprinted by permission, with changes.)

Designing dynamic simplified decouplers for systems with more than two inputs and two outputs is cumbersome. Generally, static simplified decouplers are used when the process is larger than 2 × 2. Another way to design a static decoupler is to simply invert the process gain matrix. This approach works well for systems with more than two inputs and two outputs and the gain of the process, as seen by each controller, is 1.0.

A 5 × 5 static decoupler is used to control the riser temperatures in an ammonia reform furnace (Otto and Nieman, 1979). The schematic of the furnace shown in Figure 7.27 is a view of the end of the furnace. There are five rows of catalyst tubes, $T_1$ through $T_5$, whose temperature is to be controlled by six burners. The furnace is highly interactive and has long time constants, and thus it is difficult to control with multiloop control when the process is upset by a change in ambient conditions, production rate changes, and changes in the heating value of the fuel. The control problem is to maintain the five catalyst tube temperatures with the six fuel valves. In reality, the manipulated valves shown in the diagram are setpoints to fuel pressure regulators. In order to remove one degree of freedom, control of the third and fourth valves is combined. Rather than develop a dynamic model (deemed difficult), a static gain-only model was developed from empirical furnace data. The furnace gain matrix is

$$\mathbf{T} = \mathbf{K} \times \mathbf{MV}$$

$$\begin{bmatrix} T_1 \\ T_2 \\ T_3 \\ T_4 \\ T_5 \end{bmatrix} = \begin{bmatrix} 1 & 1 & 0 & 0 & 0 \\ 0 & 1 & 1 & 0 & 0 \\ 0 & 0 & 2 & 0 & 0 \\ 0 & 0 & 1 & 1 & 0 \\ 0 & 0 & 0 & 1 & 1 \end{bmatrix} \begin{bmatrix} \mathrm{MV}_1 \\ \mathrm{MV}_2 \\ \mathrm{MV}_3 \\ \mathrm{MV}_4 \\ \mathrm{MV}_5 \end{bmatrix} \tag{7.20}$$

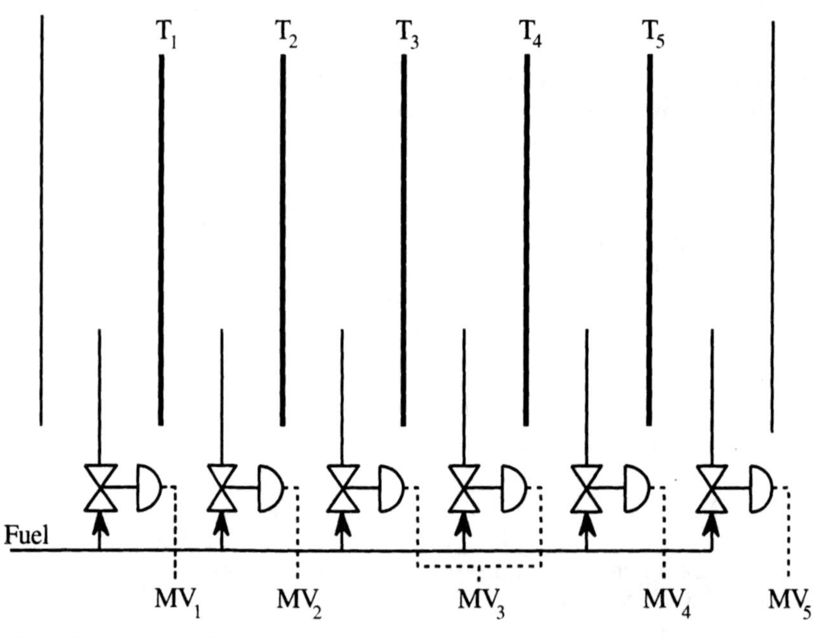


**FIGURE 7.27.** Schematic drawing of the reform furnace. (Reprinted from *Industrial Process Design*, with changes. Copyright © 1979, American Institute of Chemical Engineers.)

The zero entries represent a negligible temperature response to a change in the respective manipulated variable. A static decoupler is constructed by inverting the gain matrix in (7.20),

$$\mathbf{D} = \mathbf{K}^{-1} = \begin{bmatrix} 1 & -1 & 0.5 & 0 & 0 \\ 0 & 1 & -0.5 & 0 & 0 \\ 0 & 0 & 0.5 & 0 & 0 \\ 0 & 0 & -0.5 & 1 & 0 \\ 0 & 0 & 0.5 & -1 & 1 \end{bmatrix} \tag{7.21}$$

Each of the five temperature controllers is a simple P controller whose gain is less than 1, to minimize the influence of modeling errors due to process nonlinearities. The control algorithm is executed once every 5 min. This multivariable controller drove the temperatures to their setpoints within 100 min *the first time* it was placed into automatic operation. With minor retuning, the temperatures could be brought under control within 30 min.

Later, Otto (1984) used all six manipulated variables, yielding a $5 \times 6$ gain matrix similar to the $5 \times 5$ gain matrix in (7.21). In this case the decoupler is obtained from the pseudoinverse of the gain matrix, yielding a $6 \times 5$ decoupler. The results are similar as for the $5 \times 5$ decoupler. However, Otto notes that it is imperative that the gain matrix be accurate, or system instability will result.

## 7.4 SINGULAR-VALUE DECOMPOSITION

Singular-value decomposition (SVD) is a numerical algorithm important to linear algebra but with important applications in multivariable process control. As a tool, SVD obtains quantitative information about process controllability, sensor placement, loop pairing, and decoupler design. Use of SVD as a tool to support multivariable control design is emphasized in this section.

### 7.4.1 Definition of SVD

The SVD of a matrix is a numerical algorithm developed to minimize computational errors in large matrix operations. The SVD of a matrix **K** results in three matrices,

$$\mathbf{K} = \mathbf{U}\boldsymbol{\Sigma}\mathbf{V}^{\mathrm{T}} \tag{7.22}$$

where

**K** is an $n \times m$ matrix

**U** is an $n \times n$ orthonormal matrix whose columns are called the "left singular vectors" (an orthonormal matrix is one in which each column is a vector of length 1 and orthogonal to the other columns, $u_i \cdot u_j = 0,\ i \neq j$)

**V** is an $m \times m$ orthonormal matrix whose columns are called the "right singular vectors"

$\boldsymbol{\Sigma}$ is an $n \times m$ matrix whose diagonal elements are called the "singular values," arranged in descending order, $\sigma_1 \geq \sigma_2 \geq \sigma_3 \geq \cdots \geq \sigma_m \geq 0$.

Mathematically, both **U** and **V** represent a coordinate rotation. The *condition number* (CN) is the ratio of the largest singular value to the smallest singular value,

$$\text{CN} = \frac{\sigma_1}{\sigma_m} \tag{7.23}$$

For systems larger than 2 × 2, intermediate condition numbers may be defined as

$$\text{CN}_j = \frac{\sigma_1}{\sigma_j} \quad \text{for } 1 < j \leq m \tag{7.24}$$

The SVD is designed to determine the condition and rank of a matrix and other information about the matrix that can help avoid numerical computation errors. The SVD algorithm is discussed in many texts (Forsythe et. al., 1977) and are available in standard numerical packages, like MATLAB.

When applied to multivariable control problems, SVD provides physical insight into the nature of the problems. Each part of the decomposition can be understood in terms of some physical aspect of the process as follows:

Matrix **K** is the matrix of steady-state process gains. It is important that the variables be scaled as they are seen by the controller. In other words, the sensors and manipulated variables should be scaled to 0–100% and not to the physical units (e.g., lb/hr or °C). Unlike the RGA, the SVD is sensitive to scaling. Therefore, the results are properly interpreted only if the gain matrix is scaled as the process is perceived by the controller.

Matrix $\mathbf{U} = [U_1 \; U_2 \; \ldots \; U_n]$, whose columns, $U_1, U_2, \ldots$, the left singular vectors, indicate the best way in which to view the sensors. The first column is the best "sensor direction," the best combination of sensors, that is, the combination of sensors that is most easily influenced by changes in the manipulated variables. The second column is the next easiest combination of sensors. And so on.

Matrix $\mathbf{V} = [V_1 \; V_2 \; \ldots \; V_m]$, whose columns, $V_1, V_2, \ldots$, the right singular vectors, indicate the best way in which to view the manipulated variables. The first column is the best "process direction," the combination of manipulated variables that have the strongest effect on the process. The second column is the next best combination of manipulated variables. And so on.

Matrix $\boldsymbol{\Sigma} = \text{diag}(\sigma_1 \; \sigma_2 \; \ldots \; \sigma_m)$, the singular values, indicates the decoupled gain of the open-loop process. The condition number CN is a measure of the control difficulty of the decoupled multivariable problem. A large condition number indicates that good control will be difficult, even with decoupling. The intermediate condition numbers $\text{CN}_j$ are used to determine the appropriate number of sensors and manipulated variables.

As a simple example of SVD analysis, consider the blending process of Example 7.1 at three different operating points, $A_{1_1} = 0.2$, $A_{1_2} = 0.5$, and $A_{1_3} = 0.8$. The three gain matrices and their SVDs are as follows:

$$\mathbf{K}_1 = \begin{bmatrix} 0.8 & -0.2 \\ 1 & 1 \end{bmatrix}$$

$$\mathbf{U}_1 = \begin{bmatrix} 0.3606 & -0.9327 \\ 0.9327 & 0.3606 \end{bmatrix} \begin{matrix} A_1 \\ F_1 \end{matrix} \quad \mathbf{\Sigma}_1 = \begin{bmatrix} 1.494 & 0 \\ 0 & 0.6694 \end{bmatrix}$$

$$\mathbf{V}_1 = \begin{bmatrix} 0.8174 & -0.5760 \\ 0.5760 & 0.8174 \end{bmatrix} \begin{matrix} V_A \\ V_S \end{matrix}$$

$$\mathbf{K}_2 = \begin{bmatrix} 0.5 & -0.5 \\ 1 & 1 \end{bmatrix}$$

$$\mathbf{U}_2 = \begin{bmatrix} 0 & -1 \\ 1 & 0 \end{bmatrix} \begin{matrix} A_1 \\ F_1 \end{matrix} \quad \mathbf{\Sigma}_2 = \begin{bmatrix} 1.414 & 0 \\ 0 & 0.707 \end{bmatrix} \quad \mathbf{V}_2 = \begin{bmatrix} 0.707 & -0.707 \\ 0.707 & 0.707 \end{bmatrix} \begin{matrix} V_A \\ V_S \end{matrix}$$

$$\mathbf{K}_3 = \begin{bmatrix} 0.2 & -0.8 \\ 1 & 1 \end{bmatrix}$$

$$\mathbf{U}_3 = \begin{bmatrix} -0.3606 & -0.9327 \\ 0.9327 & -0.3606 \end{bmatrix} \begin{matrix} A_1 \\ F_1 \end{matrix} \quad \mathbf{\Sigma}_3 = \begin{bmatrix} 1.494 & 0 \\ 0 & 0.6694 \end{bmatrix}$$

$$\mathbf{V}_3 = \begin{bmatrix} 0.5760 & -0.8174 \\ 0.8174 & 0.5760 \end{bmatrix} \begin{matrix} V_A \\ V_S \end{matrix}$$

For all three operating conditions, the condition number is approximately 2. The $\mathbf{U}$ matrix indicates best sensor direction. The $\mathbf{V}$ matrix indicates best manipulated variable direction. For $\mathbf{K}_1$, the indicated pairings are $F_1$–$V_A$ (largest element of the first column of $\mathbf{U}_1$ is paired with the largest element of the first column of $\mathbf{V}_1$) and $A_1$–$V_S$ (largest element of the second column of $\mathbf{U}_1$ is paired with the largest element of the second column of $\mathbf{V}_1$). For $\mathbf{K}_3$, the indicated pairings are $F_1$–$V_S$ and $A_1$–$V_A$, the reverse of the pairings for $\mathbf{K}_1$. These pairings make physical sense since the overall flow should be controlled by the valve that has the largest effect on the combined flow. For small $A_1$, $V_A$ has the most effect, and for large $A_1$, $V_S$ has the most effect. Examining the SVD for $\mathbf{K}_2$ indicates that this system will be hard to control in its present form since all gains in $\mathbf{V}_1$ have the same magnitude and thus there is no clear rationale for choosing pairs.

### 7.4.2 Picking Sensors/Manipulated Variables

The magnitude of the condition number in combination with the singular values provides insight into potential problems of the multivariable control system. As will be shown later, the singular values are a direct measure of the decoupled gains of the process. That is, if a perfect decoupler removed the interactions, the singular values would be the open-loop gains of the noninteracting system. Very large or very small singular values thus indicate a poorly scaled system and should be avoided. A small singular value indicates that the manipulated variable does not have much influence on the process variable and is a serious practical problem. Moore (1989) suggests that singular values less than or equal to the magnitude of the sensor noise (including analog-to-digital conversion resolution) should be assumed degenerate. For example, if the SVD analysis on a $4 \times 4$ system yields two singular values that are less than the noise magnitude, then only a $2 \times 2$ system should be considered for the controller. Large singular values are a less common problem, but they indicate a large gain in the system, which means that a process variable is very sensitive to changes in the manipulated variable. For this reason, singular values that are equal to or greater than the reciprocal of the valve resolution should be avoided (Moore, 1989). A smaller dimensioned system should be selected in this case.

The magnitude of the components of the vectors of **U** and **V** also provides insight into the design of the multivariable control system. The $U_1$ vector points in the most sensitive sensor direction; $U_2$ points in the next most sensitive direction; and so on. It is reasonable to expect that the location of the principal component (one with the largest magnitude) of each $U$ vector is a good choice for a relatively sensitive, yet independent sensor. The same reasoning applies to the vectors of **V**, that is, the principal components of the $V_1, V_2, \ldots$ vectors are the best choice of manipulated variables that have the strongest effect on the process, but are relatively independent. This selection process is also called *principal component analysis*.

Based on previous observations of SVD, the procedure to pick the sensors and manipulated variables for a system is as follows:

1. Calculate the SVD for the system with all possible sensors and manipulated variables. Find the condition number for the full system as well as for systems with a smaller number of sensors and manipulated variables. The appropriate system dimension is one in which the condition number is not too high and the singular value is not too low. A high condition number indicates a poorly behaved system. Unfortunately, it is hard to quantify the value of a high condition number. Sometimes, when the condition number takes a relatively large jump from one system dimension to the next, the lower dimension is the proper one. This point will be illustrated by examples.
2. Select the particular manipulated variables by calculating a SVD on all possible combinations of the proper number of manipulated variables, selected from the set of all possible manipulated variables. The combination that has the lowest condition number is the appropriate one. If the condition numbers for more than one combination are close in value, then save all of these for further analysis.
3. For the set of manipulated variables chosen in step 2, examine the **U** vectors and, in each column, pick the sensor with the highest magnitude element. If more than one combination is viable, then hold them for further consideration.
4. Check the condition number of the system with the selected manipulated variables and sensors and verify that it is not too large. If too large, then use alternate manipulated variables for step 2 or alternate sensors selected from step 3.

This procedure is illustrated by the following examples.

**Example 7.13.** The tunnel dryer of Figure 7.28 (Moore, 1989) has four manipulated variables and eight measurements and the gain matrix is

$$\mathbf{K} = \begin{bmatrix} 0.00471 & 0.00522 & 0.00713 & 0.01066 \\ 0.01711 & 0.01903 & 0.02606 & 0.02352 \\ 0.03944 & 0.04315 & 0.05072 & 0.02724 \\ 0.05795 & 0.06315 & 0.07074 & 0.02720 \\ 0.06509 & 0.07091 & 0.06038 & 0.02254 \\ 0.06021 & 0.06334 & 0.04802 & 0.01599 \\ 0.04834 & 0.04676 & 0.03157 & 0.01006 \\ 0.03452 & 0.02976 & 0.01858 & 0.00528 \end{bmatrix}$$

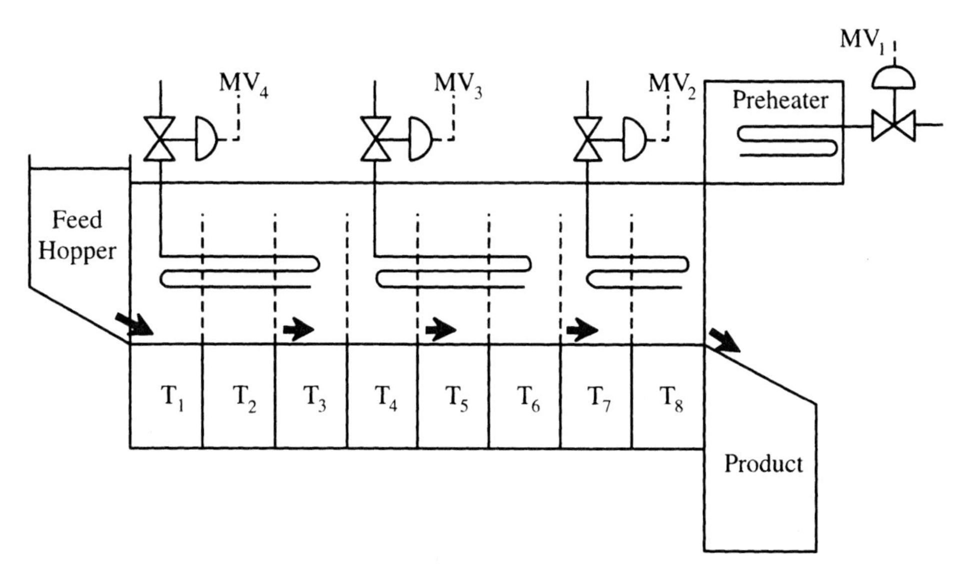


**FIGURE 7.28.** Tunnel dryer. (Copyright © 1989, ISA, all rights reserved. Reprinted by permission, with changes.)

Determine the appropriate sensors and manipulated variables for this system.

Solution.#An SVD analysis on the gain matrix **K** yields the following singular values (the diagonal elements of the $\boldsymbol{\Sigma}$ matrix):

$$\boldsymbol{\Sigma} = \begin{bmatrix} \sigma_1 & 0 & 0 & 0 \\ 0 & \sigma_2 & 0 & 0 \\ 0 & 0 & \sigma_3 & 0 \\ 0 & 0 & 0 & \sigma_4 \end{bmatrix} = \begin{bmatrix} 0.2291 & 0 & 0 & 0 \\ 0 & 0.0316 & 0 & 0 \\ 0 & 0 & 0.0108 & 0 \\ 0 & 0 & 0 & 0.0038 \end{bmatrix}$$

From Equation (7.24), the condition numbers for the full $4 \times 4$ system and smaller dimensioned systems are calculated as:

| Dimension | Condition Number |
|---|---|
| $4 \times 4$ | 59.57 |
| $3 \times 3$ | 21.25 |
| $2 \times 2$ | 7.25 |

The appropriate dimension of this system is difficult to determine. The best condition number is for the $2 \times 2$ system, though the $3 \times 3$ system is acceptable. The singular value for the $3 \times 3$ system is almost too low. The singular value for the $4 \times 4$ system is probably too low. The final decision as to whether a $2 \times 2$ or a $3 \times 3$ system is best will probably depend on the process dynamics and possibly other factors. In general, as many manipulated variables as possible should be used, so start with a $3 \times 3$ system.

To select the appropriate manipulated variables for a 3 × 3 controller, an SVD is done for every combination of three manipulated variables and the condition numbers with the results as follows:

| Manipulated Variables | Condition Number |
|---|---|
| $MV_2$, $MV_3$, $MV_4$ | 19.61 |
| $MV_1$, $MV_3$, $MV_4$ | 17.18 |
| $MV_1$, $MV_2$, $MV_4$ | 38.45 |
| $MV_1$, $MV_2$, $MV_3$ | 55.52 |

The choice of $MV_1$, $MV_3$, and $MV_4$ as the manipulated variables is the best one, although the choice of $MV_2$, $MV_3$, and $MV_4$ is also acceptable. Another way to choose the manipulated variables is to examine the first three columns of **V** and pick the largest element in each column (as will be done for loop pairing later). Using this method, the best manipulated variables are $MV_2$, $MV_3$, and $MV_4$. This result is acceptable, but each possible combination of the manipulated variables should be examined in order to determine the best one. If one examines the system, elimination of $MV_2$ is the most logical choice, since $MV_1$ controls the preheater. Hence, the use of $MV_2$, $MV_3$, and $MV_4$ is probably not wise.

To choose the proper sensors, examine the first three columns of **U** from the SVD of the system whose manipulated variables are $MV_1$, $MV_3$, and $MV_4$. The first three columns of **U**, with the maximum sensor magnitudes circled, are:

| $U_1$ | $U_2$ | $U_3$ | |
|---|---|---|---|
| 0.0593 | 0.2256 | (0.4378) | $T_1$ |
| 0.1932 | (0.4915) | 0.5681 | $T_2$ |
| 0.3710 | 0.4174 | 0.0499 | $T_3$ |
| (0.5123) | 0.2822 | −0.5635 | $T_4$ |
| 0.4943 | −0.1214 | −0.0691 | $T_5$ |
| 0.4216 | −0.3265 | 0.0630 | $T_6$ |
| 0.3091 | −0.4192 | 0.2673 | $T_7$ |
| 0.2035 | −0.3957 | 0.2922 | $T_8$ |

If the system is constructed with $T_1$, $T_2$, $T_4$ as the sensors and $MV_1$, $MV_3$, $MV_4$ as the manipulated variables, the condition number is 122.59, which is high. Therefore, the 2 × 2 system is also examined. Examining the condition number of all possible combinations of two manipulated variables, the following results are obtained:

| Manipulated Variables | Condition Number |
|---|---|
| $MV_1$, $MV_2$ | 31.47 |
| $MV_1$, $MV_3$ | 7.71 |
| $MV_1$, $MV_4$ | 5.47 |
| $MV_2$, $MV_3$ | 9.43 |
| $MV_2$, $MV_4$ | 6.06 |
| $MV_3$, $MV_4$ | 8.49 |

Using $MV_1$ and $MV_4$ is the best choice of manipulated variables and physically makes the most sense. To choose the proper sensors, the first two columns of **U** from the SVD of the system whose manipulated variables are $MV_1$ and $MV_4$ is examined. The first two columns of **U**, with the maximum sensor magnitudes circled, are:

| $U_1$ | $U_2$ | |
|---|---|---|
| 0.0598 | 0.3267 | $T_1$ |
| 0.1773 | 0.6244 | $T_2$ |
| 0.3380 | 0.4416 | $T_3$ |
| 0.4630 | 0.1744 | $T_4$ |
| 0.4991 | −0.1004 | $T_5$ |
| 0.4489 | −0.2727 | $T_6$ |
| 0.3531 | −0.3217 | $T_7$ |
| 0.2472 | −0.3001 | $T_8$ |

The best sensors are $T_2$ and $T_5$. For the system constructed with $T_2$ and $T_5$ as the sensors and $MV_1$ and $MV_4$ as the manipulated variables, the condition number is 4.67, which is good. In conclusion, the $2 \times 2$ system will be much easier to control than a $3 \times 3$ or $4 \times 4$ system and should be used in this case, unless other considerations override this decision.

As another example of determining the location of sensors, Moore (1989) considers the problems of determining the appropriate tray temperatures to use for the control of an ethanol–water distillation column. The objective is to control two tray temperatures by manipulating the distillate flow and the steam flow. The particular column has 50 trays and the two best tray temperatures from the control point of view need to be chosen. By examining the column temperature profile, trays 18 and 19 have the largest temperature drop and thus are the most sensitive. However, since they are adjacent trays, they are not independent, and changes in the column operation will affect both of them about equally. One would expect that trays at either end of the column are relatively independent, but the temperature sensitivity is too low for control.

An SVD analysis of the gain matrix for the 50-tray distillation column yields a $50 \times 2$ **U** matrix. The value of each component of the $U_1$ and $U_2$ vectors, the two columns of **U**, are plotted versus the tray number in Figure 7.29. The largest element for $U_1$ corresponds to tray 18 and the largest element for $U_2$ corresponds to tray 13. Since these trays are close to each other, there will be some interaction between them. This interaction is evident in Figure 7.29 since at the peak of $U_1$ and at the peak of $U_2$ the other vector coefficient is nonzero. The trade-off between sensitivity and interaction is shown in Figure 7.30, which plots the difference between the absolute values of the $U$ vector components. The positive peak indicates the position of the principal component in the first column and the negative peak indicates the position of the principal component in the second column. This graph suggests that trays 13 and 18 are good choices, even with the interaction.

### 7.4.3 Determine Loop Pairings for Multiloop Control

Loop pairings can be determined by a SVD analysis by pairing the principal components. The sensor associated with the principal (largest component of $U_1$ is paired with the manipulated variable associated with principal (largest) component of $V_1$. The largest component of $U_2$ is paired with the largest component of $V_2$. This process continues until all pairings are determined. In most cases, the SVD pairing procedure gives results

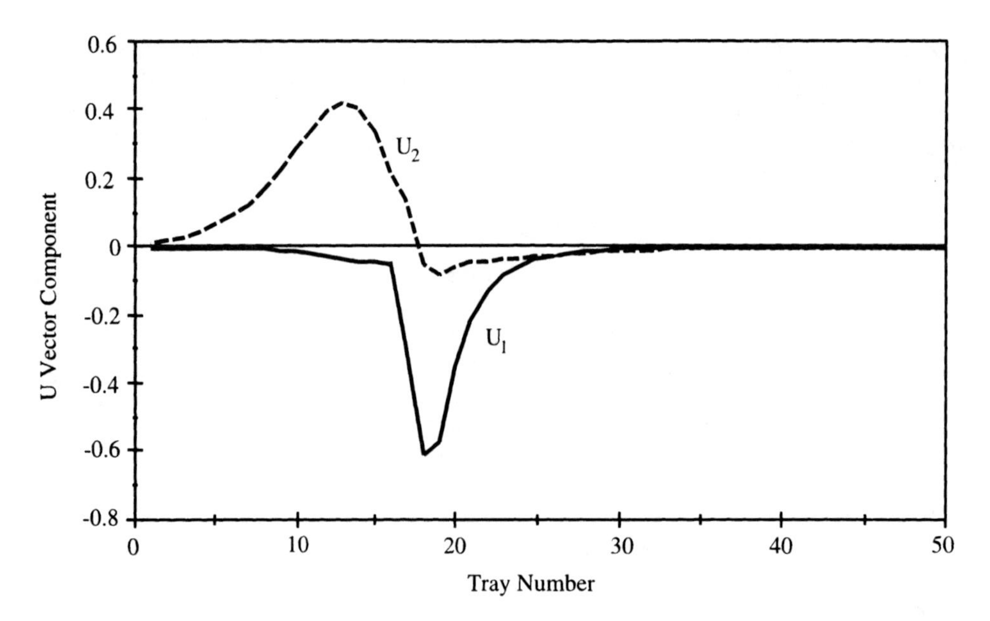


**FIGURE 7.29.** U vector plots for ethanol–water column.

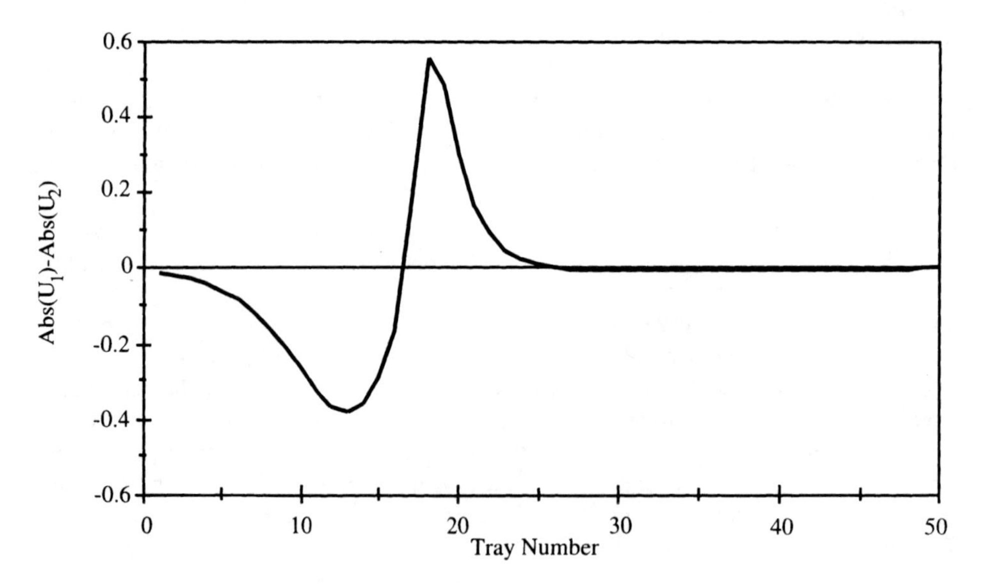


**FIGURE 7.30.** Difference in principal components.

consistent with the RGA. In cases where the pairings are different, both analyses should be considered, since the proper choice depends on the situation.

To illustrate this pairing process, RGA analysis examples are repeated here with the SVD technique, and the differences in results are discussed.

**Example 7.14.** For the Wood–Berry distillation column of Example 7.9, determine the loop pairings using SVD analysis and compare them with the RGA pairings from Example 7.6.

***Solution.*** Note that for proper results the gain matrix must be properly scaled. Hence, the gain matrix is taken from (7.13). The SVD analysis is as follows:

$$\mathbf{U} = \begin{bmatrix} 0.7219 & -0.6920 \\ 0.6920 & 0.7219 \end{bmatrix} \begin{matrix} X_D \\ X_B \end{matrix} \qquad \mathbf{\Sigma} = \begin{bmatrix} 8.577 & 0 \\ 0 & 0.8645 \end{bmatrix}$$
$$\mathbf{V} = \begin{bmatrix} 0.3220 & 0.9468 \\ -0.9468 & -0.3220 \end{bmatrix} \begin{matrix} F_R \\ F_S \end{matrix} \tag{7.25}$$

The pairings suggested by the SVD analysis are $X_D$–$F_S$ (first column of **U** and **V**) and $X_B$–$F_R$ (second column of **U** and **V**), which is different from the RGA result (Example 7.6). However, note that the relative magnitude of the elements in **U** is approximately equal, indicating that the sensors are very interactive and there is not a strong "sensor direction."

**Example 7.15.** For the heat exchanger system of Example 7.7, determine the loop pairings using SVD analysis and compare them with the RGA pairings.

***Solution.*** The SVD analysis results, with the pairings indicated, is as follows:

$$\mathbf{U} = \begin{bmatrix} -0.3068 & -0.0164 & \boxed{-0.9516} \\ -0.1450 & \boxed{0.9890} & 0.0297 \\ \boxed{-0.9407} & -0.1471 & 0.3058 \end{bmatrix} \begin{matrix} T_1 \\ \theta_2 \\ \theta_4 \end{matrix}$$
$$\mathbf{\Sigma} = \begin{bmatrix} 1.0713 & 0 & 0 \\ 0 & 0.0536 & 0 \\ 0 & 0 & 0.0263 \end{bmatrix}$$
$$\mathbf{V} = \begin{bmatrix} \boxed{0.0006} & 0.0254 & 0.09154 \\ 0.0209 & 0.2301 & \boxed{-0.9729} \\ -0.0212 & \boxed{-0.9728} & -0.2305 \end{bmatrix} \begin{matrix} MV_1 \\ MV_2 \\ MV_3 \end{matrix}$$

The suggested pairings are $\theta_4$–$MV_1$, $\theta_2$–$MV_3$, and $T_1$–$MV_2$ and are the same as for the RGA.

**Example 7.16.** For the crude tower system of Example 7.8, determine the loop pairings using SVD analysis and compare them with the RGA pairings.

***Solution.*** The results of the SVD analysis, with the pairings indicated, is as follows:

$$\mathbf{U} = \begin{bmatrix} 0.0407 & -0.4228 & 0.9000 & 0.0350 & -0.0913 \\ 0.8216 & 0.0915 & 0.0617 & -0.3892 & 0.4015 \\ 0.5680 & -0.0682 & -0.1394 & 0.5458 & -0.5962 \\ 0.0255 & -0.6839 & -0.2836 & 0.4027 & 0.5376 \\ 0.0031 & -0.5835 & -0.2938 & -0.6222 & -0.4313 \end{bmatrix} \begin{matrix} T_1 \\ T_2 \\ T_3 \\ T_4 \\ T_5 \end{matrix}$$

$$\begin{bmatrix} \sigma_1 \\ \sigma_2 \\ \sigma_3 \\ \sigma_4 \\ \sigma_5 \end{bmatrix} = \begin{bmatrix} 0.5227 \\ 0.3843 \\ 0.1749 \\ 0.0543 \\ 0.0197 \end{bmatrix}$$

$$\mathbf{V} = \begin{bmatrix} 0.0882 & -0.7286 & 0.6790 & -0.0078 & 0.0136 \\ -0.0020 & -0.4694 & -0.4865 & 0.4150 & -0.6089 \\ -0.1560 & -0.4246 & -0.4493 & 0.1109 & 0.7624 \\ -0.1866 & -0.2564 & -0.2571 & -0.8900 & -0.2030 \\ -0.9659 & 0.0525 & 0.1853 & 0.1525 & -0.0814 \end{bmatrix} \begin{matrix} MV_1 \\ MV_2 \\ MV_3 \\ MV_4 \\ MV_5 \end{matrix}$$

The suggested pairings, indicated with the solid circles, are $T_2$–$MV_5$, $T_4$–$MV_1$, $T_1$–$MV_2$, $T_5$–$MV_4$, and $T_3$–$MV_3$. However, checking these pairings with the Niederlinkski stability theorem shows these pairings to be unstable. Therefore, $MV_1$ and $MV_2$ are swapped (dashed circles) and this set of pairings is considered. A check with the Niederlinski stability theorem shows these pairings are not unstable. This result is different from the RGA pairing, as indicated below,

| Sensor | RGA Pairing | SVD Pairing |
|---|---|---|
| $T_1$ | $MV_1$ | $MV_1$ |
| $T_2$ | $MV_5$ | $MV_5$ |
| $T_3$ | $MV_2$ | $MV_3$ |
| $T_4$ | $MV_3$ | $MV_2$ |
| $T_5$ | $MV_4$ | $MV_4$ |

The RGA pairings are preferred in this case since instability will result if the $T_4$–$MV_2$ loop is opened.

### 7.4.4 Design Decouplers

The results of a SVD analysis can be used to design static decouplers. Two decouplers are designed, a sensor decoupler and a manipulated variable decoupler. Consider the SVD of a gain matrix,

$$\mathbf{K} = \mathbf{U}\boldsymbol{\Sigma}\mathbf{V}^{\mathrm{T}}$$

When $\mathbf{U}^T$ is used as a sensor decoupler and $\mathbf{V}$ is used as a manipulated variable decoupler, the process, as perceived by the controller, becomes

$$\mathbf{U}^T(\mathbf{U\Sigma V}^T)\mathbf{V} = \mathbf{\Sigma} \tag{7.26}$$

Since $\mathbf{U}$ and $\mathbf{V}$ are orthonormal, $\mathbf{U}^T\mathbf{U} = \mathbf{V}^T\mathbf{V} = \mathbf{I}$. As a block diagram, the system with sensor and valve decouplers appears as Figure 7.31*a*. From (7.26), the block diagram of the system with the decouplers appears as in Figure 7.3*b*. The sensor decoupler redefines the sensors into structured process variables and redefines the manipulated variables as structured manipulated variables. To the controller, the process appears as noninteracting loops, with the singular values as the gains. The sensor decoupler can also be viewed as a coordinate rotation that aligns the structured process variables to the relative strengths and weaknesses of the sensors. The valve decoupler does the same to the manipulated variables. When the PID controllers are combined with the decouplers, the control system appears as in Figure 7.32. Note that the setpoints, which are the desired values of the process variables, must be redefined to match the structured process variables.

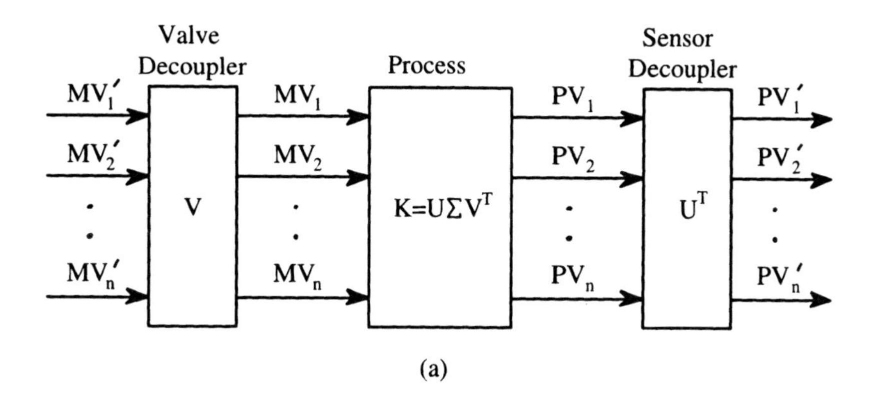


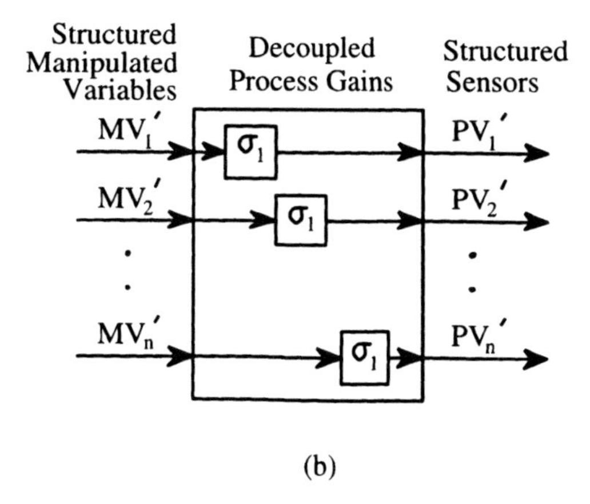


**FIGURE 7.31.** Block diagram of system with SVD decouplers: (*a*) system with valve and sensor decouplers; (*b*) equivalent decoupled system.

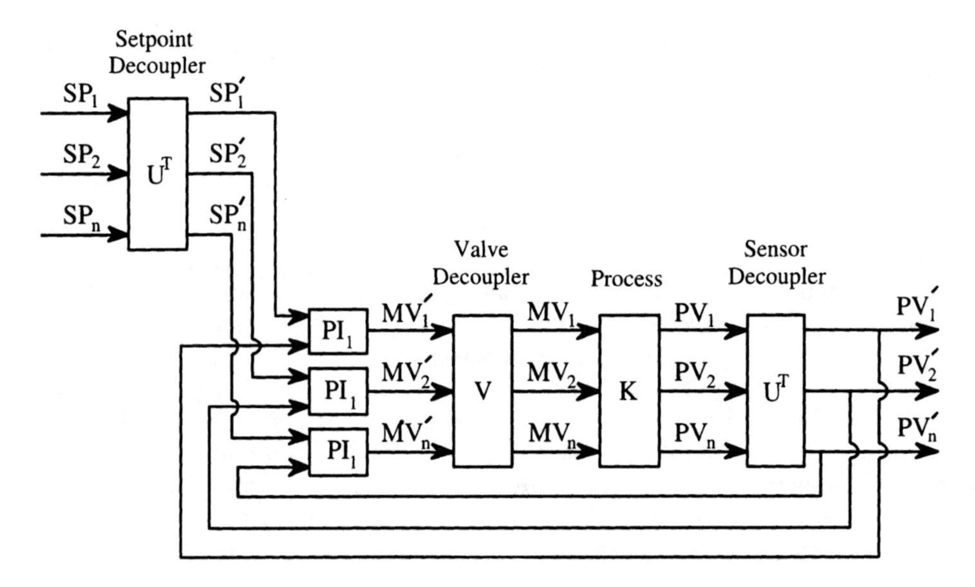


**FIGURE 7.32.** Multivariable control system with SVD decouplers.

**Example 7.17.** For the Wood–Berry distillation column of Example 7.9, design decouplers based on the SVD analysis and determine the response of the process variables to a feed flow disturbance.

***Solution.*** Using the SVD analysis from Example 7.14, (7.25), the sensor decoupler is

$$\begin{bmatrix} X'_D \\ X'_B \end{bmatrix} = \begin{bmatrix} 0.7219 & 0.6920 \\ -0.6920 & 0.7219 \end{bmatrix} \begin{bmatrix} X_D \\ X_B \end{bmatrix}$$

and the valve decoupler is

$$\begin{bmatrix} F_R \\ F_S \end{bmatrix} = \begin{bmatrix} 0.3220 & 0.9468 \\ -0.9468 & -0.3220 \end{bmatrix} \begin{bmatrix} F'_R \\ F'_S \end{bmatrix}$$

If an approximate first-order-plus-deadtime model is found for each process, as perceived by the controller, and the Fertik disturbance tuning is used, the $\mathrm{PV}'_1$–$\mathrm{MV}'_1$ PI controller parameters are $K_P = 0.44$ and $T_I = 13.0\,\mathrm{min}$ and the $\mathrm{PV}'_2$–$\mathrm{MV}'_2$ tuning parameters are $K_P = 3.8$ and $T_I = 4.3\,\mathrm{min}$. However, these parameters lead to a fairly oscillatory response to a change in either setpoint. Thus, the $K_P$ for the $\mathrm{PV}'_1$–$\mathrm{MV}'_1$ controller is decreased to 0.25 and the $K_P$ for the $\mathrm{PV}'_2$–$\mathrm{MV}'_2$ controller is decreased to 1.9. The responses to a +15% (0.34 magnitude) feed flow disturbance are shown in Figure 7.33. Compared with the multiloop controller of Example 7.9 (Figure 7.17), the responses are worse. However, they are better than the static decoupler of Example 7.12 (Figure 7.25).

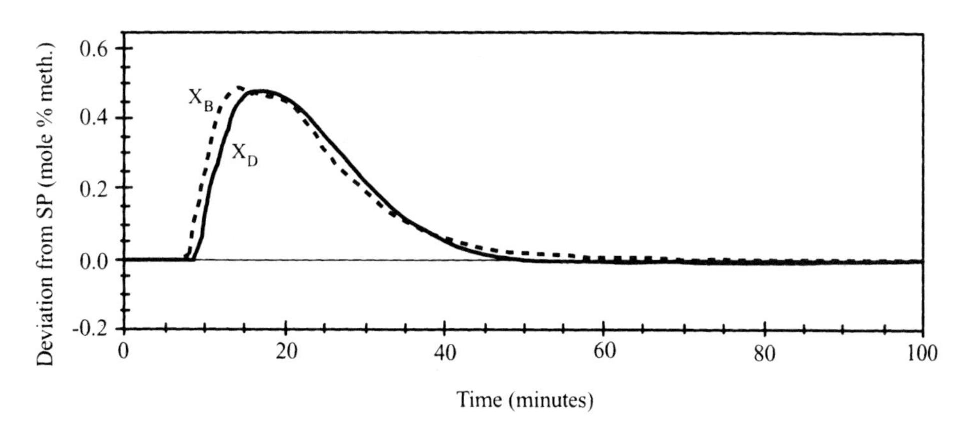


**FIGURE 7.33.** Response of Wood–Berry distillation column with SVD decouplers.

## 7.5 MULTIVARIABLE MODEL PREDICTIVE CONTROL

In contrast to the multiloop methods presented in earlier sections of this chapter, multivariable model predictive control (MPC) is a full multivariable controller. All measurements are used simultaneously to calculate the manipulated variable. It is a common form of multivariable control being used in chemical process industries. Multivariable MPC is a straightforward extension of the single-loop model predictive controllers presented in Chapter 6.

Both model predictive controllers introduced in Chapter 6, the dynamic matrix controller and the forward modeling controller, are extended to multivariable systems. The FMC is a single-point MPC algorithm and is conceptually simpler than DMC, and so it is presented first. The application of each controller is illustrated on the Wood–Berry distillation column. The extension of these controllers to handle process constraints is briefly discussed. In addition, their ability to handle mismatch between the model and the process is illustrated with an example.

### 7.5.1 Multivariable Forward Modeling Control

Unlike other model predictive controllers, the forward modeling controller calculates the controller action to minimize the error at only one future sample (the output horizon) of the predicted outputs. Other model-based predictive controllers minimize the error at multiple samples of the predicted outputs. Therefore, the matrix that must be inverted in the FMC algorithm is much smaller than the matrix used in the DMC algorithm (Cutler and Ramaker, 1980). In contrast to MAC, FMC does not require the on-line solution of an optimization problem or the solution of an off-line Ricatti difference equation for processes with nonminimum phase (Rouhani and Mehra, 1982). The multivariable single-point FMC also does not require the derivation of an approximate process inverse, as in IMC (Garcia and Morari, 1985). The response models are used directly in the calculation of the controller outputs. In contrast to the simplified model predictive control (Arulalan and Deshpande, 1987), the multivariable FMC does not use an off-line optimization procedure to determine the tuning parameters. Multivariable FMC is also similar to the single-step control of Brosilow and Zhao (1988) but does not factor out the process deadtime and then add filtering to get less vigorous control action. The aim of this section is to present the multivariable FMC for unconstrained processes. Feher (1993) and Feher and Erickson (1993) extend the FMC to processes with soft and hard constraints.

The feature of the multivariable FMC that makes it unique among other model-based predictive controllers is the use of the output horizon as the only major tuning parameter. Each controlled variable may have a different output horizon, depending on how tightly each is controlled. No external filter factors or move suppression factors are used.

*7.5.1.1 Algorithm Development.* The development of the multivariable FMC is very similar to the development of the SISO version in Chapter 6, except that all measurements and manipulated variables are now vectors, and scalar coefficients are now matrices. Formally, the process has $q$ outputs and $r$ inputs and is modeled by a discrete moving-average (MA) model with $n$ samples:

$$\mathbf{y}(k) = \mathbf{H}_1\mathbf{m}(k-1) + \mathbf{H}_2\mathbf{m}(k-2) + \cdots + \mathbf{H}_n\mathbf{m}(k-n) + \hat{\mathbf{d}}(k) \tag{7.27}$$

where

$\mathbf{y}(k) = [y_1\ y_2\ \ldots\ y_q]^{\mathrm{T}}$, a $q \times 1$ vector of process variable measurements at the $k$th sample;

$\mathbf{m}(k) = [m_1\ m_2\ \ldots\ m_r]^{\mathrm{T}}$, an $r \times 1$ vector of the manipulated variables at the $k$th sample measured as close to the process as practical;

$\mathbf{H}_l$ is a $q \times r$ matrix of the $l$th MA model coefficient; $(h_{ij})_l$ is the $l$th MA model coefficient between the $j$th manipulated variable and the $i$th process variable; and

$\hat{\mathbf{d}}(k) = [d_1\ d_2\ \ldots\ d_q]^{\mathrm{T}}$, a $q \times 1$ vector of the current discrepancies between the model and the measurement at the $k$th sample.

With the same assumptions as for the SISO version, the multivariable single-point FMC computes a prediction of the process output $\mathbf{T}(k)$, into the future with the following assumptions:

1. The manipulated variable is held constant into the future: $\mathbf{m}(k) = \mathbf{m}(k+1) = \mathbf{m}(k+2) = \cdots$.
2. The future discrepancies remain at their current value: $\hat{\mathbf{d}}(k) = \hat{\mathbf{d}}(k+1) = \hat{\mathbf{d}}(k+2) = \cdots$.

The equations for the process variable predictions are developed in the same manner as for the SISO FMC [(6.12) and (6.13)] and the predicted process variables at the $j$th sample in the future, given the information of the current, $k$th, sample, $\mathbf{T}_j(k)$:

$$\begin{aligned}\mathbf{T}_j(k) &= (\mathbf{H}_1 + \mathbf{H}_2 + \cdots + \mathbf{H}_j)\mathbf{m}(k) + \mathbf{H}_{j+1}\mathbf{m}(k-1) + \cdots + \mathbf{H}_n\mathbf{m}(k-n+j)\hat{\mathbf{d}}(k) \\ &= \mathbf{A}_j\mathbf{m}(k) + \mathbf{H}_{j+1}\mathbf{m}(k-1) + \cdots + \mathbf{H}_n\mathbf{m}(k-n+j) + \hat{\mathbf{d}}(k)\end{aligned} \tag{7.28}$$

where $\mathbf{A}_j = \sum_{i=1}^{j} \mathbf{H}_i$ is the $j$th coefficient of the process discrete-time step response. Similar to the SISO case,

$$\mathbf{T}_j(k) = \mathbf{T}_{j+1}(k-1) + \mathbf{A}_j\,\Delta\mathbf{m}(k) + \Delta\hat{\mathbf{d}}(k) \tag{7.29}$$

where $\Delta\mathbf{m}(k) = \mathbf{m}(k) - \mathbf{m}(k-1)$ and $\Delta\hat{\mathbf{d}}(k) = \hat{\mathbf{d}}(k) - \hat{\mathbf{d}}(k-1)$. Expressed in words, the prediction of the $j$th sample in the future is updated from the previous prediction of the $(j+1)$st sample by the change in the unmeasured disturbances and the changes in the manipulated variables.

The controller is designed that yields at sample $k$ a controller output $\mathbf{m}(k)$ that, if it were kept constant from here on, would minimize $\mathbf{V}$, the Euclidean norm of the error between the prediction and the setpoint at $P$ samples in the future:

$$\mathbf{V} = [\mathbf{s}_P(k) - \mathbf{T}_P(k)]^{\mathrm{T}}[\mathbf{s}_P(k) - \mathbf{T}_P(k)] = [\mathbf{e}_P(k)]^{\mathrm{T}}[\mathbf{e}_P(k)] \tag{7.30}$$

where $\mathbf{s}_P(k)$ and $\mathbf{e}_P(k)$ are the setpoint vector and error vector, respectively, at $P$ samples in the future. The details of the actual computation are in Erickson and Otto (1991) and involve the pseudo inverse of a matrix for nonsquare systems (unequal numbers of process inputs and outputs). Assuming the controller manipulated variable is calculated to drive the prediction after $P$ sample intervals in the future to the setpoint $\mathbf{s}_P(k)$ minus an error vector $\mathbf{e}_P(k)$ yields

$$\begin{aligned}\mathbf{T}_P(k) &= \mathbf{s}_P(k) - \mathbf{e}_P(k)\\ &= \mathbf{A}_P\mathbf{m}(k) + \mathbf{H}_{P+1}\mathbf{m}(k-1) + \cdots + \mathbf{H}_n\mathbf{m}(k-n+P) + \hat{\mathbf{d}}(k)\end{aligned}$$

For systems where the number of manipulated variables is greater than or equal to the number of controller variables ($r \geq q$), it is theoretically possible to calculate a controller output that makes the error $\mathbf{e}_P(k)$ zero. However, for systems where $r < q$ or for constrained systems where $r \geq q$, it may not be possible to achieve the desired setpoint. For convenience, the setpoint is redefined as $\mathbf{s}(k) = \mathbf{s}_P(k) + \mathbf{e}_P(k)$. The value of $\mathbf{s}(k)$ thus represents the setpoint that can be achieved given the limitations of the system. In conventional controller terms (Figure 6.47), the controller transfer function matrix $\mathbf{G}_c(z)$ is given as

$$\mathbf{G}_c(z) = [\mathbf{A}_P + \mathbf{H}_{P+1}z^{-1} + \cdots + \mathbf{H}_n z^{-n+P} - (\mathbf{H}_1 z^{-1} + \cdots + \mathbf{H}_\mathbf{n} z^{-n})]^{(-1)} \tag{7.31}$$

and is developed in a manner similar to the SISO FMC. The notation $[\cdot]^{(-1)}$ indicates the left pseudoinverse of a nonsquare matrix. Of course, if the matrix is square, the left pseudoinverse is the usual matrix inverse. One way to examine stability of the closed-loop system is to examine the closed-loop system poles. Obviously, this task is not trivial for multivariable systems. However, by converting the controller to the internal model configuration (Figure 6.48), stability analysis is simpler. It can be shown that when exact process modeling is assumed, the internal model controller $\mathbf{G}_{cp}(z)$ is given as

$$\mathbf{G}_{cp}(z) = [\mathbf{A}_P + \mathbf{H}_{P+1}z^{-1} + \cdots + \mathbf{H}_n z^{-n+P}]^{(-1)} \tag{7.32}$$

For system stability, the controller poles are given by the roots of the equation

$$\det[\mathbf{A}_P z^{n-P} + \mathbf{H}_{P+1} z^{n-P-1} + \mathbf{H}_{P+2} z^{n-P-2} + \cdots + \mathbf{H}_n] = 0 \tag{7.33}$$

Thus the loop is stable if and only if the roots of (7.33) are strictly inside the unit circle. The order of (7.33) is $(n - P) \times \max(q, r)$ and can be high since typically $n \geq 20$. Also, the roots tend to cluster close to $z = 1$, making the task of accurately finding the roots difficult.

As an alternative, the stability of (7.33) can be assessed using Ngo's stability test (Ngo and Erickson, 1997) without fiinding any equation roots. The test is as follows. Define the matrices $\mathbf{B}$ and $\mathbf{B}_T$ as

$$\begin{aligned} \mathbf{B} &= [\mathbf{H}_{P+1} \quad \mathbf{H}_{P+2} \quad \cdots \quad \mathbf{H}_n] \\ \mathbf{B}_T &= [\mathbf{H}_{P+1}^{\mathrm{T}} \quad \mathbf{H}_{P+2}^{\mathrm{T}} \quad \cdots \quad \mathbf{H}_n^{\mathrm{T}}] \end{aligned}$$

where the matrix $\mathbf{B}$ is $q \times (n - P)r$ and the matrix $\mathbf{B}_T$ is $r \times (n - P)q$. The system (7.33) is asymptotically stable if either

$$\|\mathbf{B}\|_\infty < \frac{1}{\|\mathbf{A}_P^{-1}\|_\infty} \tag{7.34}$$

or

$$\|\mathbf{B}_T\|_\infty < \frac{1}{\|\mathbf{A}_P^{-1}\|_1} \tag{7.35}$$

where $\|\cdot\|_\infty$ denotes the $\infty$-norm of a matrix, the maximum row sum of the absolute values of the matrix elements, and $\|\cdot\|_1$ denotes the 1-norm of a matrix, the maximum column sum of the absolute values of the matrix elements. For nonsquare systems $(q > r)$, the ordinary matrix inverse in (7.34) and (7.35) is replaced by the left pseudoinverse. The smallest value of $P$ that satisfies either (7.34) or (7.35) is the minimum value of $P$ that guarantees stability. Note that if $P$ is set equal to $n$ (steady-state control), the theorem shows that the loop is stable; hence stability is guaranteed if the modeling is exact. The results of the Ngo stability test tend to be conservative (Ngo and Erickson, 1997) but are more representative of actual practice, since a fair amount of modeling error can be tolerated.

The preceding development assumed that $P$, the output horizon, was the same for each controlled variable. However, each controlled variable may have a different value of $P$, depending on how tightly each one is to be controlled. To accommodate different values of $P$ for each process output, $\mathbf{T}_P(k)$ is redefined as

$$\mathbf{T}_P(k) = [T^{P_1} \quad T^{P_2} \quad \cdots \quad T^{P_q}]^{\mathrm{T}}$$

where $P_i$ is the output horizon for the $i$th process variable. Similarly $\mathbf{A}_P$ is redefined as

$$\mathbf{A}_P = \begin{bmatrix} a_{11}^{P_1} & a_{12}^{P_1} & \cdots & a_{1r}^{P_1} \\ a_{21}^{P_2} & a_{22}^{P_2} & \cdots & a_{2r}^{P_2} \\ \vdots & \vdots & \cdots & \vdots \\ a_{q1}^{P_q} & a_{q2}^{P_q} & \cdots & a_{qr}^{P_q} \end{bmatrix} \tag{7.36}$$

where $a_{jk}^{P_i}$ is the step response model coefficient at output horizon $P_i$ of the model between the $k$th process input and the $j$th process output. The expression for the controller transfer function (7.32) becomes similarly complex and is not shown here.

*7.5.1.2* ***Filtering for Robustness.*** If modeling errors exist, stability cannot be guaranteed even for $P_i = n$. Furthermore, even with exact modeling, a policy of setting $P_i = n$ may produce a controller that moves the manipulated variables too vigorously when large amounts of noise are present in the measurements. The controller needs to be modified to produce robustness (tolerance to modeling errors) and noise rejection. Garcia and Morari (1985) have shown that if a filter $\mathbf{F}(z)$ is added to the controller input, the closed-loop system can be made stable for arbitrarily large modeling errors (other than the wrong sign on the model gains) by filtering heavily enough. The filter $\mathbf{F}(z)$ is a diagonal matrix of the form

$$\mathbf{F}(z) = \text{diag}\left\{\frac{1}{1 + f_i - f_i z^{-1}}\right\}$$

where $f$ is picked as

$$f_i = \begin{cases} 0 & \text{for } P_i \le n \\ \frac{1}{3}(P_i - n) & \text{for } P_i > n \end{cases}$$

When a particular $P_i$ is larger than the length of the step response model, the $i$th controller input is filtered and provides robustness to modeling error. Therefore, adjustment of the output horizon $P_i$ can move the controller from high-performance control to noise-rejecting, sluggish, robust control.

*7.5.1.3* ***Feedforward Control.*** Feedforward control is easily accommodated by the MIMO FMC. Given a model of the effect of a vector of *measured* disturbance $\mathbf{d}$ on the process outputs of the same form as the discrete-time MA model used for the process,

$$\mathbf{y}(k) = \mathbf{H}_1'\mathbf{d}(k-1) + \mathbf{H}_2'\mathbf{d}(k-2) + \cdots + \mathbf{H}_n'\mathbf{d}(k-n) \tag{7.37}$$

where the matrices $\mathbf{H}_j'$ are the appropriate dimensions. As done in the SISO case, with the additional assumption that the measured disturbances remain at their present values, the prediction vector at the $j$th sample in the future is

$$\mathbf{T}_j(k) = \mathbf{T}_{j+1}(k-1) + \mathbf{A}_j\,\Delta\mathbf{m}(k) + \mathbf{A}_j'\,\Delta\mathbf{d}(k) + \Delta\hat{\mathbf{d}}(k) \tag{7.38}$$

where $\mathbf{A}'_j = \sum_{i=1}^{j} \mathbf{H}'_i$ is the $j$th coefficient of the measured disturbance step response. As for the SISO case, the only change to the control algorithm is to update the process variable predictions with the expected effect of the measured disturbance. No other change to the control algorithm is necessary.

*7.5.1.4* ***Controller Tuning.*** The multivariable FMC has only two types of adjustments. The above development focused on the output horizon for each process variable $P_i$, which smoothly takes the control action from

1. extremely aggressive control by setting each $P_i$ to its minimum value to
2. steady-state control where the controller moves the manipulated variables only to statically compensate the process to
3. extremely sluggish, noise-rejecting, robust control that should be stable in all practical situations.

In addition to the output horizon, the controller has only one other tuning parameter, the control interval $\Delta t$. The number of points in the model, $n$, is usually fixed. For ease of communication with the engineer commissioning the loop and with operating personnel, the two types of parameters can be renamed:

1. the *closed-loop settling time* (CLST) of the controlled variable, $P_i\,\Delta t$, and
2. the *open-loop settling time* (OLST) of the process, $n\,\Delta t$.

In practice, the OLST of the process is rarely changed. The CLST is approximate (except for $P_i = n$ with exact modeling) but is useful for explaining the trade-off between loop performance and robustness. Decreasing the CLST tends to make the system less tolerant to modeling errors. This feature has strong intuitive appeal and is well received by operating personnel.

*7.5.1.5* ***Algorithm Implementation.*** As for the SISO version, the controller is implemented to recursively process a prediciton of the future trends of the process variables. When a predictor is used within the control loop, the structure of the controller with feedforward changes to the configuration shown in Figure 6.50. The algorithm details are in Erickson and Otto (1991).

**Example 7.18.** For the Wood–Berry distillation column of Example 7.9, design a multivariable FMC, with and without feedforward, and determine the response of the process variables to a feed flow disturbance.

***Solution.*** The multivariable FMC is implemented using a sampling period of 1 min. The transfer function model (7.13) is converted to matrices of MA models of 60 samples each. Using stability analysis (Jury, 1964) on the poles of $\mathbf{G}_{cp}(z)$ in (7.32), the minimum output horizons are $P_{1,\min} = 2$ and $P_{2,\min} = 4$. At these output horizons, no modeling error can be tolerated. Using the Ngo stability test, the minimum output horizons are $P_{1,\min} = 15$ and $P_{2,\min} = 16$. Obviously, they are conservative but more representative of the parameters used in practice since modeling error can be tolerated. The manipulated variable moves are calculated with the algorithm in Erickson and Otto (1991). With $P_1 = P_2 = 5$ and no

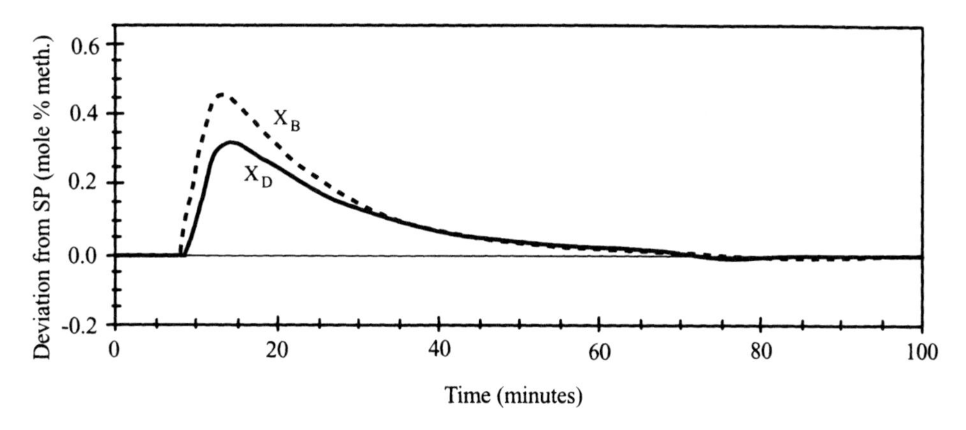


**FIGURE 7.34.** Response of Wood–Berry distillation column with FMC and no feedforward.

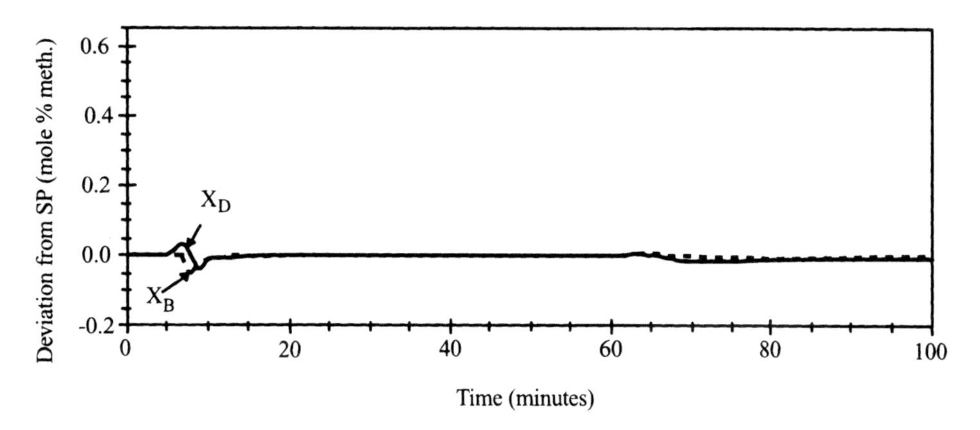


**FIGURE 7.35.** Response of Wood–Berry distillation column with FMC and feedforward.

feedforward, the responses of the column to a 15% feed flow disturbance are shown in Figure 7.34. Compared with the multiloop control of Example 7.9 (Figure 7.17), the distillate response is slightly worse while the bottom composition is much better. Overall, the IAE is only slightly higher. Compared with the dynamic decoupler of Example 7.12 (Figure 7.25), the response is only slightly higher. However, the FMC is not tuned to its more aggressive settings. With feedforward, the responses of the column to a 15% feed flow disturbance are shown in Figure 7.35, clearly showing that the feedforward nearly cancels the disturbance. Feedforward control is difficult to design for the multiloop or decoupling controllers.

### 7.5.2 Multivariable Dynamic Matrix Control

The multivariable DMC algorithm is developed as an extension of the single-loop DMC in Chapter 6. The process is modeled by a discrete MA model (7.27). The prediction assumes more than one manipulated variable move in order to bring the process variable back to its setpoint. The prediction is developed assuming $M$ future manipulated variable moves and

$N$ predicted process variable measurements ($N < n$, the step response or MA model length). The process has $q$ process variable measurements and $r$ manipulated variables. The prediction between the $j$th manipulated variable and the $i$th process variable is expressed in the same form as for the SISO DMC (6.27),

$$\begin{bmatrix} \Delta T'_{i_1}(k) \\ \Delta T'_{i_2}(k) \\ \vdots \\ \Delta T'_{i_M}(k) \\ \vdots \\ \Delta T'_{i_N}(k) \end{bmatrix} = \begin{bmatrix} a_{ij_1} & 0 & \cdots & 0 \\ a_{ij_2} & a_{ij_1} & \cdots & 0 \\ \vdots & \vdots & & \vdots \\ a_{ij_M} & a_{ij_{M-1}} & \cdots & a_{ij_1} \\ \vdots & \vdots & & \vdots \\ a_{ij_N} & a_{ij_{N-1}} & \cdots & a_{ij_{N-M+1}} \end{bmatrix} \begin{bmatrix} \Delta m_j(k) \\ \Delta m_j(k+1) \\ \vdots \\ \Delta m_j(k+M-1) \end{bmatrix} \tag{7.39}$$

$$\Delta \mathbf{T}'_i(k) = \mathbf{A}_{ij}\ \Delta \mathbf{m}_j(k)$$

The entire prediction of the $i$th process variable is

$$\Delta \mathbf{T}_i(k) = \sum_{j=1}^{r} \mathbf{A}_{ij}\ \Delta \mathbf{m}_j(k) \tag{7.40}$$

As for the SISO DMC, the prediction vector (7.40) is set equal to the vector of predicted errors $\hat{\mathbf{e}}_i$,

$$\hat{\mathbf{e}}_i = \sum_{j=1}^{r} \mathbf{A}_{ij}\ \Delta \mathbf{m}_j(k) \tag{7.41}$$

Note that $\hat{\mathbf{e}}_i$ only contains the effect of past moves and not the prediction of future moves. Since the system of equations in (7.41) is overdetermined, the error of the solution is minimized,

$$\mathbf{e}_i^* = \hat{\mathbf{e}}_i - \sum_{j=1}^{r} \mathbf{A}_{ij}\ \Delta \mathbf{m}_j(k) \tag{7.42}$$

By defining

$$\Delta \mathbf{T}(k) = \begin{bmatrix} \Delta \mathbf{T}_1(k) \\ \Delta \mathbf{T}_2(k) \\ \vdots \\ \Delta \mathbf{T}_q(k) \end{bmatrix} \qquad \Delta \mathbf{m}(k) = \begin{bmatrix} \Delta \mathbf{m}_1(k) \\ \Delta \mathbf{m}_2(k) \\ \vdots \\ \Delta \mathbf{m}_r(k) \end{bmatrix} \qquad \mathbf{A} = \begin{bmatrix} \mathbf{A}_{11} & \cdots & \mathbf{A}_{1r} \\ \vdots & & \vdots \\ \mathbf{A}_{q1} & \cdots & \mathbf{A}_{qr} \end{bmatrix}$$

the entire prediction of the process can be written concisely as

$$\Delta \mathbf{T}(k) = \mathbf{A}\ \Delta \mathbf{m}(k) = \mathbf{A}\ \Delta \mathbf{m} \tag{7.43}$$

where $\mathbf{A}$ is commonly called the *dynamic matrix* and has $Nq$ rows and $Mr$ columns. The equation is in the same form as for the SISO DMC (6.28). The set of moves $\Delta \mathbf{m}(k)$ is calculated to produce changes in the predicted output $\Delta\mathbf{T}(k)$ that force the prediction of the error to zero at each future sample. As for the SISO DMC, $\Delta\mathbf{m}(k)$ is calculated to minimize the sum of the squares of the solution error vector components plus a penalty term for the manipulated variables. In addition, since the relative importance of each process variable is different, a penalty term is added to the error term. The manipulated variable moves are calculated to minimize

$$\min_{\Delta\mathbf{m}}\left[\sum_{i=1}^{q}\gamma_i\left(\sum_{j=1}^{N}(e^*_{i_j})^2\right)+\sum_{i=1}^{r}\left(\lambda_i\sum_{j=1}^{M}[\Delta m_i(k+j-1)]^2\right)\right] \tag{7.44}$$

where

$\lambda_i$ are called the *move suppression factors*, one for each manipulated variable;
$\gamma_i$ are called the *control weighting factors*, one for each process variable; and
$(e^*_{i_j})$ is the $j$th component of the vector $\mathbf{e}^*_i$ in (7.42).

In matrix form, (7.44) is written as

$$\min_{\Delta\mathbf{m}}[(\hat{\mathbf{e}}-\mathbf{A}\,\Delta\mathbf{m})^{\mathrm{T}}\Gamma(\hat{\mathbf{e}}-\mathbf{A}\,\Delta\mathbf{m})+(\Delta\mathbf{m})^{\mathrm{T}}\Lambda(\Delta\mathbf{m})] \tag{7.45}$$

The unconstrained solution to (7.45) is

$$\begin{aligned}\Delta\mathbf{m}&=(\mathbf{A}^{\mathrm{T}}\Gamma\mathbf{A}+\Lambda)^{-1}\mathbf{A}^{\mathrm{T}}\Gamma\hat{\mathbf{e}}\\&=\mathbf{S}\hat{\mathbf{e}}\end{aligned} \tag{7.46}$$

As for the SISO DMC, in practice, only the first current move for each manipulated variable is implemented.

To complete the controller, the prediction must be updated with the actual process variable measurement. The update is accomplished in the same manner as for the FMC (7.38).

Feedforward control is accomplished in the same manner as for the multivariable FMC. The prediction is updated with the expected effect of the disturbance before the manipulated variable move is calculated.

In addition to the sample period and the step response model length, the multivariable DMC has four adjustments:

$N$—The *output horizon*, the future time over which the control performance is evaluated. Typically, this parameter is 20–50 samples. The closed-loop system should approach steady state in $N$ samples.

$M$—The *manipulated variable* (input) *horizon*, the number of future manipulated variable moves considered in the calculation. Typically, this parameter is $\frac{1}{4}$ to $\frac{1}{3}$ of $N$.

$\gamma$—The *control weighting factor*, one for each process variable. Increasing this value for a process variable tends to reduce its deviation but at the expense of increasing the deviation of the other process variables.

$\lambda$—The *move suppression factor*, one for each manipulated variable. Increasing this value tends to slow the manipulated variable move, which degrades control loop performance. However, increasing the value also improves the robustness of the system to model mismatch.

**Example 7.19.** For the Wood–Berry distillation column of Example 7.9, design a multivariable DMC, with and without feedforward, and determine the response of the process variables to a feed flow disturbance.

***Solution.*** The multivariable DMC is implemented using a sampling period of 1 min. The transfer function model (7.13) is converted to matrices of MA models of 60 samples each. The controller parameters are $N = 25$, $M = 10$, $\gamma_1 = \gamma_2 = 1.0$, and $\lambda = 0.1$. With no feedforward, the responses of the column to a 15% feed flow disturbance are shown in Figure 7.36. With feedforward, the responses of the column to a 15% feed flow disturbance are shown in Figure 7.37. The responses are comparable to the FMC, although the IAE is slightly smaller.

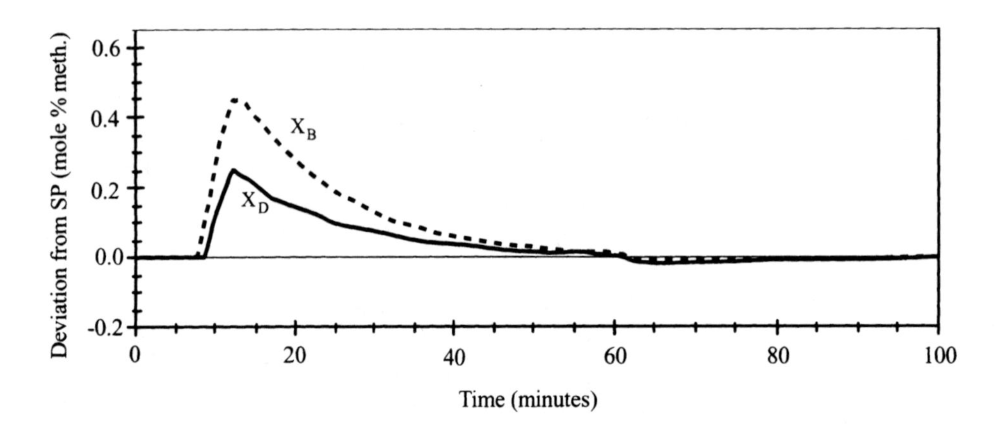


**FIGURE 7.36.** Response of Wood–Berry distillation column with DMC and no feedforward.

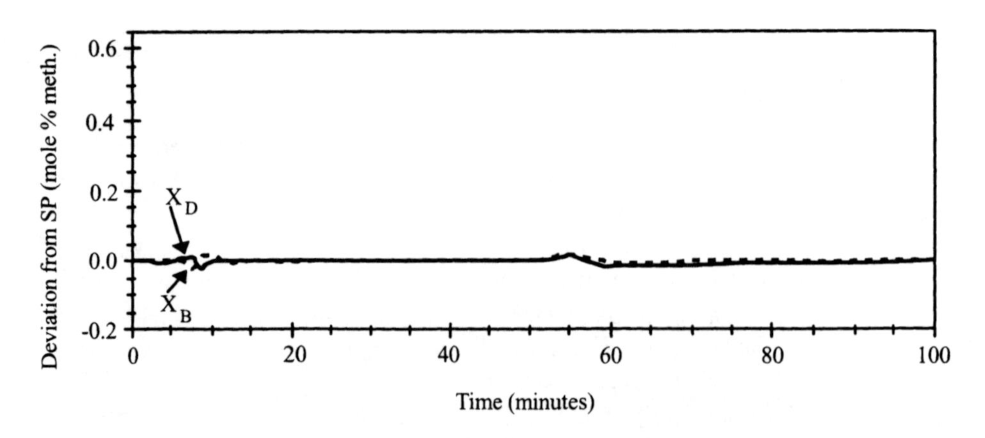


**FIGURE 7.37.** Response of Wood–Berry distillation column with DMC and feedforward.

### 7.5.3 Constraints

In many cases, the control system behavior is limited by constraints. These constraints may be a limitation of the physical equipment (e.g., a valve can be no further open that 100%) or may be imposed by the process engineer (e.g., feed flow should not go above 200 l/min). Unfortunately, economical operation of a unit often compels operating points that lie on or near one or more process constraints. Two types of constraints need to be considered: *hard constraints*, where no violations of the bounds are allowed at any time, and *soft constraints*, where violations of the bounds are tolerated in order to satisfy other criteria. Hard constraints generally represent those bounds set by safety or equipment considerations. On the other hand, soft constraints represent operating bounds that may be relaxed if the process is operating abnormally, for example, one or more of the manipulated variables is already at its hard constraint boundary.

The model-based approach to control was originally proposed in order to handle such constraints. Both FMC and DMC may be extended to calculate the control action in order to minimize the original objective, (7.30) or (7.45), while observing the process constraints. When hard constraints are added to the problem, *quadratic programming* (QP) is the method of optimization. When soft constraints are added to the problem, a *nonlinear quadratic programming* algorithm must be used. Model predictive controller formulations for constraints are given in Garcia and Morshedi (1986), Ricker (1985), and Feher and Erickson (1993).

### 7.5.4 Sensitivity to Mismatch between Model and Process

Both FMC and DMC can tolerate mismatch between the process model and the actual process. Mathematical analysis of the closed-loop system in order to determine stability is difficult, and generally the controllers are just detuned to account for the modelling error. The performance degradation is illustrated by the following example.

**Example 7.20.** For the model predictive controllers designed in the previous two examples for the Wood–Berry distillation column, suppose the tuning parameters of the controller for the reflux flow rate change and the perceived dynamic gain of the first

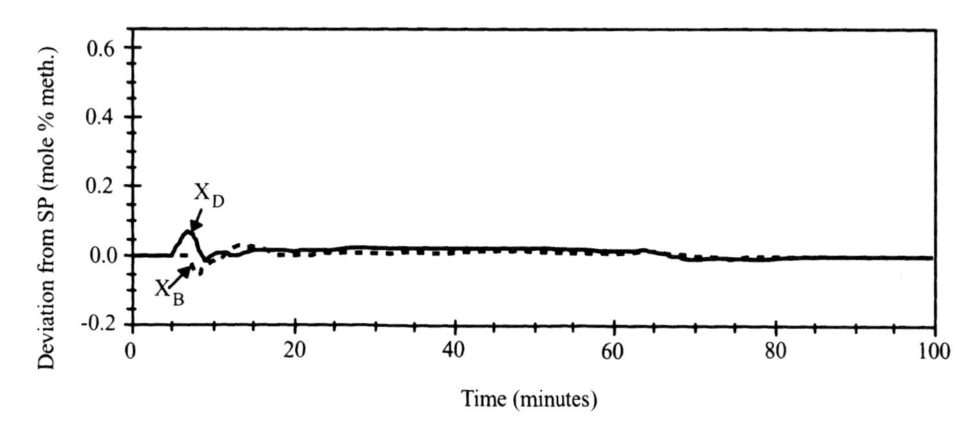


**FIGURE 7.38.** Response of Wood–Berry distillation column with FMC and feedforward when the dynamic gain of $MV_1$ is doubled.

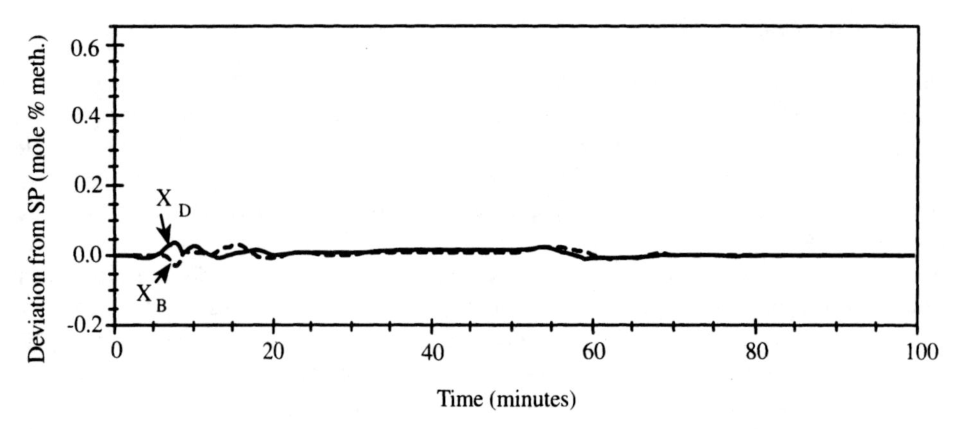


**FIGURE 7.39.** Response of Wood–Berry distillation column with DMC and feedforward when the dynamic gain of $MV_1$ is doubled.

manipulated variable doubles. Assess the effect of this process change on the response of the system when FMC or DMC are used.

***Solution.*** Multivariable FMC is the same as implemented in Example 7.18 and multivaraible DMC is the same as designed in Example 7.19. With feedforward, the responses of the column with FMC and DMC to a 15% feed flow disturbance are shown in Figures 7.38 and 7.39, respectively. Compared with Examples 7.18 and 7.19, the responses show some increased deviation from the setpoint and IAE. However, the responses are stable.

## 7.6 USER INTERFACE

Depending on the type of multivariable control, the user interface will vary. For a multiloop control scheme, the operator interface is not much different from the typical single-loop or cascade control faceplate. A decoupling controller needs to present summary information about the process variables and manipulated variables in addition to the usual interface to the single-loop controllers. This summary is generally simply a table with two columns, the first showing the current process variable values (and possibly the setpoints) and the second showing the current manipulated variable values. An example summary display for the Wood–Berry distillation column decoupling controller (Example 7.12) is shown in Figure 7.40. The process variable and setpoint in the first column are also

METHANOL COLUMN CONTROL SUMMARY

| Process Variable | | SP | Man. Variable | | Actual | |
|---|---|---|---|---|---|---|
| Top Concentration (mole % meth.) | 96.35 | 96.25 | Reflux Flow (lb/min.) | 1.90 | 1.95 | Detail |
| Bottom Concentration (mole % meth.) | 0.45 | 0.50 | Steam Flow (lb/min.) | 1.70 | 1.71 | Detail |

**FIGURE 7.40.** Decoupling controller summary display for Wood–Berry distillation column.

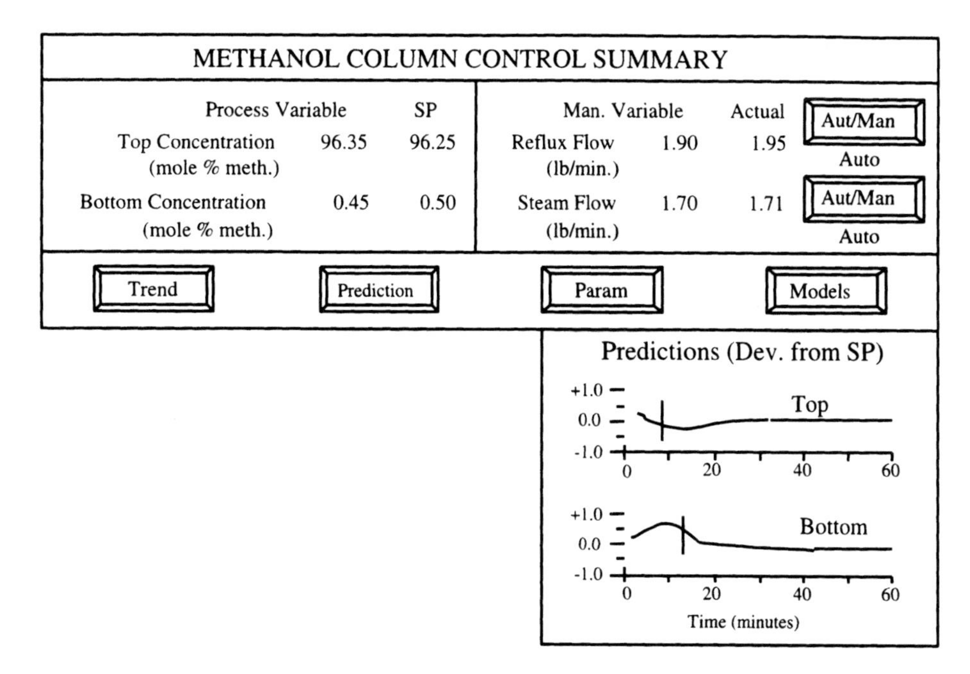


**FIGURE 7.41.** FMC summary display for Wood–Berry distillation column.

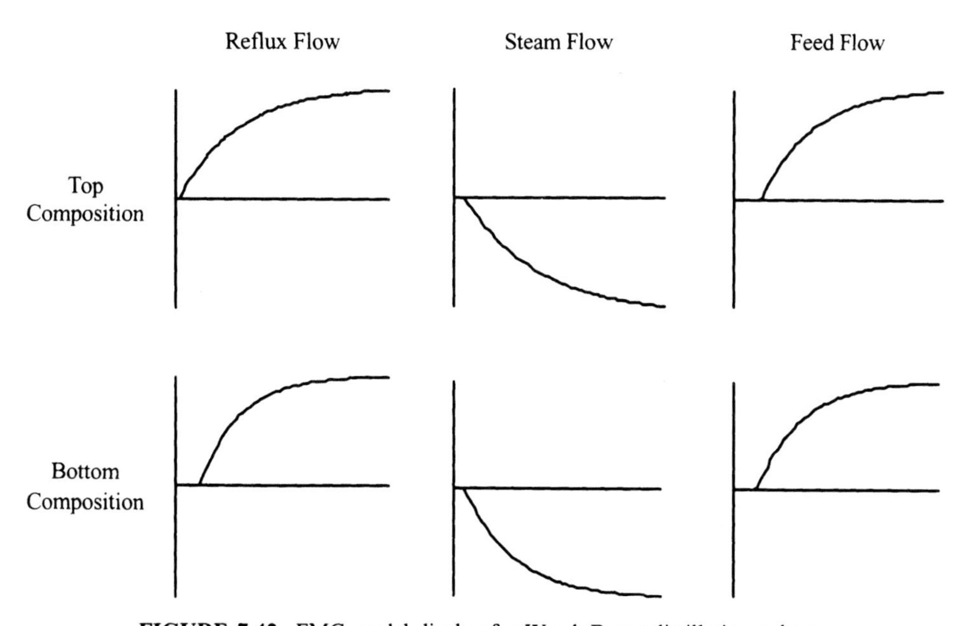


**FIGURE 7.42.** FMC model display for Wood–Berry distillation column.

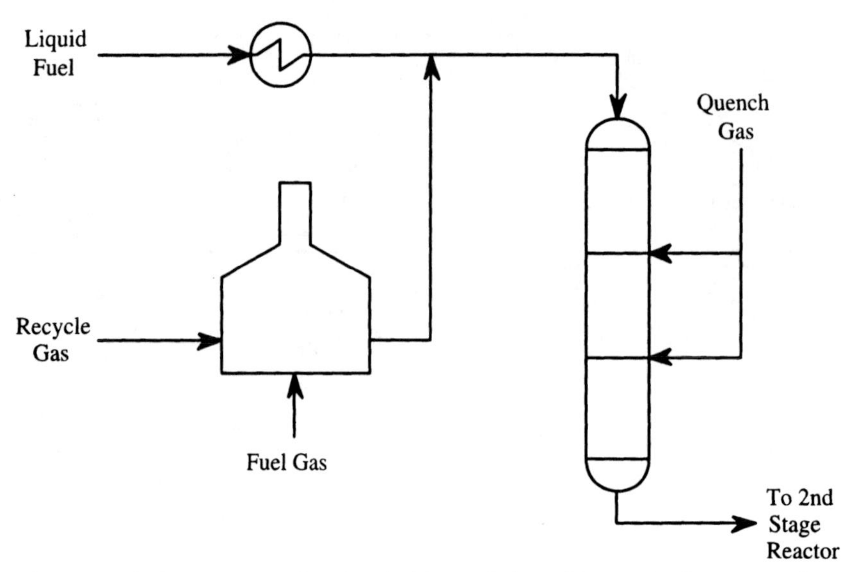


**FIGURE 7.43.** Hydrotreater unit. (Reprinted with permission of the AACC from *Proceedings of the 1987 American Control Conference.* Copyright © 1987, American Automatic Control Council.)

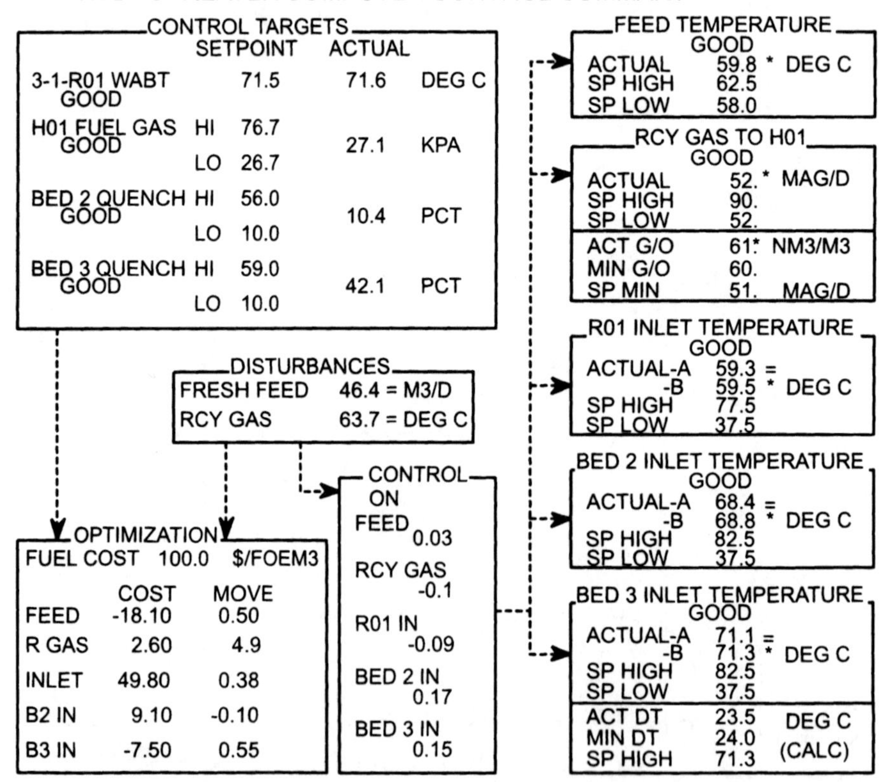


**FIGURE 7.44.** Summary display for hydrotreater unit. (Reprinted with permission of the AACC from *Proceedings of the 1987 American Control Conference.* Copyright © 1987, American Automatic Control Council.)

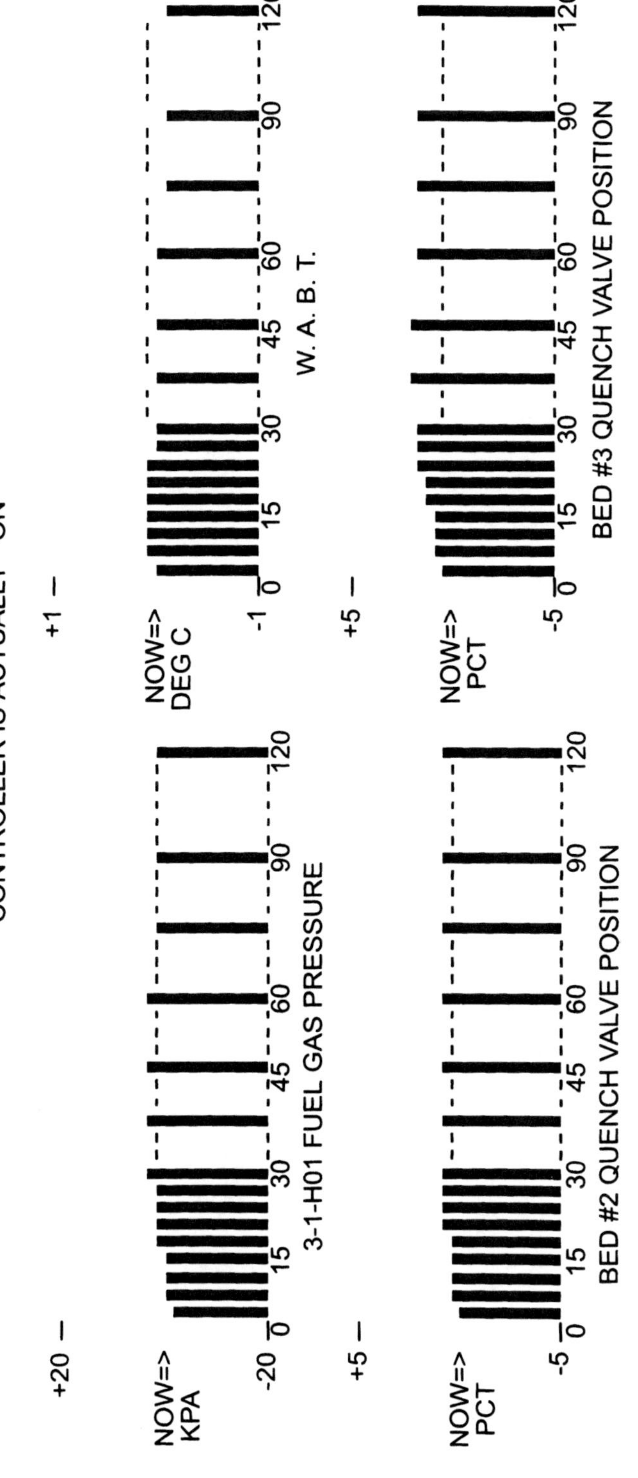


**FIGURE 7.45.** Prediction display for hydrotreater unit. (Reprinted with permission of the AACC from *Proceedings of the 1987 American Control Conference*. Copyright © 1987, American Automatic Control Council.)

distinguished by the text color. The individual controller faceplates are accessed by selecting the appropriate button next to the manipulated variable display.

A full multivariable controller has different information that must be presented. For FMC and DMC, additional information includes the process variable predictions, the process model, and optimization information. Figure 7.41 shows an example user interface for FMC when used with the Wood–Berry distillation column (Example 7.18). As for the single-loop FMC, the prediction window is generally not part of the faceplate and must be invoked by the operator. A display of the models may be displayed as shown in Figure 7.42.

Cutler and Hawkins (1987) show a typical user interface for a hydrotreater reactor. A simplified diagram of the unit is shown in Figure 7.43. The fresh oil feed is preheated by second-stage reactor effluent in a series of heat exchangers and is sent to the hydrotreater reactor. Recycle gas is heated in a gas-fired furnace and mixed with the liquid feed before it enters the reactor. The reactor converts the sulfur and nitrogen in the feed to hydrogen sulfide and ammonia and saturates the unsaturated hydrocarbons. The reactions are exothermic, so cold quench gas is added between each of the beds to maintain control of the temperature. The effluent from this reactor goes directly to the second-stage reactor. The objective of the controller is to hold the hydrotreater reactor weighted average bed temperature (WABT) to an operator-specified setpoint while minimizing energy usage and staying within the specified limits on the variables. For this unit, the process variables are WABT, fuel header pressure, and the two quench valve positions. The recycle gas temperature and the liquid feed flow rate are treated as disturbances. The manipulated variables are feed temperature, recycle gas flow, reactor inlet temperature, bed 2 inlet

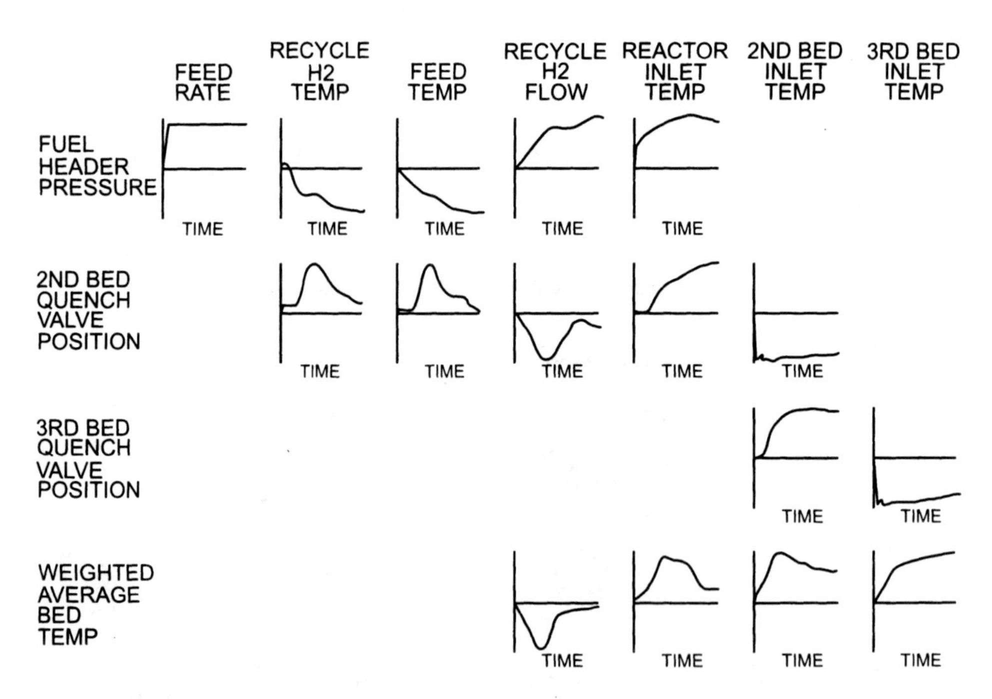


**FIGURE 7.46.** Model display for hydrotreater unit. (Reprinted with permission of the AACC from *Proceedings of the 1987 American Control Conference*. Copyright © 1987, American Automatic Control Council.)

temperature, and bed 3 inlet temperature. The overview display for this unit is shown in Figure 7.44. The prediction and model displays are shown in Figures 7.45 and 7.46, respectively.

## REFERENCES


Arulalan, G. R., and P. B. Deshpande, "Simplified Model Predictive Control," *Ind. Eng. Chem. Res.*, **26**, 347–356 (1987).

Bristol, E., "On a New Measure of Interaction for Multivariable Process Control," *IEEE Trans. Automatic Control*, **AC-11**, 133–134 (1966).

Brogan, W. L., *Modern Control Theory*, 3rd ed., Prentice-Hall, Englewood Cliffs, NJ, 1991.

Brosilow, C., and G. Q. Zhao, "A Linear Programming Approach to Constrained Multivariable Process Control," in Leondes, C. T., Ed., *Control and Dynamic Systems*, Vol. 27, Academic, New York, 1988, pp. 141–181.

Cutler, C. R., and R. B. Hawkins, "Constrained Multivariable Control of a Hydrotreator Reactor," in *Proceedings of the 1987 American Control Conference*, Minneapolis, MN, American Automatic Control Council, Green Valley, AZ, Vol 2, 1987, pp. 1014–1020.

Cutler, C. R., and B. L. Ramaker, "Dynamic Matrix Control—A Computer Control Algorithm," paper WP5-B, in *Proceedings of the 1980 JACC*, Vol. 1, San Francisco, CA, American Automatic Control Council, Green Valley, AZ, 1980.

DiBiano, R., "Importance of Versatile Control Strategy on Crude Unit," *Chem. Eng. Prog.*, **77**(2), 56–64 (1981).

Edgar, T. F., and D. Himmelblau, *Optimization of Chemical Processes*, McGraw-Hill, New York, 1988.

Erickson, K. T., and R. E. Otto, "Development of A Multivariable Forward Modeling Controller," *Ind. Eng. Chem. Res.*, **30**, 482–490 (1991).

Feher, J. D., "Forward Modeling Control with Soft Constraints," Ph.D. Dissertation, University of Missouri-Rolla, 1993.

Feher, J. D., and K. T. Erickson, "Solving the Model Predictive Control Problem with Soft Constraints," in *Proceedings of the 1993 American Control Conference*, San Francisco, CA, American Automatic Control Council, Evanston, IL, 1993, pp. 377–378.

Forsythe, G. F., M. A. Malcolm, and C. Moler, *Computer Methods for Mathematical Computations*, Prentice-Hall, Englewood Cliffs, NJ, 1977.

Garcia, C. E. and M. Morari, "Internal Model Control. 2. Design Procedures for Multivariable Systems," *Ind. Eng. Chem. Process Des. Dev.*, **24**, 472–484 (1985).

Garcia, C. E., and A. Morshedi, "Quadratic Programming Solution of Dynamic Matrix Control (QDMC)," *Chem. Eng. Commun.*, **46**, 73–87 (1986).

Grosdidier, P., M. Morari, and B. R. Holt, "Closed-Loop Properties from Steady-State Information," *Ind. Eng. Chem. Fund.*, **24**, 221–235 (1985).

Johnson, C. R. "Hadamard Products of Matrices," *Linear Multilinear Alg. I*, 295–307 (1974).

Jury, E. I., *Theory and Application of the Z-Transform Method*, Huntington, New York, 1964.

Marlin, T. E., *Process Control: Designing Processes and Control Systems for Dynamic Performance*, McGraw-Hill, New York, 1995.

McAvoy, T. J., "Connection Between Relative Gain and Control Loop Stability and Design," *AIChE J.*, **27**, 613–619 (1981).

McAvoy, T. J. *Interaction Analysis–Principles and Applications*, Instrument Society of America, Research Triangle Park, NC, 1983.

Moore, C. F., "Singular Value Analysis," in Deshpande, P. B. (Ed.), *Multivariable Process Control*, Instrument Society of America, Research Triangle Park, NC, 1989.

Ngo, K. T., and K. T. Erickson, "Stability of Discrete-Time Matrix Polynomials," *IEEE Trans. Automatic Control*, **42**, 538–542 (1997).

Niederlinski, A., "Heuristic Approach to the Design of Linear Multivariable Control Systems," *Automatica*, **7**, 691–701 (1971).

Otto, R. E., "Decouplers for Furnace Control," *ISA Trans.*, **23**(4), 55–57 (1984).

Otto, R. E., and G. R. Nieman, "Computer Control of an Ammonia Reform Furnace," *Industrial Process Control*, Proceedings of AIChE Workshop, American Institute of Chemical Engineers, New York, 1979.

Ricker, N. L., "Use of Quadratic Programming for Constrained Internal Model Control," *Ind. Eng. Chem. Process Des. Dev.*, **24**, 925–938 (1985).

Rouhani, R., and R. K. Mehra, "Model Algorithmic Control (MAC): Basic Theoretical Properties," *Automatica*, **18**, 401–414 (1982).

Rugh, W. J., *Linear System Theory*, 2nd ed., Prentice-Hall, Upper Saddle River, NJ, 1996.

Seborg, D. E., T. F. Edgar, and D. A. Mellichamp, *Process Dynamics and Control*, Wiley, New York, 1989.

Shinskey, F. G., *Controlling Multivariable Processes*, Instrument Society of America, Research Triangle Park, NC, 1981.

Shinskey, F. G., *Process Control Systems*, 3rd ed., McGraw-Hill, New York, 1988.

Stanley, G., M. Marino-Galarraga, and T. J. McAvoy, "Shortcut Operability Analysis. 1. The Relative Disturbance Gain," *Ind. Eng. Chem. Process Des. Dev.*, **24**, 1181–1188 (1985).

Stephanopoulos, G., *Chemical Process Control*, Prentice-Hall, Englewood Cliffs, NJ, 1984.

Wood, R. K., and M. W. Berry, "Terminal Composition Control of a Binary Distillation Column," *Chem. Eng. Sci.*, **28**, 1707–1717 (1973).

# 8 Discrete Control

**Chapter Topics**

- Basic discrete control
- Interlock control
- Discrete-control modules
- Unit supervision discrete control

*Scenario:* There is more to plantwide process control than continuous control.

At his first employer, the young engineer was involved in control systems for batch, continuous, and discrete manufacturing. Glass manufacturing starts with a material handling system that prepares batches of raw material by weight according to a prescribed formula and distributes them to the various glass furnaces. The glass furnaces melt the glass material and feed the container-forming machines. After forming, the containers are annealed, coated, and checked for defects. The process ends with packing the glass containers in cases and cartons on the packaging lines.

In the batching process cell, he gained experience with large-scale relay and PLC systems. This plant system is an example of a purely discrete control application. There were no loops in the batch control system. The process cell included rail car and truck unloading as well as formulation of the glass container raw materials. The raw materials were weighed in batches by weigh scales and delivered to the furnace bins by conveyors, elevators, and distributors. The process management control function included electronic recipe management and batch scheduling. In one system alone there were over 2000 relays that were coordinated by a computer system to perform the batch management function. He learned to appreciate the control requirements in this type of application.

His next employer was valve-centric in its view of control. When they tried to develop a DCS to do plantwide control they did not include ladder logic capability to handle discrete control. At one point, he was told that only the "on" state of motors was valid and the "off" state was an alarm condition. The young engineer was convinced that without some type of discrete-control capability the DCS would be unusable in batch applications. During a heated discussion the young engineer asked them to visualize a typical control panel. He asked them to describe what they saw. It was then that they understood the potential for DCS control if they could handle discrete control along with regulatory control. Every control panel discussed included lights and switches to handle the on–off equipment. In some cases, such as batching process units, the ratio of discrete to regulatory panel space was much greater than 1. He eventually won part of his case, for the DCS vendor added the necessary discrete-control capability to its offering. However, ladder logic capability was not included. The vendor was still too valve-centric to see the total discrete-control picture.

## 8.1 INTRODUCTION

Generally, discrete control is the control type that is used to prevent process upsets and enforce correct equipment operation in a stepwise manner. It is often referred to as "on-off" control and includes interlock and sequence control. The interlock has served the industry for years as the control logic to prevent upsets. Sequence control is the industrial automation workhorse to enforce the correct operation of process equipment in a stepwise fashion.

As discussed above, discrete control is misunderstood and underrated in plantwide process control. The term process control is often limited to loop control in continuous applications. But, as defined by the reference model of Chapter 3, discrete control is an integral part of the process control activity. It is also an integral part of unit supervision control.

The purpose of this chapter is to define discrete control and its role in plantwide control systems. Basic discrete-control concepts are discussed and sequential function charts (SFCs) are introduced. The interlock control type structure and design specifications are then presented. Next, discrete device control is described and illustrated with examples. Lastly, unit sequence control, usually reserved for batch control, is applied to nonbatch situations and also illustrated with examples.

## 8.2 BASIC DISCRETE CONTROL

Although basic discrete control is similar to loop control, it differs in the nature of its control objective. A loop control objective is to regulate a process around an operating setpoint. The control objective of discrete control is to set equipment in definable discrete states. The difference lies in the statewise nature of the measurements, control algorithm, and process actions.

To illustrate the difference, consider the forced air heating and cooling system in the average home. This system is controlled by a thermostat. The family decides the operating temperature and the thermostat maintains the temperature at this setting. It achieves this control objective by turning the heating (or cooling) on and off as the temperature cycles around the desired temperature setting. The control action is on–off. One could argue that this is a discrete-control example. The key is in the control objective that is to maintain the temperature around the setpoint. This is a regulatory application and thus an example of loop control, although the distinction in this case is not clear and could be argued either way. When the control objective is regulation, the loop is the basic control type. When the control objective is discrete, setting equipment to discrete states that can be labeled, the device is the basic control type.

Discrete control may stand alone as a separate device control module or it may be a part of another control module such as an interlock action within a loop control module. Examples of basic discrete-control applications include safety shutdown systems, block-and-bleed valve logic, and motor control circuits.

Discrete-control measurements provide state information about process equipment. As such, the basic measurement device is the binary state limit switch used to detect process stream events like flow, no-flow, low-level, or high-level conditions.

Discrete-control logic solvers perform an action based on the measured state. For instance, a high-level interlock will close an inlet block valve to prevent a spill.

The user interface with discrete-control modules includes the commands to change the state (e.g., start/stop, open/close) of process equipment either individually or in a prescribed sequence.

The traditional expression of basic discrete-control logic is the ladder logic diagram. This format has been around for many years and has even survived the emergence of the computer in control technology.

The basic concept of ladder logic is in the power flow across a ladder of rungs with power on the left and neutral on the right. Figure 8.1 illustrates a typical motor start–stop circuit. It is a single rung on the power ladder. The start push-button (Start PB), stop push-

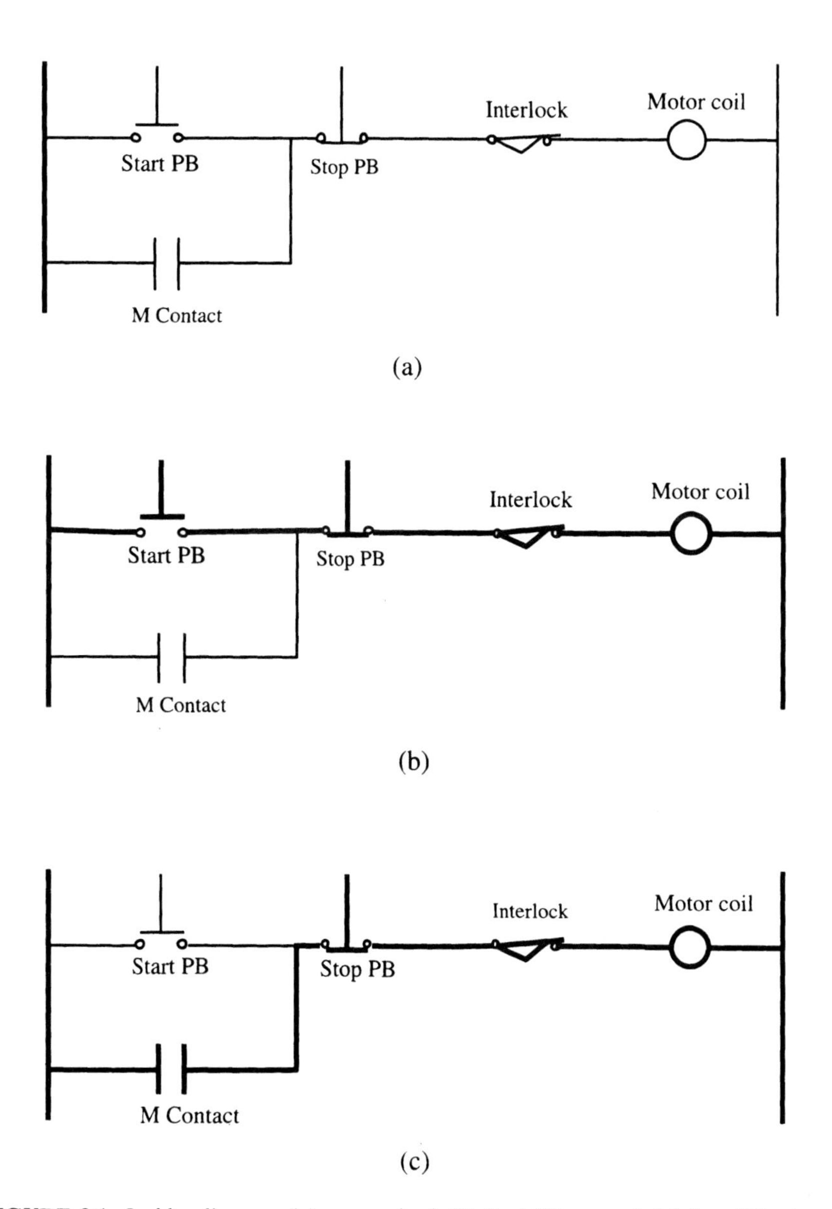


**FIGURE 8.1.** Ladder diagram: (*a*) unenergized; (*b*) Start PB pressed; (*c*) Start PB released.

button (Stop PB), and interlock switch are wired in series. The coil is part of a relay that energizes the motor contactor to start the motor. The M contact indicates whether or not the motor contactor has been energized. When closed, it implies that the motor is running. For this reason the M contact is sometimes called the run contact. This contact is in parallel with Start PB and acts as a seal-in to maintain the energized coil when the Start PB is released.

Figure 8.1*a* illustrates that the power flow in the motor is off because the Start PB switch and the M contact are normally open. The Stop PB switch and interlock are shown as normally closed. By depressing Start PB with no interlock condition, the power will flow to the coil and energize the contactor (Figure 8.1*b*). If Start PB is held in long enough—usually less than a second—the M contact will close and seal in the motor start circuit, maintaining power to the coil (Figure 8.1*c*). The motor is now running. When running, the motor will stop if Stop PB is depressed or the interlock condition opens the interlock switch.

The ladder rung logic of Figure 8.1 may be expressed in Boolean form. The true–false statement of the coil is a function of the logic expression:

Coil := (Start PB OR M contact) AND NOT Stop PB AND NOT Interlock

The coil will hold the value of the logical expression combining the states of the Start PB switch, M contact, Stop PB switch, and the interlock contact. If the expression is true, the motor is on; if false, the motor is off.

In the early 1970s, the ladder logic model of graphic rungs and Boolean logic was introduced in the form of a dedicated, computer-based device called the programmable logic controller (PLC). The PLC has been a workhorse in discrete-control applications. The PLC execution is basically the same as it was when first introduced. The ladder program is scanned to resolve the logic of the ladder. After this scan the process input–output (I/O) is serviced followed by any communications on a control data highway. There are many variations of this cycle, but it has served the world of basic discrete control for many years. A brief introduction to ladder logic is contained in Appendix B. Modern PLCs have added real number processing, preemptive task execution, and other performance enhancements as the application software demanded more sophisticated control logic.

### 8.2.1 Sequence Control Logic

Many control functions in a plant include discrete steps that force a process to an end condition. When an endpoint is reached, it is time to repeat the steps or to start another set of discrete steps. The detection of a successful endpoint and initiating the next set of control actions are common control functions in batch processing. Sequencing is the time-oriented step logic of process control. Sequencing is also common in discrete-parts manufacturing and packaging, where the operations are expressed as a sequence of steps. Though traditionally not associated with process control, many otherwise continuous processes require sequential control for startup and shutdown.

Actually, the simplest form of sequence control is the interlock. The interlock is designed to detect an abnormal process condition and take action to prevent an undesirable or hazardous event. After the condition clears, the consequences of the action (e.g., valve closure, loop in manual) must be reset to continue routine operation. The interlock drives

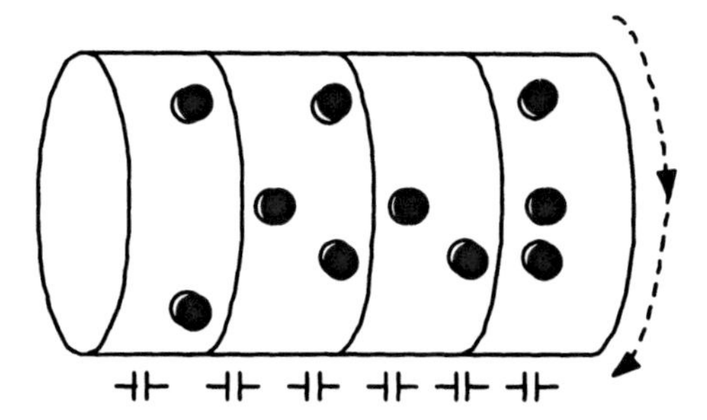

**FIGURE 8.2.** Drum sequencer.

process equipment through a cycle of states. This cycle of states is the essence of sequence control.

The classical way of describing discrete-control logic in sequential control applications is the truth table. This is often pictured in the form of a drum sequencer, depicted in Figure 8.2. A mechanical drum sequencer is a cylinder with pegs strategically located to make or break contacts. The contact applies power to some discrete logic or final control element. The drum advances a row at a time triggered by a discrete event. As the drum advances, it makes or breaks control circuits in a pattern prescribed by the position of each peg in a row. As a simple example, the drum sequencer peg could depress the motor Start PB of Figure 8.1 and advance when the M contact closes.

This mechanical drum sequencer illustrates the primary control objective of sequence control, enforcing a predefined set of control actions on process equipment in an ordered set of steps. The basic nature of sequencing is a cycle that is repeated as long as the process is operating.

Like ladder logic, there are software implementations of the drum sequencer in the form of a bit matrix that emulates the actions of the mechanical drum sequencer. Unlike ladder logic, this form of computer-based programming has not enjoyed widespread use.

### 8.2.2 Sequential Function Charts

The emerging IEC 1131 standard (International Electrotechnical Commission, 1993) defines a method of programming sequence control. It is derived from the IEC 848 function chart standard (International Electrotechnical Commission, 1988) that has been used to define sequential control logic. This format has emerged as a major programming tool in modern control systems. Appendix C summarizes the format and operation of SFCs.

The SFC is a diagram of interconnected steps, actions, and transitions as illustrated in Figure 8.3. A SFC begins with an initial step (box enclosed in double line) followed by an ordered set of (numbered) steps configured to perform the desired sequential control scheme. Associated with each step is a transition condition, shown as a horizontal bar. If the step is active and the transition is true, the logic advances to the next step or steps. Each step is labeled for reference. The stepwise flow continues until the end of the diagram. At this point, the sequencing ends or it may recycle back to the initial step waiting to resume the next cycle. There may be one or more action boxes attached to a step. The action box is used to perform a process action such as opening a valve, starting a motor, or calculating an endpoint for the transition condition.

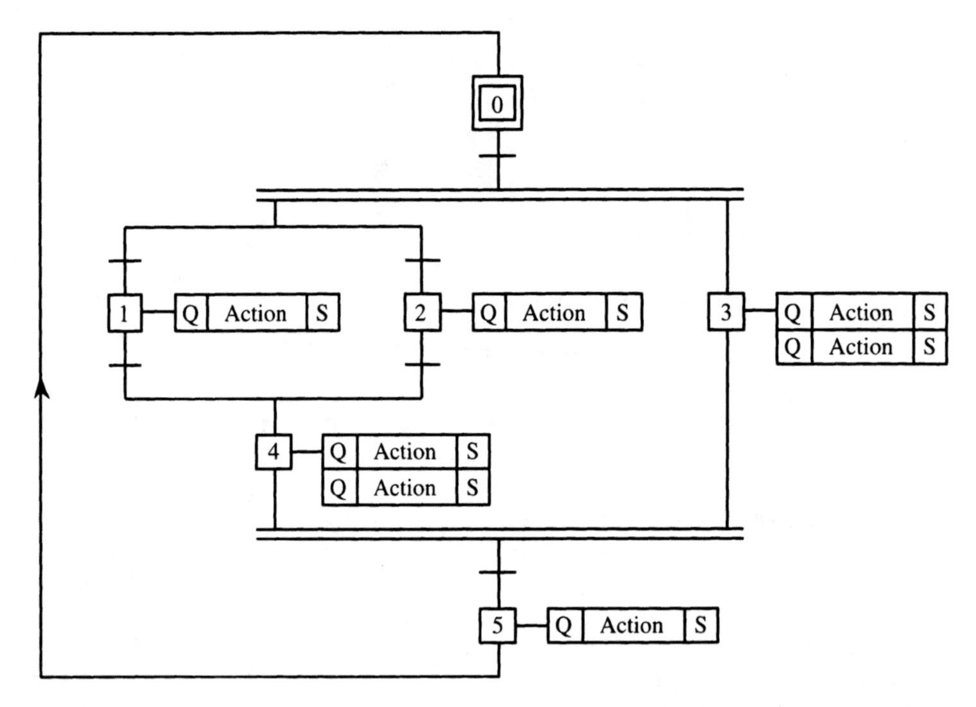


**FIGURE 8.3.** Example sequential function chart (SFC).

As defined, a SFC may advance in a diverging or converging manner (Figure 8.4). When the next step is one of many based on the transition conditions of each of the succeeding steps, it is referred to as an exclusive divergent-path diagram. If a set of steps are initiated in parallel, the construct is called a simultaneous divergent-path diagram. Likewise, multiple steps are converged with similar constructs as either an exclusive or simultaneous path operation.

**Example 8.1.** A simplified diagram of a machine that fills cereal boxes is shown in Figure 8.5. The operation of the machine is described as follows: Upon initial startup, the infeed conveyor motor, EX101, runs until an empty box is in the proper position, which is

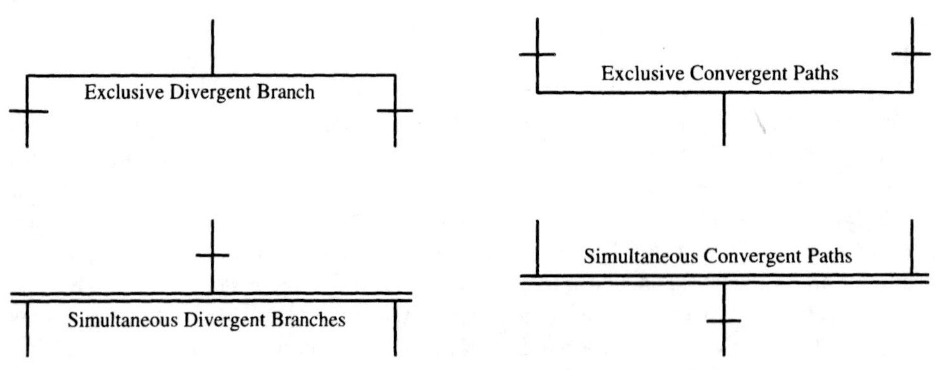


**FIGURE 8.4.** Divergent and convergent paths on SFC.

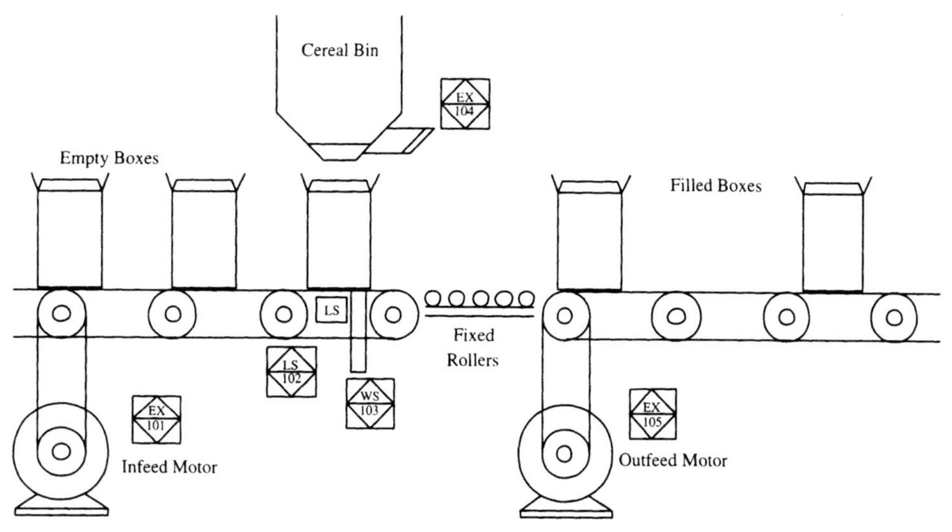


**FIGURE 8.5.** Cereal box filler process. (Reprinted by permission. Copyright © 1996, Kelvin T. Erickson.)

detected by the limit switch, LS102. After a 0.3-sec wait, the box is filled (by activating the solenoid, EX104) until the weight switch WS103 detects a full box. A 0.5-sec wait after the box is filled ensures that the cereal from the bin has fallen into the box. After the wait is completed, the full box is moved out and then the operation repeats.

In order to keep the example simple, issues of stopping, pausing, and interlocking are not considered. A SFC of the control for this machine is shown in Figure 8.6. The last step, though not in the text description, is necessary since the limit switch does not change its state until the box has moved out of the filling position. Motor EX105 runs continuously (unless paused) and does not appear on the SFC.

## 8.3 INTERLOCK CONTROL

Interlock control is the most basic form of discrete control, in the same manner as PID control is the most basic form of regulatory control. To many control engineers, when discussing discrete control, the first thought is interlock control. One has to be careful in discussing interlock control, however, because many people use the term to mean safety interlocks. In practice, there are many discrete-control applications that require the use of interlock logic but are not safety situations (e.g., Example 8.5). To avoid the stigma associated with the word *interlock*, interlocks in situations not related to safety are often labeled differently, for example, process actions and operational interlocks. It is important to understand the context of discrete control. For this reason, the S84.01 standard (ISA, 1996) deals with classifying interlock control for safety situations.

### 8.3.1 Safety Instrumented System

The S84.01 standard defines the control technology for automatic safety protection, called the Safety Instrumented System (SIS). As depicted in Figure 8.7, the SIS is required to be

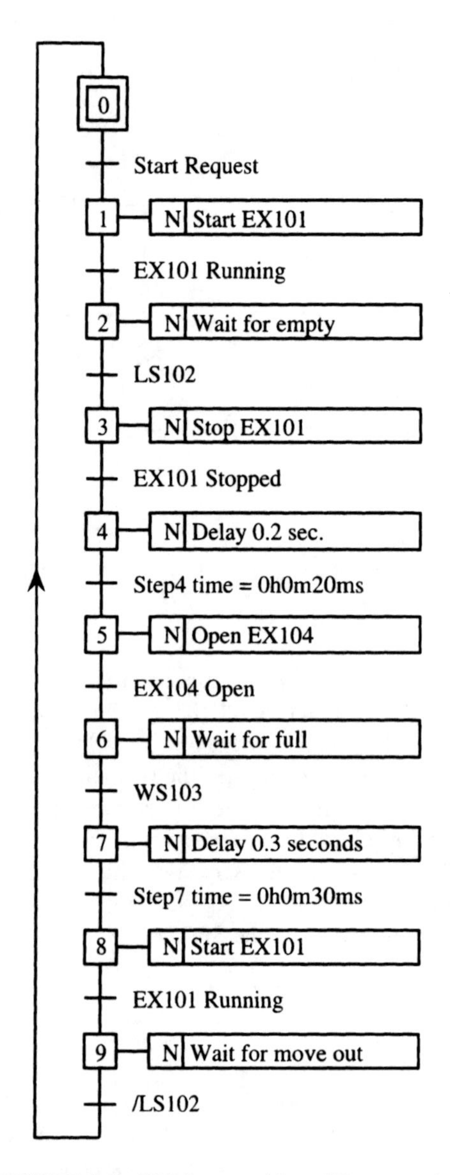


**FIGURE 8.6.** SFC for cereal box filler control.

a totally separate control technology from the rest of the process control technology. The S84.01 standard defines the non-safety-system technology as the Basic Process Control System (BPCS). According to the S84.01 standard, associated with the SIS is a user interface that provides the SIS annunciation and data presentation. The SIS user interface must also be separate from the BPCS. However, the S84.01 standard recognizes the need to share safety system status with the BPCS. Figure 8.7 thus shows the general system architecture recommended by the S84.01 standard for the SIS.

According to the S84.01 standard, the SIS is composed of the sensors and final control elements and a logic solver, depicted in Figure 8.8. Sensors and final elements include the typical limit switches and on–off valves used to enforce hazard mitigation in the process

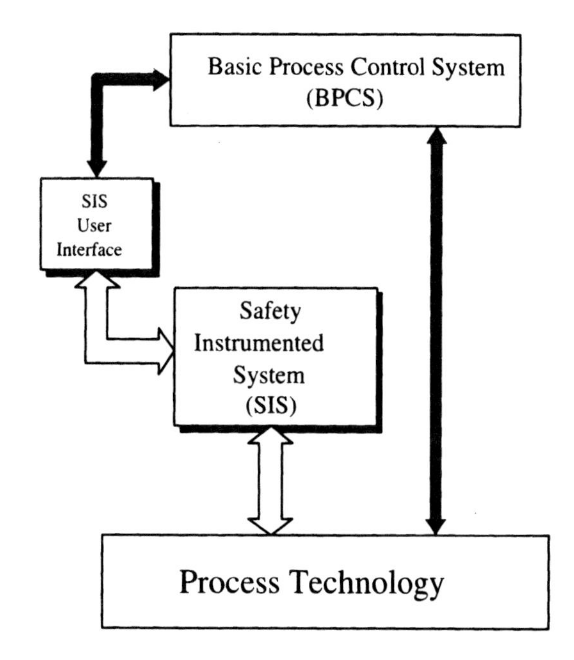


**FIGURE 8.7.** Relationship of SIS to rest of process control technology.

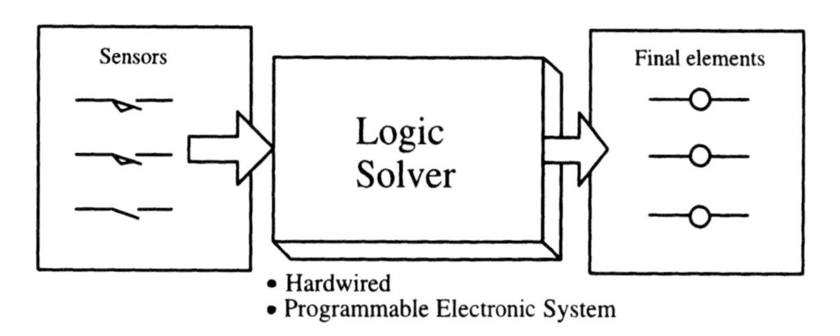


**FIGURE 8.8.** Structure of SIS system.

technology. The logic solver is the control mechanism that is triggered by the measured demand signals from the sensors to take action with the final elements. The logic solver can be electrical/electronic hardware or a programmable electronic system (PES). The former is commonly referred to as "hard-wired interlocks." A PES is required to be a hardened, high-reliability PLC with its associated I/O hardware connected to the sensors and final controls. The PES includes the embedded and applications software that performs the discrete control logic using interlock control types. Many SIS implementations are combinations of these technologies.

### 8.3.2 Interlock Management

The purpose of the interlock logic solver is to mitigate the possibility of a hazardous event. The hazardous operability study (HAZOP) performed by the capital project team is the first step in identifying and classifying the hazards in a process design. The mitigation of the hazard may be accomplished by changing the design, by using dikes or other noncontrol technology methods, or by utilizing automation. The automated hazard prevention strategy is captured in the form of an interlock description. As a result, a very important part of the control engineering activity on a capital project is to develop the interlock descriptions.

The interlock description identifies each interlock in the SIS and defines the demand signals and associated control actions that must be taken to prevent a hazardous event. It also defines the reset condition, or what happens when the interlock demand signal returns to normal. The demand signals are the switches and analog sensors used to measure the presence of an impending hazardous condition. In extremely hazardous situations, the SIS sensors must be redundant and of diverse technologies to ensure the highest possible system reliability. Also, extremely hazardous situations require redundant logic solvers.

The sections of a typical interlock description include:

- Hazard description
- Prevention strategy
- Safety integrity level
- Cause/action specification

The interlock descriptions provide the control requirements for the SIS. These requirements are used to develop the SIS technology specification. The specification may call for hard-wired interlocks in some situations and PES technology in other interlocks. The specification is used to procure and implement the SIS.

The SIS is implemented like any control system except it is subject to more stringent functional testing than the BPCS to meet enterprise and government regulations. The functional testing includes the prestartup checkout and periodic retesting during the lifecycle of the SIS. These tests must be supported by clearly documented test procedures.

Typical SIS functional tests include:

- Sensors, final elements, and logic solver inspection
- Fault injection and results recording of SIS modules
- PES application software functional test
- Fault injection and results recording of the entire SIS
- Process hazardous event simulation

The interlock descriptions and SIS system specifications are excellent candidates for electronic data management. In a data base, this information is easier to manage than paper systems. In addition, government regulation may require that the management of interlock changes be enforced by a data base.

The typical contents of an interlock data base include:

- Interlock description information
- One or more interlock demand specifications
- One or more interlock actions specifications
- Interlock logic description

- Reset actions and condition
- Plant cell or area
- SIS controller
- Interlock test procedure (may reference a document)

The SIS user interface indicates the state of the SIS, which is very important. The user interface is designed to meet the same reliability requirements as the SIS logic solver. The display of an interlock trip is critical to good operating practice in a plant. The main display requirements for both the SIS and BPCS interlock status presentation include:

- Process graphic display indications
- Interlock trip annunciation

The SIS performs a vital service in a process plant. As such, operating personnel closely monitor it. For PES-style SIS applications, on-line monitoring is possible to ensure the integrity of the SIS. The monitoring system is completely separate from the SIS. The SIS monitoring system captures interlock events and saves the events in a data history file. From this file, various reports and analyses can be obtained by operating personnel. The functions of on-line interlock monitoring include:

- Monitor and record interlock events (demand and action occurred)
- Compare event with interlock data base
- Verify interlock action performed by application logic
- Verify final control element action
- Store interlock event in data repository
- Provide predefined and temporary reports for human review and approval

## 8.4 DISCRETE-CONTROL MODULES

Discrete-control modules are composed of two subclasses: status and device. The status object indicates a discrete state process variable either as a direct measurement or as a derived measurement. Derived measurements include signal selection and multicontact Boolean logic variables. There are no outputs associated with a status object at this level. Statuses may be used in higher level objects to perform supervisory, sequential, or batch control.

The device object includes the traditional on–off valve and motor control. Also included are metering and weighing stations.

**Example 8.2: Typical Motor Device.** The bulk of this example will concern the control aspects of this device. However, in order to present a complete picture of this device, the control objective and concept that lead to the control strategy are described.

**Control Objective:** To move material from tank T-100 to tank T-101.

**Control Concept:** Because tank T-101 is at a higher level than T-100, a pump must be employed to move the material. Figure 8.9 shows the device tag EX100 associated with the pump. The pump needs to be started and stopped automatically by sequence step or manually by the operator.

**Control Strategy:** To start and stop the motor in the Manual or Automatic control state. Stop the motor on failure conditions. Generate overload fail alarm, auxiliary fail alarm, and HOA-switch-not-in-auto indications.

The device has two control states, Manual and Automatic, shown in Figure 8.10. The control state determines the source of the commands to start and stop the motor. In the Manual control state, the motor may only be started or stopped by the operator. In the Automatic control state, the motor may only be started or stopped by steps in automatic sequences. Switching between the Manual and Automatic control states should not change the operating state (Running or Stopped) of the motor. The operational states of the equipment are shown in Figure 8.11. The Failed state is entered from any of the other states.

The auxiliary fail alarm must not be generated until 20 sec has elapsed after the motor contactor has been closed and the auxiliary contact has not closed. Provisions must be made to allow the operator to reset the auxiliary fail alarm indication.

**Process System Items**

Motor starter (contactor or motor control center)
Hand-Off-Auto (HOA) switch
Overload indicator
Auxiliary contact

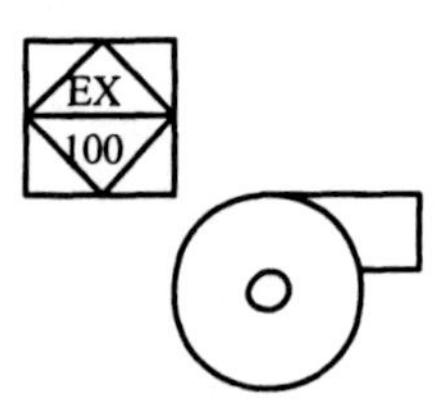


**FIGURE 8.9.** P&ID symbol for motor device control associated with pump.

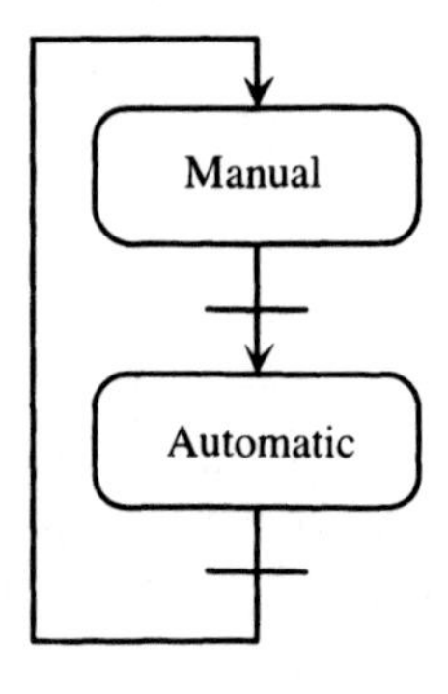


**FIGURE 8.10.** Control states of motor and valve devices.

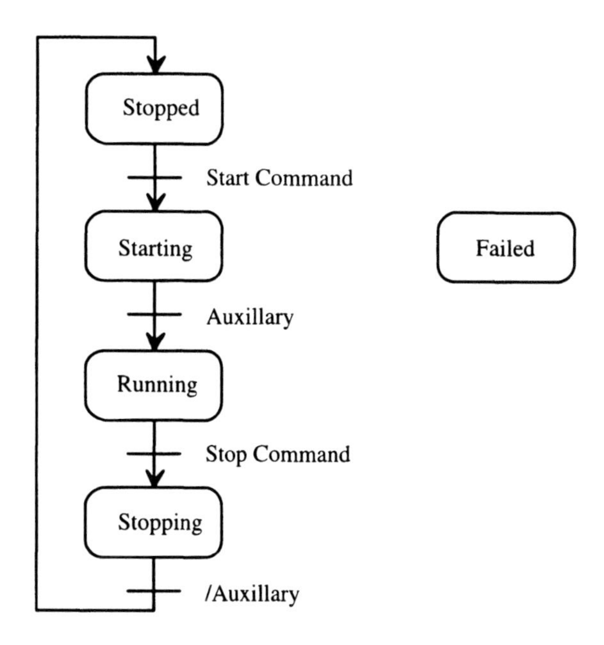


**FIGURE 8.11.** Operational states of motor device.

**Detail Design:**

- Physical inputs
  - Auxiliary contact (on when motor running)
  - Overload trip (on when motor overloaded)
  - HOA switch auto contact (on when auto contact is closed)
- Physical outputs
  - Starter (on to start/run motor)
- Operator commands
  - Manual Start
  - Manual Stop
  - Manual/Automatic control state
  - Reset alarm
- Operator indication
  - Run status
  - Failure alarm
  - Auxiliary fail alarm
  - Overload alarm
  - HOA-switch-not-in-auto alarm
- Automatic sequence commands
  - Automatic Start generated by steps of one or more sequences
  - Automatic Stop generated by steps of one or more sequences

Internal storage

Start command, starts motor in the Manual or Automatic control state

Stop command, stops motor in the Manual or Automatic control state

The IEC 1131-3 ladder logic code for the control of a motor with tag EX100 is shown in Figure 8.12. The first rung controls the physical output that drives the motor contactor (or motor starter). The motor is turned off on the first scan of the ladder or when any failure occurs. The start and stop internal coils to drive the motor contactor are determined by the second and third rungs. In these two rungs, the operator generates the start and stop commands when the control is in the Manual state. When the control is in the Automatic state (not the Manual state), the motor is started and stopped by steps in the various sequences (function charts). For the ladder shown in Figure 8.12, the motor is started in steps 2 and 14 of the startup sequence and stopped in step 7 of the startup sequence and step 9 of the shutdown sequence. The Manual control state, shown here as specific to this motor, may be the state for a grouping of equipment. For example, the Manual control state may be defined for the group of pumps and valves associated with a reactor. The fourth rung generates the auxiliary fail alarm when the auxiliary contact drops out (turns off) and it cannot be generated until 20 sec after the motor is started. The auxiliary fail alarm must be latched since this failure will cause the output to the starter to be turned off, thus disabling the conditions for this alarm. The fifth and sixth rungs generate the overload fail alarm and the indication that the HOA switch is not in the auto position. The seventh rung generates one summary failure indication that would appear on an alarm summary screen. The last rung resets the auxiliary failure alarm so that another start attempt is allowed. As for the Manual/Automatic control state, the reset may be for a group of equipment rather than specific to each device.

**Implementation:** The motor control is implemented in ladder logic within a PLC, as shown in Figure 8.12. A possible operator interface faceplate for this device is shown in Figure 8.13. In a typical system, this faceplate is invoked from an operator screen that shows all of the devices in a unit, for example, a packaging machine.

**Example 8.3.** Typical discrete valve device. As for the pump in the previous example, the bulk of this example will concern the control aspects of this device. However, to present a complete picture of this device, the control objective and concept that lead to the control strategy are described.

**Control Objective:** To drain the products from reactor R-103 to storage tank T-104.

**Control Concept:** Because the reactor is vented and at a higher level than T-104, gravity will be employed to move the material. Figure 8.14 shows the device tag XV100 associated with the valve. The valve needs to be opened and closed automatically by sequence step or manually by the operator.

**Control Strategy:** To open and close the valve in the Manual or Automatic control state. Generate fail-to-open and fail-to-close alarms.

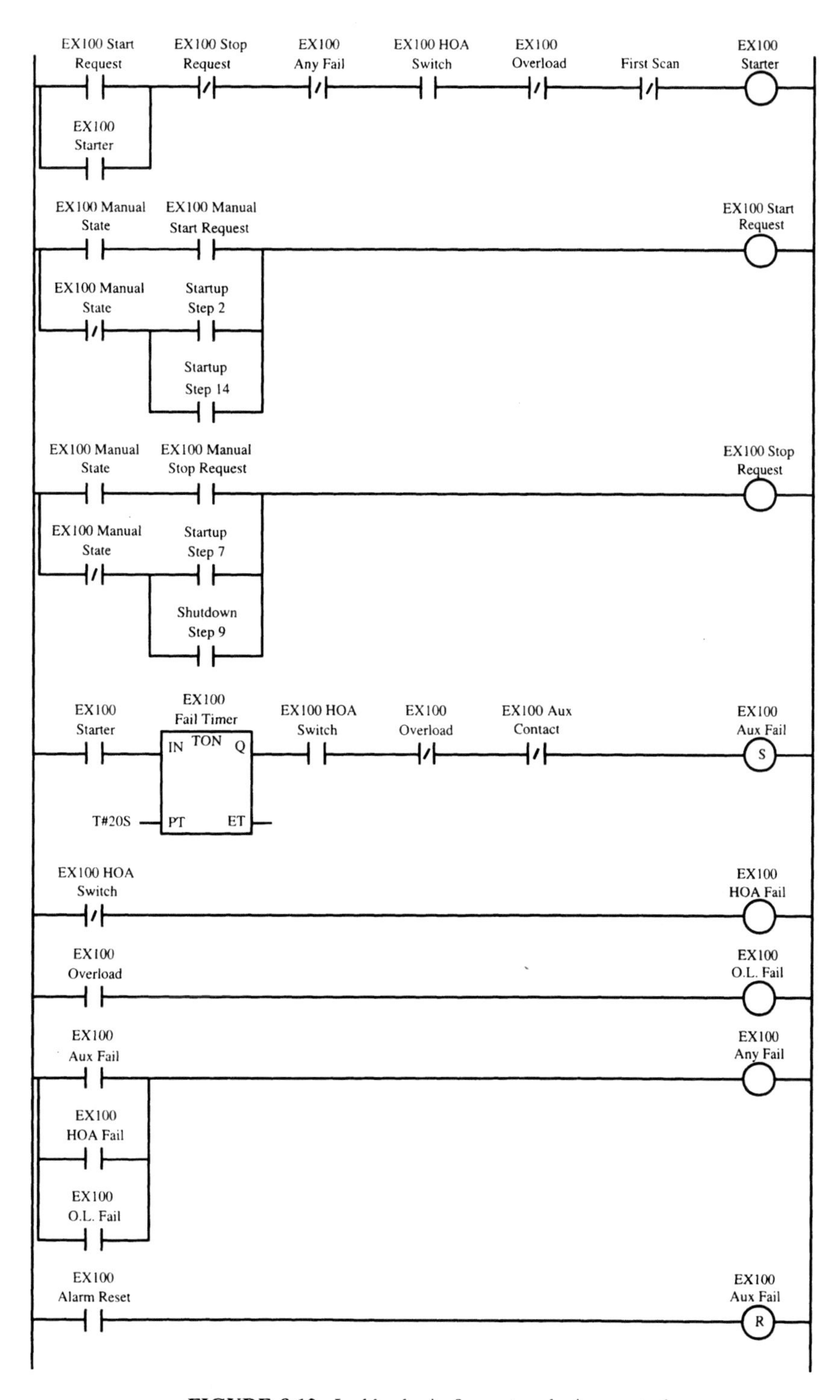


**FIGURE 8.12.** Ladder logic for motor device control.

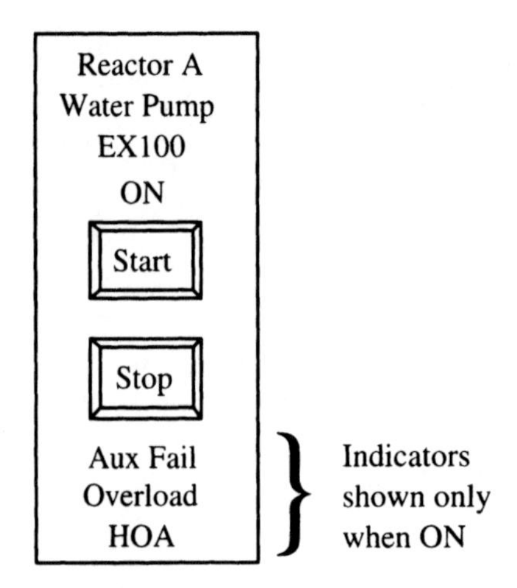


**FIGURE 8.13.** Faceplate for motor device control.

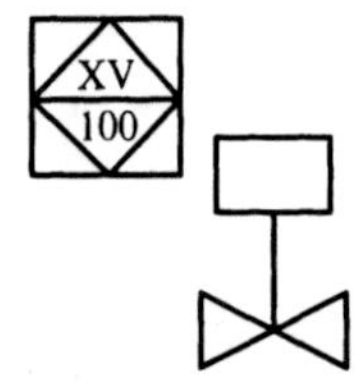


**FIGURE 8.14.** P&ID symbol for discrete-valve device control.

The device has two control states, Manual and Automatic, shown in Figure 8.10. The control state determines the source of the commands to open and close the valve. In the Manual control state, the valve may only be opened or closed by the operator. In the Automatic control state, the valve may only be opened or closed by steps in automatic sequences. Switching between the Manual and Automatic control state should not change the operating state (Opened or Closed) of the valve. The operational states of the valve are shown in Figure 8.15. The Failed state is entered from any of the other states.

The fail-to-open and fail-to-close alarms must not be generated until 20 sec has elapsed after the valve has been commanded to open or close and the appropriate limit switch has not closed. Provisions must be made to allow the operator to reset the alarm indications.

### Process System Items

Valve solenoid
Valve open limit switch
Valve closed limit switch

### Detail Design

Physical inputs
  Valve open limit switch (on when valve fully open)
  Valve closed limit switch (on when valve fully closed)

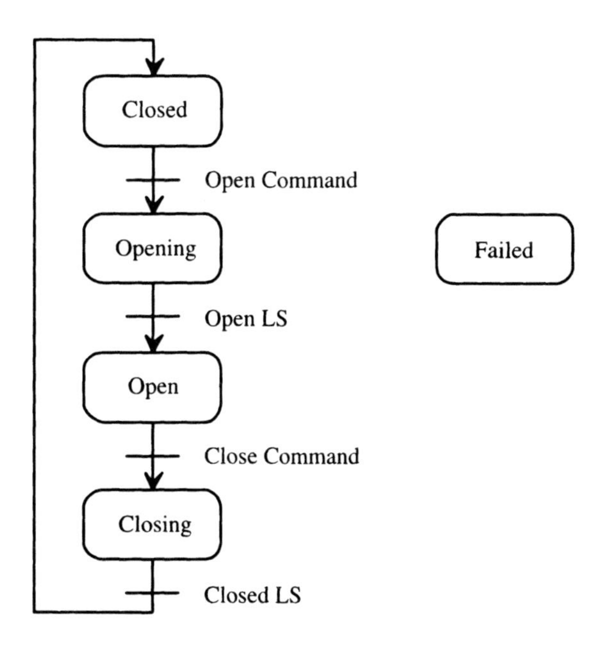


**FIGURE 8.15.** Operational states of discrete-valve device.

Physical outputs
  Valve solenoid (on to open the valve)
Operator commands
  Manual Open
  Manual Close
  Manual/Automatic control state
  Reset alarms
Operator indication
  Open status
  Failure alarm
  Fail-to-open alarm
  Fail-to-close alarm
Automatic sequence commands
  Automatic Open generated by steps of one or more sequences
  Automatic Close generated by steps of one or more sequences

The IEC 1131-3 ladder logic code for the control of a valve with tag XV100 is shown in Figure 8.16. The first two rungs control the physical output that drives the valve solenoid coil. Latching outputs are used, in contrast to the motor control in the previous example, because failures do not automatically close or open the valve. In these two rungs, the operator generates the open and close commands when the control is in the Manual state. When the control is in the Automatic state (not Manual state), the valve is opened and closed by steps in the various sequences (function charts). For the ladder shown in Figure

8.16, the valve is opened in step 5 of the startup sequence and closed in step 2 of the shutdown sequence and step 1 of the emergency shutdown sequence. As for the motor control example, the Manual state shown here is specific to this valve but may be the state for a grouping of equipment. The third rung generates the failure alarms, but only after the condition persists for 20 sec. This delay allows time for the valve to change state when

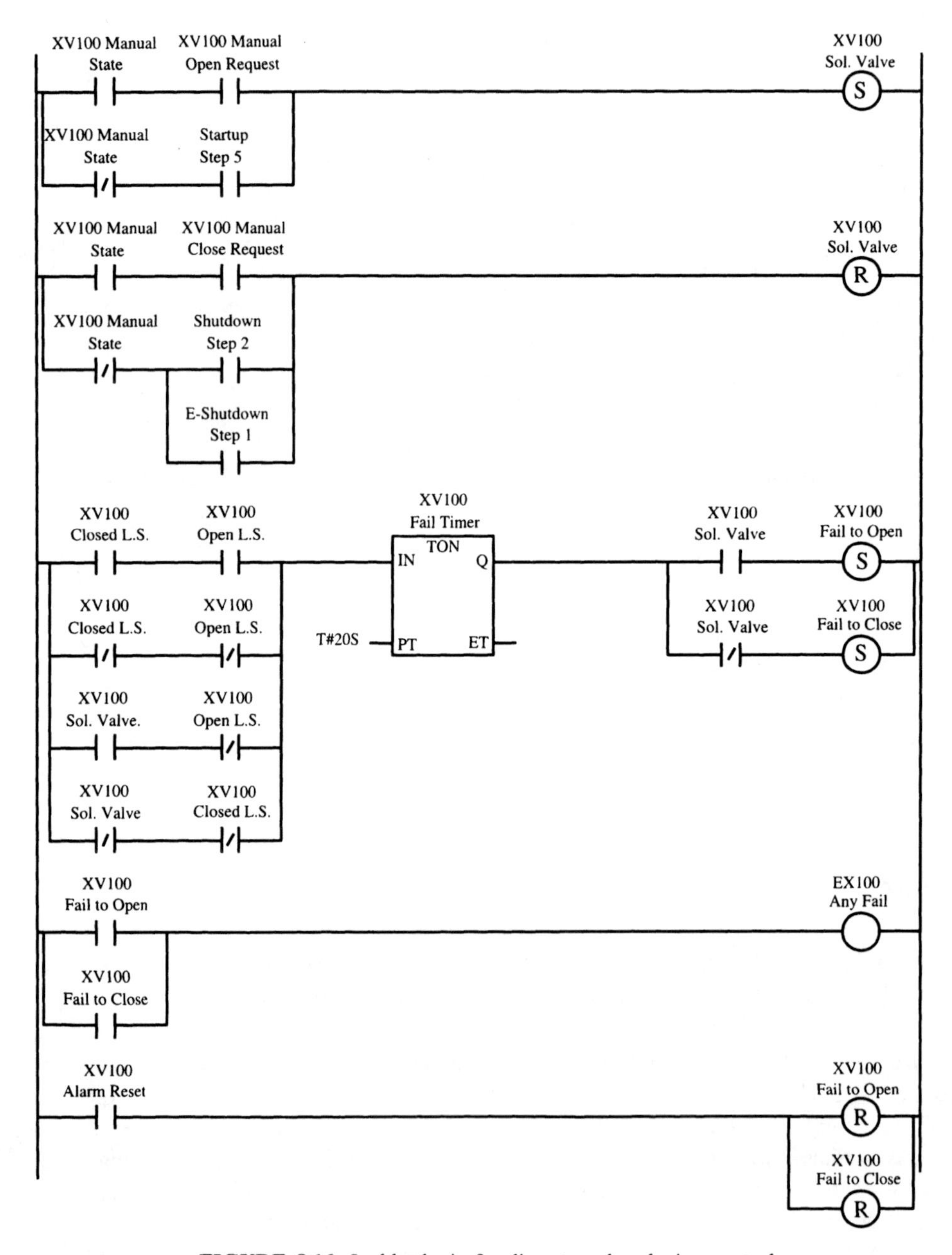


**FIGURE 8.16.** Ladder logic for discrete-valve device control.

neither limit switch will be closed. A failure is assumed when one of four conditions persists for 20 sec:

Both limit switches closed
Both limit switches open
Valve commanded to open and open limit switch not closed
Valve commanded to close and closed limit switch not closed

The fourth rung generates one summary failure indication that would appear on an alarm summary screen. The last rung resets the failure alarms. As for the Manual/Automatic control state, the reset may be for a group of equipment rather than specific to each device.

**Implementation:** The valve control is implemented in ladder logic within a PLC, as shown in Figure 8.16. A possible operator interface faceplate for this device is shown in Figure 8.17 and is similar to the motor control faceplate. In a typical system, this faceplate is invoked from an operator screen that shows all of the devices in a unit, for example, a reactor.

**Example 8.4.** Block and bleed discrete valve device. As for the previous examples, the bulk of this example will concern the control aspects of this device. However, to present a complete picture of this device, the control objective and concept that lead to the control strategy are described.

**Control Objective:** Mixing upstream material with downstream material is undesirable. There is intermittent use of the stream in the downstream plant system. The upstream material is a caustic for clean-in-place operations and would ruin the downstream material if it contacts the downstream material during normal production.

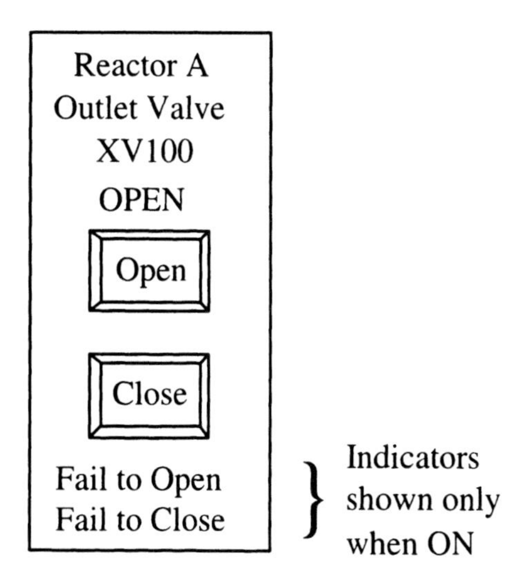


**FIGURE 8.17.** Faceplate for discrete-valve device control.

**Control Concept:** Figure 8.18 shows the equipment needed to fulfill the control objective. Two valves are used to ensure that the upstream material does not leak when the valves are closed. In addition, any material between the valves is bled off. The device tag XV101 references the entire control device. The individual valves have tags but are not individually manipulated. The device needs to be opened and closed.

**Control Strategy:** Discrete-control module to enforce the opening and closing of the block valves coordinated with the bleed. The bleed valve is open when the two block valves are closed. When opening the block valves, the bleed valve must first be closed. When closing the block valves, the bleed valve must be opened after the block valves are closed. As for the discrete valve device (Example 8.3), the device has two control states, Manual and Automatic, shown in Figure 8.10. The control state determines the source of the commands to start and stop the valve. In the Manual control state, the valve may only be opened or closed by the operator. In the Automatic control state, the valve may only be opened or closed by steps in automatic sequences. Switching between the Manual and Automatic control states should not change the operating state (Opened or Closed) of the valve. The operational states of the valve are shown in Figure 8.15. The Failed state is entered from any of the other states.

**Detail Design**

Physical inputs
- Valve open limit switch (one for each valve)
- Valve closed limit switch (one for each valve)

Physical outputs
- Valve solenoid (one for each valve)

Operator commands
- Manual Open
- Manual Close

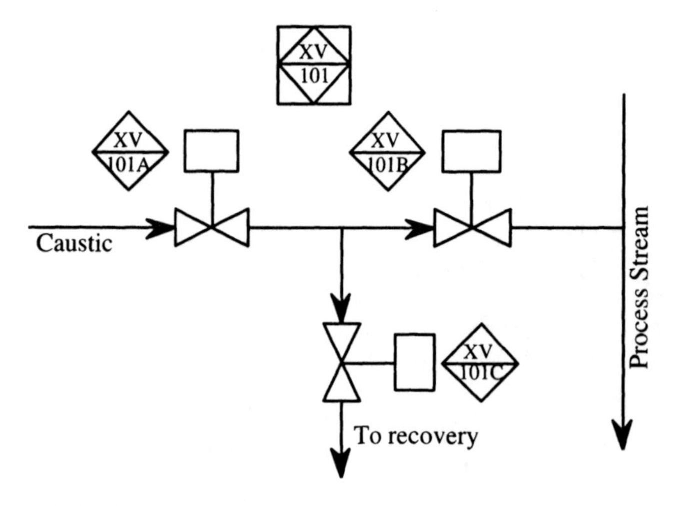


**FIGURE 8.18.** P&ID for block and bleed valve device control.

- Manual/Automatic control state
- Reset alarms

Operator indication

- Open status
- Failure alarm
- Fail-to-open alarm (one for each valve)
- Fail-to-close alarm (one for each valve)

Automatic sequence commands

- Automatic Open generated by steps of one or more sequences
- Automatic Close generated by steps of one or more sequences

The ladder logic for each valve is similar to the previous example. The only real difference is that this discrete-control device coordinates the operation of these three valves. A SFC is used to describe the coordination that must take place when the device is commanded to open (Figure 8.19) or to close (Figure 8.20).

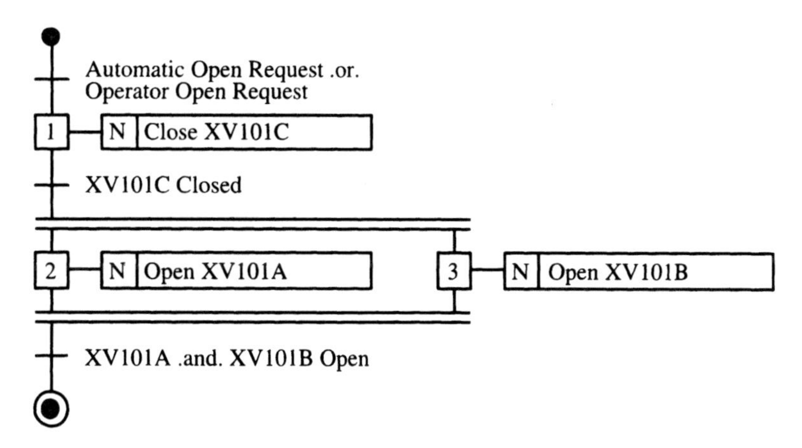


**FIGURE 8.19.** SFC for opening of block and bleed device.

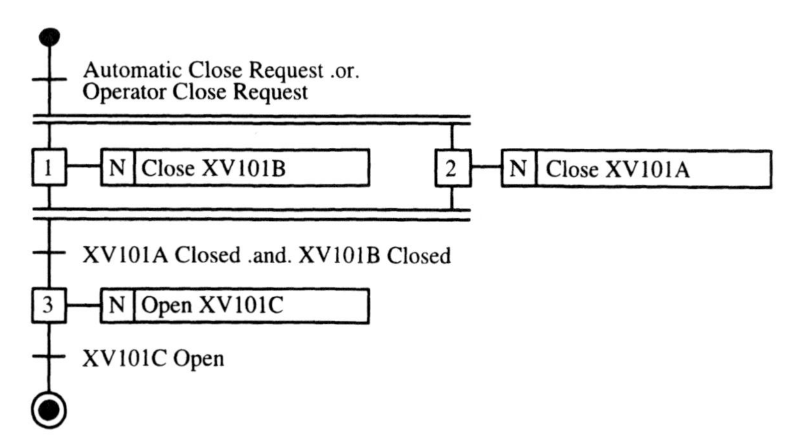


**FIGURE 8.20.** SFC for closing of block and bleed device.

**Implementation:** The valve control is implemented as a combination of SFC and ladder logic within a PLC. The SFCs are as shown in Figures 8.19 and 8.20. The ladder logic for each individual valve is implemented as in Example 8.3, without the open and close commands from the operator. A possible operator interface faceplate for this device is identical to that for Example 8.3 (Figure 8.17), except that there are more error indications. In a typical system, this faceplate is invoked from an operator screen that shows all of the devices in a unit, for example, a reactor.

## 8.5 UNIT SUPERVISION DISCRETE CONTROL

Sequence control is often discussed in the context of driving a set of process equipment in a coordinated cycle to make a batch. This type of sequence control manipulates device and loop setpoints and sets parameters to accomplish the prescribed control objective. However, this type of sequence control can be employed in non-batch process control applications. For example, sequence control is used in the startup and shutdown of a continuous process.

Figure 8.21 illustrates sequence control using the terminology of S88 (ISA, 1995) as defined in Chapter 3. In the figure, the process is depicted as a state diagram and the sequence control as a simplified SFC. The initial state of a process unit is Idle (i.e., not making product). When it begins operation it is considered in the Operating state (i.e., making product).

In Figure 8.21, sequence control is depicted as a step cycle. The SFC initial step is often called the "home" step in a sequential control cycle. A manual or automatic event causes

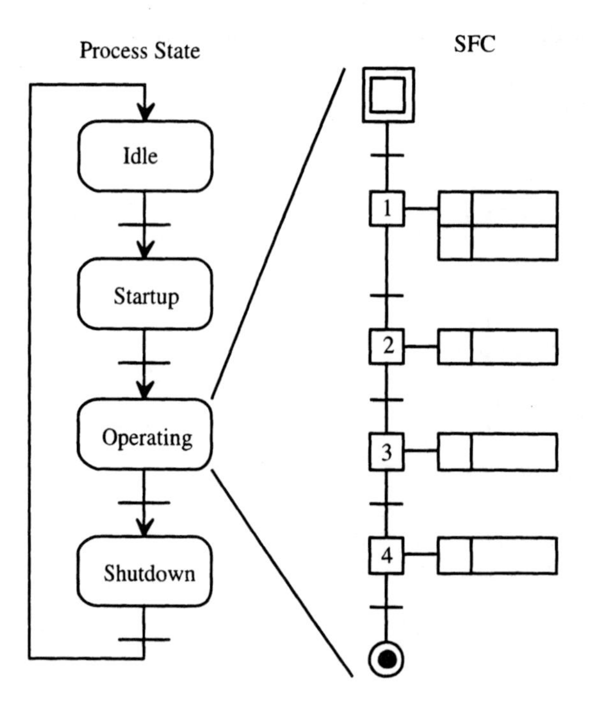


**FIGURE 8.21.** Sequence control using the terminology of S88.

the sequencer to move from the home step through its sequence logic. At the end of the sequence logic the sequencer returns to the home step and awaits the next start command. If the process is not idle when the sequence is at the home step, the sequencer must be synchronized to the process.

### 8.5.1 Unit versus Device

The control objective of a unit is very similar to the objective of a device. Normally, the unit manager coordinates the operation of the devices and loops. However, the grouping of unit control in a monolithic form is undesirable in times of process upset, in checkout, or in routine maintenance. If full automatic sequencing cannot be performed, operating personnel must be allowed to use the loop and device objects to complete a sequence or bring the process down gracefully. The proper allocation of device, loop, and unit control is a critical consideration in design for operability. The devices and loops perform their task at the lowest level of operator access.

### 8.5.2 Unit Procedure

The unit manager provides operation support, performance monitoring, and control enforcement for a group of process equipment. Figure 8.22 depicts the unit object class and associated phases. The relationship with the operator, loop, and device objects and the activity are also depicted.

Each unit manager has one or more sequential control schemes, called unit procedures. The individual unit procedures are subdivided into operations and the operations are subdivided into phases. This hierarchical relationship between unit procedure, operation, and phase is shown in Figure 3.9. Figure 8.23 depicts a unit procedure for each of two ion exchange units. The unit procedure has states and state transition logic, as shown in Figure 8.24. When activated, the unit procedure changes from the IDLE to the RUNNING state. While in the RUNNING state, the operator or some other event may force the unit procedure into a HOLD state, where normal automatic operation may be temporarily suspended. If the operator stops the unit or the procedure finishes execution, the unit will branch automatically to the STOPPING state. An emergency stop is often handled differently and is shown as a separate state. When the unit procedure ends, it returns to the IDLE state.

An operation is associated with a unit operational state. The operation performs the control actions in a unit procedure state. For example, the RUNNING, RE-STARTING, STOPPING, and E-STOPPING states in Figure 8.24 each have an operation performing the control actions.

Each operation is divided into one or more phases. The typical phase is composed of multiple steps. The control instructions in the steps manipulate the process unit equipment through the loop and device objects. Usually, the first step, or set of steps, performs the actions required to set up the process unit for the rest of the phase cycle, such as setting the initial loop and device setpoints and control states. More complex setups include operator actions such as selecting raw-material tanks, pumps, and material amounts. The operator may be requested to enter laboratory analysis, set manual valves, and start auxiliary equipment around the unit. Setup includes checking permissives that must be satisfied before proceeding.

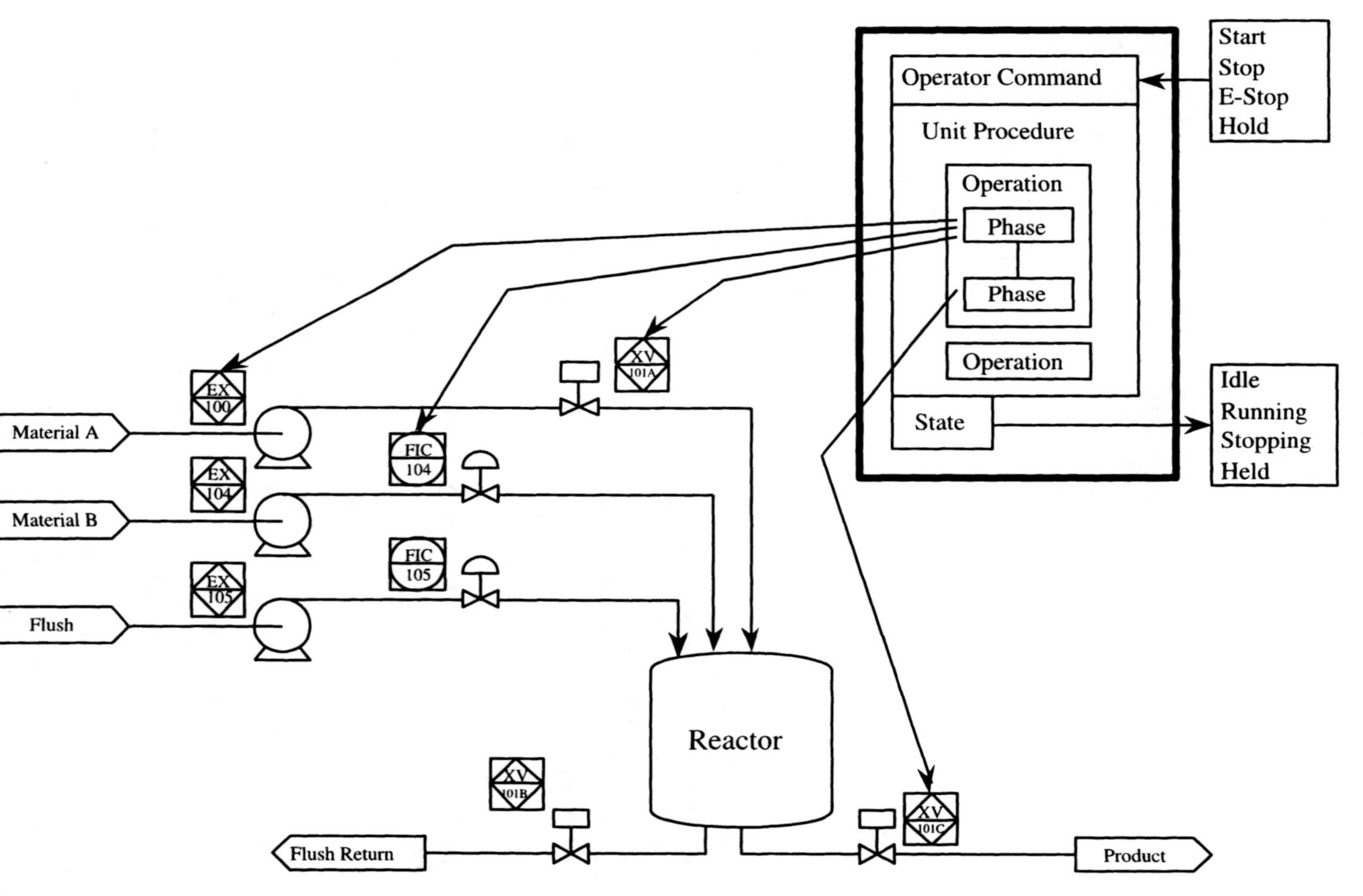


**FIGURE 8.22.** Unit object and relationship with devices. (Reprinted by permission. Copyright © 1996, Automation and Control Technologies, Inc.)

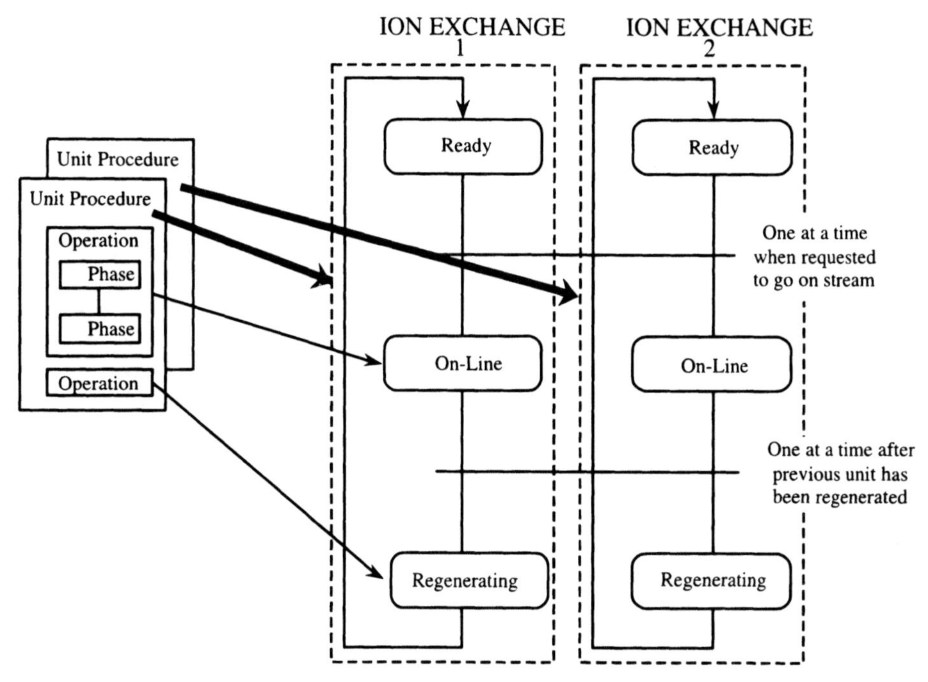


**FIGURE 8.23.** Operations and phases as part of a unit procedure. (Reprinted by permission. Copyright © 1996, Automation and Control Technologies, Inc.)

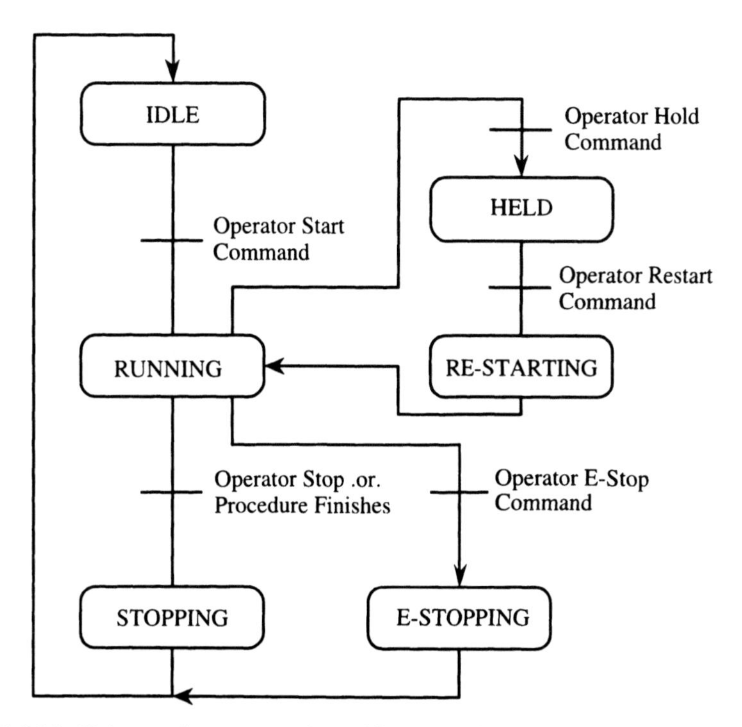


**FIGURE 8.24.** Unit procedure states and transitions. (Reprinted by permission. Copyright © 1997, Automation and Control Technologies, Inc.)

After setup, the phase performs the sequence control actions to operate the process unit. This is the operating part of the phase and includes control instructions such as metering, weighing, and profiling. Some of these control actions may continue through multiple steps as concurrent tasks. The length of a phase varies from a few minutes to many days.

The last part of a phase is the setdown step(s), which prepares the process equipment for the next cycle. Typical setdown control tasks include unit clean-out, pump stops, loop and device deactivation, and vessel drain-out or purge. As in the case of setup, this can be quite elaborate and involve operator action responses.

Each phase also has states. When activated by the unit scheduler, the phase changes from the INACTIVE to the EXECUTING state. While in the EXECUTING state, the phase may be in one of several substates, as shown in Figure 8.25. When routine step logic is being performed, the phase is in the NORMAL state. The normal state has a substate called HOLD that is activated from external commands—the operator or activity—and should only be valid at the beginning of a phase. If a failure occurs or the operator stops the unit object, the phase will branch automatically to the FAIL state. When the phase ends, it returns to the INACTIVE state. At this point, the operation may activate another phase or return to the home step. When implemented, a typical phase has two distinct parts, the routine step logic and the fail action and recovery logic, as shown in Figure 8.26.

### 8.5.3 Simple versus Complex Phase Logic

A simple phase is a fixed cycle with end conditions that are serial and that perform a single process control task. Serial logic means the step end conditions occur essentially at the same time or one at a time. Complex phase logic involves concurrent multiple control

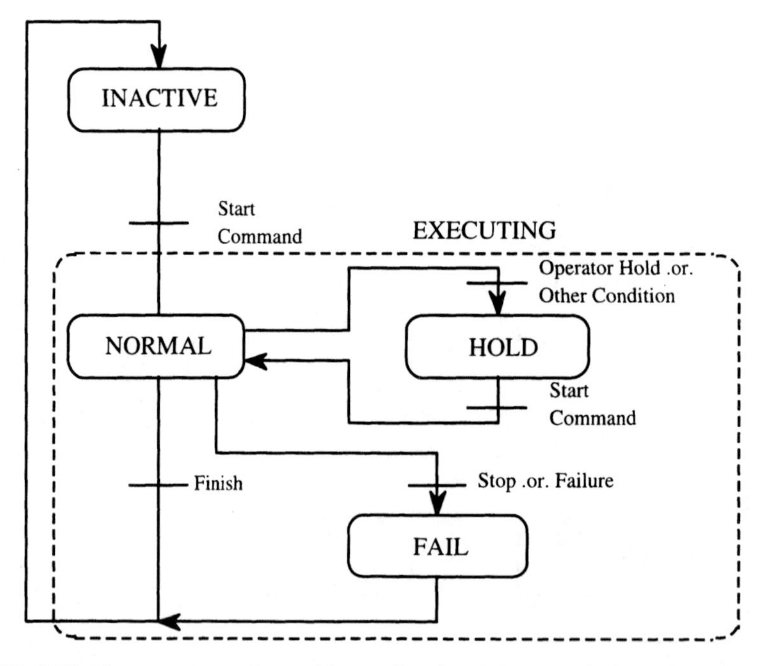


**FIGURE 8.25.** Phase states and transitions. (Reprinted by permission. Copyright © 1997, Automation and Control Technologies, Inc.)

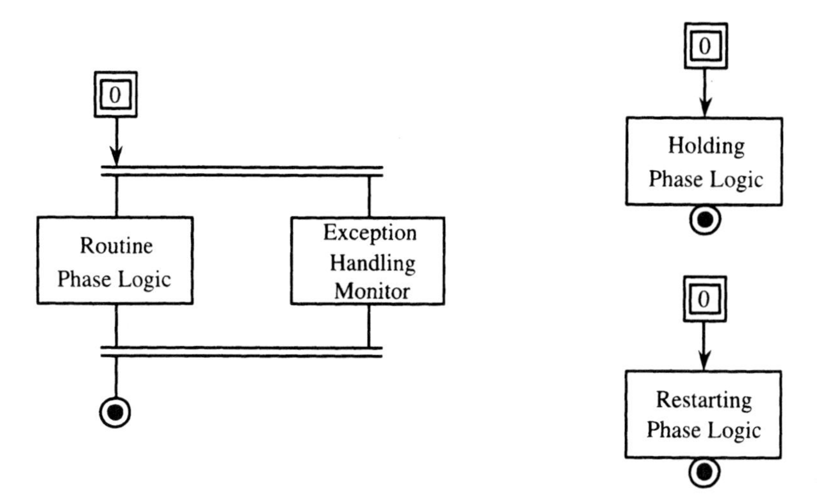


**FIGURE 8.26.** Phase logic.

tasks (i.e., more than one unit control task executes at the same time) where the end conditions may not happen in unison or one at a time. The charge of multiple reactants into a reactor without knowing which reactant charge will finish first is an example of a complex sequence control task.

### 8.5.4 Phase Fail Action and Recovery

Phases include fail action and recovery functions because things can go wrong in routine plant operations. During plant operation, batches can be saved if the proper fail action and recovery are implemented. However, fail actions can change a simple sequence into a very complex one.

Fail actions are triggered by fail conditions that are monitored by the unit procedure or operation if the condition is unitwide or by specific phases if the condition is only of concern during a particular phase. Once a fail action is triggered, appropriate loop and device objects are placed in a safe position. It is then the operator's job to troubleshoot the problem. Supporting the operator in the diagnosis with accurate information is a critical aspect of good unit object design. Once the cause is identified, the recovery logic should allow safe restart or a proper abort path.

Aborting the phase returns control to the unit where the cycle may be terminated, necessitating a manual operation to finish the cycle. A restart recovery action returns the phase to the scheduler, which moves the phase backward, forward, or to the last step, resuming normal automatic control (i.e., recovery).

### 8.5.5 Examples

**Example 8.5.** Material transfer system unit. As for the previous examples, the bulk of this example will concern the control aspects of this device. However, to present a complete picture of this device, the control objective and concept that lead to the control strategy are described.

**Control Objective:** To transfer raw material (grain, sand, cement, etc.) from rail cars to a storage silo.

**Process Description:** A simplified P&ID for this unit is shown in Figure 8.27. Material is unloaded from one or two rail cars into a hopper below. Gates G-101 and G-102 regulate the flow onto small conveyors (C-103 and C-104) that direct the material onto the C-201 conveyor. Material is transferred to the C-202 bucket conveyor and then to the C-203 transfer conveyor, which moves the material to the storage bin. A blower (B-301) collects dust from the conveyors into a bin. After settling, the dust is reintroduced into the system at the C-203 conveyor. The operating states of this unit are shown in Figure 8.28.

**Process System Items**

West rail dump sliding gate (G-101)
East rail dump sliding gate (G-102)
West rail dump transfer conveyor (C-103)
East rail dump transfer conveyor (C-104)
Rail-dump-to-bucket transfer conveyor (C-201)
Bucket conveyor (C-202)
Bucket-to-bin transfer conveyor (C-203)
Dust collector bin (T-301)
Dust collector blower (B-301)
Rotary airlock valve (L-302)
Raw-material storage silo (T-401)

**Control Concept:** Provide automatic startup and shutdown for the transfer of raw material from either or both rail dump stations.

The control of the rail dump gates and the conveyors below the gates will be under manual operator control. The transfer system will normally be operated automatically, with the operator requesting startup or shutdown. However, there are provisions for operator control for each control module. Normally, the operator requests startup when the system is stopped, and when startup is complete, the system is running and the conveyors and gates at the rail dump hoppers may be manually operated by the operator. To shut down the system, the operator requests a shutdown. An E-stop state is requested by the operator and results in immediate shutdown of all of the material transfer equipment. The Hold state is requested by the operator in order to gain manual control of the modules. When the operational state is Shutdown or E-stop the operator may request manual control of the control modules.

**Control Strategy:** When the system is started, the equipment must be started in backward order. For example, the last conveyor in the transfer system must be started first. When the system is shut down, the equipment must be stopped so that all material on a conveyor is removed before that conveyor is stopped. Interlocking must be provided so that the backward startup and forward shutdown must be enforced even when manually controlled by the operator. Shutdown must be automatically initiated upon equipment failure.

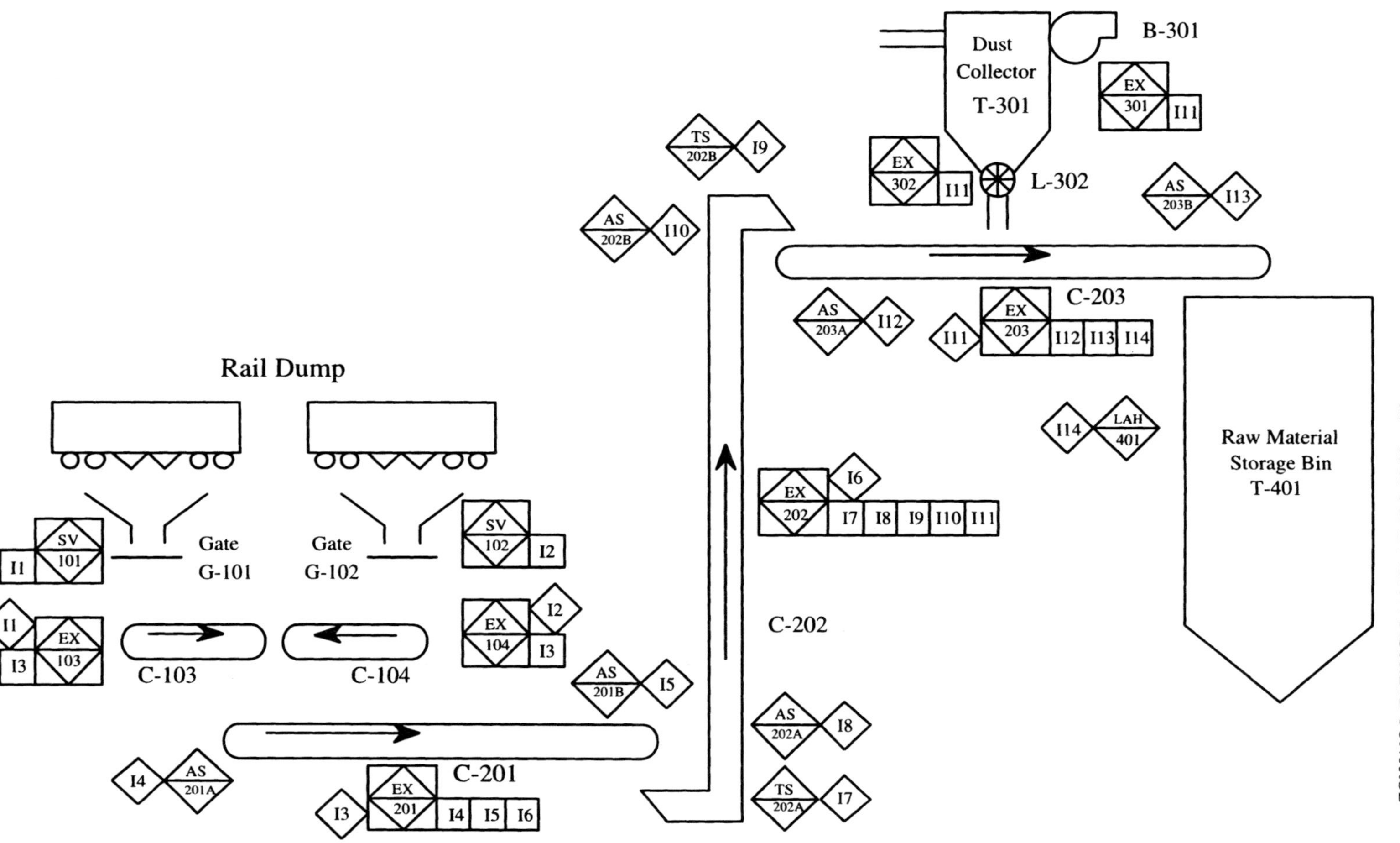


**FIGURE 8.27.** P&ID of material transfer unit.

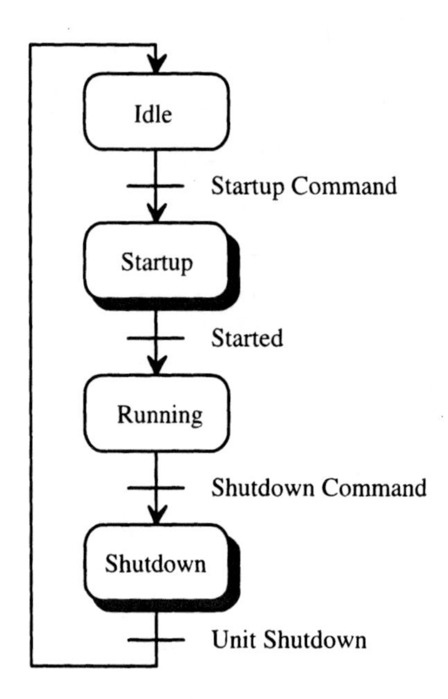


**FIGURE 8.28.** Operational states of transfer unit.

Interlocking must be provided so that if a downstream conveyor is stopped because of a failure, the upstream conveyor must be E-stopped. Likewise, interlocking must ensure the entire conveyor system is running before either rail dump transfer conveyor is running. In addition, the respective rail dump transfer conveyor must be running before the rail dump gate is allowed to be opened by the operator. The rail dump slide gates are normally under control of the operator. However, if the system is shut down, the slide gates are immediately closed.

**Detail Design:** The unit procedure states are the same as the unit operational states in Figure 8.28. The operation associated with each state has only one phase. The Startup, Shutdown, and E-stop phase steps are shown in Figures 8.29–8.31. When in the Automatic control state, the interlocks are enforced by the SFCs. Many of the branches during Shutdown enforce a particular interlock, namely that when a downstream conveyor fails, the upstream conveyors must be stopped immediately. During Startup, delays are added after the start of each conveyor to allow the operator to take action, for example, to take the system to the HOLD state after a conveyor starts but does not sound like it is operating correctly.

Skip delay of C-103 when there is C-103 failure or C-201 failure or C-202 failure or C-203 failure or T-301 failure or LAH401.

Skip delay of C-104 when there is C-104 failure or C-201 failure or C-202 failure or C-203 failure or T-301 failure or LAH401.

Skip delay of C-201 when there is C-201 failure or C-202 failure or C-203 failure or T-301 failure or LAH401.

Skip delay of C-202 when there is C-202 failure or C-203 failure or T-301 failure or LAH401.

Skip delay of C-203 when there is C-203 failure or T-301 failure or LAH401.

Skip delay of T-301 when there is C-203 failure or T-301 failure or LAH401.

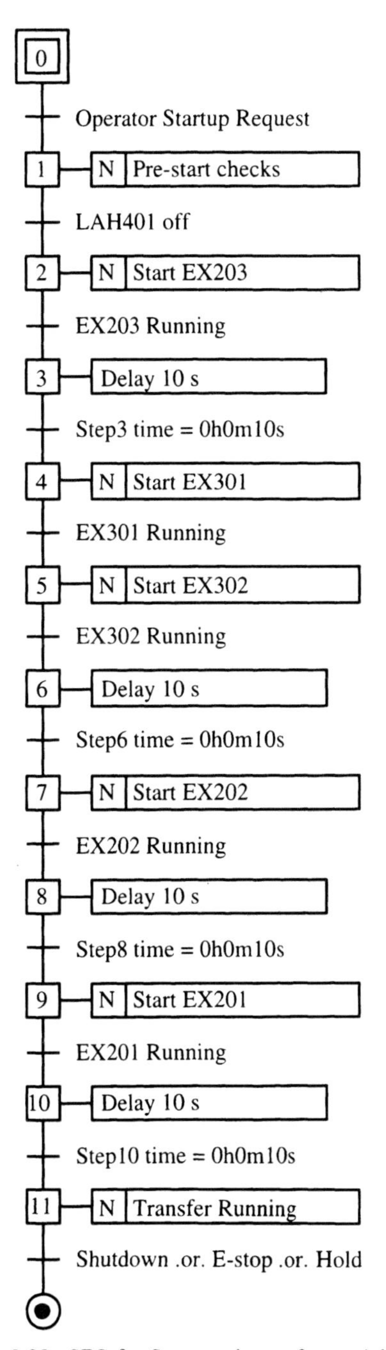


**FIGURE 8.29.** SFC for Startup phase of material transfer unit.

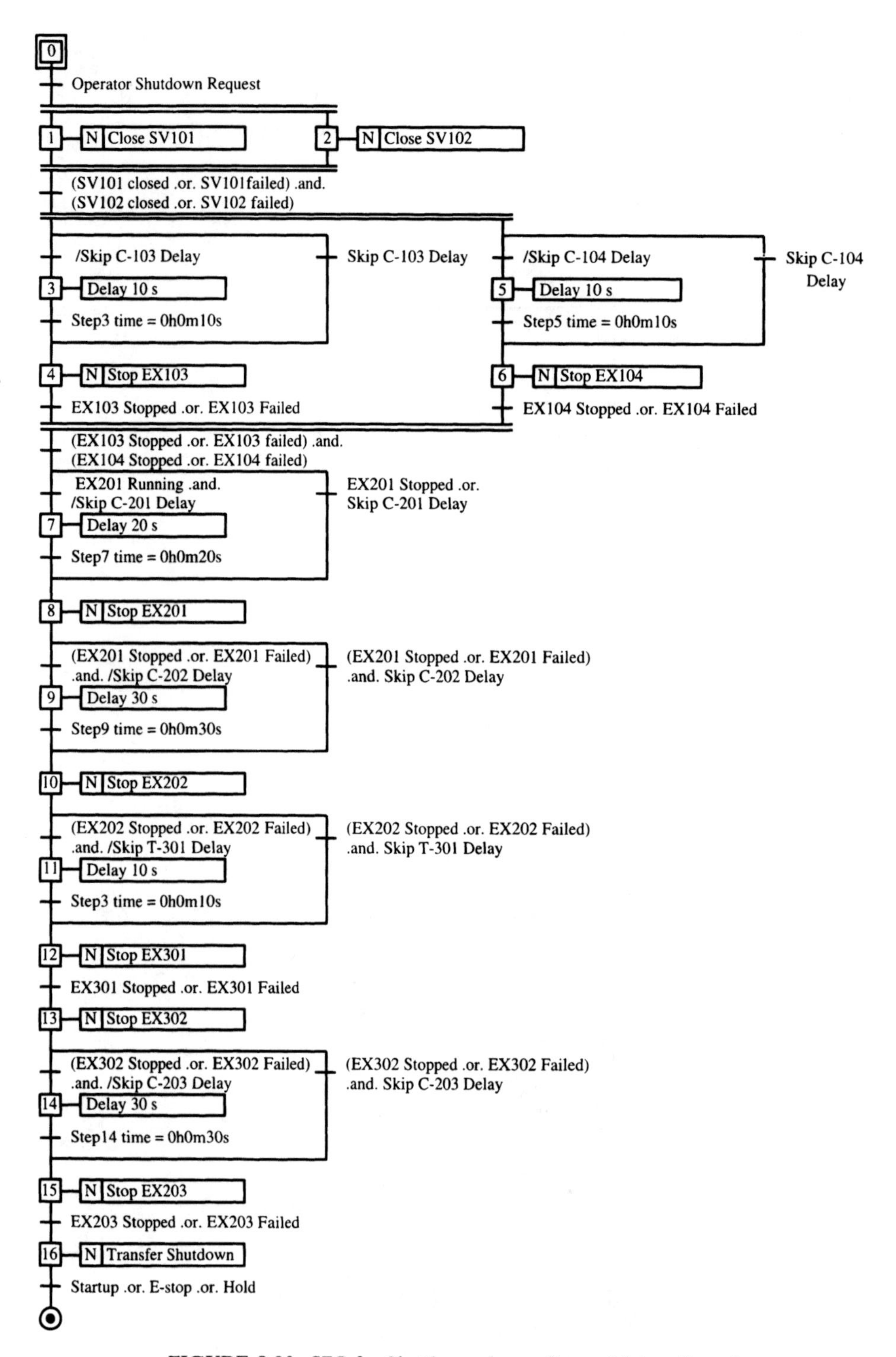


**FIGURE 8.30.** SFC for Shutdown phase of material transfer unit.

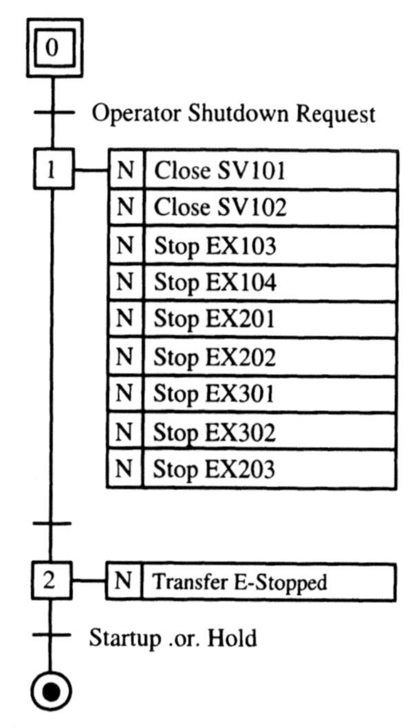


**FIGURE 8.31.** SFC for E-stop phase of material transfer unit.

**Failures**

| Failure Condition | Action |
|---|---|
| EX103 failed for 1 sec | Shutdown |
| EX104 failed for 1 sec | Shutdown |
| EX201 failed for 1 sec | Shutdown |
| AS201A indicates C-201 belt misalignment for 1 sec | Shutdown |
| AS201B indicates C-201 belt misalignment for 1 sec | Shutdown |
| EX202 failed for 1 sec | Shutdown |
| TS202A indicates C-202 bearing temperature high for 1 sec | Shutdown |
| AS202A indicates C-202 belt misalignment for 1 sec | Shutdown |
| TS202B indicates C-202 bearing temperature high for 1 sec | Shutdown |
| AS202B indicates C-202 belt misalignment for 1 sec | Shutdown |
| EX203 failed for 1 sec | Shutdown |
| AS203A indicates C-203 belt misalignment for 1 sec | Shutdown |
| AS203B indicates C-203 belt misalignment for 1 sec | Shutdown |
| LAH401 indicates high storage bin level for 1 sec | Shutdown |
| EX301 failed for 1 sec | Shutdown |
| EX302 failed for 1 sec | Shutdown |

**Interlocks in the Manual Control State**

| Interlock Number | Demand | Action | Reset |
|---|---|---|---|
| 1 | EX103 stopped | Close SV101 | Automatic |
| 2 | EX104 stopped | Close SV102 | Automatic |
| 3 | EX201 stopped | Stop EX103<br>Stop EX104 | Automatic |
| 4 | AS201A indicates C-201 belt misalignment | Stop EX201 | Automatic |
| 5 | AS201B indicates C-201 belt misalignment | Stop EX201 | Automatic |
| 6 | EX202 stopped | Stop EX201 | Automatic |
| 7 | TS202A indicates C-202 bearing temperature high | Stop EX202 | Automatic |
| 8 | AS202A indicates C-202 belt misalignment | Stop EX202 | Automatic |
| 9 | TS202B indicates C-202 bearing temperature high | Stop EX202 | Automatic |
| 10 | AS202B indicates C-202 belt misalignment | Stop EX202 | Automatic |
| 11 | EX203 stopped | Stop EX202<br>Stop EX301<br>Stop EX302 | Automatic |
| 12 | AS203A indicates C-203 belt misalignment | Stop EX203 | Automatic |
| 13 | AS203B indicates C-203 belt misalignment | Stop EX203 | Automatic |
| 14 | LAH401 indicates high storage bin level | Stop EX203 | Automatic |
| 15 | EX301 failed | Stop EX203 | Automatic |
| 16 | EX302 failed | Stop EX203 | Automatic |

**Implementation:** A PLC is programmed to directly execute the SFCs (Figures 8.29–8.31). Alternatively, the SFCs may be translated into ladder logic code, which is executed by a PLC. The operator interface contains a graphic representation of the process (Figure 8.27).

**Example 8.6.** Ion exchange bed unit. As for the previous examples, the bulk of this example will cover the control aspects of this unit. Even though this is a continuous process, it has a significant amount of sequence control for automated switchover and regeneration.

**Control Objective:** To remove solid waste from a continuous feed stream.

**Process Operation:** The P&ID for this unit is shown in Figure 8.32. The feed stream is passed through one or more beds that contain material that collects undesirable components. Over time, a bed accumulates enough undesirable material that it must be taken off-line and regenerated, removing the waste material. The need for regeneration may be detected by differential pressure across the bed, by an in-stream measurement, or by laboratory analysis. In any case, the spent bed must be switched out and replaced with a fresh bed. After the spent bed is taken off-line, the bed material is regenerated through a sequence of water, acid, and caustic washes. After regeneration, the bed is ready for the next switchover to replace another spent bed.

The operating states of the ion exchange bed units are shown in Figure 8.33. Because of the coordination that must take place, the states of all of the beds are shown on the same diagram.

**Process System Items**

Ion exchange tank 1 (T-101)
Ion exchange tank 2 (T-102)
Ion exchange tank 3 (T-103)
Feed pump (P-100)
Caustic regeneration pump (P-104)
Flush water pump (P-105)
Downstream pump (P-110)

**Control Concept:** Provide automatic startup, shutdown, and regeneration for the ion exchange beds. The operator can request that a particular bed be started up or shut down.

Only one bed may be started or shut down at a time. The process is designed so that only one bed may be regenerated at time, because the three beds share common water and caustic supply headers. The feed stream processing and regenerating may be performed as concurrently executing tasks, but not with the same bed.

**Control Strategy:** The Startup, Shutdown, and Regenerating unit states have a corresponding operation. The Startup and Shutdown operations consist of only one phase and are shown as function charts in Figures 8.34 and 8.35 for ion exchange bed 1. Depending on the vendor, it may be possible to define one Startup procedure shared among the ion exchange units. For example, the SFC can use aliases instead of actual equipment tag names. To execute the procedure for a particular ion exchange unit, the Startup procedure is called with a table of the actual tags to use in place of the aliases. The phases for the Regeneration operation is shown in Figure 8.36. The function charts for the Flush phase and Regen phase are shown in Figures 8.37 and 8.38, respectively. Depending on the cycle chosen by the operator, the Flush phase may be called only once, or the Regen and Flush phases may be called multiple times.

There must be semaphores that prevent other ion exchange units from simultaneously executing the Startup, Shutdown, and Regeneration operations.

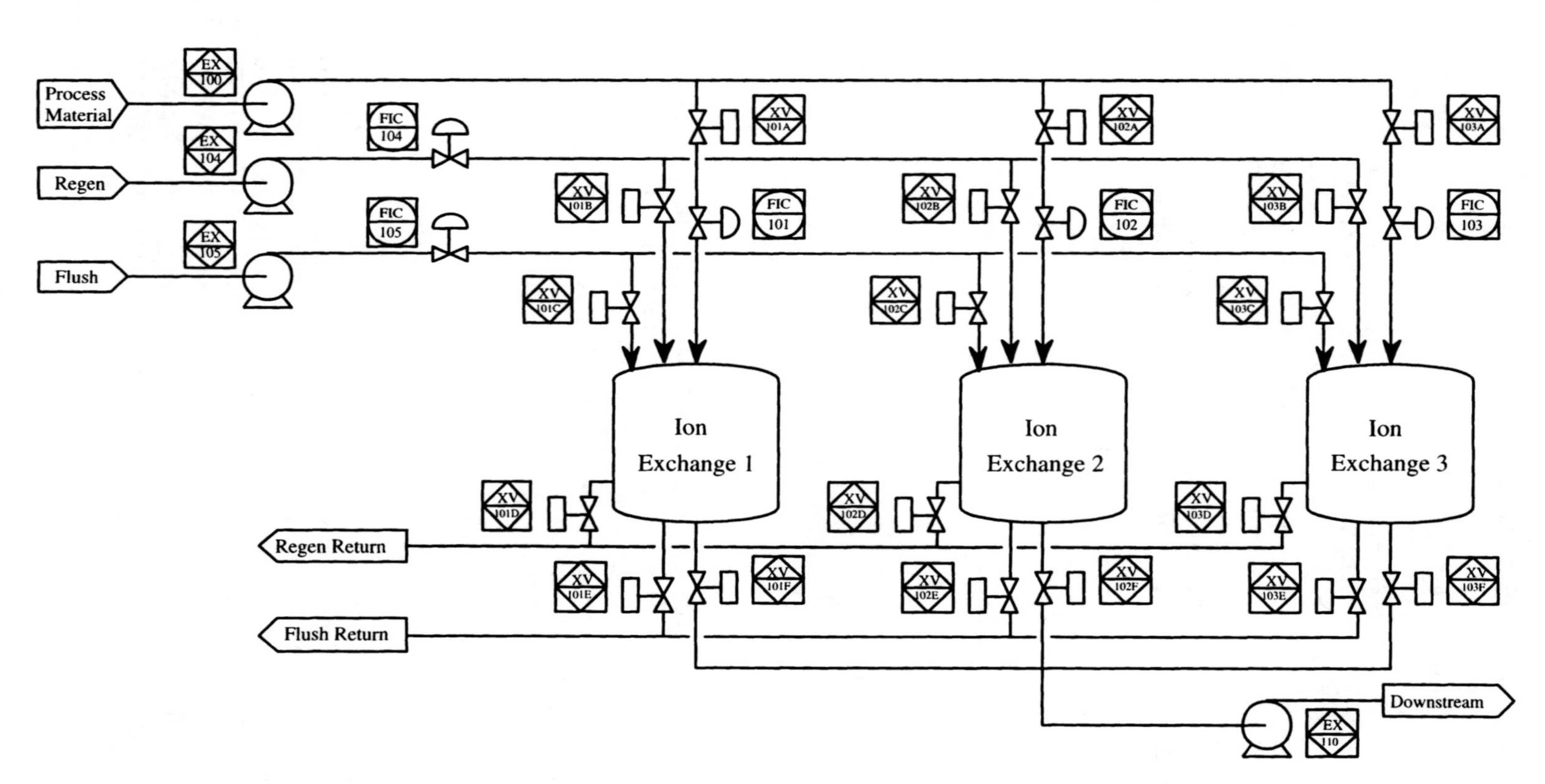


**FIGURE 8.32.** P&ID of ion exchange cell. (Reprinted by permission. Copyright © 1996, Automation and Control Technologies, Inc.)

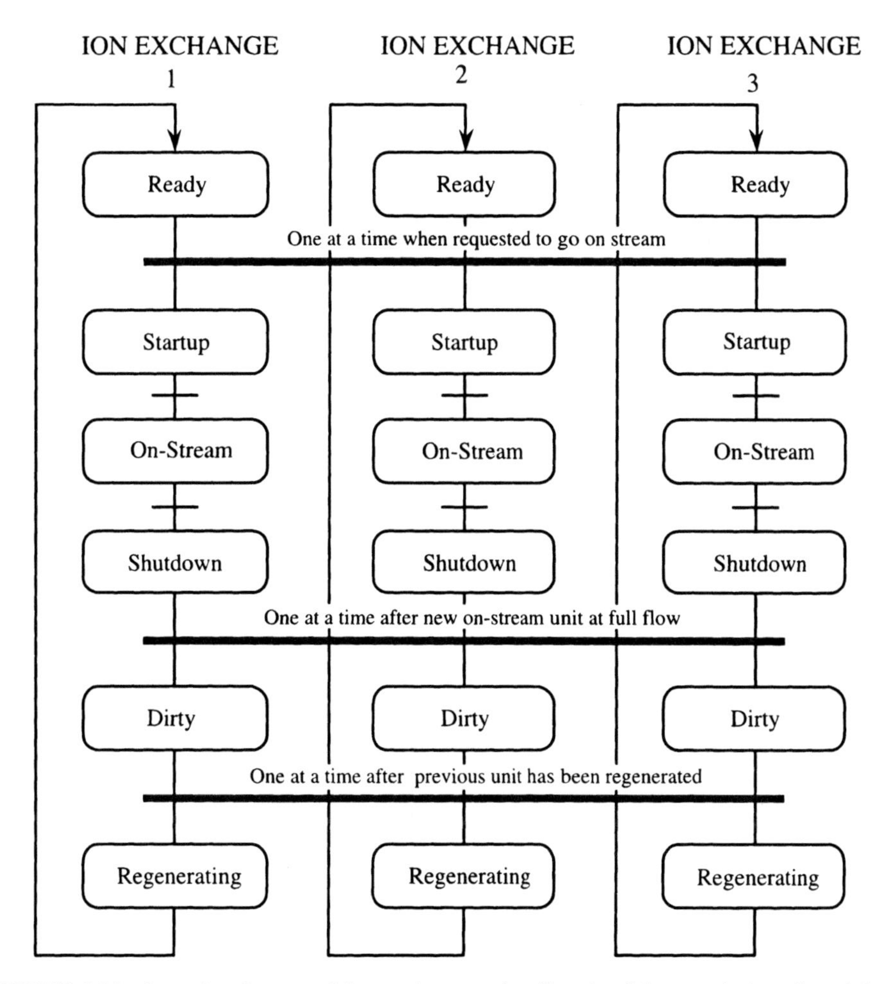


**FIGURE 8.33.** Operational states of ion exchange units. (Reprinted by permission. Copyright © 1996, Automation and Control Technologies, Inc.)

**Implementation:** A PLC/DCS is programmed to directly execute the SFCs (Figures 8.34–8.38). Alternatively, the SFCs may be translated into ladder logic code, which is executed by a PLC. The operator interface contains a graphic representation of the process (Figure 8.32).

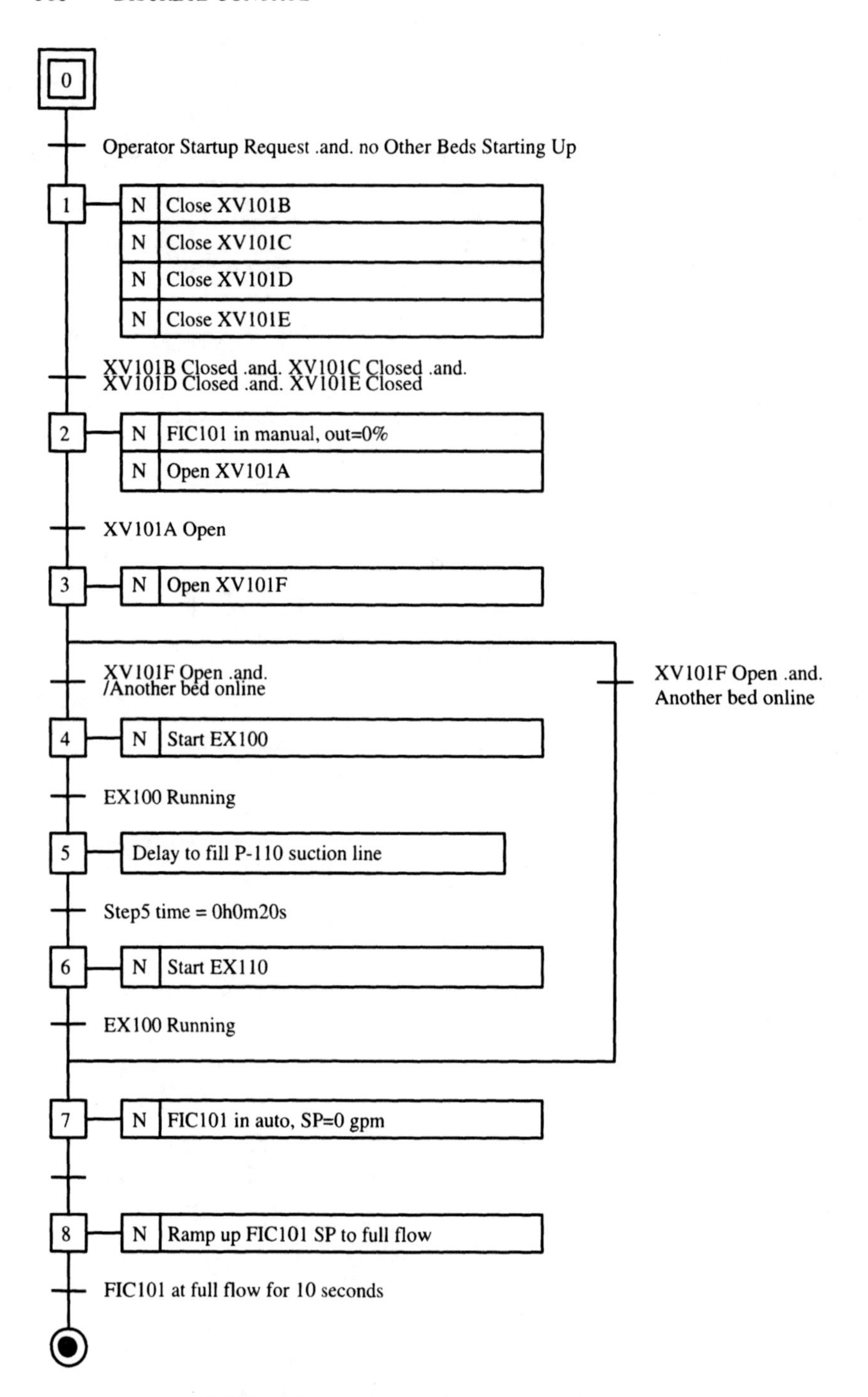


**FIGURE 8.34.** SFC for Startup phase of ion exchange 1 unit.

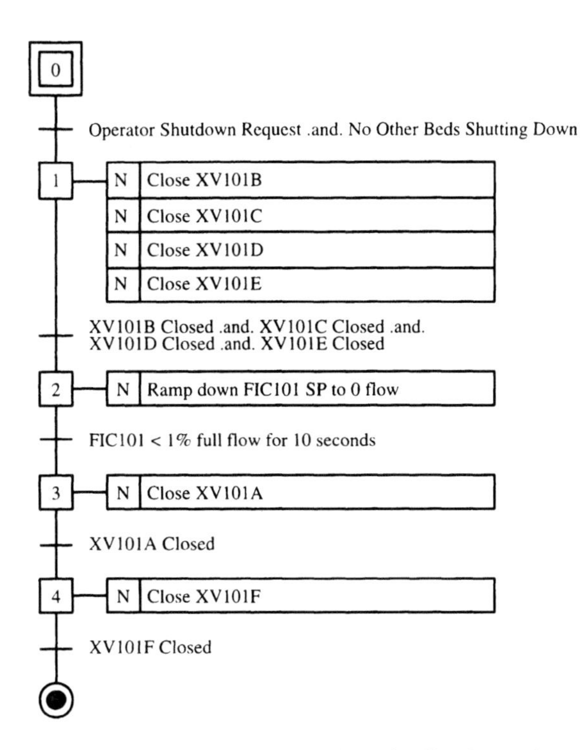


**FIGURE 8.35.** SFC for Shutdown phase of ion exchange 1 unit.

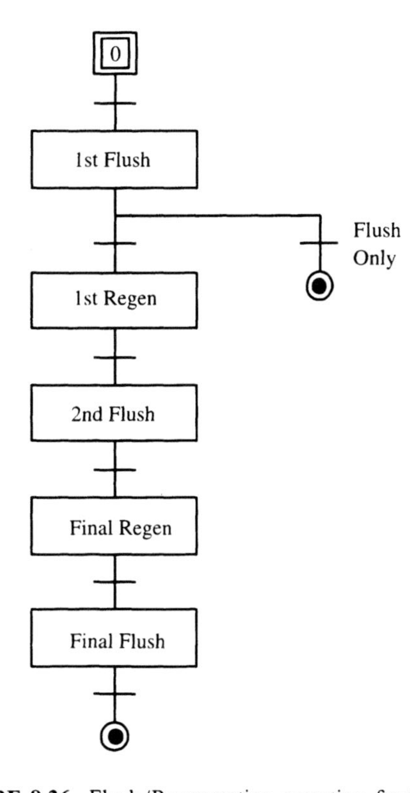


**FIGURE 8.36.** Flush/Regeneration operation for ion exchange 1.

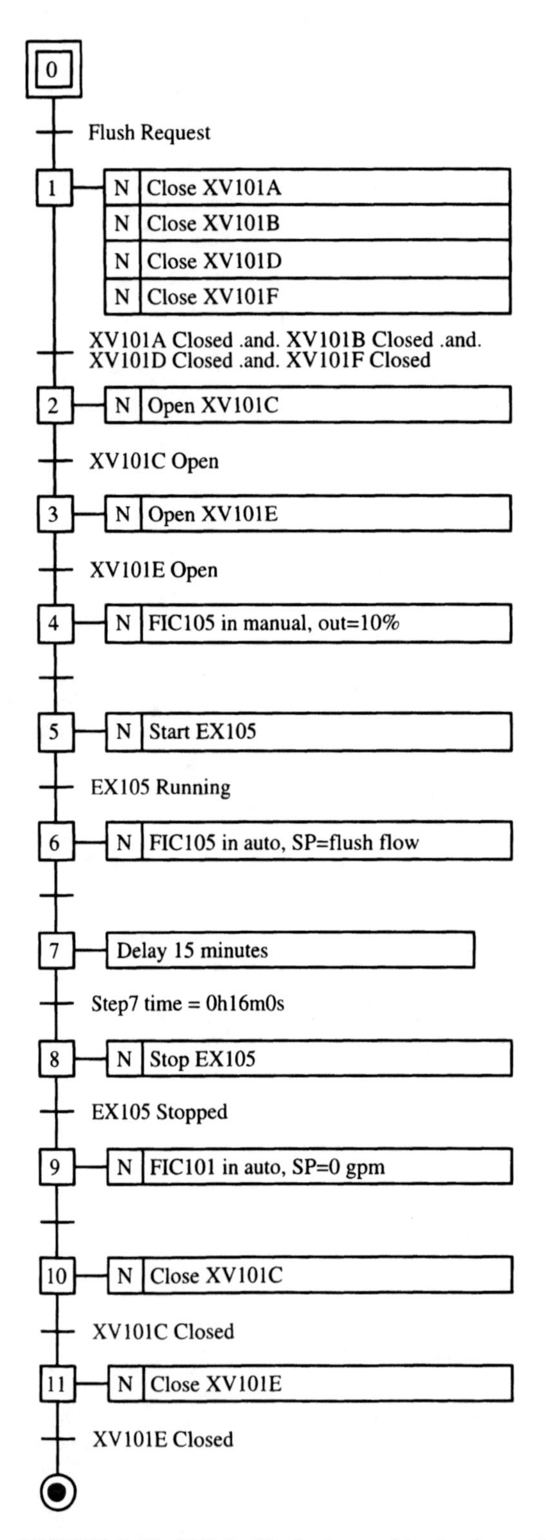


**FIGURE 8.37.** SFC for Flush phase of ion exchange 1.

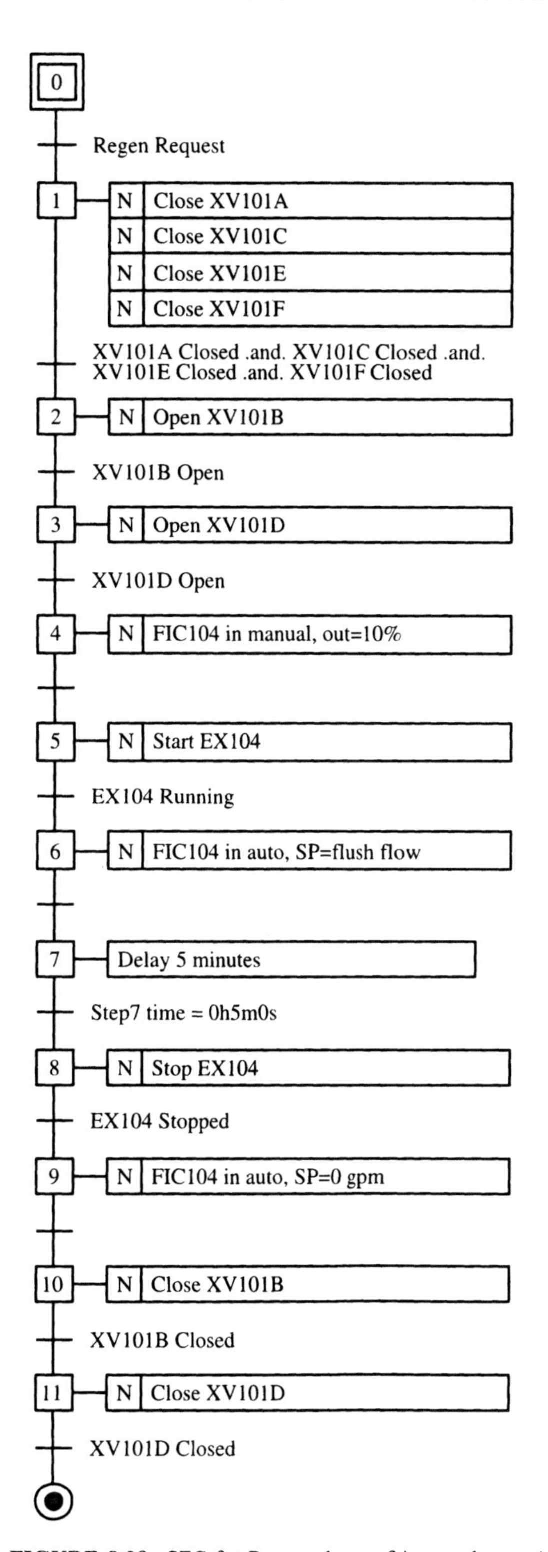


**FIGURE 8.38.** SFC for Regen phase of ion exchange 1.

## REFERENCES

Instrument Society of America, *ISA-S88.01, Batch Control, Part 1: Models and Terminology*, Instrument Society of America, Research Triangle Park, NC, 1995.

Instrument Society of America, *ISA-S84.01, Application of Safety Instrumented Systems for the Process Industries—1996*, Instrument Society of America, Research Triangle Park, NC, 1996.

International Electrotechnical Commission, *IEC 848: Preparation of Function Charts for Control Systems*, International Electrotechnical Commission, Geneva, Switzerland, 1988.

International Electrotechnical Commission, *IEC 1131-3: Programmable Logic Controllers—Part 3: Programming Languages*, International Electrotechnical Commission, Geneva, Switzerland, 1993.

# 9 Batch Control

**Chapter Topics**

- Batch procedural control
- Coordination control
- Recipe control
- Batch campaigns and runs
- Batch data collection
- Batch user interface for setup and operation
- Batch project engineering example

*Scenario:* The instrument vendor realized that batch is more than making PID work in a chemical batch reactor application.

Over his experience in batch control, the young engineer soon became frustrated with his new employer's attitude. His batch control experience was largely discounted by some of his new colleagues because it was not relevant to "real" batch control. After all, there were no loops in a glass container batch house and, even worse, there were no control valves. He soon learned about the syndrome of the elephant and the blind men: His new employer only touched the world around the control valve. The young engineer was convinced of a significant market potential for batch recipe control but his new employer's preoccupation with the control valve was diverting attention from a wider business opportunity. He realized that the very first lesson he learned about process control had not been learned by his new employer. Control technology must fit the nature of the process. Conventional analog controllers of the time were designed for linear, steady-state operation, not for batch. He continually wondered when they would they realize this and develop control products that could handle the broader requirements of batch processing. The young engineer eventually gave up and left the company, before they realized the true magnitude of the batch control opportunity. He did not leave them empty handed, however, for he left a legacy of batch control software that eventually proved very lucrative to the former employer. To this day, the young engineer is very proud of his contribution to the batch control industry.

## 9.1 INTRODUCTION

It is no surprise that the world is involved in batch control everyday. In the kitchen, a birthday cake or a special stew is made. On a hot July day, homemade ice cream is carefully prepared on the back porch. In this case, the batch of ice cream is actually started in the kitchen with premixing and cooking activities that prepare a batch of ingredients.

This mixture is carefully—and lovingly—transferred to the back porch for finishing and final disposition. Yes, everyday someone follows a recipe with its prescribed list of ingredients and sequence of steps to make a batch of something. The kitchen (and back porch) is capable of making many different products using a variety of recipes. This is the essence of batch processing.

The control of batch processes requires the integration of regulatory, discrete, and sequential control. As described in Chapter 8, all plants require some degree of discrete control. Discrete control ranges from simple interlock logic to complex sequential control strategies that automate the operation of entire process units. Batch plant automation requires a high level of discrete control. Also, some batch applications require complex regulatory control schemes such as temperature control in batch reactors. It is the recipe control requirement that distinguishes batch processes from continuous processes.

The challenge in the design and implementation of plantwide process control for batch processing is to handle the complexity imposed by this mixture of diverse control types. This complexity has impeded the progress of batch control technology. After many years of trial and error and evolution, the batch control technology is now available to handle the diversity of control techniques in an integrated manner. Batch automation has never been easier than it is today. This chapter describes the considerations for plantwide process control in batch processing.

## 9.2 BATCH PLANT OPERATION COMPLEXITY

A batch plant system is commissioned to prepare process materials in discrete quantities. The plant system may require weighing and mixing a formula of materials. There may be some sort of chemical or biological conversion of the mixture. The batch of material may require physical separation to remove the final product from the by-products. The final batch may be blended or it may be segregated to ensure its use can be traced in the marketplace. A batch process system is operated in repeated cycles, day after day, during active production. The processing of discrete quantities of product is fundamental to batch processing.

The intermittent flow of material within the plant is another defining characteristic of a batching operation. In other words, unlike a continuous process, there is an interruption in the flow of material and energy streams among the various batch operating modules. This up-and-down nature has created a challenge for control technologists.

As defined in Chapter 3, the plant operating module that makes a batch is called a process cell. In the simplest case, the batch may be completely prepared in a single unit of the process cell. On the other hand, the production of a batch may require passing it through many process units within a process cell. According to the S88.01 standard (ISA, 1995), each stop along the batch path is called a stage. The stage is part of the S88.01 batch process model. The number of different paths and stages in a process cell is a measure of the batch control complexity.

Another control complexity factor in batch processing is the number of different products a process cell must be able to manufacture. Some batch plants are designed to make a single product with minor adjustments from batch to batch—like the ice cream maker. The ice cream maker can only make frozen desserts with minor variations in flavoring and additives (e.g., vanilla, chocolate, chocolate chips, strawberries). The more complex situation is the multiproduct batch plant. This type of plant system is required to handle more than one type of product. It may be one product at a time or several different products at the same time, much like the kitchen stove. While a cake is being baked, a stew can be simmering to completion on the stovetop. The most complex batch processing involves many parallel process lines arranged in multi-unit configurations. In this case, more than one batch is prepared at the same time in a completely asynchronous operation. This plant configuration is called a multiproduct plant.

The complexity of batch operations may be classified by the following criteria:

Single-stage versus multistage operation requirements
Single-product versus multiproduct requirements
Single-path versus multipath requirements

The simplest batch operations are single-stage, single-product, and single-path plant systems. The most complex are multistage, multiproduct, and multipath situations. Batch process classification helps identify the requirements for the control technology and is similar to classifying regulatory control by the number and interaction of loop control modules. Classifying a batch process is very helpful as a first step on a capital project involving batch processing.

## 9.3 BATCH OPERATION MODEL

The control reference model described in Chapter 3 is based on the S88.01 control activity model. As discussed earlier in this book, the control reference model provides the frame of reference for the Control Concept in process design. It also provides a model to discuss the automation of batch plant operations. This section presents the reference model in specific batch operation terminology as defined by the S88.01 standard.

The S88.01 standard control activity model provides the common structure for any batch plant operation. It is a means of expressing batch operations in terms of what activity is performed and where it fits in the other batch operations.

Instead of using the S88.01 control activity diagram in the standard, the authors are compelled to present the S88.01 activity model in terms of an IDEF0 diagram (National Institute of Standards and Technology, 1993). This method is more precise than the S88.01 model and will prove useful to future STEP data exchange efforts as described in Chapter 2. The AIChE's pdXi consortium plans to extend the process plant activity model to include operating activities and batch control.

The process plant lifecycle IDEF0 diagram in Figure 2.2 focused on the plant design aspects of a capital project. The batch operations activity model presented in Figure 9.1 is part of the *Manage, Operate and Maintain Plant* activity (node A5 of Figure 2.2).

In the IDEF0 diagram of Figure 9.1, the operation of a batch process is segmented into six activities as defined by the S88.01 standard. The nodes are:

Node A1: Perform Recipe Management
Node A2: Perform Production Planning and Scheduling
Node A3: Perform Production Information Management
Node A4: Perform Process Management
Node A5: Perform Unit Supervision
Node A6: Perform Process Control

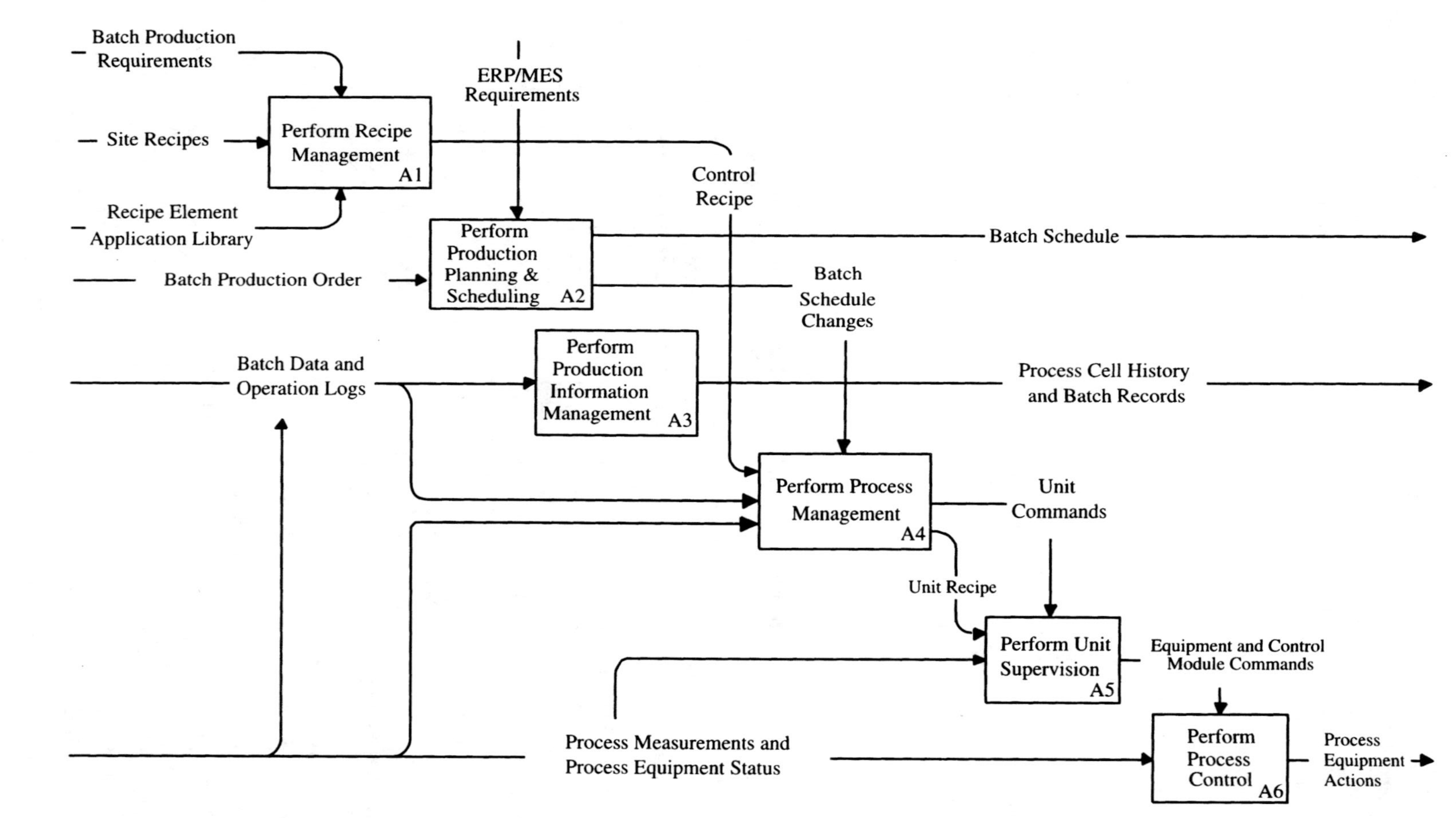


**FIGURE 9.1.** S88.01 control activity model; IDEF0 format. (Reprinted by permission. Copyright © 1997, Automation and Control Technologies, Inc.)

The primary focus of the S88.01 standard is on nodes A1, A3, A4, and A5. These activities are the main areas of concern for plantwide, batch recipe control systems.

According to S88.01, recipe management is the plant activity that develops the master recipes for use in actual production. A master recipe is based on the requirements of the enterprise and site recipes developed by the enterprise. The master recipe is formulated by converting batch production knowledge into a library of master recipes. The batch production cycle begins when one of the master recipes is converted to a control recipe and used in the process management activity to make a batch of product. Following good engineering and operation practice, S88.01 defines the need for a recipe element library to construct master recipes. These elements include generic procedural control elements, formula parameters, and equipment information.

Process management is the batch operation activity that executes the control recipe, coordinates the batch path, and collects batch data for batch end reports and future reference.

The unit supervision activity executes the recipe control elements according to the control recipe and is coordinated by process management. It includes unit coordination control and data collection.

Production information management in a batch plant is focused on reporting time-series data, events, and operator inputs associated with each batch cycle.

This batch operation model applies in both simple and complex batch operating environments. It applies to both manual and automated situations.

## 9.4 PROCEDURAL CONTROL IN BATCH PROCESSING

As discussed in Chapters 3 and 8, procedural control is the automatic control type used to correctly operate process equipment in a stepwise fashion. The procedural control type is involved in operating a single item of plant equipment or it may be involved in the coordination of many batches in a process cell. This is the essence of batch control. One of the most significant contributions made by the S88.01 standard is the differentiation between recipe procedural control and equipment procedural control.

Often, process designers develop operating modules that are used in a repeatable manner. Assume the designer had a particular operation method in mind. In other words, he or she embedded a particular design for the control scheme in a process unit. This design includes both the basic controls as well as the procedural control. The plant system must be operated in a specific manner to produce the desired outcome as envisioned by the designer. In this case, the process management and unit supervision control activities are fixed by the design of the plant systems. The sequence control for this type of plant system is called equipment procedural control. The process type may be batch or continuous.

This type of fixed control is prevalent in single-product batching applications. Examples of processes of this type are found in chemical batch reactors and batch distillation plant systems. The reactor or column makes the same product over and over, day in and day out. Adjustments are made to control parameters over time and as raw-material assays change, but the procedure is usually fixed for long periods. The procedural control is fixed by the unit design. In these cases the procedural control remains stable for long periods with changes made infrequently and very carefully.

When operation flexibility is required, such as that found in a multiple-product batch plant, the process management and/or the unit supervision needs recipe procedural control. Section 9.6 discusses recipe control in more detail. While a batch application may not include recipe procedural control, there is always some degree of equipment control. The key to the design and implementation of batch control is determining the line between recipe and procedural control. At some point in the control hierarchy, the equipment control stops and recipe control is employed. If there is too much recipe control, the process may be unsafe to operate. If there is too much equipment control, the process will not be agile enough for flexible manufacturing.

### 9.4.1 S88.01 Procedural Control

According to the U.S. Patent Office (1996), "Sequential control is where the output signals of a system are sequentially or selectively applied to control a process in a repetitive, orderly manner according to a predetermined schedule of operation." This definition can be applied to any application from interlock logic to running a whole plant, in any type of industry. In the S88.01 standard, batch process management, unit supervision, and equipment module control are all called procedural control. According to S88.01 (ISA, 1995, p. 15) procedural control is "control that directs equipment-oriented actions to take place in an ordered sequence in order to carry out some process-oriented task." The S88.01 procedural control model is composed of procedural control elements. There are four types of elements, shown in Figure 9.2. As depicted, all procedural elements have three primary attributes as defined by S88.01: state, mode, and operator command. The procedural elements are also defined in a four-tiered structure (Figure 9.3), each with its own attributes.

The basic element is called a phase. The phase is the sequence control logic that applies the prescribed order of outputs to the control modules (devices and loops) as a function of process state measurements and operator commands. The control modules apply output signals to the process. This is a two-tier architecture and conforms to the U.S. Patent Office definition. Phases are parameterized in recipe control applications or fixed in equipment control applications.

The S88.01 equipment modules contain a single phase that is executed on demand or in an auto-cycle sequence. The demand signal is generated from an external command

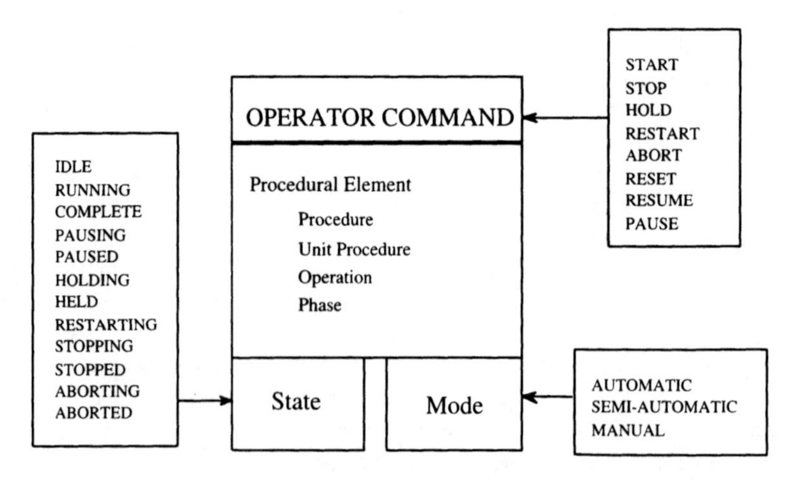


**FIGURE 9.2.** Procedural control elements. (Reprinted by permission. Copyright © 1996, Automation and Control Technologies, Inc.)

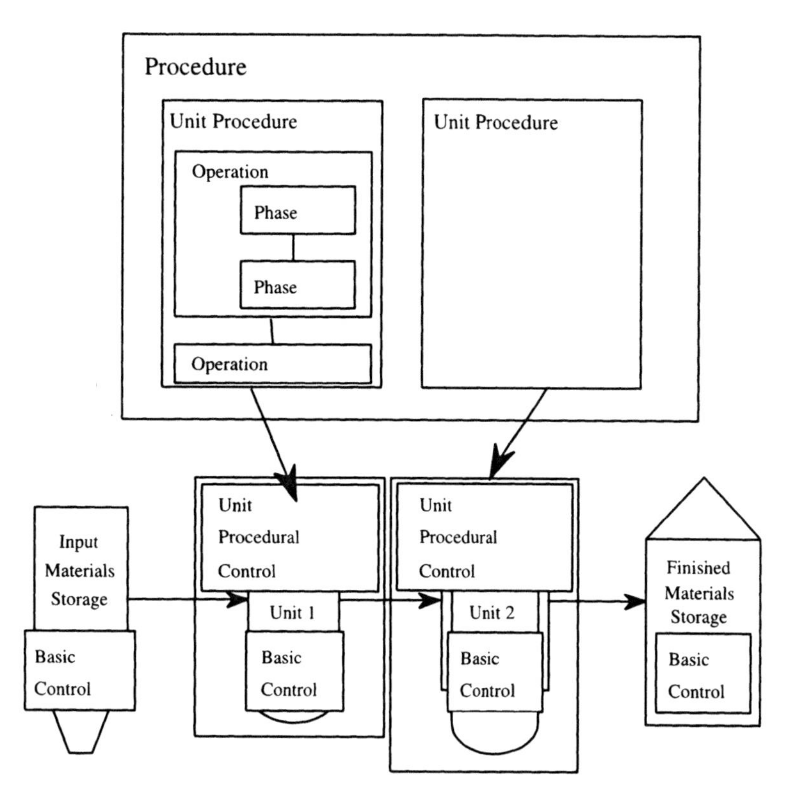


**FIGURE 9.3.** Procedural control model. (Reprinted by permission. Copyright © 1996, Automation and Control Technologies, Inc.)

originating from the operator or from another procedural control element. The auto-cycle is a type of phase that repeats its logic in a continuous cycle. Equipment module control is found in shared equipment sequencing and complex transfer system applications.

At the other end of the procedural control hierarchy is the procedure, the sequential control logic used to make a complete batch of product. It is an essential element of the S88.01 recipe. In a recipe, the procedure executes the lower level procedural element in the prescribed order. The lower level elements are at least phases and sometimes operations and unit procedures.

The unit procedure defines the sequential control logic to operate a unit. It is a subdivision of the procedure and is usually associated with fulfilling the control of a batch process stage in a multistage batch application. The unit procedure provides the instructions to operate a process unit in conjunction with the unit coordination control and data collection functions of the unit equipment entity. The unit procedure coupled with the unit's recipe parameters and equipment information is sometimes called the unit recipe.

In some applications, the phase logic is fixed (i.e., equipment procedural control), but each recipe may require a different set of phases to make a batch. In this case, it is convenient to further subdivide the unit procedure into operations. The operations initiate and monitor the appropriate set of phases for a particular step in the unit procedure.

This procedural element flexibility is called collapsibility. Collapsibility means that S88.01 batch control requires only procedures with a sequence of phases. The need for unit procedures and operations is arbitrary and is left for the system implementers.

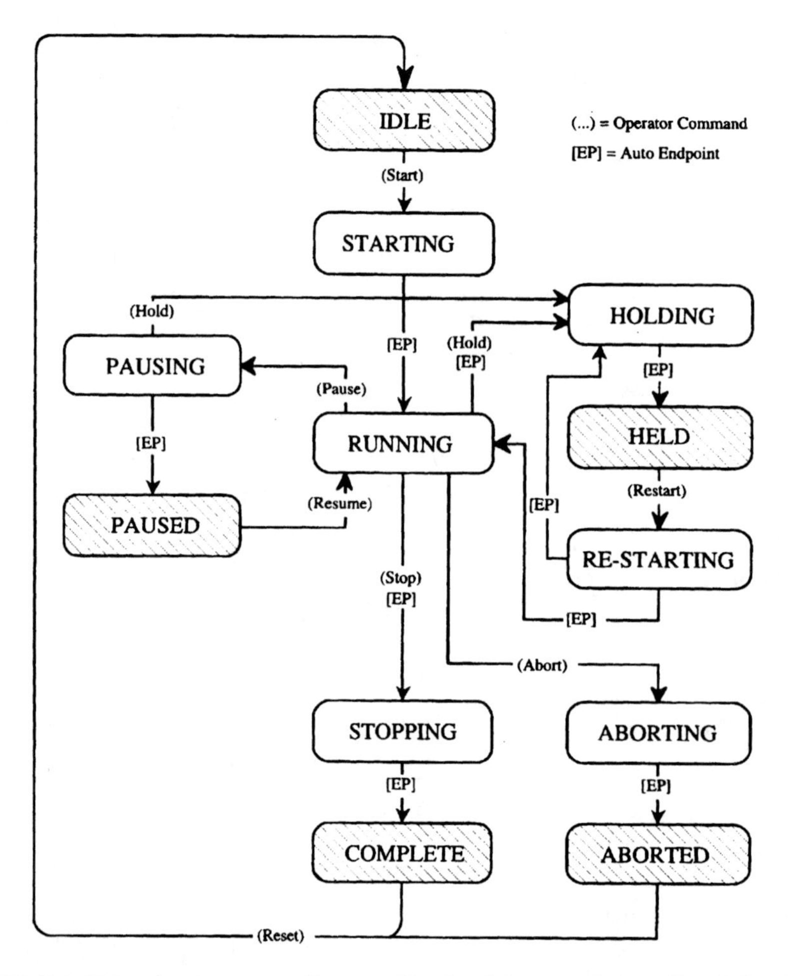


**FIGURE 9.4.** Unit procedure state diagram. (Reprinted by permission. Copyright © 1997, Automation and Control Technologies, Inc.)

### 9.4.2 Procedural Control State

Each procedural control entity possesses a state. Figure 9.4 defines the unit procedure state transition logic defined by S88.01. Operations and phases have a simpler state diagram, without the Pausing, Paused, Aborting, and Aborted states. The process normally starts in the IDLE state. The shaded states are final states. The valid states are described as:

IDLE. The procedural element is waiting for a Start command that will cause a transition to the STARTING state.

STARTING. The procedural element executes startup logic. Once complete, the procedural element automatically transitions to the RUNNING state.

RUNNING. Normal operation. The operation, phase, or procedure is executing. The procedural element remains in this state until an endpoint is reached or an operator command or an abnormal condition forces it to another state.

STOPPING. The normal operation has run to completion, and any special logic needed to bring the process to a setdown state is executed. The operator can also issue a Stop command. If no sequencing is required, then the procedural element or equipment entity transitions immediately to the COMPLETE state.

COMPLETE. The procedural element has completed its STOPPING logic and is waiting for a Reset command to transition to IDLE.

PAUSING. The procedural element has received a Pause command. The procedural element stops at the *next* defined state or stable stop location in its normal RUNNING logic. Once stopped, the state automatically transitions to PAUSED.

PAUSED. Once the procedural element has paused at the defined stop location, the state changes to PAUSED. This state is normally used for short-term stops. A Resume command causes a transition to the RUNNING state, resuming normal operation immediately following the defined stop location.

HOLDING. The procedural element receives a Hold command from the operator or an abnormal process condition forces a transition to the HOLDING state. The HOLDING logic places the procedural element or equipment entity into a known state. If no sequencing is required, then the procedural element transitions immediately to the HELD state.

HELD. Once the procedural element has completed its HOLDING logic and is at the known state, the state changes to HELD. This state is normally used for a long-term stop. A Restart or Abort command causes a transition out of this state.

RESTARTING. The procedural element has received a Restart command while in the HELD state. If no sequencing is required, the procedural element transitions immediately to the RUNNING state; otherwise restart logic is executed.

ABORTING. The procedural element has received an Abort command and is executing its ABORTING logic, which facilitates a quicker, and not generally controlled, emergency shutdown. Generally, no sequencing is required, and the procedural element or equipment entity transitions immediately to the ABORTED state.

ABORTED. The procedural element has completed its ABORTING logic and is waiting for a Reset command to transition to IDLE.

Within the procedural element states, the state logic provides the sequential control selected for the automation of the batch application activity. The state logic is subdivided into one or more steps. A step is a node of a sequential function chart (SFC). Each SFC node represents the actions to be performed. As such, within the state logic the steps may be serial or parallel. The typical actions at each step may be one or more of the following:

- Wait for a step end condition.
- Change control module setpoints or internal parameters.
- Ask the operator to perform a manual activity.
- Save batch data parameters.
- Perform a calculation.
- Start a lower level procedural element.

Each SFC step may have zero, one, or more actions.

The STARTING, STOPPING, and RUNNING states are composed of the batch procedural control for routine batch production. It also includes the logic to detect procedural element exception conditions. If an exception condition is detected, the state is changed to HOLDING and eventually results in a HELD state. The HOLDING logic is sometimes called fail action logic. From HOLDING, the operator may attempt to restart the procedural element. The RESTARTING state logic is designed to recover from an exception condition once it is corrected and to continue the RUNNING logic. Dividing the running logic into steps will help facilitate the restart at a safe restart point. The ABORTING state logic is used to place the batch plant system in a state that can be reset to the IDLE state and allow the system to be used for the next batch. The PAUSING and PAUSED state logic allows operating personnel the ability to temporarily pause the RUNNING logic. The RUNNING logic stops at a safe step transition point. The exception condition monitoring continues uninterrupted during the PAUSING and PAUSED states.

## 9.5 COORDINATION CONTROL

Another consideration in batch automation is the coordination of plant systems to make the batches. In multiple-stage batch applications where batches are to be automatically transferred between process units, there is a need to make sure the transfer is done correctly. Preventing material contamination in both the transfer systems and the downstream units is a common requirement.

There are two basic ways to implement batch transfer control. One way is to make the transfer plant system (e.g., the pumps, on–off valves, flow control, and piping manifold) an equipment module and enforce a one-path-only routing. The other way is to set up cooperating transfer phases in the cooperating unit procedural control.

In some plants the process units and cells must share plant systems. This situation requires coordination control to make sure the shared systems are not misused or overloaded. This may be as simple as a shared utility such as steam or water or as complex as batch day tanks that are in themselves batch units requiring procedural control.

The most common method of implementing shared resource coordination control is by using a technique employed in many computer operating systems to coordinate the use of printers. This technique assigns a semaphore to each shared resource. Each program task that requires the use of a printer, for instance, must acquire the resource and then release it for other tasks to use. If a shared plant system's control logic manages the requests for its use, various resource arbitration schemes can be utilized for coordination control of shared resources.

## 9.6 RECIPE CONTROL

As defined in Section 9.3, executing a recipe makes a batch. The parts of the recipe that define the method to make a batch are the formula, procedure, and equipment information. Some batch products may vary only in the formula of materials. The procedural control is fixed. In more complex batch processing applications, each product may require a different recipe with its own formula and procedure. Sometimes the same formula and procedure must accommodate different plant system characteristics such as equipment sizes, paths, and processing parameters. The complexity of recipe control is dictated by the flexibility of these parts of the recipe.

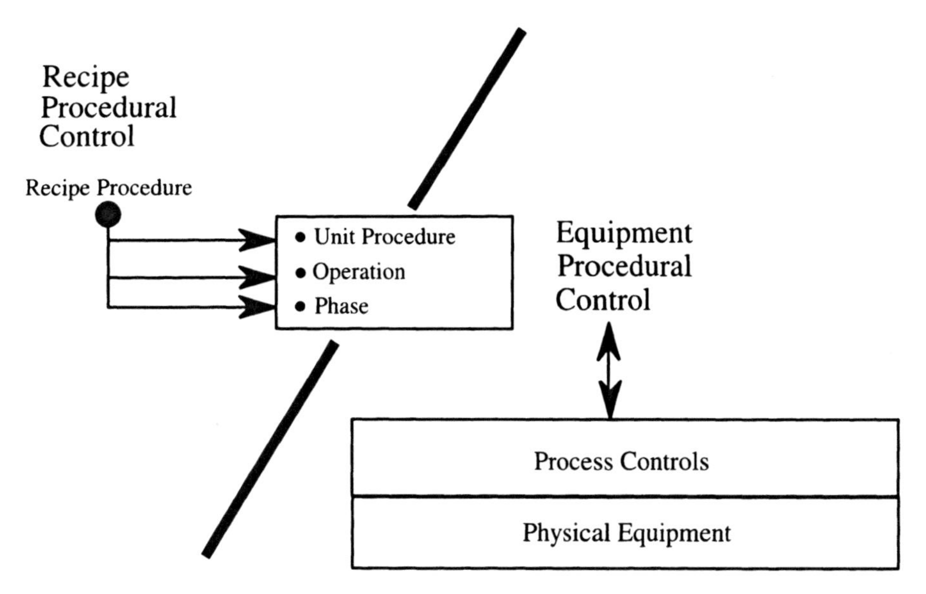


**FIGURE 9.5.** Recipe/equipment batch control. (Reprinted by permission. Copyright © 1996, Automation and Control Technologies, Inc.)

As described earlier in this chapter, recipe procedural control is the control type that provides flexibility in the operation of plant production to produce products. The recipe control must allow for the degree of flexibility required to fulfill the designed purpose of the plant or process unit. In addition, there is always some degree of equipment control. The key to the design and implementation of batch control is determining the line between recipe and procedural control, shown in Figure 9.5. At some point in the control hierarchy, the equipment control stops and recipe control is employed. Too much recipe control and the process may be unsafe to operate. Too much equipment control and the process will not be agile enough for flexible manufacturing.

The required flexibility, therefore, is defined by the recipe. There may be a need for flexibility in the recipe formula. The application may require both flexible formula and procedures. In addition, the application may require flexible batch path selection.

The S88.01 recipe model is composed of four types of recipes: (1) general, (2) site, (3) master, and (4) control, shown in Figure 9.6. The general recipe describes how to make a batch without regard to the equipment available at a specific site. At the plant site, the general recipe is converted into a site-specific version by naming specific process cells and equipment entities at the site. The process operations and actions are converted to site-specific terms. The basic structure of general and site recipes is the same.

The site recipe management personnel transform the site recipe into master recipes. The master recipe is the specific instructions to make a batch of product. The master recipe is converted to a control recipe when the time to make a batch arrives. As a result, the master and control recipes have the same structure. The control recipe can be considered a copy of the master recipe but one that is linked directly to the basic controls and selected process equipment that will be used to make a batch. The data recorded about the batch and the information contained in the control recipe form the batch history record.

The S88.01 master recipe, and therefore the control recipe, consists of the following information:

*Header.* The data that describe the recipe. It also contains version information, authorship, and other important information.

*Formula.* The list of recipe parameters that define the values of product inputs, processing parameters, and product outputs for use in making a batch according to the procedure.

*Procedure.* The operating sequence to make a batch.

*Equipment Requirements.* The information about batch sizing, equipment constraints, and other information for use in making a batch according to the recipe's formula and procedure.

*Other Information.* Anything not covered in the previous information but pertaining to making a batch of product.

There is considerable flexibility in the definition of the master recipe. As a result, the concept of S88.01 compliance is not really possible. This model is more for guidance in developing technology than in assuring interoperable systems. The real value of the S88.01

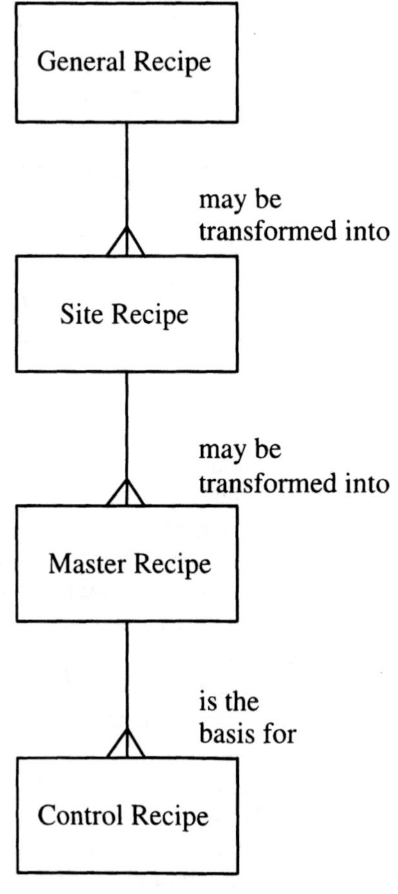


**FIGURE 9.6.** Recipe types. (Copyright © 1995, ISA, all rights reserved. Reprinted by permission with changes. From S88.01.)

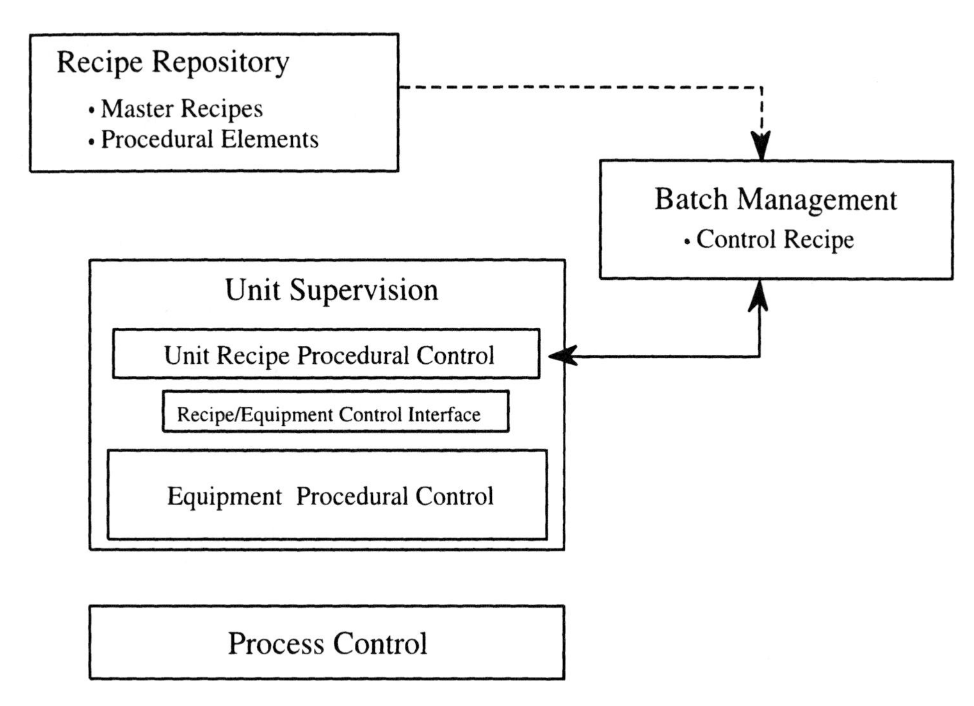


**FIGURE 9.7.** Recipe control. (Reprinted by permission. Copyright © 1996, Automation and Control Technologies, Inc.)

standard is in establishing a common frame of reference so that the industry can communicate better about the requirements of batch control systems. The users can express their requirements more precisely and suppliers can deliver open batch control software products. As of this writing, the results are encouraging.

The S88.01 standard envisions a recipe management system that provides the tools to develop master recipes from the requirements expressed in general and site recipes, shown in Figure 9.7. Part of this management system includes a repository of predefined procedural elements. This repository provides phases, operations, and unit procedures that could be invoked to enforce proven techniques in making a batch. The user could start with a site recipe and construct a master recipe from the procedural element library, modifying or adding as required to fulfill the site recipe intent. This vision has been realized in many commercial, batch recipe control packages. The only limitation is the imagination of software engineers who are commissioned to implement the S88.01 standard.

## 9.7 BATCH CAMPAIGN MANAGEMENT

From a plantwide control viewpoint, batch products are made to fill a production run. A plant production run is the operation of a plant system to manufacture a quantity of product to fill a production order. The production order is the request that originates in the business unit to define the requirements to execute a run.

There are many types of plant production runs. Continuous, batch, or discrete process systems may execute runs. In many applications a run may be executed on a mixture of these system types.

Runs are scheduled for one of the following reasons: make-to-order, make-to-inventory, or some combination of these. Make-to-order runs fill a specific customer sales order while make-to-inventory orders produce product for general market demand requirements.

Some processes are designed to handle only one production run at a time. Oil refineries, for instance, usually make gasoline during part of the year and switch to heating oil for the remainder. Other examples of long, single-product production runs may be found in commodity, agricultural, and intermediate chemical manufacturing. In these cases, run plan management seeks to meet whole market requirements rather than specific customer orders. Run plans are composed of the required production rates and operating conditions to maintain production yield to build inventory that meets the expected market demand.

Other plants handle a diverse set of runs simultaneously on parallel production lines. The runs may change frequently, even several times per shift on each line. This run management requirement is common in plastic resin compounding, food, beverage, specialty chemicals, and consumer product manufacturing.

In a batch process, a run is referred to as a campaign. In some cases, the overall campaign is subdivided into smaller runs called lots. Generally speaking, a batch lot is a portion of a campaign that is filled by one or more batches of product in a process cell using the same control recipe.

The lifecycle of a run can be defined in terms of its state. For example, a run starts in the Backlog state. Once the physical preparations are begun, it is considered Committed. This means that aborting the run may cost the enterprise in lost material, labor, or production. From the Committed state the run proceeds to the Active state on a line where actual product manufacturing is performed. Once the order requirements are fulfilled, it is considered Complete. A record of the completed run is saved for future reference.

Executing a run to fill a production order is usually the responsibility of plant operations. Run management may be totally manual or supported by computer automation. Run management automation is emerging as a need in today's global and agile manufacturing environment.

A run plan is the specification of production requirements, such as formulation, procedure, equipment, and/or schedule to fill a production order. Executing a run plan performs a plant production run. In batch processing the run plan is called the recipe.

The typical run planning activities in a plant are:

- Prepare the run plan and update schedule.
  - Determine formulation to fill the production order quality requirements.
  - Determine production rate to fulfill the production order timing requirements.
  - Determine the plant equipment requirements with current constraints.
  - Determine availability of on-hand raw material inventory levels.
  - Determine product inventory storage availability.
  - Determine run schedule.
- Release run plan for execution.
- Periodically update the in-process run plan.
- Report run plan status to the business unit.

Batch campaign planning is similar but may require the preparation of new recipes to handle new product requirements in a multiproduct plant.

## 9.8 BATCH DATA MANAGEMENT

The collection of performance data is an important consideration in any plant system. In continuous plants the data consist of time-stamped events and time-series trends. Since many batch plants have continuous process units, this type of data collection is also required. In batch plants, however, the collection of batch records is important. This collection may occur before, during, and after the actual batch of material is produced. The batch history, or repository, is a collection of events, trends, and process data that represent the history of a batch that was produced, as shown in Figure 9.8. In addition to the trends and events from the continuous process control functions, batch information must be collected from the batch management and unit supervision functions. The typical batch data collection consists of batch timing events, recipe parameter values, phase profiles, and operator comments. These data are used to produce a batch end report and, in modern control systems, an electronic batch record.

## 9.9 BATCH USER INTERFACE FOR SETUP AND OPERATION

In batch control there are several user interface windows. At the lowest level of procedural control there may be a need to operate equipment modules individually and separate from the batch units. The batch units require a separate user interface. The other windows include the batch list for the process cell and the campaign user interface to initiate and control the execution of campaigns. The user interface is composed of equipment entity

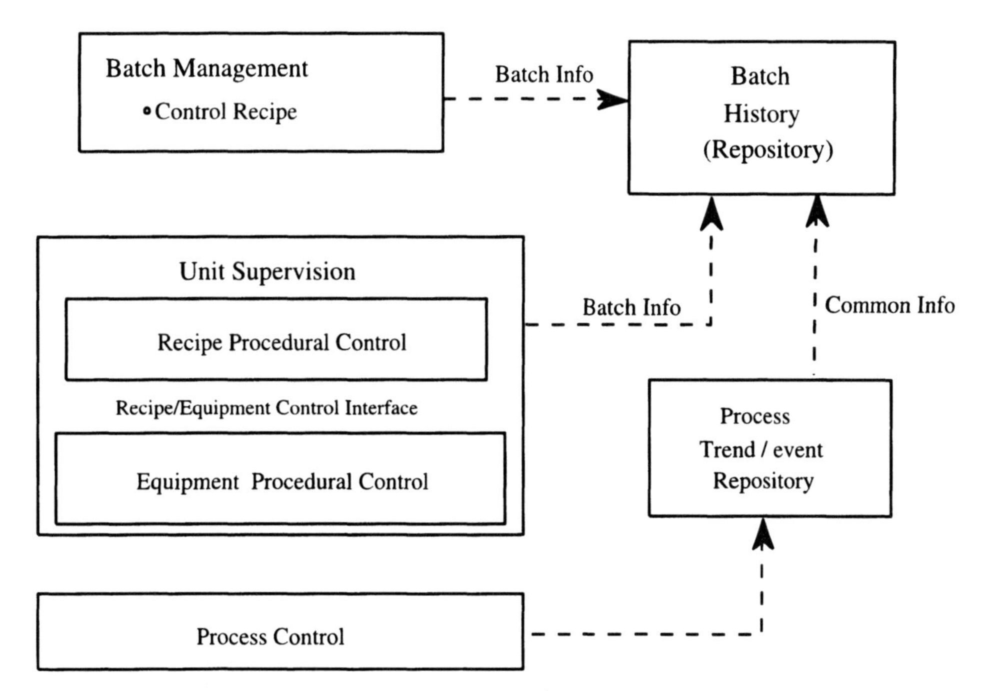


**FIGURE 9.8.** Batch data management. (Reprinted by permission. Copyright © 1996, Automation and Control Technologies, Inc.)

state variables and commands to manually operate the equipment module, unit, or process cell.

## 9.10 BATCH CONTROL EXAMPLE

The batch control example is documented as a control definition in the manner set forth in Chapters 2 and 3. First, the system is generally described and the control objectives are stated. As part of the control objectives, the operating states are defined and described. The Control Concept, which documents what is being controlled, describes the control as a hierarchy and the various automated/manual functions that must be performed in each layer. The control strategy delineates the control entities. Following the control strategy, more detailed control information is presented for the equipment modules up through the unit managers.

The plant system is a single process cell with three batch units, shown as a simplified process flow diagram in Figure 9.9. The process water, cooling water, reactant, material, and batch transfer headers are shown with more detail in the P&ID in Figure 9.10. A P&ID for Reactor 1 is shown in Figure 9.11. The other two reactors are similar.

### 9.10.1 Reactor Operation

The recipe formula parameters are:

Material 1 charge amount
Starting batch temperature

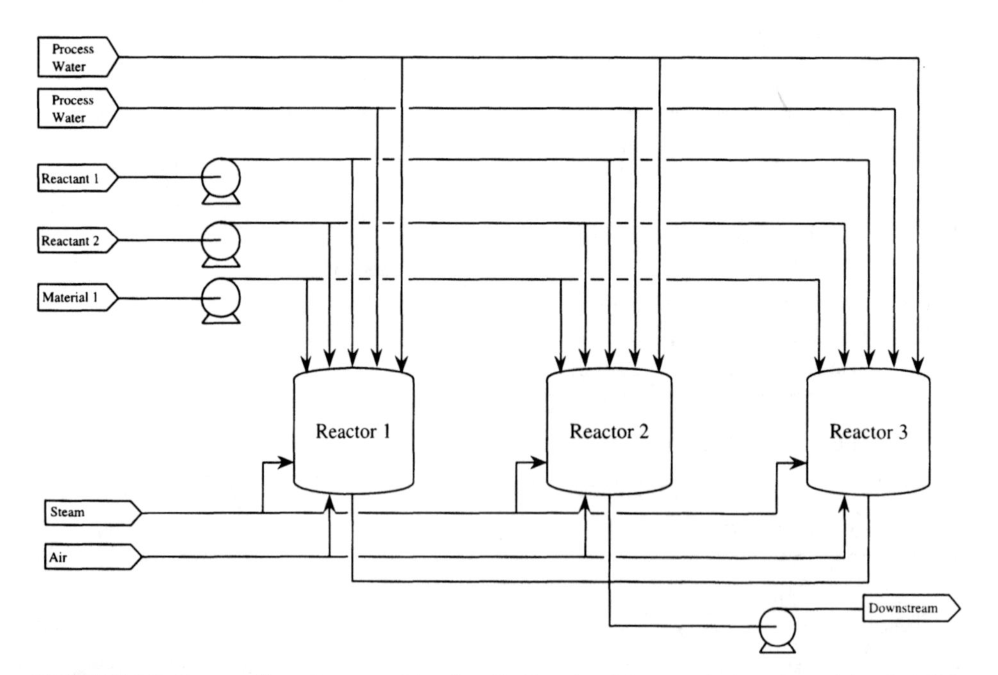


**FIGURE 9.9.** Process flow diagram of batch cell. (Reprinted by permission. Copyright © 1996, Automation and Control Technologies, Inc.)

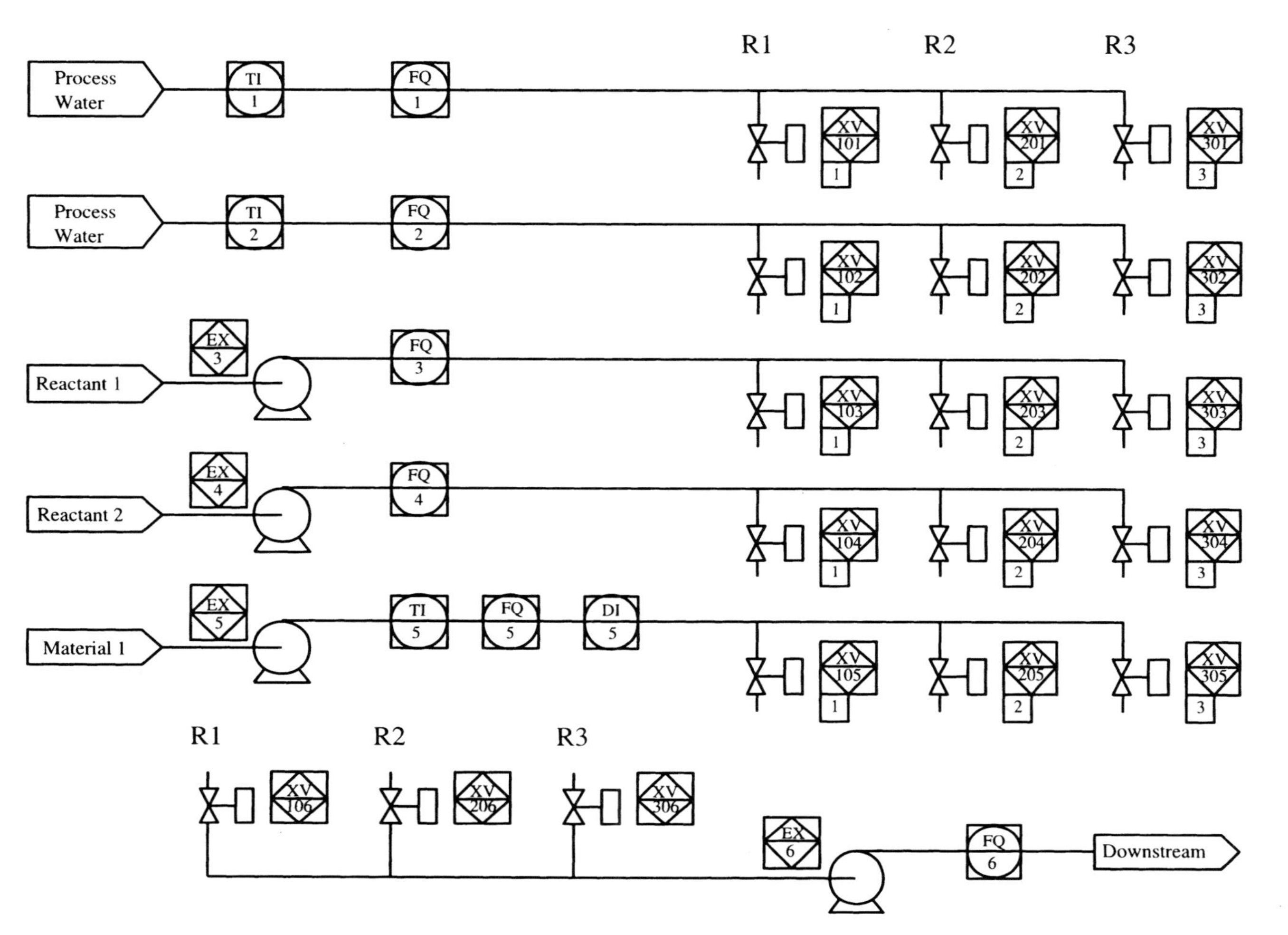


**FIGURE 9.10.** Batch cell headers P&ID. (Reprinted by permission. Copyright © 1996, Automation and Control Technologies, Inc.)

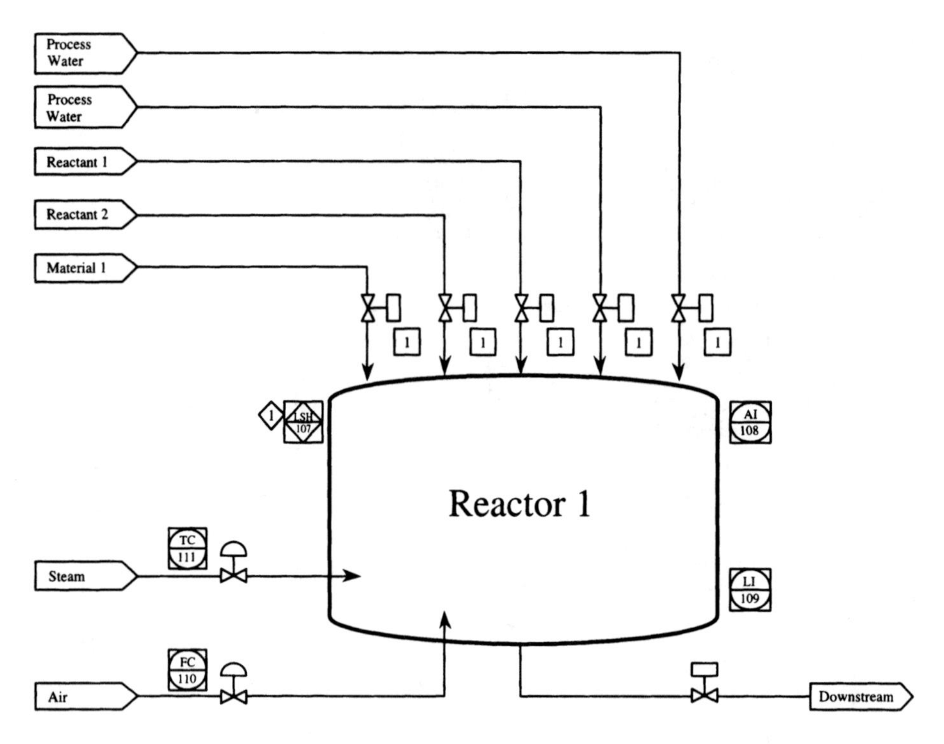


**FIGURE 9.11.** Reactor 1 P&ID. (Reprinted by permission. Copyright © 1996, Automation and Control Technologies, Inc.)

Batch size
Air flow setpoint for mixing
Reactant charge amount

The operating states for the reactor cell are shown pictorially in Figure 9.12 and are described as follows:

**Prestart State Activity**

Batch sheet data from laboratory
Material 1 charge amount
  Adjust for temperature and specific gravity.
Starting batch temperature (after first charge)
Total batch size
Air flow setpoint for mixing
Titration values before and after reactant charge
Transition: reactant charge amount

**Charging Material 1 State Activity**

Charge first materials.
If batch cooling is required, charge chilled water (usually required in summer).

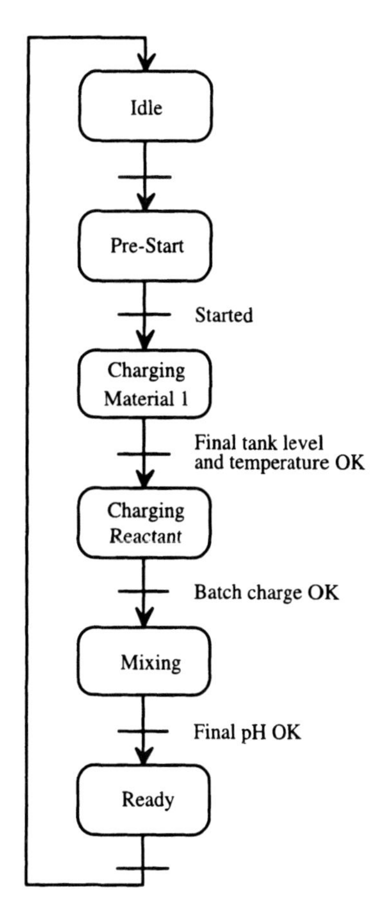


**FIGURE 9.12.** Operating states for reactor cell.

Record actuals and make sure Material 1 is charged.

If batch heating is required, use steam to heat the batch (usually required in winter).

Transition: final tank level and temperature OK.

### Charging Reactant State Activity

Do initial titration and record.

Charge reactant and agitate with air.

Record actual reactant amount.

After reactant is charged, do another titration and record.

Record initial pH.

Transition: batch charge OK.

### Mixing State Activity

Mix with air agitation.

Record batch sheet data.

Transition: final pH OK.

**Ready State Activity**

Transfer all or part of batch to product separating and finishing.

### 9.10.2 Control Concept

The Control Concept describes the automatic and manual control functions of the process control, unit supervision, and process management layers of the control activity model shown in Figure 3.18:

**Process Control**

Interlocks

Loop control

Manual operation from operator interface

  Open/close automated block valves.

  Start/stop pumps.

  Operate loop control.

  View process indicators and trends.

Semi-automatic batching (equipment modules)

  Set amounts and start/stop individual material charges.

  Set temperature endpoint and start/stop batch heatup.

  Set pH endpoint and start/stop automated mixing.

  Set amount and source reactor and start/stop batch transfer.

**Unit Supervision**

Perform full automatic caustic batch cycle.

Electronic batch sheet is provided with operator data entry.

Operator starts batch cycle on each reactor.

Adjust batch sheet charge amount as a function of measured specific gravity and temperature.

Use chilled-water charge if process water is too warm to achieve the required batch temperature.

Use heatup phase if process water is too cold to achieve the required batch temperature.

Charge Material 1 and water; add reactant.

Monitor transfer of reactor to downstream finishing.

Save and display batch sheet data versus actual for current cycle; reset at start of next cycle.

**Process Management**

Semiautomatic batching

  Set amounts and start/stop individual material charges.

  Set amount and source reactor and start/stop batch transfer.

### 9.10.3 Control Strategy

An overall diagram of the entities comprising the control strategy is shown in Figure 9.13. Note that the equipment modules are shared among all of the unit managers.

**Control Modules** (Also shown on P&ID)

Process water charge valves: XV101, XV201, XV301
Chilled-water charge valves: XV102, XV202, XV302
Reactant 1 charge valves: XV103, XV203, XV303
Reactant 2 charge valves: XV104, XV204, XV304
Material 1 charge valves: XV105, XV205, XV305
Batch transfer valves: XV106, XV206, XV306
Reactant 1 charge pump: EX3
Reactant 2 charge pump: EX4
Material 1 charge pump: EX5
Batch transfer pump: EX6

**Equipment Modules**

Charge process water
Charge chilled water
Charge Material 1
Charge Reactant 1
Charge Reactant 2
Batch transfer

**Unit Managers**

Reactor 1
Reactor 2
Reactor 3

The unit procedure for each of the three unit managers will be identical, except for the heatup and mix phases. These two phases are individual for semiautomatic control.

### 9.10.4 Detailed Procedural Control

The procedural control is detailed for the equipment modules, unit managers, and process manager.

***9.10.4.1 Equipment Modules.*** The equipment modules have a state diagram similar to the state diagram for unit managers in Figure 9.4, except that there is no PAUSING state. The general logic for each equipment module is described as follows:

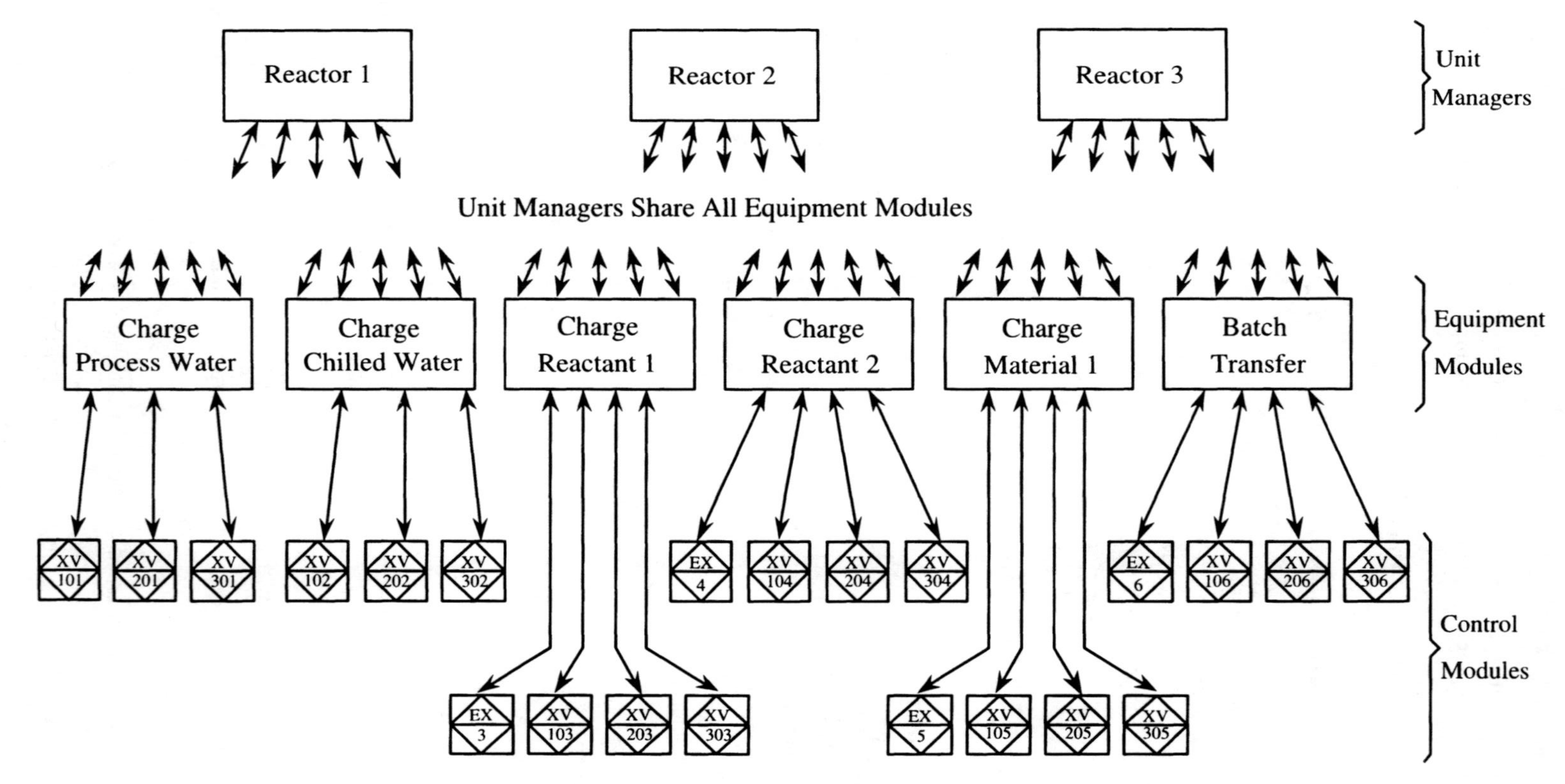


**FIGURE 9.13.** Overall diagram of the control strategy.

### General Equipment Module State Logic

- IDLE logic
  Initialize/reset operating parameters.
- RUNNING logic
  Set up and manipulate control modules.
  Wait for endpoint.
  Detect failure conditions.
- HOLDING logic → HELD
  Failure condition action.
- RESTARTING logic → RUNNING
  Failure condition recovery.
- STOPPING logic → COMPLETE
  Shut down control modules.

The state logic for each of the equipment modules that charge material into a reactor is very similar, so only the logic for the Charge Material 1 equipment module is described. The primary difference is in the tags of the control modules that are manipulated. In addition, the Charge Process Water and Charge Cooling Water equipment modules do not start a pump. A SFC of the RUNNING and STOPPING logic for the Charge Material 1 equipment module is shown in Figure 9.14.

### Charge Material 1

- IDLE logic
  Reset totalization and failure status.
- RUNNING logic
  Open reactor block valve, start pump and totalization.
  Endpoint: totalization $\geq$ Material 1 amount endpoint.
  Fail conditions: level too high detected for 5 sec; no flow detected for 5 sec
- HOLDING logic
  Close block valves, stop pump, and alert operator.
- RESTARTING logic
  Reset failure status and continue.
- STOPPING logic
  Stop pump, totalization, and close block valve.

The state logic for the Transfer Batch equipment module is described as:

### Transfer Batch

- IDLE logic
  Reset totalization, reactor, and failure status.
- RUNNING logic
  Open transfer block valve.

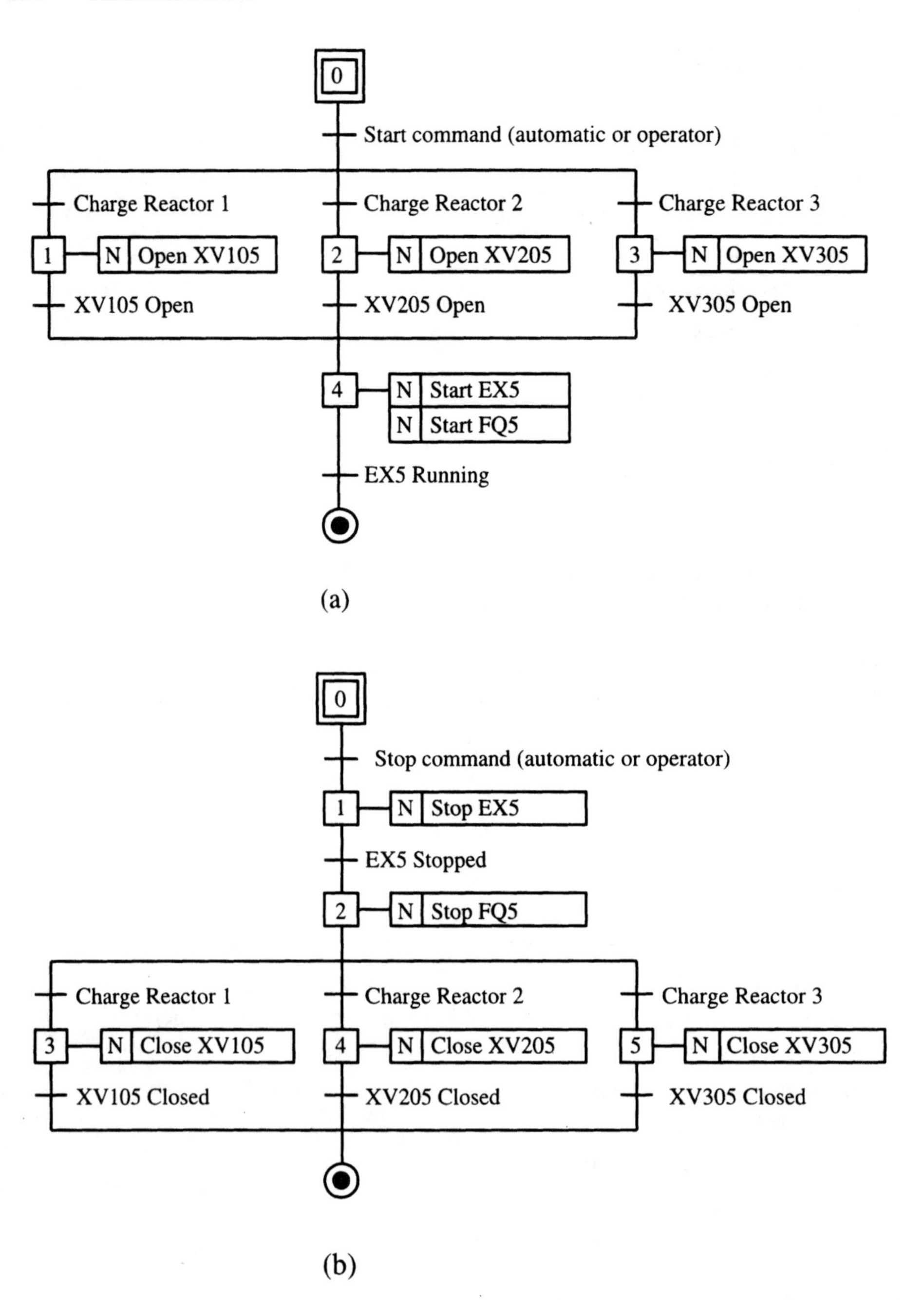


**FIGURE 9.14.** SFC for Charge Material 1 (*a*) RUNNING; (*b*) STOPPING.

Start pump and totalization.

Endpoint: totalization $\geq$ transfer amount.

Fail conditions: source reactor not ready;
no change in reactor level detected for 15 sec,
no flow detected for 5 sec.

- HOLDING logic
  Stop pump and alert operator.
- RESTARTING logic
  Reset failure status and continue.
- STOPPING logic
  Stop pump and totalization.

*9.10.4.2 Unit Managers.* There are three unit managers, one for each reactor. For a given batch type, the unit procedure used by each of the three unit managers will be identical, except for the heatup and mixing phases. These two phases are individual for semiautomatic control. The three unit managers must also coordinate the shared equipment modules. The three reactor unit managers have the state diagram shown in Figure 9.4. The general logic for each unit manager is described as follows:

**General Unit Manager State Logic**

- IDLE logic
  Clear previous cycle operating data.
- STARTING logic → RUNNING
  Do prestart checks and set up actions.
- RUNNING logic
  Execute operation logic and monitor for failures.
- PAUSING logic → PAUSED
  Running logic at step change and pause requested.
- HOLDING logic → HELD
  Perform step fail actions.
- RESTARTING logic → RUNNING (at retry step)
  Perform failure recovery actions.
- ABORTING logic → ABORTED
  Perform abort actions.
- STOPPING logic → COMPLETE
  Do setdown actions.

The STARTING and RUNNING logic is described as an operation in terms of the phases and steps that are executed. These operations are described below, followed by a description of the state logic common to all steps. Pictorially, the Running operation is shown as phases in Figure 9.15. The first three phases of the Running operation are shown as SFCs in Figures 9.16–9.18.

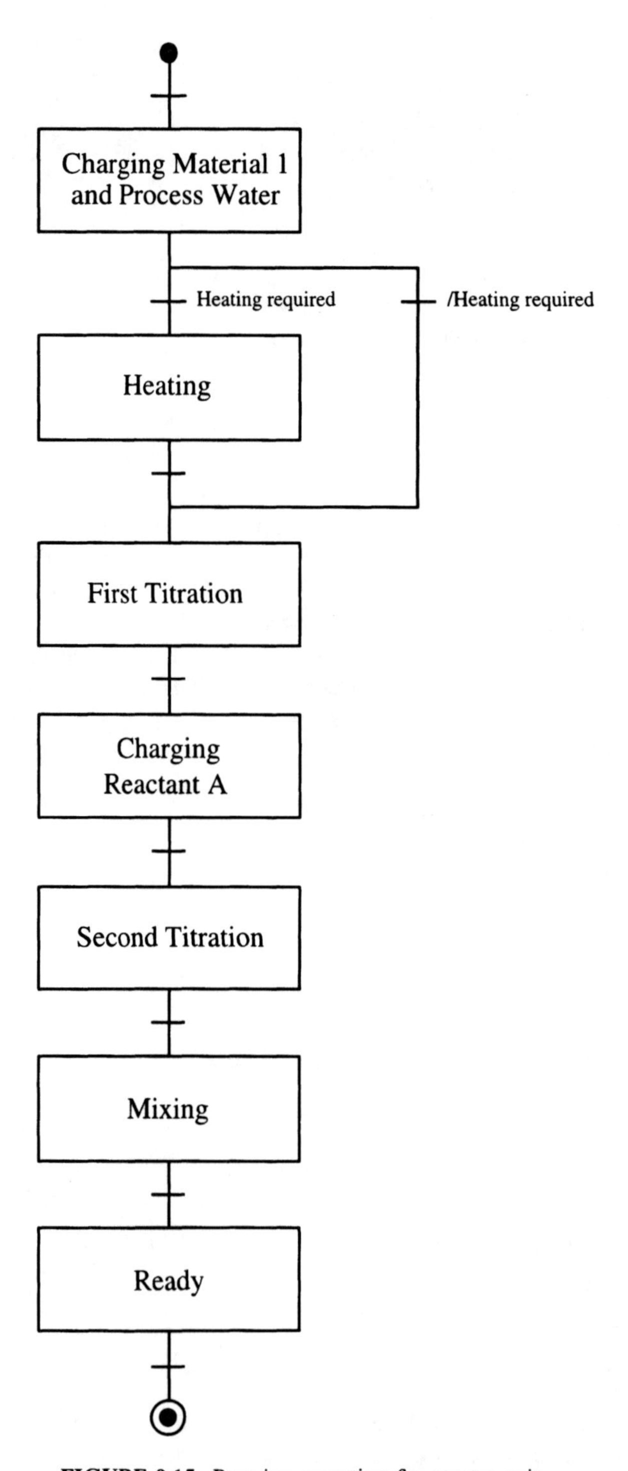


**FIGURE 9.15.** Running operation for reactor unit.

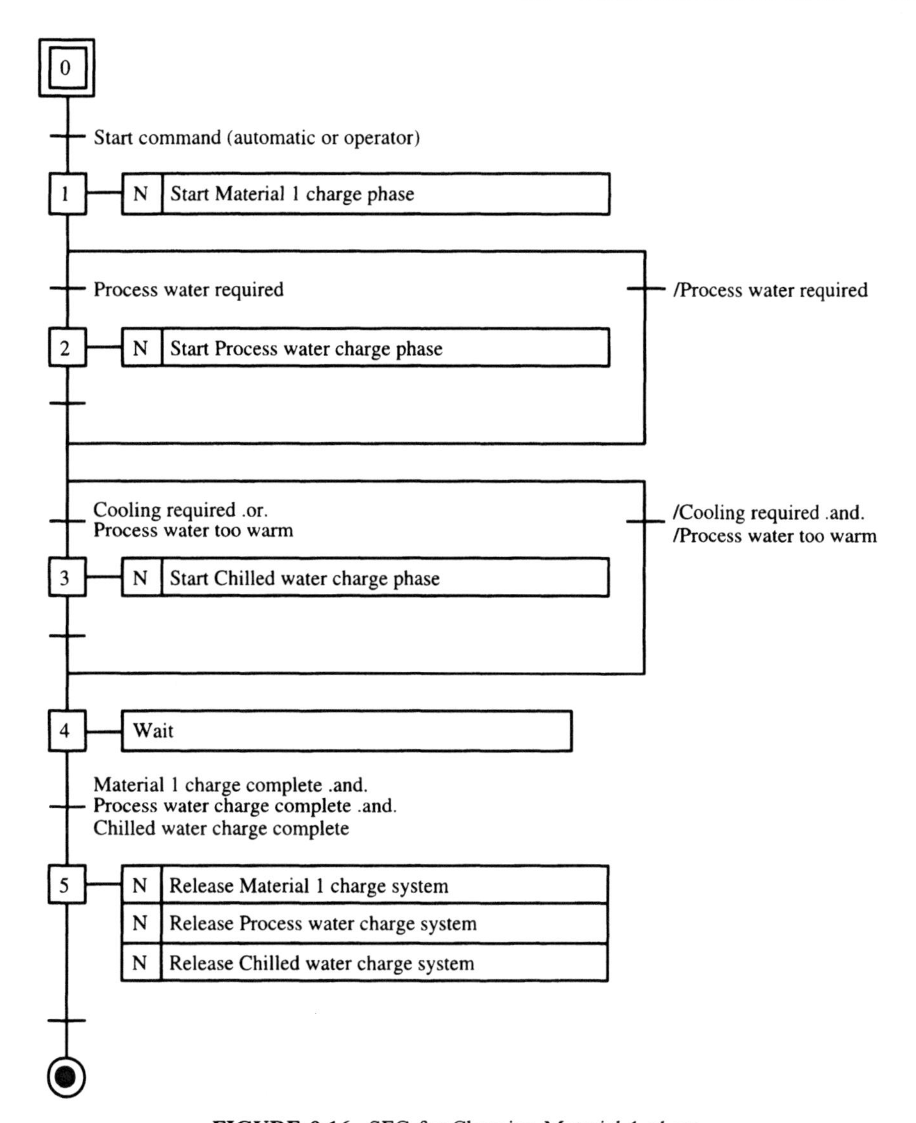


**FIGURE 9.16.** SFC for Charging Material 1 phase.

## Operation: Starting

*Step 1*

Adjust Material 1 charge amount for temperature and specific gravity.

Calculate required water temperature to achieve desired batch temperature after mixing with Material 1.

If process water is too warm, calculate amount of chilled water to charge with process water (if still too hot, use chilled water only).

If process water is too cold, use heatup phase after water and Material 1 are charged.

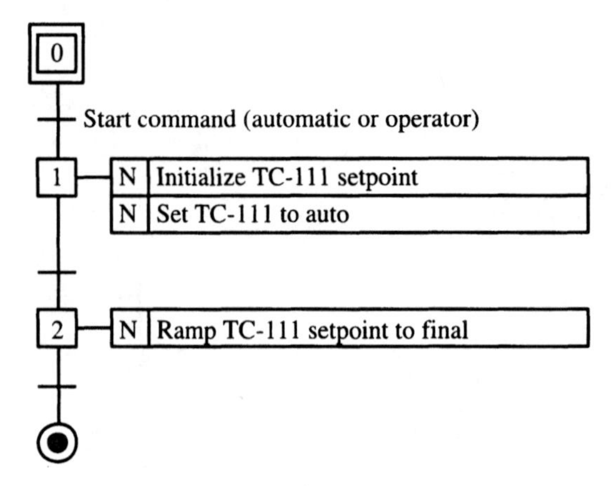


**FIGURE 9.17.** SFC for Reactor 1 Heating phase.

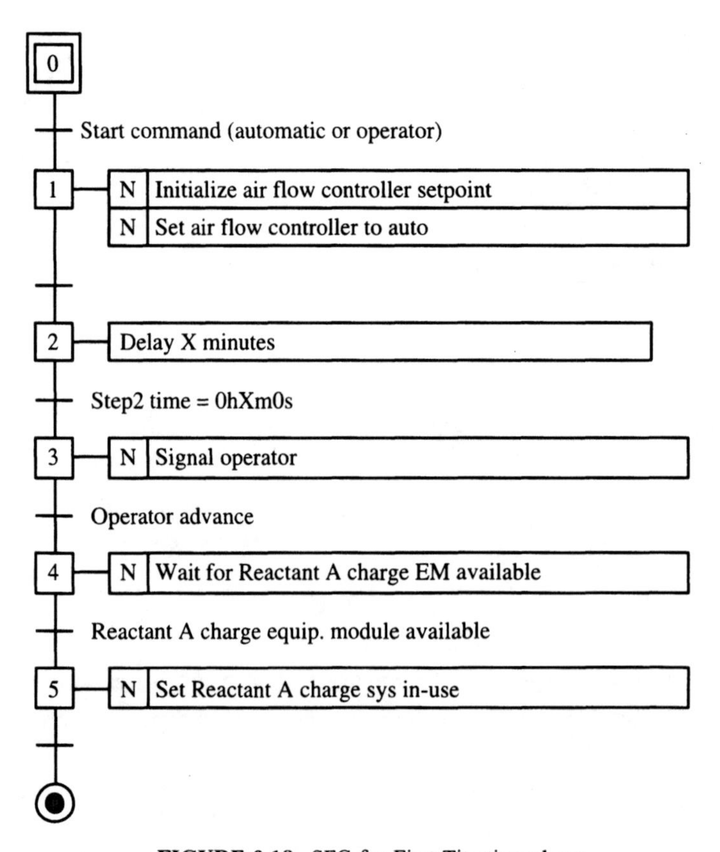


**FIGURE 9.18.** SFC for First-Titration phase.

*Step 2*

Wait until Material 1 and water charge systems are available.

Set Material 1 and water systems in-use.

Initialize equipment modules and control modules.

### Operation: Running

*Step 1: Charging Material 1 and Process Water*

Start Charge Material 1 phase.

If required, start Charge Process Water phase.

If cooling is required or process water is too warm to use, start Chilled Water phase.

Monitor batch temperature, calculate endpoint, and indicate to operator.

Monitor Material 1 charge and alert operator if it will not finish before final level is reached.

When equipment modules are complete, release Material 1 and water charge systems.

Fail conditions:

phases in HELD state.

*Step 2: Heating*

Skip if no heating required.

Initialize and start Unit Heatup phase.

Fail conditions:

phase in HELD state.

*Step 3: First Titration*

Mix for $X$ minutes.

Signal operator and wait for operator advance.

Wait until Reactant A (1 or 2) charge system is available.

Set Reactant A system in-use.

Fail conditions:

none.

*Step 4: Charge Reactant A*

Initialize and start Charge Reactant A phase.

Fail conditions:

phase in HELD state.

*Step 5: Second Titration*

Start Reactor Mix Batch phase.

When complete, signal operator and wait for operator advance.

Fail conditions:

phase in HELD state.

*Step 6: Mixing*

Release Reactant A charge system.

Wait until pH $\leq$ warning level; alert operator.

Wait until pH $\leq$ batch endpoint; do next step.

Fail conditions:

phase in HELD state.

*Step 7: Ready*

Enable batch transfer for this unit.

Monitor reactor level during transfer; if it does not change in $X$ seconds, alert operator.

Endpoint: Wait for operator advance; go to STOPPING state.

**Reactor Unit Procedure State Logic common to all steps**

- PAUSING logic
  Alert operator of failure and wait for operator resume.
- HOLDING logic
  Alert operator and save step for restart.
- RESTARTING logic
  If required, restart appropriate phases.
  Reset appropriate unit variables.
  Return to restart step.
- ABORTING logic
  Release any charge systems in use.
  Reset any HELD phases.
- STOPPING logic
  Go to COMPLETE state.

***9.10.4.3 Process Manager.*** The process manager carries out the batch management functions, including the following:

User interface

Batch data

Batch initiation

Batch recipe entry

The operator does the following functions:

Initiate the batch.

Enter batch sheet recipe data, one per reactor (option to use previous batch formula data).

Data entry during batch:

Laboratory analysis values

Comments

**Automated batch data collection:**

Unit manager collects batch data.

Batch sheet data.

Previous batch data reset at beginning of next cycle.

## REFERENCES

Instrument Society of America, *ISA-S88.01, Batch Control, Part 1: Models and Terminology*, Instrument Society of America, Research Triangle Park, NC, 1995.

National Institute of Standards and Technology, *Integration Definition for Function Modeling (IDEF0)*, Federal Information Processing Standards Publication 183, December 21, Washington, D.C., 1993.

U.S. Patent Office, Washington, D.C., copied from http://www.uspto.gov, 1996.

# 10 Case Study: Pulp and Paper Mill

This chapter presents a control requirements definition (CRD) for an actual pulp and paper plant. Because of space limitations, it is not complete but gives the flavor of the type of documentation needed for a plant. In particular, the descriptions are generally shorter than one would use in the front-end design of an actual control project, as described in Chapter 2. The organization is complete down to the unit level for most of the pulp mill area. In addition, control details are presented for some of the units. The information furnished in this case study is believed to be accurate and reliable; however, no responsibility is assumed for any errors.

As outlined in Chapter 2, the CRD has three main parts: the process operation description (POD), the Control Concept, and the control strategy. The CRD is organized by process cell. For a process cell, the POD is done in a top-down manner, with the cell POD described first, followed by a POD for each unit in the cell. Each POD contains a description of the operating states that describe what the entity (cell or unit) is doing when that state is active. Each operating state is described with the following subheadings:

*Routine Activities*

If the activities contain steps, the steps are listed and their sequence is indicated. If a unit is being described, these steps often become the operations and/or phases in the unit procedure.

*Exception Handling*

Abnormal conditions and the response (e.g., shutdown if temperature $> 200°C$).

*Primary Control Objectives*

The control objective is the primary objective of the manual and/or automated control strategies. It is the goal of a process action in terms of physical property quantities (e.g., product composition, material amounts, molecular structure, specific gravity) and/or time-domain quantities (e.g., mixing time of 1 hr, rinse for 20 min). Performance tolerances are sometimes used (e.g., maintain product composition of 30% A and 70% B within $\pm 2\%$). A control objective may apply to more than one operating state.

*Performance Information*

The key plant information quantities that indicate past, present, or future performance of an operating module. These are used to determine the process capability for the operating state or for the operating module as a whole. These include quality and production rate measurements whether taken manually or automatically.

*State End Conditions*

Conditions that cause a transition to the next operating state.

After the POD, the Control Concept is described. The Control Concept includes the cellwide process management and the unit supervision for each unit. The Control Concept is described in a top-down manner, like the POD. However, it is usually written in a bottom-up fashion. That is, the Control Concept for the units is usually developed before the Control Concept for the cell is complete. The Control Concept defines the extent of automation in terms of control requirements for the operating modules (units, cells). The Control Concept for a cell or unit is described in outline form with the following subheadings:

*Extent of Automation*

Operating module activities and control objectives that are automated.

*Flexibility of Automation*

Multiple products, frequent equipment changes, and so on.

*Control Activity Coordination*

Concurrent operations and shared resources.

*Interaction with Operating Personnel*

Data presentation, equipment selections, control commands, and data entry.

The control strategy defines the control types and interrelationships to fulfill the control requirements. For process cells, any high-level control, such as multiple-unit coordination or scheduling, is detailed. For units, any advanced control algorithms are described. Loops and devices are described in terms of how they are to be operated (e.g., start/stop in certain unit procedure phases and maintain pressure during a certain unit phase). Status and indicator devices are described. Interlocks are also documented in the control strategy. The control strategy information is organized in a top-down fashion but may be easier to formulate bottom-up. In other words, it may be easier to start describing the control strategy in terms of the individual loops and devices and then proceed to the unit supervision.

## Site

A typical site of a paper production enterprise is shown in Figure 10.1 divided into plant areas and process cells (Biermann, 1996; MacDonald and Franklin, 1969, 1970). The site is divided into four plant areas: pulp mill, paper mill, powerhouse, and warehouse. In addition, the pulp mill area is subdivided into the process cells of woodyard, brown pulp mill, groundwood mill, caustic, and bleach. An overall process flow diagram (PFD) of the pulp mill area is depicted in Figure 10.2. The paper mill area is subdivided into two paper machine cells and a packaging cell. The site is broken down into areas, cells, and units as follows:

| Plant Area | Cell | Unit |
|---|---|---|
| Pulp Mill | Woodyard | Chip Unloading<br>Chip Screening<br>Log Handling |
| | Brown Pulp Mill | #1–#13 Digesters<br>Softwood Brown Stock Washer<br>Hardwood Brown Stock Washer<br>Blow Heat Evaporator |
| | Caustic | Green Liquor<br>White Liquor<br>Lime Mud<br>#1, #2 Lime Kilns |
| | Bleach | Softwood Bleach<br>Hardwood Bleach<br>Chlorine Dioxide Generator |
| | Groundwood Mill | Grinding<br>Screening<br>Cleaning |
| Paper Mill | #1 Paper Machine | Stock Preparation<br>Wet End<br>Dryer<br>Supercalender<br>Coater |
| | #2 Paper Machine | Stock Preparation<br>Wet End<br>Dryer<br>Extruder |
| | Packaging | #1, #2 Slitters<br>Wrapper |

yyyymmdd hh:mm | Pulp and Paper Co | Page 1

## Site

| Plant Area | Cell | Unit |
|---|---|---|
| Powerhouse | Power<br>Recovery<br>Bark Boiler | #1–#3 Power Boilers<br>#1, #2 Recovery Boilers<br>Bark Boiler |
| Warehouse | | |

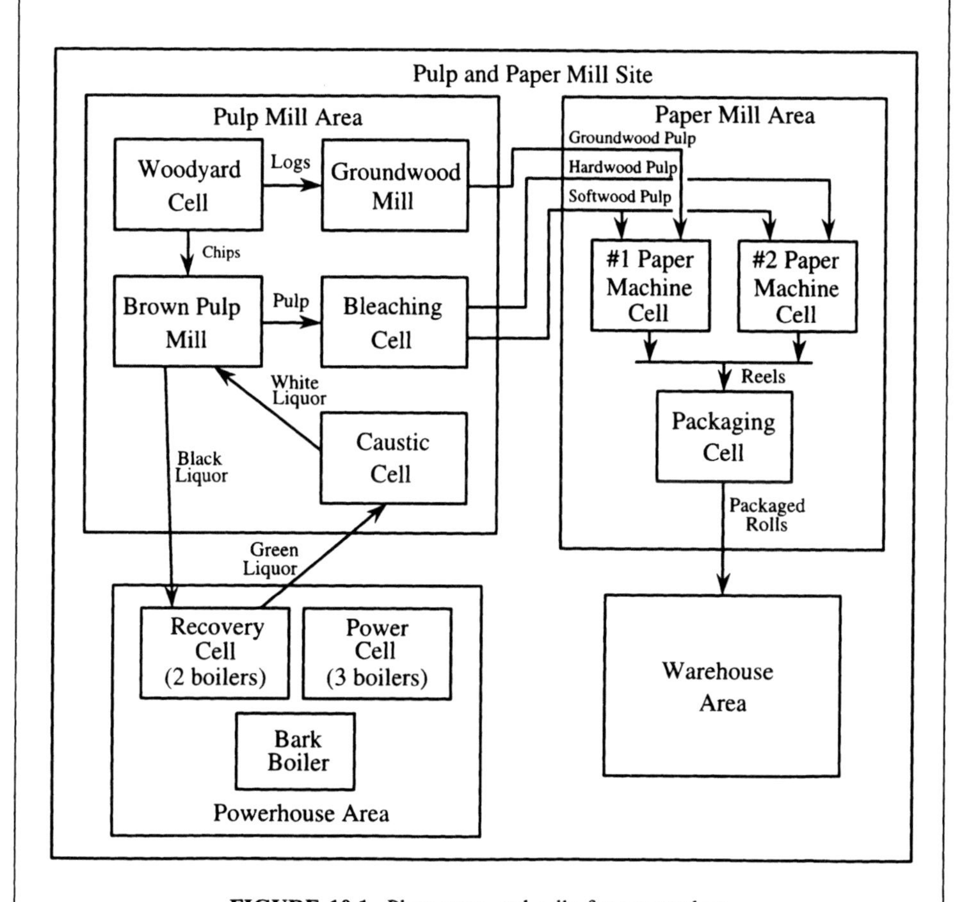


**FIGURE 10.1.** Plant areas and cells for paper plant.

| POD | Area: Pulp Mill Cell: Unit: |
|---|---|

**PULP MILL AREA**

The pulp mill area is subdivided into the process cells of woodyard, brown pulp mill, groundwood mill, caustic, and bleach (Figure 10.2). The POD, Control Concept, and control strategy are developed for all of these areas, except for the groundwood mill. For the groundwood mill, only the POD and Control Concept at the cell level are developed.

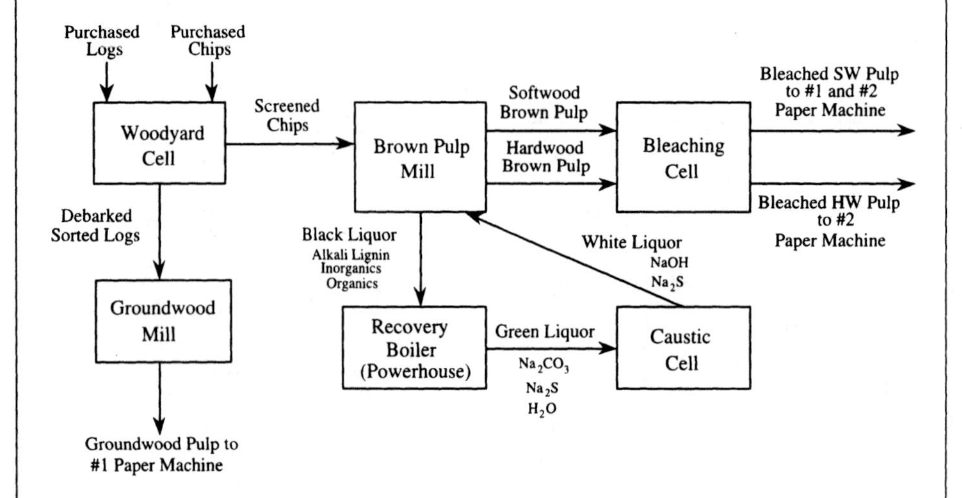


**FIGURE 10.2.** PFD for pulp mill area.

| POD | Area: Pulp Mill Cell: Woodyard Unit: |
|---|---|

**Woodyard Cell**

As shown in Figure 10.3, the woodyard handles logs for the groundwood cell and chips for the brown pulp mill cell. Purchased logs are removed from trucks and stacked for temporary storage before being fed into a barker. From the barker, the logs are sorted and stored in small stacks before being conveyed to the groundwood cell. Purchased chips are unloaded from one of two truck dumps or a rail dump. The east truck dump is used only for hardwood chips and the west truck dump is used only for pine chips. The chips from the railcar dump can be conveyed to either chip pile. In addition, a bypass path exists for the chips to be conveyed directly to the chip screening unit. The screening unit processes chips by ensuring that they fall within an acceptable size range before being stored in silos. Large chips are broken down into an acceptable size before being stored in a silo.

The operating states of the woodyard cell are shown in Figure 10.4:

Idle

Running

The various units operate relatively autonomously and so there is no special logic to transition between states. If any of the units are not idle, the cell is running. If all units are idle, the cell is idle.

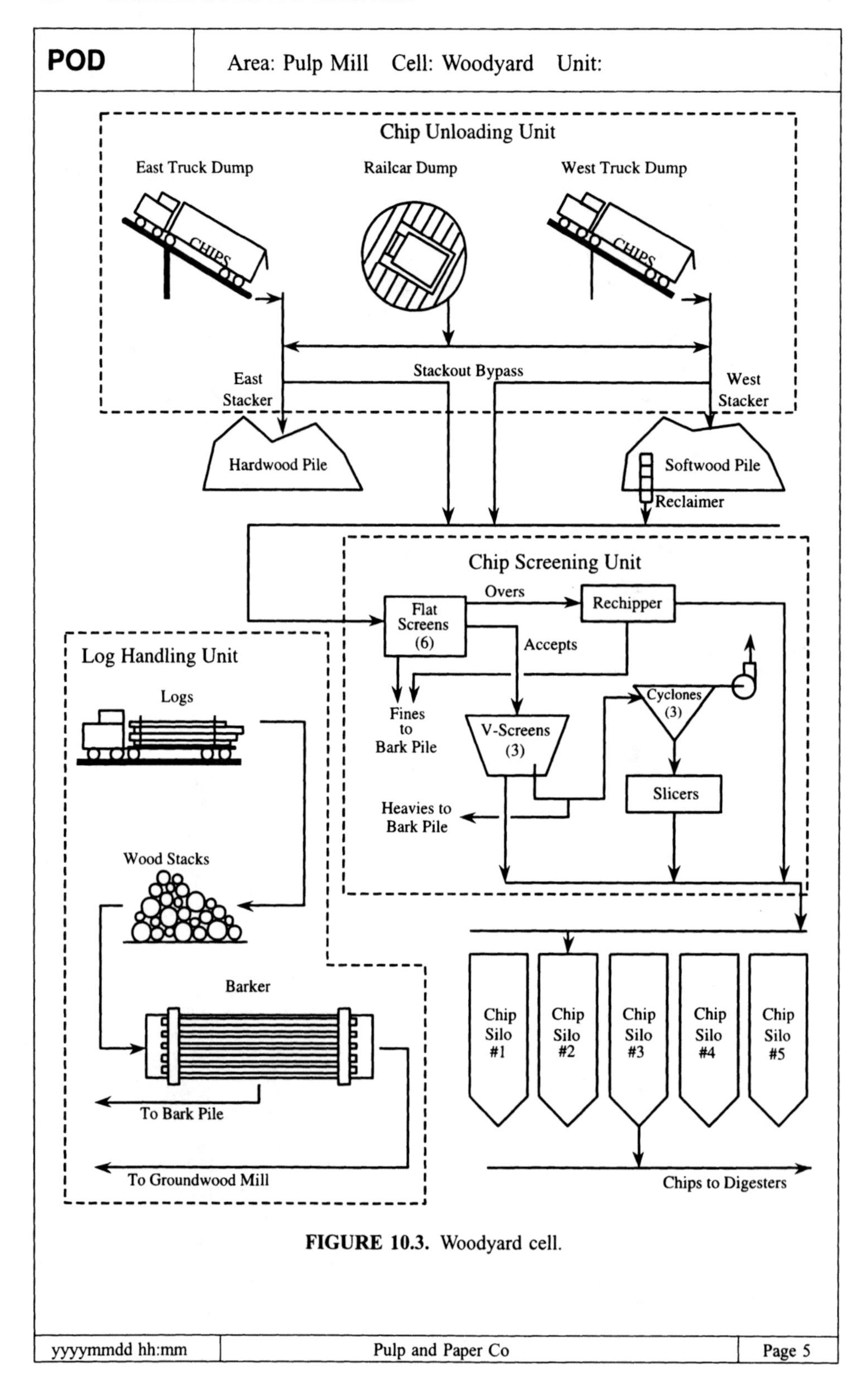


**FIGURE 10.3.** Woodyard cell.

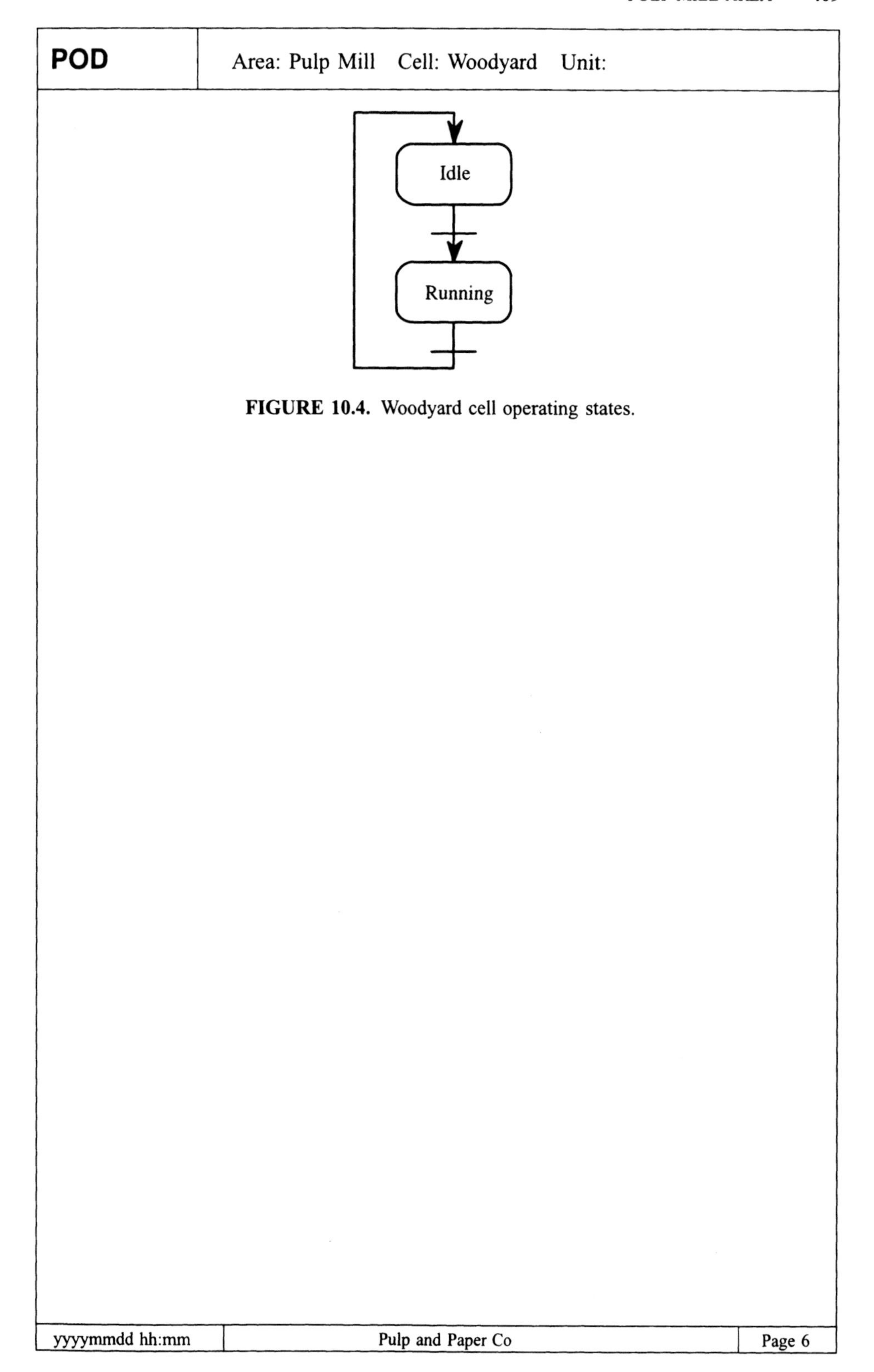

**POD** | Area: Pulp Mill Cell: Woodyard Unit:



**FIGURE 10.4.** Woodyard cell operating states.

yyyymmdd hh:mm | Pulp and Paper Co | Page 6

| POD | Area: Pulp Mill Cell: Woodyard Unit: Chip Unloading |
|---|---|

*Chip Unloading Unit.* In operation, this unit resembles the material transfer unit of Example 8.5 with the additional complication that there are multiple conveyor paths. Purchased chips are unloaded from one of two truck dumps or a rail dump. The truck dumps operate by tilting the truck-trailer, allowing the chips to fall out into a hopper, where the chips are conveyed to a destination (pile or chip screening unit). The rail dump is a rotary dumper that rotates the wood chip car, causing the chips to fall into a dump pit, and the chips are conveyed to the destination. The east truck dump is used only for hardwood chips and the west truck dump is used only for softwood chips. The chips from the railcar dump can be conveyed to either chip pile. In addition, a bypass path exists for the chips to be conveyed directly to the screening unit. The operating states of the chip unloading unit are shown in Figure 10.5:

Idle
Starting
Running
Shutting Down

Normal operation is continuous in order to supply the chip screening unit (if running) or the softwood and hardwood piles. The states are described as follows:

**Operating State Name: Idle**

*Routine Activities*

Clean equipment as needed.
Perform inspection.

*Exception Handling*

None.

*Primary Control Objectives*

None.

*Performance Information*

None.

*State End Conditions*

Startup requested.

| POD | Area: Pulp Mill Cell: Woodyard Unit: Chip Unloading |
|---|---|

**Operating State Name: Starting**

*Routine Activities*

Prestart checks.

Start conveyors for a source/destination.

*Exception Handling*

Shutdown if:

Conveyor motor fails.

Conveyor belt misaligned.

*Primary Control Objectives*

None.

*Performance Information*

None.

*State End Conditions*

Conveyors for a source/destination path started.

**Operating State Name: Running**

*Routine Acitivities*

Starting source/destination paths.

*Exception Handling*

Shutdown if:

Conveyor motor fails.

Conveyor belt misaligned.

High level on a dump station.

High level on chip pile.

*Primary Control Objectives*

None.

*Performance Information*

Chips unloaded per shift.

*State End Conditions*

Shutdown requested.

**Operating State Name: Shutting Down**

*Routine Activities*

Shutdown all source/destination path conveyors.

| POD | Area: Pulp Mill Cell: Woodyard Unit: Chip Unloading |
|---|---|

*Exception Handling*

None.

*Primary Control Objectives*

None.

*Performance Information*

None.

*State End Conditions*

All conveyors stopped.

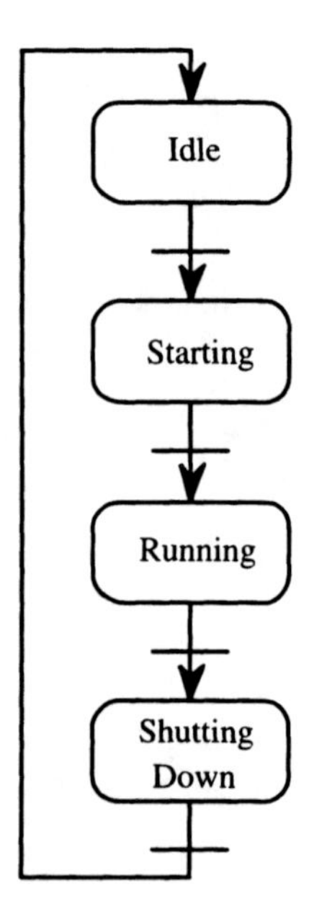


**FIGURE 10.5.** Operating states.

| POD | Area: Pulp Mill Cell: Woodyard Unit: Chip Screening |
|---|---|

*Chip Screening Unit.* Chips to be screened come either from the hardwood or softwood piles or directly from the dump stations. The screening unit processes chips by ensuring that they fall within an acceptable size range before being stored in silos. A set of flat screens initially sorts the chips into three size ranges. Fine chips are conveyed to the bark pile to be used as boiler fuel. Large chips ($> 2\frac{1}{2}$ in.) are sent through a rechipper to break them down to an acceptable size before being stored into the selected silo. Accepts from the flat screens are screened further by a set of V screens. Chips larger than 6 mm that are too dense are assumed to be bark and are conveyed to the bark pile. Chips larger that 6 mm and not too dense are conveyed to a slicer to reduce the chip size before being conveyed to the selected silo. The operating states of the screening unit are shown in Figure 10.5:

Idle
Starting
Running
Shutting Down

Normal operation is continuous in order to maintain adequate inventories of softwood and hardwood chips in the chip silos that supply the digesters in the brown stock cell. The Idle state is described in a similar manner as for the chip unloading unit and is thus not described. The other states are described as follows:

**Operating State Name: Starting**

*Routine Activities*

Prestart checks.
Start conveyors for a destination silo.
Start flat screens.
Start rechipper.
Start cyclone blowers.
Start V screens.
Start slicers.

*Exception Handling*

Shutdown if:

All flat screen motors fail.
Rechipper motor fails.
All V-screen motors fail.
All cyclone blowers fail.
Slicer motor fails.
Conveyor motor fails.
Conveyor belt misaligned.

| POD | Area: Pulp Mill Cell: Woodyard Unit: Chip Screening |
|---|---|

*Primary Control Objectives*

None.

*Performance Information*

None.

*State End Conditions*

All equipment started.

**Operating State Name: Running**

*Routine Activities*

Startup/shutdown individual flat screens.

Startup/shutdown individual V screens.

Startup/shutdown individual cyclone blowers.

*Exception Handling*

Shutdown if:

All flat screen motors fail.

Rechipper motor fails.

All V-screen motors fail.

All cyclone blowers fail.

Slicer motor fails.

Conveyor motor fails.

Conveyor belt misaligned.

High level on dump station when being fed from dump station.

High level in destination silo.

*Primary Control Objectives*

Produce $x$ tons per hour of screened hardwood chips.

Produce $x$ tons per hour of screened softwood chips.

*Performance Information*

Tons of hardwood chips screened per shift

Tons of softwood chips screened per shift

*State End Conditions.*

Shutdown requested.

**Operating State Name: Shutting Down**

*Routine Activities*

Stop all conveyors.

| **POD** | Area: Pulp Mill Cell: Woodyard Unit: Chip Screening |
|---|---|

Stop flat screens.
Stop rechipper.
Stop V screens.
Stop cyclone blowers.
Stop slicers.

*Exception Handling*

None.

*Primary Control Objectives*

None.

*Performance Information*

None.

*State End Conditions*

All equipment stopped.

| **POD** | Area: Pulp Mill Cell: Woodyard Unit: Log Handling |
|---|---|

*Log Handling Unit.* Purchased logs are removed from trucks and stacked for temporary storage before being fed into a barker. From the barker, the logs are conveyed to the groundwood cell. A small pile of debarked logs is maintained in order to provide some inventory for starting up the groundwood mill. The bark is conveyed to the powerhouse area for boiler fuel.

The states of the log processing unit are shown in Figure 10.5:

Idle
Starting Up
Running
Shutting Down

Normal operation is continuous in order to supply barked logs to the groundwood mill. The Idle state is described in a similar manner as for the chip unloading unit and is thus not described. The other states are described as follows:

**Operating State Name: Starting**

*Routine Activities*
Prestart checks.
Start barker.
Start conveyors.

*Exception Handling*
Shutdown if:
Barker motor fails.
Conveyor motor fails.
Conveyor belt misaligned.

*Primary Control Objectives*
None.

*Performance Information*
None.

*State End Conditions*
Conveyors started.

**Operating State Name: Running**

*Routine Activities*
None.

| POD | Area: Pulp Mill Cell: Woodyard Unit: Log Handling |
|---|---|

*Exception Handling*

Shutdown if:

Barker motor fails.

Conveyor motor fails.

Conveyor belt misaligned.

Groundwood mill shuts down.

*Primary Control Objectives*

Produce $x$ tons per hour of barked logs.

*Performance Information*

Tons of barked logs produced per shift.

*State End Conditions*

Shutdown requested.

**Operating State Name: Shutting Down**

*Routine Activities*

Shutdown conveyors.

Shutdown barker.

*Exception Handling*

None.

*Primary Control Objectives*

None.

*Performance Information*

None.

*State End Conditions*

Barker stopped.

| yyyymmdd hh:mm | Pulp and Paper Co | Page 14 |
|---|---|---|

| CONTROL CONCEPT | Area: Pulp Mill Cell: Woodyard Unit: |
|---|---|

**Woodyard Cell Process Management**

*Extent of Automation*

Collect performance data from units.

*Flexibility of Automation*

None.

*Control Activity Coordination*

Chip piles are shared by the unloading and chip screening units.

*Interaction with Operating Personnel*

Operators enter chip pile heights and log inventory estimates.

Summary information from units displayed for operator.

**Chip Unloading Unit Supervision**

*Extent of Automation*

Provide control of dump stations.

Provide automatic startup, shutdown, and path changes for the transfer of chips from the dump stations.

Collect performance data (tons of softwood and hardwood chips unloaded per shift).

*Flexibility of Automation*

None.

*Control Activity Coordination*

Path selection equipment modules will interact.

*Interaction with Operating Personnel*

Operators control raising/lowering of truck dump stations at the station.

Operators control rail dump station at the station.

Chip path selection done by operator. Interlocks prevent illegal selections.

Display status information from control modules.

**Chip Screening Unit Supervision**

*Extent of Automation*

Provide automatic startup and shutdown for chip screening.

Collect performance data (tons of chips screened per shift).

*Flexibility of Automation*

Either hardwood or softwood chips are screened during running.

In order to change the type of chip, the unit is shut down, the reclaimer is moved and then the unit is started up.

**CONTROL CONCEPT** | Area: Pulp Mill Cell: Woodyard Unit:

*Control Activity Coordination*

Path selection equipment modules will interact.

*Interaction with Operating Personnel*

Operator selects source pile (hardwood or softwood) and destination silo at startup.

Operator starts/stops individual flat screens.

Operator starts/stops individual V screens.

Operator starts/stops individual cyclone blowers.

Display status information from control modules.

**Log Handling Unit Supervision**

*Extent of Automation*

Provide control of barker.

Sort debarked logs by diameter onto one of three conveyors that supply logs for the groundwood mill.

Collect performance data (logs processed/shift).

*Flexibility of Automation*

None.

*Control Activity Coordination*

Startup/shutdown is coordinated with the groundwood mill.

*Interaction with Operating Personnel*

Operators control startup/shutdown of barker at a local operator station.

**CONTROL STRATEGY** | Area: Pulp Mill Cell: Woodyard Unit: Chip Unloading

*Chip Unloading Unit.* The individual conveyor devices are controlled by the conveyor path equipment modules. The conveyor path equipment modules are:

1. East truck dump to hardwood pile
2. East truck dump to screening unit
3. West truck dump to softwood pile
4. West truck dump to screening unit
5. Railcar dump to hardwood pile
6. Railcar dump to softwood pile
7. Railcar dump to screening unit via east conveyor
8. Railcar dump to screening unit via west conveyor

Each conveyor path equipment module has a Startup and a Shutdown phase that starts/stops the individual conveyor devices. The startup and shutdown of a path are similar to Example 8.5. The conveyor paths share the individual conveyor devices.

When a path is started, the equipment is started in backward order. For example, the last conveyor in a path is started first. When the path is shut down, the equipment is stopped so that all material on a conveyor is removed before that conveyor is stopped.

Certain paths are allowed to be running simultaneously. For example, when path 1 is running, paths 3, 4, 6, or 8 are allowed to be started. For a given source, its destination may be changed, but only after sufficient time has elapsed after the end of a dump to allow the material to be cleared off the conveyors of the currently selected path.

If the chip screening unit is shut down while it is a destination, the conveyor path is immediately set to send the chips to the appropriate chip pile. For example, if path 2 is currently running, the path is immediately changed to path 1.

**Devices**

East truck dump table:
  Raise/lower by operator.
East truck dump gate:
  Open/close by operator.
East truck dump chocks:
  Raise/lower by operator.
West truck dump table:
  Raise/lower by operator.

| CONTROL STRATEGY | Area: Pulp Mill Cell: Woodyard Unit: Chip Unloading |
|---|---|

West truck dump gate:
Open/close by operator.
West truck dump chocks:
Raise/lower by operator.
Railcar dump clamps:
Raise/lower by operator.
Railcar rotary dumper:
Rotate/back by operator.
East conveyor:
Start/stop by path equipment modules or operator.
West conveyor:
Start/stop by path equipment modules or operator.
Traverse conveyor:
Start/stop by path equipment modules or operator.
Railcar dump conveyor:
Start/stop by path equipment modules or operator.
East stackout bypass conveyor:
Start/stop by path equipment modules or operator.
West stackout bypass conveyor:
Start/stop by path equipment modules or operator.

**Interlocks**

Do not allow start of path equipment module if destination pile is too high.

Do not allow start of path 2 if chip screening unit is not running or is running but screening softwood chips.

Do not allow start of path 4 if chip screening unit is not running or is running but screening hardwood chips.

Do not allow truck dump gate to open until an appropriate path is running.

Do not allow rail dump gate to open until an appropriate path is running.

Do not allow a truck dump to be raised until an appropriate path is running.

Do not allow rail dump to be rotated until an appropriate path is running.

Backward startup and forward shutdown are enforced even when manually controlled by the operator.

If a downstream conveyor is stopped because of a failure, the upstream conveyors are emergency stopped.

| CONTROL STRATEGY | Area: Pulp Mill Cell: Woodyard Unit: Chip Screening |
|---|---|

*Chip Screening Unit*

**Unit Procedure**

The unit procedure phases are the same as the operational states.

**Devices**

Flat screens (six):
Start/stop by operator.
V screens (three):
Start/stop by unit supervision or operator.
Rechipper:
Start/stop by unit supervision or operator.
Cyclone blowers (three):
Start/stop by unit supervision or operator.
Slicers:
Start/stop by unit supervision or operator.

**Interlocks**

Do not allow startup if destination silo is too full.

| CONTROL STRATEGY | Area: Pulp Mill Cell: Woodyard Unit: Log Handling |
|---|---|

*Log Handling Unit*

**Unit Procedure**

The unit procedure phases are the same as the operational states.

**Devices**

Barker:
Start/stop by unit supervision or operator.
Conveyors:
Start/stop by unit supervision or operator.

**Interlocks**

Do not allow startup unless the groundwood cell is already in the starting or running state.
Shutdown log handling if groundwood cell is shutting down.

| yyyymmdd hh:mm | Pulp and Paper Co | Page 20 |
|---|---|---|

| POD | Area: Pulp Mill Cell: Brown Pulp Mill Unit: |
|---|---|

**Brown Pulp Mill Cell**

A simplified PFD of the brown pulp cell is depicted in Figure 10.6. There are a total of 13 batch digesters that process softwood or hardwood, but only one is shown here. In a typical sequence, the batch digester is charged with wood chips, white liquor, and black liquor. The digester is then sealed and pressurized and heated with steam. After the cooking pressure is reached, it is maintained for a set time. When the cook is completed, the contents are discharged to either the softwood or the hardwood

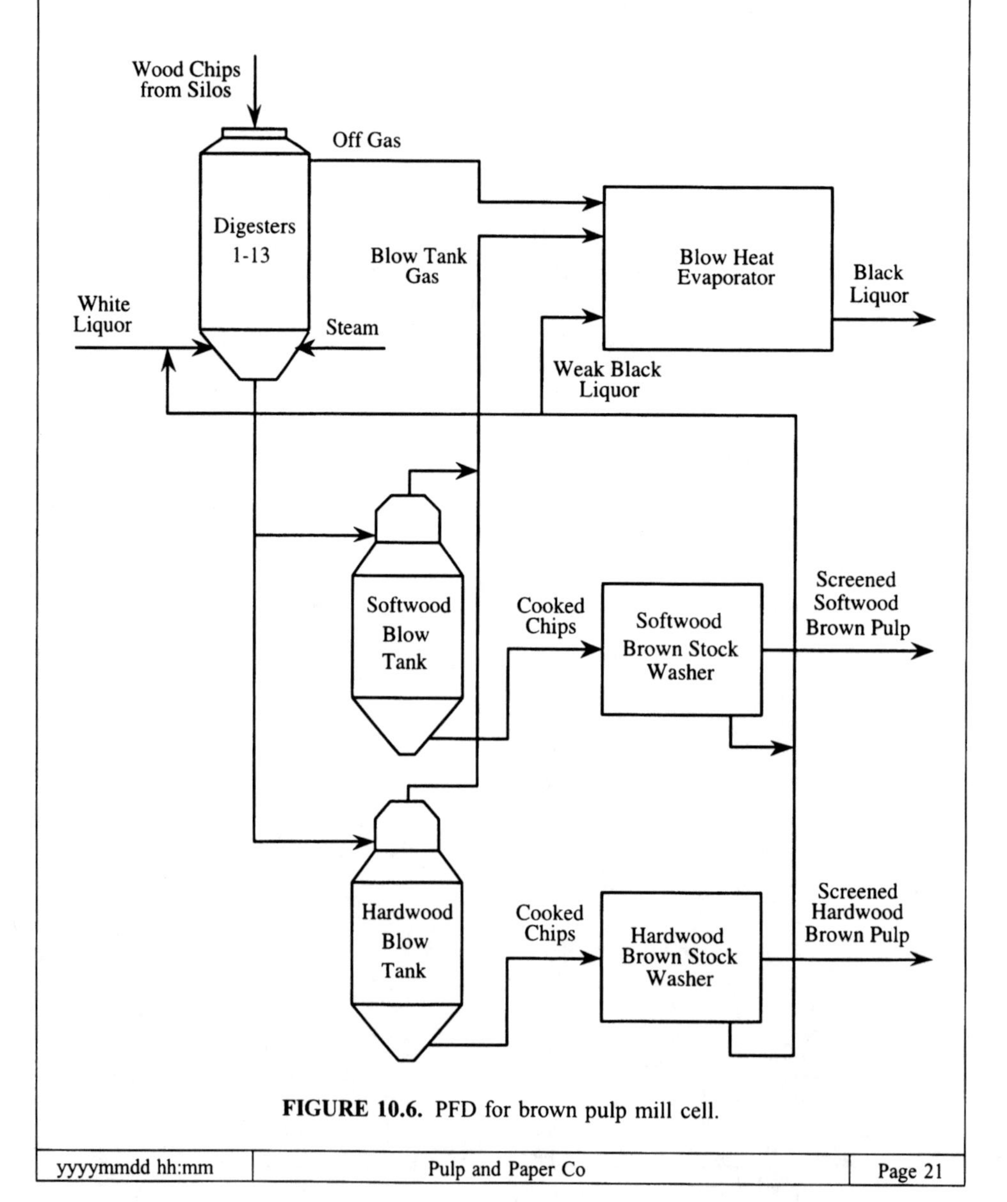


**FIGURE 10.6.** PFD for brown pulp mill cell.

| POD | Area: Pulp Mill Cell: Brown Pulp Mill Unit: |
|---|---|

blow tank. A blow tank receives the hot pulp from a digester and then the pulp is pumped to a brown stock washer unit. The brown stock washer recovers process chemicals in the form of weak black liquor. The washed and screened pulp proceeds to storage, from which it is pumped to the bleaching cell. The heat of the blow tank gas is used in the blow heat evaporator unit to concentrate the weak black liquor from the brown stock washer unit.

The operating states of this cell are shown in Figure 10.5:

Idle
Starting
Running
Shutting Down

Normal operation is continuous in order to maintain adequate inventories in the low- and high-density storage tanks that supply the bleach cell. The states are described as follows:

**Operating State Name: Starting**

*Routine Activities*

Prestart checks.
Start a digester.
Start up blow tanks.
Start up brown stock washer(s).
Start up blow heat evaporator.

*Exception Handling*

Shutdown if:

High level on low-density or high-density storage tank

*Primary Control Objectives*

None.

*Performance Information*

None.

*State End Conditions*

A digester started .and.
Blow tanks started .and.
At least one brown stock washer (softwood or hardwood) started .and.
Blow heat evaporator started

| **POD** | Area: Pulp Mill Cell: Brown Pulp Mill Unit: |
|---|---|

**Operating State Name: Running**

*Routine Activities*

Execute digester schedule.
Sample softwood washed pulp (for Kappa number).
Sample hardwood washed pulp (for Kappa number).

*Exception Handling*

Shutdown if:
High level on low-density or high-density storage tank.
Low steam pressure.
Blown seal on accumulator
Ruptured disk on digester

*Primary Control Objectives*

Level of softwood low-density storage tank above 30%.
Level of hardwood low-density storage tank above 30%.
Consistency of softwood washed pulp.
Consistency of hardwood washed pulp.
Softwood Kappa number = $x$.
Hardwood Kappa number = $x$.

*Performance Information*

Softwood Kappa number
Hardwood Kappa number
Tons produced

*State End Conditions*

Shutdown requested.

**Operating State Name: Shutting Down**

*Routine Activities:*

Shutdown all digesters.
Shutdown blow tanks.
Shutdown brown stock washers.
Shutdown blow heat evaporator.

*Exception Handling*

None.

*Primary Control Objectives*

None.

| yyyymmdd hh:mm | Pulp and Paper Co | Page 23 |
|---|---|---|

| **POD** | Area: Pulp Mill Cell: Brown Pulp Mill Unit: |
|---|---|

*Performance Information*

None.

*State End Conditions*

All units shut down.

| **POD** | Area: Pulp Mill Cell: Brown Pulp Mill Unit: Digester |
|---|---|

*Digester (Typical) Unit.* A simplified P&ID of a typical batch digester is shown in Figure 10.7. There are a total of 13 batch digesters that process softwood or hardwood, but only one is shown in Figure 10.7. In a typical sequence, the batch digester is charged with a certain amount of wood chips, white liquor, and weak black liquor. The digester is then sealed and 140 lb steam is used to pressurize and heat the vessel. When the target cooking pressure is reached, it is maintained for a target time. During the cooking time, gas is bled off to maintain vessel pressure and partly condensed in the gas-off jug. The gas-off jug condenses as much turpentine as possible. Periodically, the screen over the gas outlet is cleared of chips with a blowback. When the cook is completed and it is determined that the chips have cooked sufficiently, the contents are discharged to the appropriate blow tank.

The operating states of the unit are shown in Figure 10.8:

Idle
Starting
Filling
Pressurizing
Cooking
Blowing

Normal operation is a batch process controlled by the brown pulp cell. The overall objective is to produce pulp of a certain Kappa number (laboratory measurement). The Starting state is only used to start up the digester after it has been idle. The Idle state is similar to the Idle state for the chip unloading unit and thus is not described. The other states are described as follows:

**Operating State Name: Starting**

*Routine Activities*

Warm up with steam.

*Exception Handling*

None.

*Primary Control Objectives*

Digester temperature.

*Performance Information*

Time to warm up.

*State End Conditions*

Target digester temperature reached.

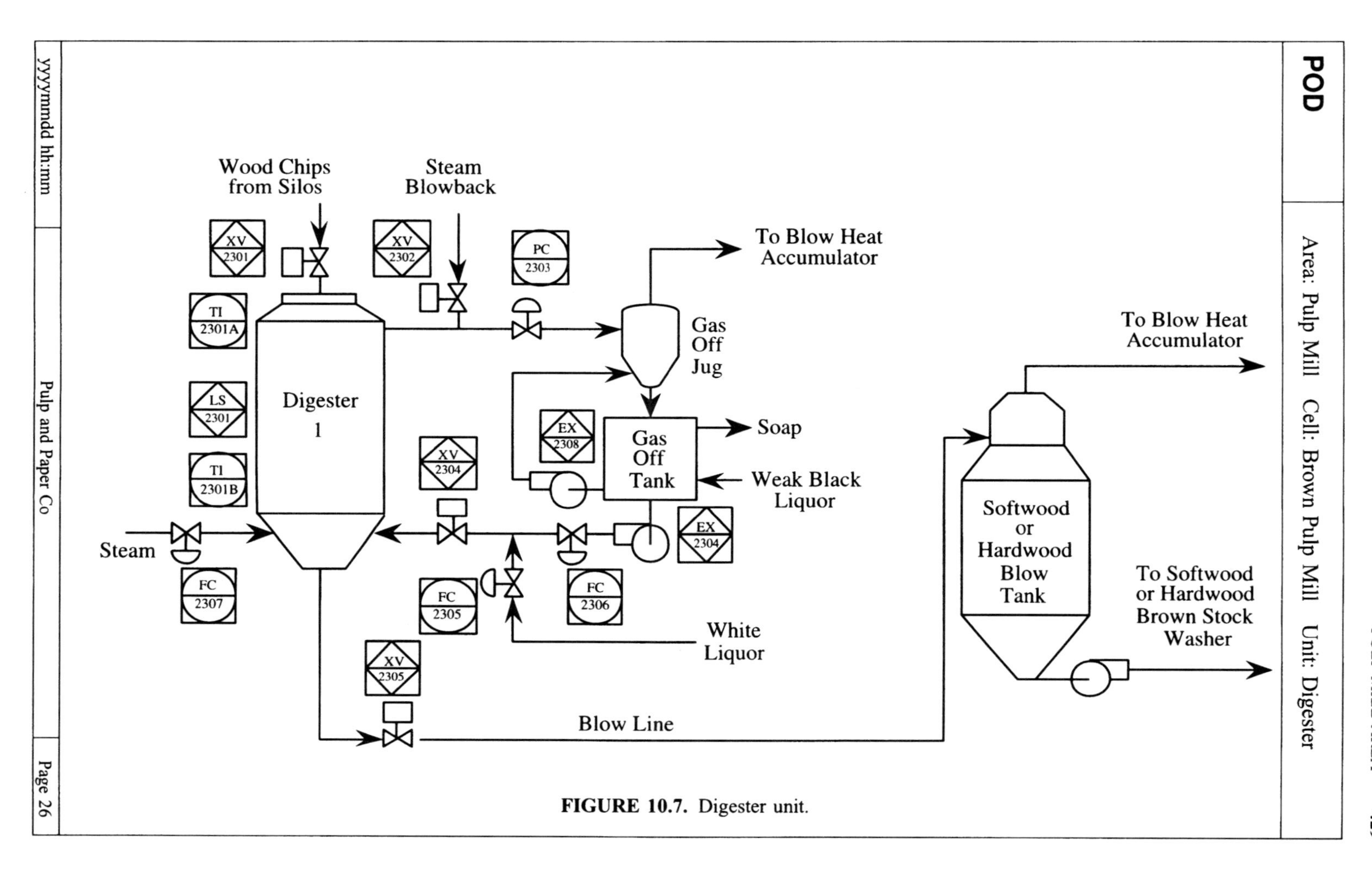


**FIGURE 10.7.** Digester unit.

| POD | Area: Pulp Mill Cell: Brown Pulp Mill Unit: Digester |
|---|---|

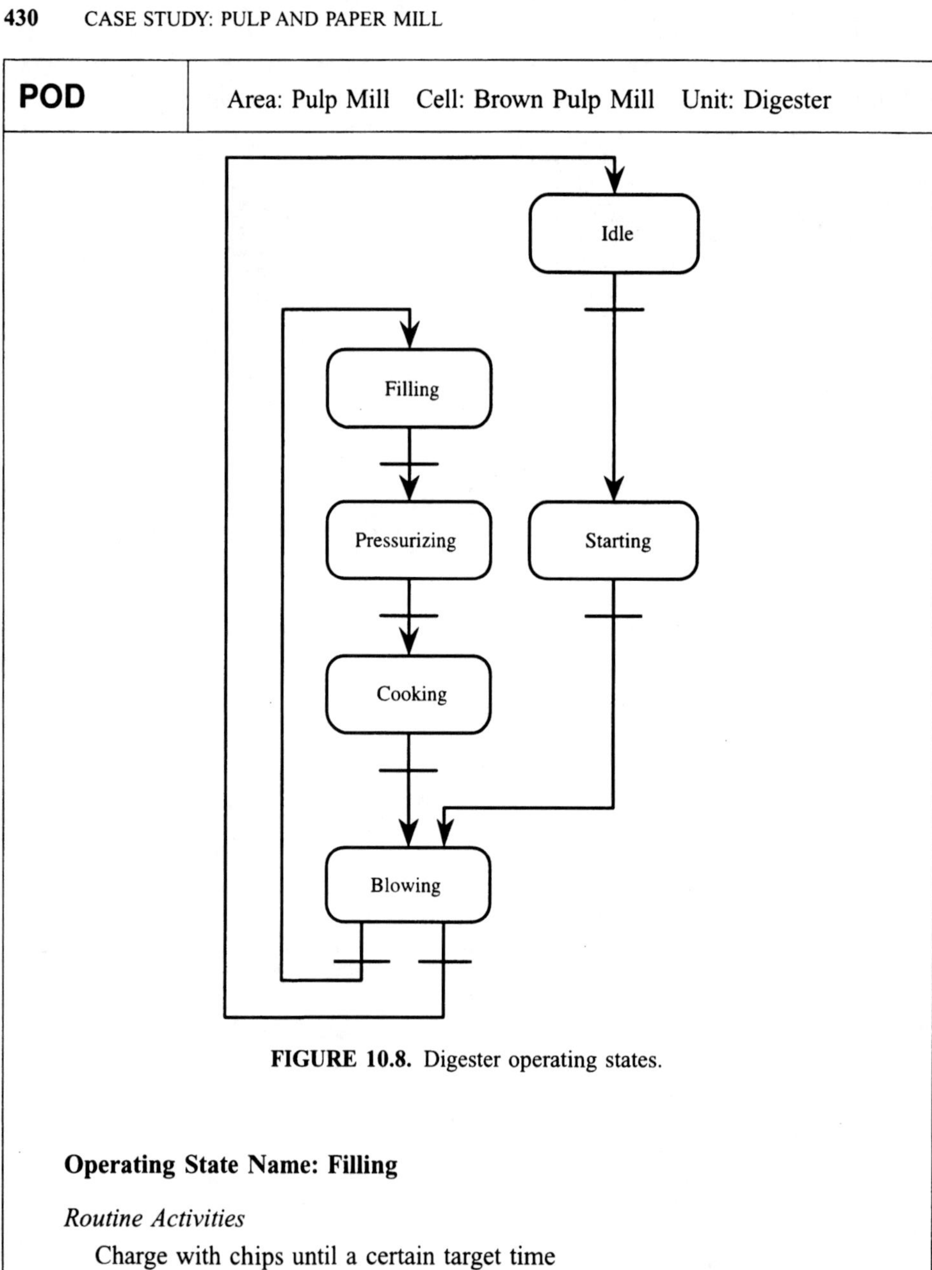


**FIGURE 10.8.** Digester operating states.

**Operating State Name: Filling**

*Routine Activities*

Charge with chips until a certain target time
Charge with white liquor
Charge with weak black liquor
Charge with chips to a certain level

*Exception Handling*

Shutdown if:

Loss of chip feed
Loss of white liquor feed

| POD | Area: Pulp Mill Cell: Brown Pulp Mill Unit: Digester |
|---|---|

Loss of weak black liquor feed

*Primary Control Objectives*

Amount of chips

Volume of liquor

White/black liquor ratio

*Performance Information*

Actual fill volume of white liquor

Actual fill volume of black liquor

*State End Conditions*

Target volume of liquor discharged into digester .or.

Target level of chips reached

**Operating State Name: Pressurizing**

*Routine Activities*

Ramp up steam flow to maximum.

Monitor temperature near top and bottom of digester and its profile.

Start digester pressure control when close to target pressure.

*Exception Handling*

Hold if:

Temperature deviates from profile by 5 degrees.

*Primary Control Objectives*

Avoid false pressure.

*Performance Information*

Pressure profile.

*State End Conditions*

Target pressure reached.

**Operating State Name: Cooking**

*Routine Activities*

Maintain vessel pressure for 70 min.

Maintain vessel temperature.

Periodically blow back gas outlet to clear chips from screen.

*Exception Handling*

Shutdown if:

Vessel pressure is too high.

**POD** Area: Pulp Mill Cell: Brown Pulp Mill Unit: Digester

*Primary Control Objectives*

Vessel pressure.

*Performance Information*

*H*-factor (integral of the relative reaction rate with respect to time).

*State End Conditions.*

Target time reached and operator allows blow.

**Operating State Name: Blowing**

*Routine Activities.*

Empty digester contents into designated blow tank.

*Exception Handling*

None.

*Primary Control Objectives*

None.

*Performance Information*

Number of blows per day.

*State End Conditions*

Low digester pressure.

| **POD** | Area: Pulp Mill Cell: Brown Pulp Mill Unit: Softwood BSW |
|---|---|

*Softwood Brown Stock Washer Unit.* A simplified P&ID of the softwood brown stock washer is shown in Figure 10.9. The device tags are not shown to minimize the diagram clutter. The hardwood brown stock washer is similar. The washer recovers process chemicals in the form of weak black liquor. Before the pulp is washed, hard, poorly cooked lumps of wood are separated from the pulp by coarse screens, called knotters. A high percentage of usable fiber occurs in the reject stream from the primary knotter; hence the rejects from the primary knotter are further treated with a secondary knotter before being removed from the pulp stream. The knots are compressed and incinerated. Four rotary vacuum drum washers are used to remove the process chemicals. The washer consists of a wire-mesh covered cylinder that rotates in a vat of the pulp slurry. As the drum contacts the slurry, a vacuum is applied through the screen to thicken the pulp and cause it to stick to the screen. The thickened pulp rotates past wash showers to displace black liquor. The vacuum is removed beyond the wash showers and the pulp mat is dislodged into a pulper where it is diluted into a slurry for the next washer stage. The pulp is screened between the third and fourth drum washer. The wash water is the evaporator condensate from the blow heat evaporator unit and flows countercurrent to the pulp flow. The black liquor concentration is highest in the effluent from the first washer. Since air is constantly being sucked into the pulp sheet, a defoaming chemical is added to the rinse water at the first and second washer. In addition, foam is removed from the first and second service tanks. Soap is skimmed from the foam and stored before being sold. The weak black liquor is stored before being concentrated in the blow heat evaporator. The weak black liquor is also part of the liquor make-up for the batch digesters. The washed and screened pulp proceeds to high-density storage. The pulp from the high-density storage is combined with water from the fourth washer to constitute the desired pulp density in the low-density storage tank, from which it is pumped to the bleaching cell.

The operating states of the unit are shown in Figure 10.5:

Idle
Starting
Running
Shutting Down

Normal operation is continuous in order to maintain adequate inventories in the low- and high-density storage tanks that supply the bleach cell. The Idle state is similar to the Idle state for the chip unloading unit and thus is not described. The other states are described as follows:

**Operating State Name: Starting**

*Routine Activities*

Prestart checks.

| yyyymmdd hh:mm | Pulp and Paper Co | Page 30 |
|---|---|---|

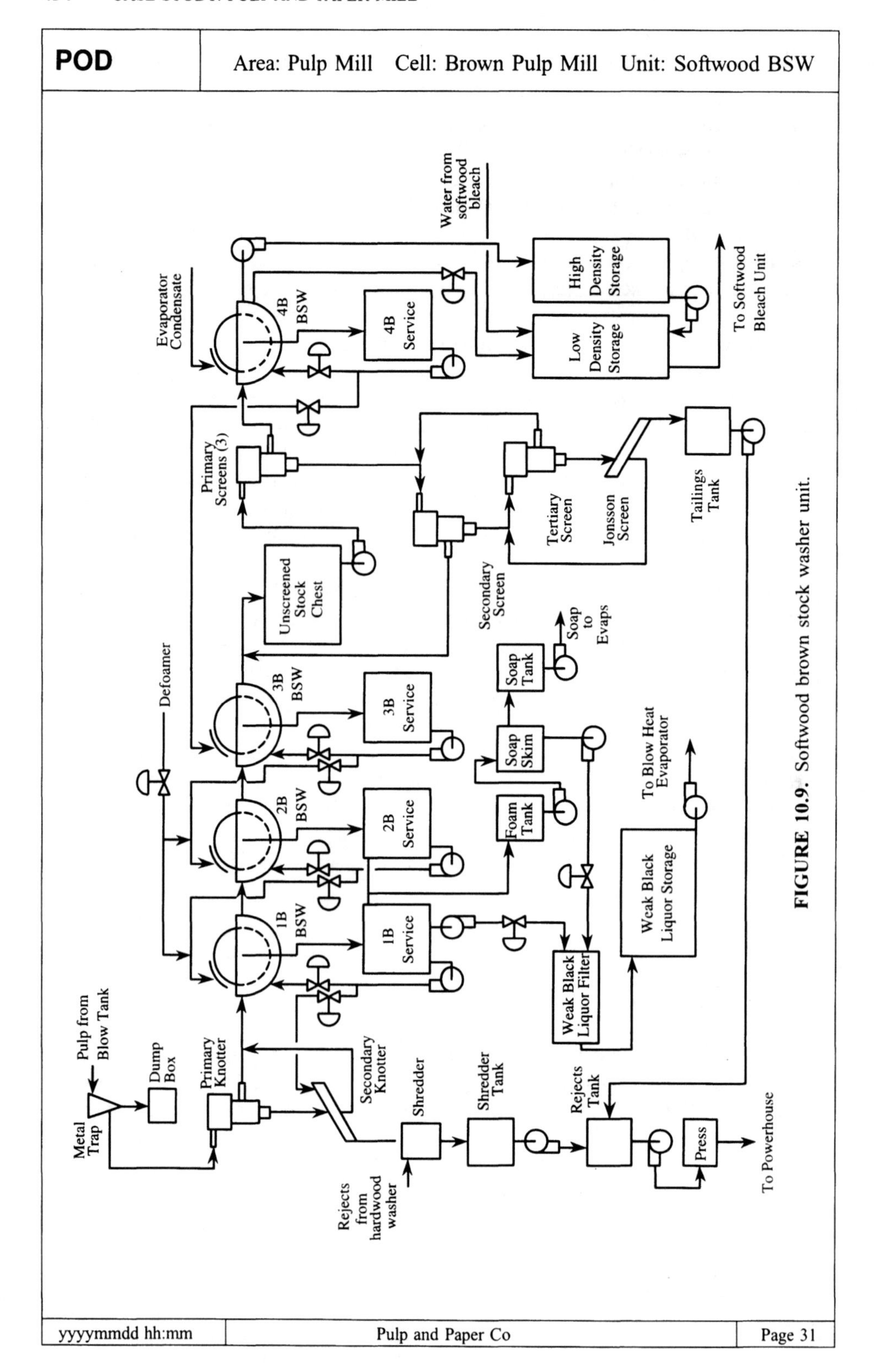


**FIGURE 10.9.** Softwood brown stock washer unit.

| POD | Area: Pulp Mill Cell: Brown Pulp Mill Unit: Softwood BSW |
|---|---|

Wait for blow heat evaporator to start.

Start evaporator condensate pump.

Start up service pumps, in reverse order of the pulp flow, to initially charge washer vats and service tanks.

Stop evaporator condensate pump and service pumps.

Wait for blow tank pump to start.

Start evaporator condensate pump.

Start service pumps, in reverse order of the pulp flow.

Start washer drum drive motors.

After some time, start pump to transfer weak black liquor to storage.

*Exception Handling*

Shutdown if:

High level on low- or high-density tank

*Primaryy Control Objectives*

Softwood brown pulp consistency.

*Performance Information*

None.

*State End Conditions*

Pump to weak black liquor storage running.

**Operating State Name: Running**

*Routine Activities*

Check blow tank level.

Check liquor storage level.

Check high-density storage level.

Run thick stock pump.

Run weak liquor pump.

Adjust showers to give correct dilution.

*Exception Handling*

Shutdown if:

High level on low- or high-density storage

High level on weak black liquor storage tanks

*Primary Control Objectives*

Brown pulp consistency in low-density storage

Washer levels

Dilution flow rate

| POD | Area: Pulp Mill Cell: Brown Pulp Mill Unit: Softwood BSW |
|---|---|

Weak black liquor concentration

*Performance Information*

Pulp consistency

Dilution factor

Pulp brightness

Pulp pH

*State End Conditions*

Shutdown requested.

**Operating State Name: Shutting Down**

*Routine Activities*

Stop evaporator condensate pump.

Empty service tanks.

Stop service pumps, in reverse order of the pulp flow.

Stop washer drum drive motors.

Stop pump to transfer weak black liquor to storage.

*Exception Handling*

None.

*Primary Control Objectives*

None.

*Performance Information*

None.

*State End Conditions*

All units shutdown.

| POD | Area: Pulp Mill Cell: Brown Pulp Mill Unit: Hardwood BSW |
| --- | --- |

*Hardwood Brown Stock Washer Unit.* Similar to softwood brown stock washer unit.

| yyyymmdd hh:mm | Pulp and Paper Co | Page 34 |
| --- | --- | --- |

| **POD** | Area: Pulp Mill Cell: Brown Pulp Mill Unit: Blow Heat Evap. |
|---|---|

*Blow Heat Evaporator Unit.* A simplified P&ID of the blow heat evaporator is shown in Figure 10.10. The device tags are not shown to minimize the diagram clutter. The evaporator uses the heat in the blow tank gas to concentrate the weak black liquor from the softwood and hardwood brown stock washer units. The concentrated black liquor is stored and then pumped to the recovery furnace. The first evaporator also condenses the turpentine from the softwood digester gas-off jugs. The vapor from the blow tanks and hardwood gas-off jugs are collected in the accumulator, from which it flows into the first effect.

The operating states of the unit are shown in Figure 10.5:

Idle
Starting
Running
Shutting Down

Normal operation is continuous in order to maintain adequate inventory of concentrated black liquor. The Idle state is similar to the Idle state for the chip unloading unit and thus is not described. The other states are described as follows:

**Operating State Name: Starting**

*Routine Activities*

Prestart checks.
Fill evaporators with weak black liquor.
Stop feed pump.
Recirculate liquor in evaporators.
Wait for accumulator pressure above 100 psi.
Start vapor flow through evaporators.
Wait for temperature in second effect to be greater than $x$ degrees.
Start feed pump.
Start cooling water recirculation pumps.
Start third-effect-to-second-effect transfer pump.
Start second-effect-to-first-effect transfer pump.
Start concentrated liquor storage transfer pump.
Stop recirculation pumps (set their ratio to low value).

*Exception Handling*

Shutdown if:

Accumulator high pressure
High level on concentrated black liquor storage tanks

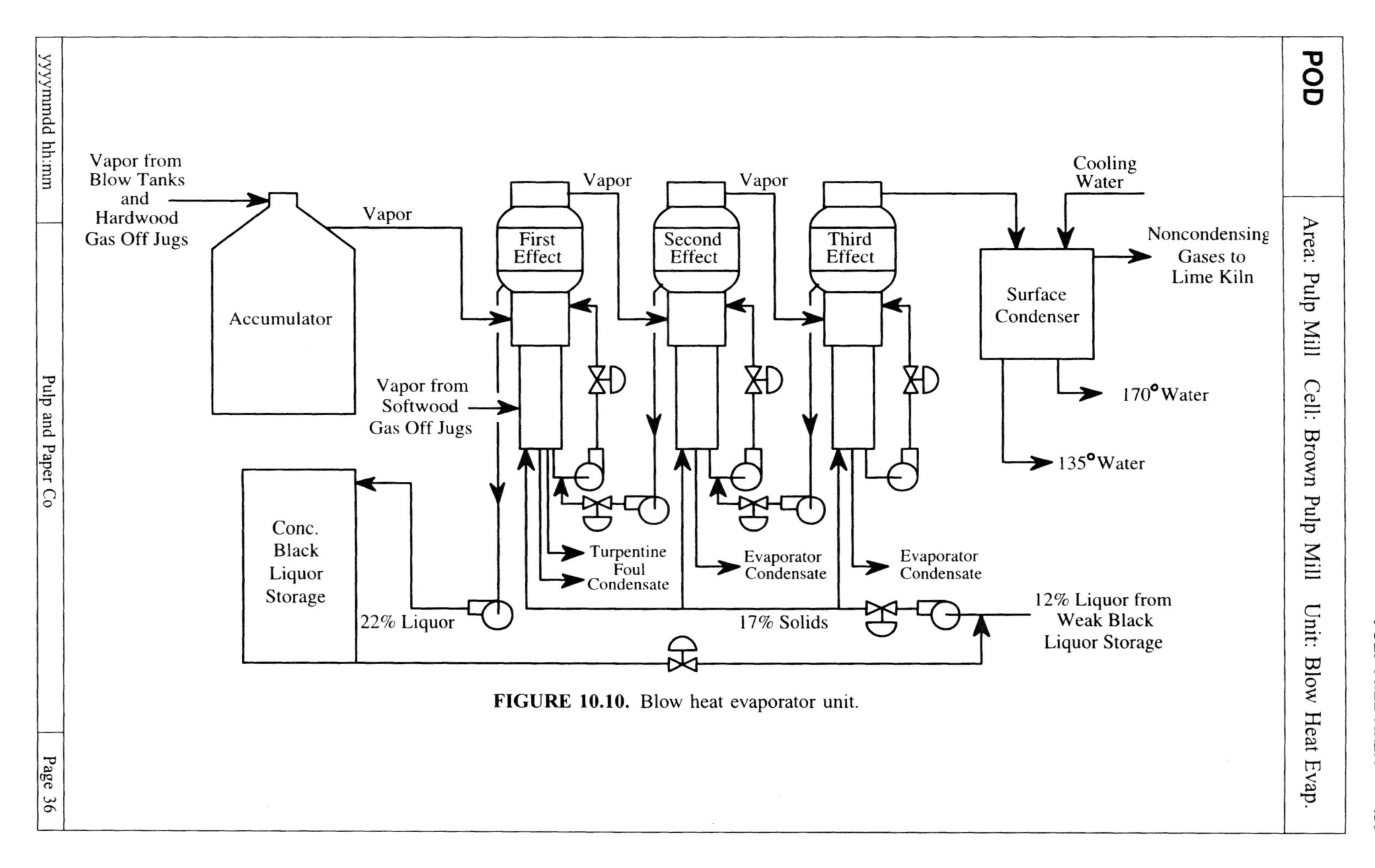


**FIGURE 10.10.** Blow heat evaporator unit.

| **POD** | Area: Pulp Mill Cell: Brown Pulp Mill Unit: Blow Heat Evap. |
|---|---|

*Primary Control Objectives*

None.

*Performance Information*

None.

*State End Conditions*

Pump to concentrated black liquor storage running.

**Operating State Name: Running**

*Routine Activities*

None.

*Exception Handling*

Shutdown if:

Accumulator high pressure

High level on concentrated black liquor storage tanks

*Primary Control Objectives*

Concentrated black liquor concentration

Maintain pressure on accumulator (do not blow seal)

*Performance Information*

Dilution factor.

*State End Conditions*

Shutdown requested.

**Operating State Name: Shutting Down**

*Routine Activities*

Stop water recirculation pumps.

Empty liquor from evaporators.

Stop liquor feed pump.

Stop transfer pumps, in the order of the liquor flow.

Stop evaporator recirculation pumps.

*Exception Handling*

None.

*Primary Control Objectives*

None.

| POD | Area: Pulp Mill Cell: Brown Pulp Mill Unit: Blow Heat Evap. |
|---|---|

*Performance Information*

None.

*State End Conditions*

All modules shut down.

**CONTROL CONCEPT** Area: Pulp Mill Cell: Brown Pulp Mill Unit:

**Brown Pulp Mill Cell Process Management**

*Extent of Automation*

Start digesters when scheduled.

Collect performance data from units.

Collect softwood and hardwood Kappa numbers.

Collect softwood and hardwood brown pulp brightness.

Collect softwood and hardwood brown pulp pH.

*Flexibility of Automation*

None.

*Control Activity Coordination*

One conveyor supplies the odd-numbered digesters.

One conveyor supplies the even-numbered digesters.

Common header supplies the weak black liquor and white liquor.

#5–#8 and #13 digesters can process either softwood or hardwood chips.

One 140-lb steam header supplies #1–#6 digester.

One 140-lb steam header supplies #7–#13 digesters.

*Interaction with Operating Personnel*

Enter a tentative digester schedule once every day. The number of digesters to be scheduled and whether they will process softwood or hardwood are determined by the expected demand of hardwood and softwood.

Operator collects sample for Kappa number hourly.

Summary information from units displayed for operator.

**Digester (Typical) Unit Supervision**

*Extent of Automation*

Control all digester activities.

Collect performance data (fill volume of white and black liquor, temperature and pressure profile, cook time, *H*-factor, blow time).

*Flexibility of Automation*

#5–#8 and #13 digesters can process either softwood or hardwood chips.

*Control Activity Coordination*

Digesters share material supply headers.

Digesters share two blow tanks.

*Interaction with Operating Personnel*

Operator is allowed to disable a digester so its operation is not scheduled.

| POD | Area: Pulp Mill Cell: Brown Pulp Mill Unit: |
|---|---|

At the end of the Cooking state, the operator must give permission for the blow to proceed.

Display digester state, primary controlled variables, and performance data.

Display status information from control modules.

**Softwood Brown Stock Washer Unit Supervision**

*Extent of Automation*

Control all aspects of unit.

Collect performance data (softwood brown pulp consistency, brightness, pH, and dilution factor).

*Flexibility of Automation*

None.

*Control Activity Coordination*

Coordinate with softwood blow tank.

*Interaction with Operating Personnel*

Accept startup and shutdown commands.

Display primary controlled variables.

Display all flows and performance data.

Display status information from control modules.

**Hardwood Brown Stock Washer Unit Supervision**

Similar to softwood brown stock washer unit supervision.

**Blow Heat Evaporator Unit Supervision**

*Extent of Automation*

Control all aspects of unit.

Collect performance data (concentrated black liquor concentration).

*Flexibility of Automation*

None.

*Control Activity Coordination*

Weak black liquor tank shared with digesters.

*Interaction with Operating Personnel*

Accept startup and shutdown commands.

Display primary controlled variables.

Display status information from control modules.

| CONTROL STRATEGY | Area: Pulp Mill Cell: Brown Pulp Mill Unit: |
|---|---|

**Brown Pulp Mill Cell**

SCHEDULING CONSTRAINTS. The 13 digesters are scheduled in order to meet constraints on the equipment modules that supply material or that accept material; they are connected to the various headers and blow tanks as shown in Figure 10.11. There are two conveyors that transfer wood chips from the silos. One conveyor supplies the odd-numbered digesters and the other supplies the even-numbered digesters. Only one odd-numbered digester can be in the process of loading chips at any given time. In a similar way, only one even-numbered digester can be in the process of loading chips at any given time. The weak black liquor and white liquor are supplied from a header common to all digesters. Only one digester may be using the liquor at any time. #1–#4 digesters only process softwood chips since they empty to the softwood blow tank. Similarly, #9–#12 digesters only process hardwood chips. #5–#8 and #13 digesters can go to either blow tank. There are two 140-lb steam headers, one supplying #1–#6 digesters and the other supplying #7–#13 digesters. For a particular header, no more than two digesters can be pressurizing simultaneously.

The digester scheduling will be done by an engineer once every day. The number of digesters to be scheduled and whether they will process softwood or hardwood, are determined by the expected demand of hardwood and softwood.

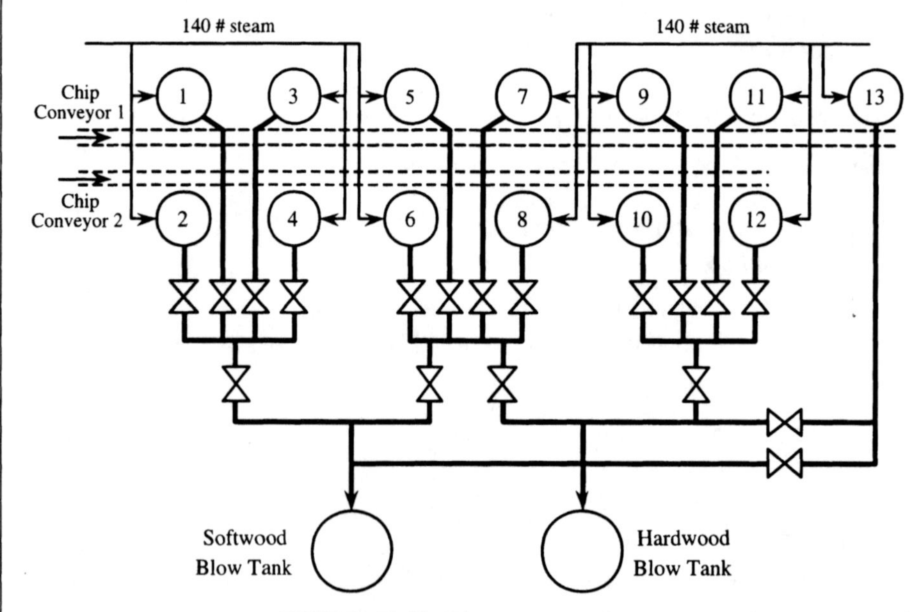


**FIGURE 10.11.** Digester connections.

| **CONTROL STRATEGY** | Area: Pulp Mill Cell: Brown Pulp Mill Unit: Digester |
|---|---|

*Digester Unit (Typical).* The cell controller normally schedules the start of a cook cycle and determines the path to the blow tank (for #5–#8 and #13 digesters).

At the end of the Cooking state, the operator must give permission for the blow to proceed. The operator approves the blow when the $H$-factor (enthalpy) is above a target value and the blow heat evaporator accumulator pressure is below a certain value.

**Unit Procedure**

The unit procedure phases are the same as the operational states depicted in Figure 10.8.

**Loops**

*Digester Temperature.*

Pressurizing phase: After the first 5 min, this loop is set to the Automatic state and manipulates the steam flow loop setpoint. The steam flow is held constant as long as the vessel high and low temperatures follow the reference high- and low-temperature profiles. If either temperature deviates a certain amount from the reference profile, then the temperature loop is set to the Manual state and the vessel switches to pressure control. When the vessel pressure gets close to the hold (cooking) pressure, the vessel always switches to pressure control, even when the temperatures are at their target values.

*Digester Pressure*

Cooking phase: Maintain vessel pressure by venting to the gas-off jug. Temporarily close vent valve when steam blowback is active.

Pressurizing phase: Maintain a target vessel pressure if either (low or high) digester temperature deviates a certain amount from the reference profile, bleeding off gas. When the vessel pressure gets close to the hold pressure, the vessel always switches to pressure control, even when the temperatures are at their target values.

*Liquor Flow and Ratio*

Filling phase: After the initial 8 min, maintain total liquor flow and ratio of white liquor to black liquor.

*Steam Flow*

Pressurizing phase: Ramp the steam flow to maximum flow during the first 5-min period. After the initial 5 min, the steam flow setpoint is controlled by the digester temperature loop.

Starting phase: Maintain constant flow.

| CONTROL STRATEGY | Area: Pulp Mill   Cell: Brown Pulp Mill   Unit: Digester |
|---|---|

**Devices**

Wood chip inlet valve
  Filling phase: Open valve.
  Other phases: Close valve.
Steam blowback
  Cooking phase: Every 5 min, open for 30 sec.
Liquor block valve
  Filling phase: Open valve after initial 8 min.
Blow block valve
  Blowing phase: Open valve.
  Other phases: Close valve.
Black liquor pump
  Filling phase: Run pump.
  Other phases: Stop pump.
Gas-off recirculating pump
  Pressurizing phase: Run pump.
  Cooking phase: Run pump.
  Other phases: Stop pump.

**Interlocks**

Blow allowed if:
  cooking target time reached .and.
  path to blow tank established .and.
  *H*-factor at or over target .and.
  blow heat accumulator pressure not too high .and.
  target blow tank level below certain value .and.
  operator allows blow.

| CONTROL STRATEGY | Area: Pulp Mill | Cell: Brown Pulp Mill | Blow Tank Equip Modules |
|---|---|---|---|

*Digester Blow Path Equipment Modules*

**Possible Paths**

1. #1–#4 digesters → Softwood Blow Tank
2. #5–#8 digesters → Softwood Blow Tank
3. #5–#8 digesters → Hardwood Blow Tank
4. #9–#12 digesters → Hardwood Blow Tank
5. #13 digester → Softwood Blow Tank
6. #13 digester → Hardwood Blow Tank

**Interlocks**

Blow paths not allowed to be active simultaneously:

1 and 2
1 and 5
2 and 5
3 and 4
3 and 6
4 and 6

Blow path cannot be started unless blow tank is below certain level.

*Softwood Blow Tank Equipment Module*

**Devices**

Outlet pump
Start when the blow tank level is above 10%.
Stop when the tank level is below 5%.

*Hardwood Blow Tank Equipment Module.*
Similar to the softwood blow tank equipment module.

| yyyymmdd hh:mm | Pulp and Paper Co | Page 44 |
|---|---|---|

| CONTROL STRATEGY | Area: Pulp Mill Cell: Brown Pulp Mill Unit: Softwood BSW |
|---|---|

*Softwood Brown Stock Washer Unit.*

All flows in the cell are adjusted in accordance with the unwashed pulp inlet flow rate.

**Unit Procedure**

The unit procedure phases are the same as the operational states.

**Loops**

Flow to secondary knotter from 1B service tank:
Maintain flow to supply adequate dilution for the secondary knotter.
Flow from 1B service tank to 1B washer:
Maintain flow as a constant ratio of the flow to the secondary knotter.
Flow from 1B service tank to weak black liquor storage:
Maintain as constant ratio of the unwashed pulp inlet flow.
Flow from 2B service tank to 1B washer shower:
Maintain as constant ratio of the unwashed pulp inlet flow.
Flow from 2B service tank to 2B washer:
Maintain as constant ratio of the flow to 1B washer shower.
Flow from 3B service tank to 2B washer shower:
Maintain as constant ratio of the unwashed pulp inlet flow.
Flow from 3B service tank to 3B washer:
Maintain as constant ratio of the flow to 2B washer shower.
Flow from 4B service tank to 3B washer shower:
Maintain as constant ratio of the unwashed pulp inlet flow.
Flow from 4B service tank to 4B washer:
Maintain as constant ratio of the flow to 3B washer shower.
Flow of evaporator condensate to 4B washer shower:
Maintain as constant ratio of the unwashed pulp inlet flow.
Washer level (1B, 2B, 3B, and 4B washers):
Maintain washer water level by adjusting drum rotation speed.
Low-density storage pulp consistency:
Maintain consistency by manipulating the flow from the high-density storage to the low-density storage and the flow of water from the 4B washer.

**Devices**

Service tank (1B, 2B, 3B, 4B) outlet pumps:
Start/stop when commanded by unit supervision or operator.

| yyyymmdd hh:mm | Pulp and Paper Co | Page 45 |
|---|---|---|

| CONTROL STRATEGY | Area: Pulp Mill Cell: Brown Pulp Mill Unit: Softwood BSW |
|---|---|

High-density storage tank inlet pumps:
Start/stop when commanded by unit supervision or operator.

Foam tank outlet pump
Start when tank level is above 10%.
Stop when tank level is below 5%.

Soap skim outlet pump
Start when tank level is above 10%.
Stop when tank level is below 5%.

Soap tank outlet pump
Start when tank level is above 10%.
Stop when tank level is below 5%.

Weak black liquor storage outlet pump
Start when tank level is above 10%.
Stop when tank level is below 5%.

Tailings tank outlet pump
Start when tank level is above 50%.
Stop when tank level is below 10%.

Shredder tank outlet pump
Start when tank level is above 60%.
Stop when tank level is below 10%.

Rejects tank outlet pump
Start when tank level is above 80%.
Stop when tank level is below 10%.

**Interlocks**

Stop service tank outlet pump when tank level below 5%.

| CONTROL STRATEGY | Area: Pulp Mill Cell: Brown Pulp Mill Unit: Hardwood BSW |
|---|---|

*Hardwood Brown Stock Washer Unit.*
Similar to softwood brown stock washer unit.

| CONTROL STRATEGY | Area: Pulp Mill Cell: Brown Pulp Mill Unit: Blow Heat Evap. |
|---|---|

*Blow Heat Evaporator*

**Unit Procedure**

The unit procedure phases are the same as the operational states.

**Loops**

First-effect recirculation flow:
Maintain as a ratio of the weak black liquor inlet flow.
Second-effect recirculation flow:
Maintain as a ratio of the weak black liquor inlet flow.
Third-effect recirculation flow:
Maintain as a ratio of the weak black liquor inlet flow.
Flow from third effect to second effect:
Maintain as a ratio of the weak black liquor inlet flow.
Flow from second effect to first effect:
Maintain as a ratio of the weak black liquor inlet flow.
Flow from first effect to concentrated liquor storage:
Maintain as a ratio of the weak black liquor inlet flow.
Accumulator pressure:
Maintain at $x$ psi, so seal is not blown.

**Devices**

First-effect recirculation pump:
Start/stop by unit supervision or operator.
Second-effect recirculation pump:
Start/stop by unit supervision or operator.
Third-effect recirculation pump:
Start/stop by unit supervision or operator.
Third-effect to second-effect transfer pump:
Start/stop by unit supervision or operator.
Second-effect to first-effect transfer pump:
Start/stop by unit supervision or operator.
Concentrated liquor storage transfer pump:
Start/stop by unit supervision or operator.
Weak black liquor feed pump:
Start/stop by unit supervision or operator.

| yyyymmdd hh:mm | Pulp and Paper Co | Page 48 |
|---|---|---|

| POD | Area: Pulp Mill Cell: Caustic Unit: |
|---|---|

**Caustic Cell**

A simplified PFD of the caustic cell is shown in Figure 10.12. In the recovery boiler, the sulfur- and sodium-based inorganic materials in the black liquor are liberated and recovered as a liquid smelt that is dissolved into water to form green liquor. The white cooking liquor is formed by converting the sodium carbonate in the green liquor to sodium hydroxide using calcium hydroxide. The precipitate from the white liquor clarifier, calcium carbonate, is heated in a kiln and converted to calcium oxide (lime), which is then used to convert more green liquor.

The operating states of the cell are shown in Figure 10.5:

Idle
Starting
Running
Shutting Down

Normal operation is continuous in order to maintain adequate inventories in the white liquor storage tanks. The Idle state is similar to the Idle state for the chip unloading unit and thus is not described. The other states are described as follows:

**Operating State Name: Starting**

*Routine Activities*

Prestart checks.
Start liquor processing unit.
Start lime kiln units.

*Exception Handling*

Shutdown if:
High level on white liquor storage

*Primary Control Objectives*

None.

*Performance Information*

None.

*State End Conditions*

Liquor processing started
At least one lime kiln started

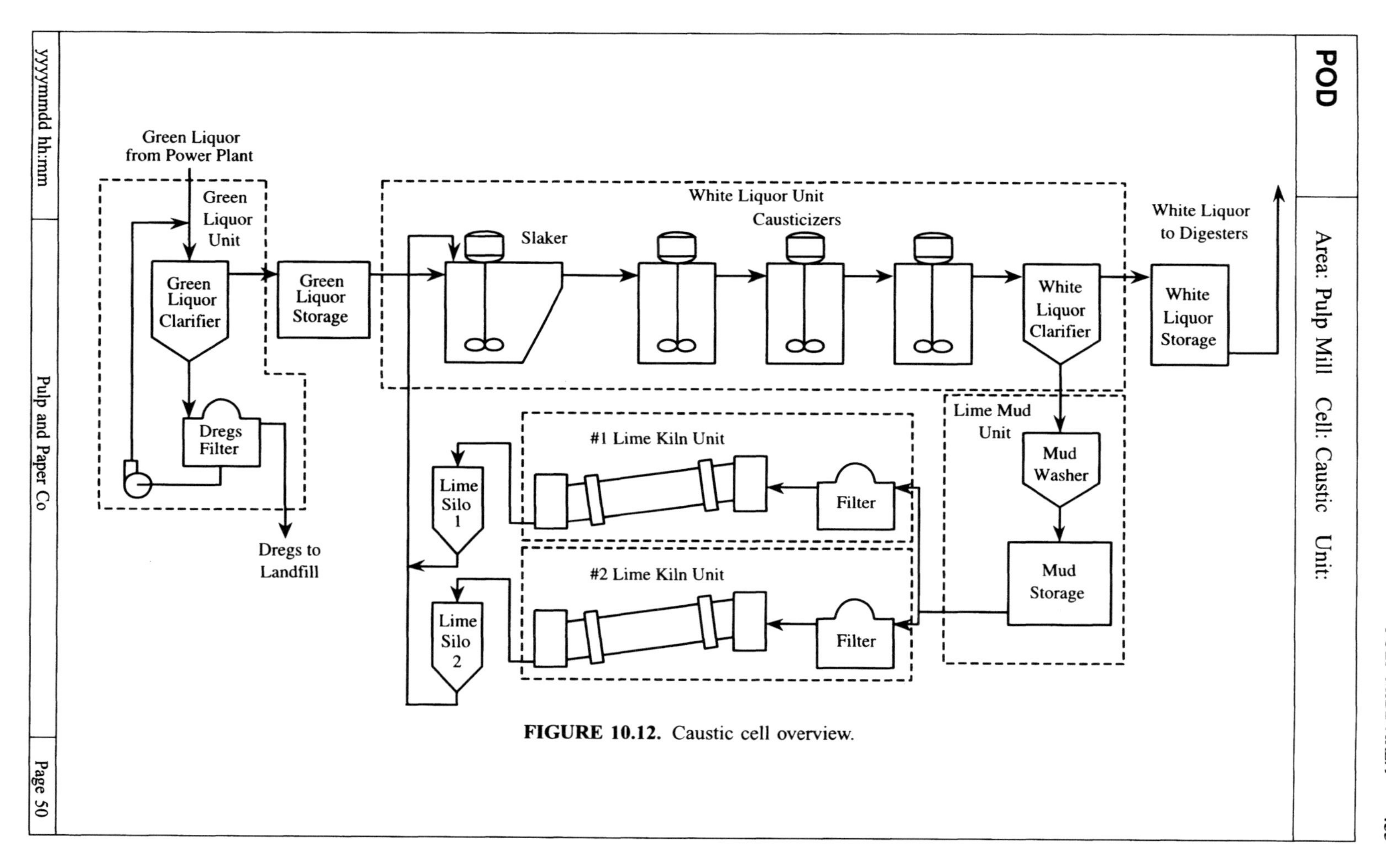


**FIGURE 10.12.** Caustic cell overview.

| POD | Area: Pulp Mill Cell: Caustic Unit: |
|---|---|

**Operating State Name: Running**

*Routine Activities*

Schedule running of lime kilns to maintain lime silo levels.

*Exception Handling*

Shut down if:

High level on white liquor storage

Low level on green liquor storage tanks

*Primary Control Objectives*

Keep inventory of white liquor tanks above $x$ gallons.

*Performance Information*

Gallons of green liquor processed per shift

Tons of lime produced per shift

Total reduced sulfur dioxide

*State End Conditions*

Shutdown requested.

**Operating State Name: Shutting Down**

*Routine Activities*

Shutdown kilns.

Shutdown liquor processing.

*Exception Handling.*

None.

*Primary Control Objectives*

None.

*Performance Information*

None.

*State End Conditions*

All units shutdown

| yyyymmdd hh:mm | Pulp and Paper Co | Page 51 |
|---|---|---|

| POD | Area: Pulp Mill Cell: Caustic Unit: Green Liquor |
|---|---|

*Green Liquor Unit.* The green liquor unit is shown in Figure 10.13, and removes the solids in the green liquor from the recovery boiler. This unit consists of a sedimentation tank and a washer. The washer is a rotary vacuum washer that removes the solids in the green liquor, which are taken to a landfill. Makeup water comes from the mill water supply. The device tags are not shown to minimize the diagram clutter.

The operating states of the unit are shown in Figure 10.5:

Idle
Starting
Running
Shutting Down

Normal operation is continuous in order to maintain adequate inventories in the green liquor storage tank. The Idle state is similar to the Idle state for the chip unloading unit and thus is not described. The other states are described as follows:

**Operating State Name: Starting**

*Routine Activities*

Prestart checks.

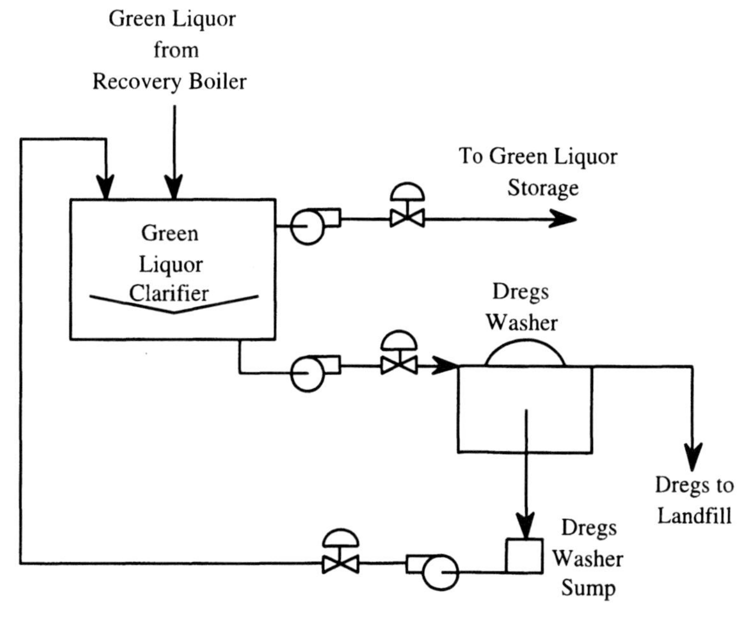


**FIGURE 10.13.** Green liquor unit.

| **POD** | Area: Pulp Mill Cell: Caustic Unit: Green Liquor |
|---|---|

Start green liquor clarifier sump pump.
Start green liquor clarifier outlet pump.
Start dregs washer drum.
Start dregs washer sump pump.

*Exception Handling*

Shutdown if:

High level on green liquor storage tank
Overflow on tanks

*Primary Control Objectives*

None.

*Performance Information*

None.

*State End Conditions*

Dregs washer sump started.

**Operating State Name: Running**

*Routine Activities*

Monitor primary control objectives.

*Exception Handling*

Shutdown if:

High level on green liquor storage
Overflow on tanks

*Primary Control Objectives*

Concentration of sodium carbonate in green liquor
Concentration of sodium sulfide in green liquor

*Performance Information*

Gallons of green liquor from recovery boiler per shift
Dregs removed

*State End Conditions*

Shutdown requested.

**Operating State Name: Shutting Down**

*Routine Activities*

Stop all pumps.

| POD | Area: Pulp Mill Cell: Caustic Unit: Green Liquor |
|---|---|

*Exception Handling*

None.

*Primary Control Objectives*

None.

*Performance Information*

None.

*State End Conditions*

All pumps stopped.

| POD | Area: Pulp Mill Cell: Caustic Unit: White Liquor |
|---|---|

*White Liquor Unit.* A simplified P&ID of the white liquor processing unit is shown in Figure 10.14. The device tags are not shown to minimize the diagram clutter. The green liquor consists of sodium carbonate, $Na_2CO_3$, and $Na_2S$ dissolved in water. In the slaker, reburned and fresh lime is combined with the green liquor. Lime silo #1 contains reburned lime and lime silo #2 contains a mixture of reburned and fresh lime. The lime, CaO, reacts with the water to form calcium hydroxide, $Ca(OH)_2$, and the calcium hydroxide reacts with the sodium carbonate to form sodium hydroxide, NaOH. The majority of the causticizing reaction occurs in the slaker. The slaking reaction is

$$CaO + H_2O \rightarrow Ca(OH)_2$$

and the causticizing reaction is

$$Ca(OH)_2 + Na_2CO_3 \leftrightarrow 2NaOH + CaCO_3(s)$$

Large, unreactive lime particles (called grits) are removed from the slaker by a classifier that uses a raking action. The chemical mixture flows to a standpipe and then it is pumped into the first of four continuous-flow, stirred reactors that are used to complete the causticizing reaction. The lime mud ($CaCO_3$) is removed from the white liquor in the white liquor clarifier, which is a settling tank. The mud is processed further by the lime mud unit before it is burned in a kiln and the white liquor goes to storage before being used by the digesters.

The operating states of the unit, are shown in Figure 10.5:

Idle
Starting
Running
Shutting Down

Normal operation is continuous in order to maintain adequate inventories in the white liquor storage tanks. The Idle state is similar to the Idle state for the chip unloading unit and thus is not described. The other states are described as follows:

**Operating State Name: Starting**

*Routine Activities*

Prestart checks.
Start green liquor feed pumps.
Start flow through causticizer.
Start agitators.
Start flow through white liquor clarifier.

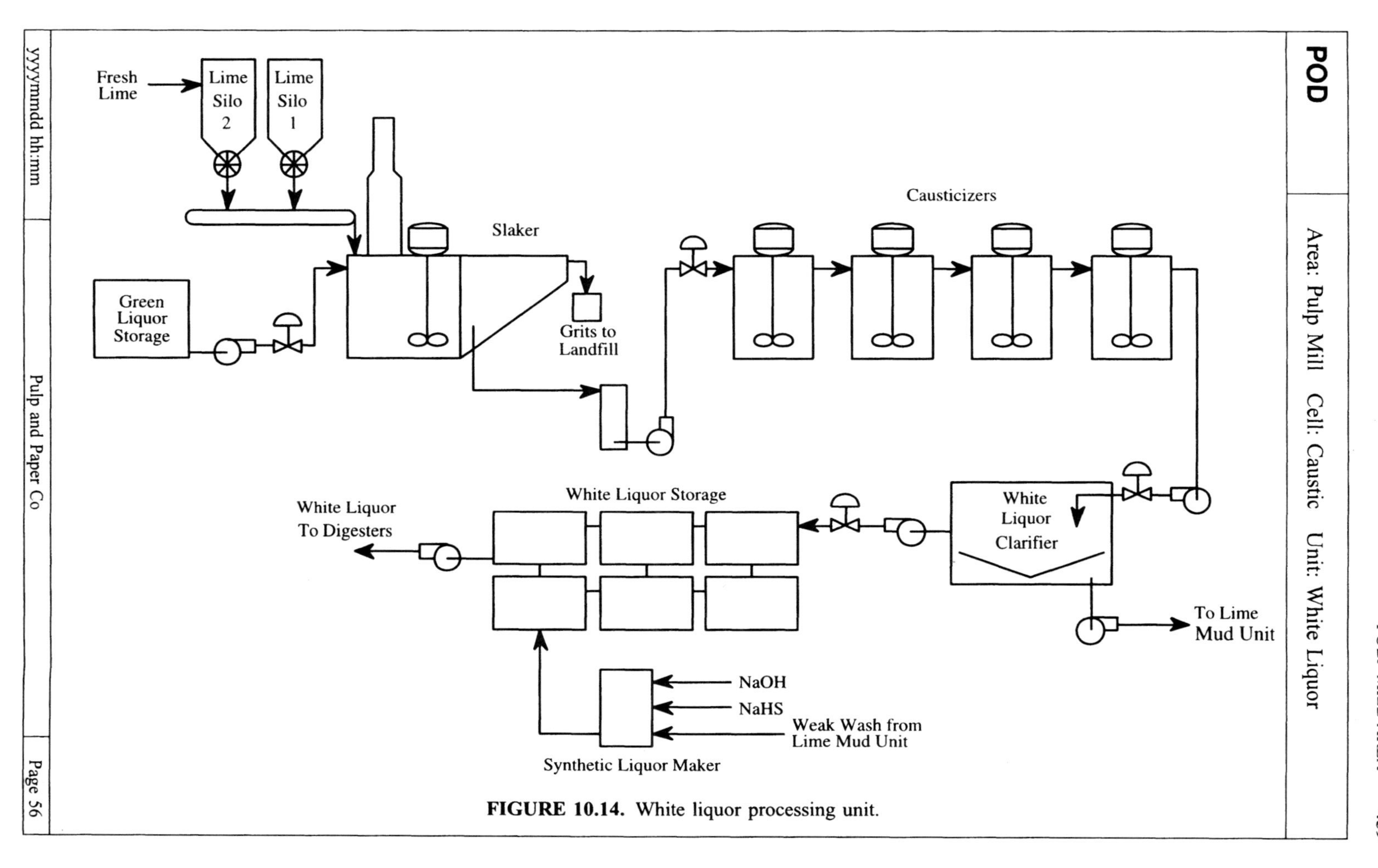


**FIGURE 10.14.** White liquor processing unit.

| POD | Area: Pulp Mill Cell: Caustic Unit: White Liquor |
|---|---|

*Exception Handling*

Shutdown if:

High level on white liquor storage

Overflow on tanks

*Primary Control Objectives*

None.

*Performance Information*

None.

*State End Conditions*

All flows started.

**Operating State Name: Running**

*Routine Activities*

Sample slaker

Sample causticizers

Sample white liquor clarifier

*Exception Handling*

Shutdown if:

High level on white liquor storage

Overflow on tanks

*Primary Control Objectives*

Keep inventory of white liquor tanks above $x$ gallons.

Excess lime to slaker, but less than 1%.

Flow through white liquor clarifier $> x$ gallons per hour

*Performance Information*

Gallons of white liquor produced per shift

Gallons of green liquor processed per shift

Pounds of grits produced per shift

*State End Conditions*

Shutdown requested.

**Operating State Name: Shutting Down**

*Routine Activities*

Stop all pumps.

Stop all agitators.

| yyyymmdd hh:mm | Pulp and Paper Co | Page 57 |
|---|---|---|

| **POD** | Area: Pulp Mill Cell: Caustic Unit: White Liquor |
|---|---|

*Exception Handling*

None.

*Primary Control Objectives*

None.

*Performance Information*

None.

*State End Conditions*

All equipment modules stopped.

| POD | Area: Pulp Mill Cell: Caustic Unit: Lime Mud |
|---|---|

*Lime Mud Unit.* This unit is shown in Figure 10.15 and removes most of the entrained sodium sulfide ($Na_2S$) in the lime mud. The unit consists of a mix tank followed by an 80-ft primary sedimentation tank and then by two smaller sedimentation tanks. The mud removed from the two smaller tanks is stored before being used by the kilns. Makeup water for the 80-ft wash tank is provided from the second set of wash tanks. The liquor removed from the first wash tank dissolves the smelt from the recovery furnace and provides makeup water for fresh white liquor. Makeup water for the two smaller mud washer tanks is provided from the mud filters

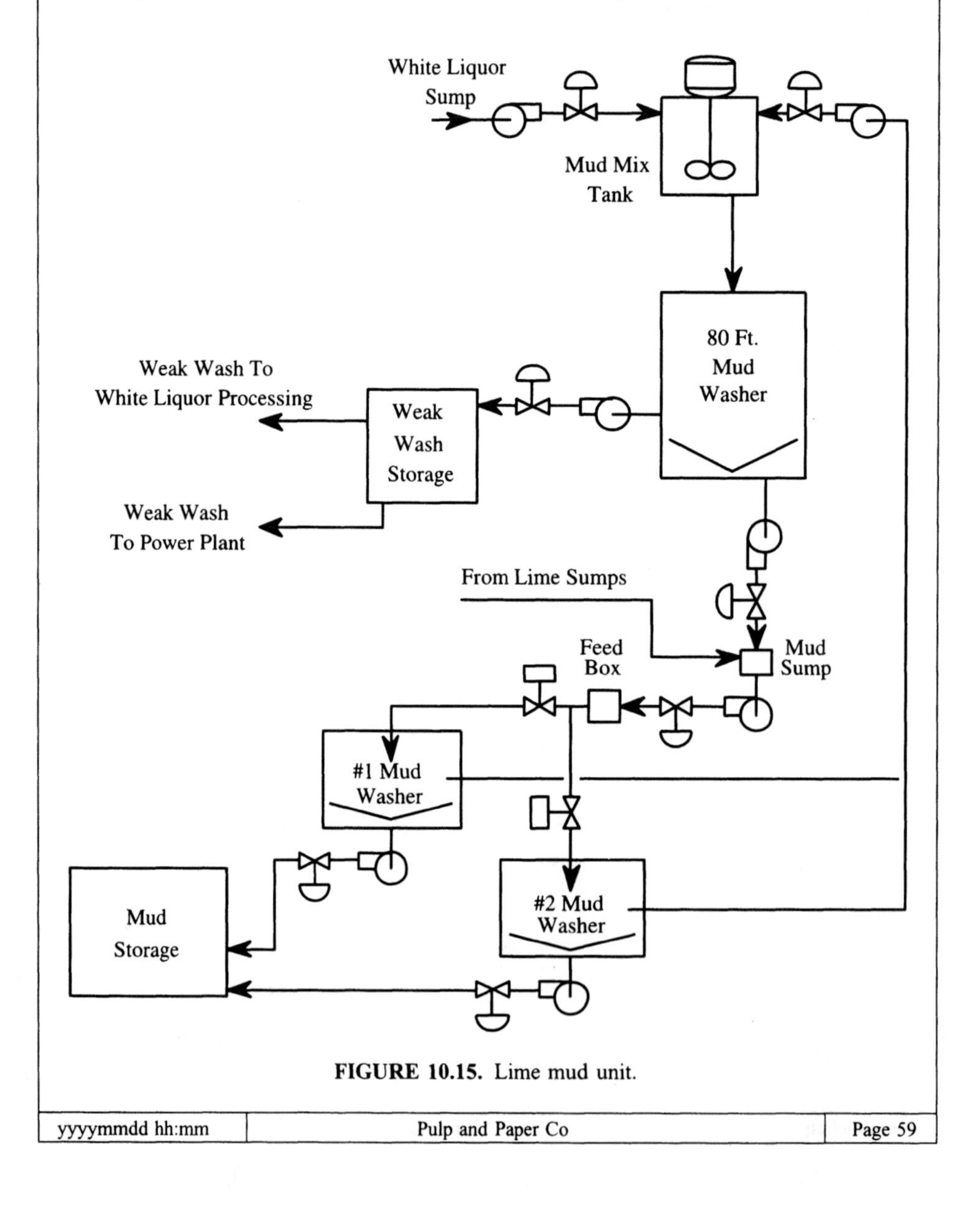


**FIGURE 10.15.** Lime mud unit.

| **POD** | Area: Pulp Mill Cell: Caustic Unit: Lime Mud |
|---|---|

associated with the lime kilns. The device tags are not shown to minimize the diagram clutter.

The operating states of the unit are shown in Figure 10.5:

Idle
Starting
Running
Shutting Down

Normal operation is continuous in order to maintain adequate inventories in the weak wash storage and mud storage tanks. The Idle state is similar to the Idle state for the chip unloading unit and thus is not described. The other states are described as follows:

**Operating State Name: Starting**

*Routine Activities*

Prestart checks.
Start white liquor clarifier sump pump.
Start mud mix tank makeup pump.
Wait for at least one kiln started.
Start sump from 80-ft mud washer.
Set valve path for selected secondary mud washer(s).
Start sump(s) from mud filter for running kiln(s).
Start sump from secondary mud washer(s).

*Exception Handling*

Shutdown if:
High level on mud storage tank
High level on weak wash storage
Overflow on tanks

*Primary Control Objectives*

None.

*Performance Information*

None.

*State End Conditions*

Sumps from secondary mud washer(s) started.

| POD | Area: Pulp Mill Cell: Caustic Unit: Lime Mud |
|---|---|

**Operating State Name: Running**

*Routine Activities*

Sample weak wash

Sample secondary mud washers

*Exception Handling*

Shutdown if:

High level on mud storage

High level on weak wash storage

Overflow on tanks

*Primary Control Objectives*

Concentration of calcium carbonate in mud to storage tank

Total mud flow to mud storage

Adequate inventory in mud storage tank

Adequate inventory in weak wash tank

*Performance Information*

Concentration of sodium sulfide in mud storage tank

Gallons of mud produced per shift

*State End Conditions.*

Shutdown requested.

**Operating State Name: Shutting Down**

*Routine Activities*

Stop all pumps.

*Exception Handling*

None.

*Primary Control Objectives*

None.

*Performance Information*

None.

*State End Conditions*

All pumps stopped.

| yyyymmdd hh:mm | Pulp and Paper Co | Page 61 |
|---|---|---|

| POD | Area: Pulp Mill Cell: Caustic Unit: #1 Lime Kiln |
|---|---|

*#1 Lime Kiln Unit.* A simplified P&ID of the #1 lime kiln unit is depicted in Figure 10.16. The device tags are not shown to minimize the diagram clutter. The mud from the mud storage tank is first passed through a mud filter. The filter is a rotary drum vacuum filter washer that does a final wash on the mud and thickens the mud to 60–70% solids before it enters the kiln. In the lime kiln, the lime mud ($CaCO_3$) is dried, heated, and converted to lime (CaO). The lime kiln is a rotary kiln 10 ft in diameter and 100 ft long. Fuel and air enter the hot end of the tube to produce the flame. Noncondensable gas (NCG) from the blow heat evaporator is also burned in the kiln. The kiln exhaust gases are passed through a Venturi scrubber that uses a water spray to trap the particulates. The lime is cooled after it exits the kiln and is then conveyed to a lime silo. Fresh lime may be added to the silo for lime kiln #2, but may not be added to lime silo #1.

The operating states of the unit are shown in Figure 10.5:

Idle
Starting
Running
Shutting Down

Normal operation is continuous in order to maintain adequate inventory in the lime silo. The Idle state is similar to the Idle state for the chip unloading unit and thus is not described. The other states are described as follows:

**Operating State Name: Starting**

*Routine Activities*

Prestart checks.
Start kiln rotation.
Start air and fuel flows and ignite flame.
Start kiln stack scrubber.
Wait for kiln operating temperature to be reached.
Start noncondensable gas flow.
Start feed pump from mud storage.
Start mud filter.
Start mud filter sump.

*Exception Handling*

Shutdown if:

High kiln hot end temperature
High level on lime silo
Low level on mud storage tank

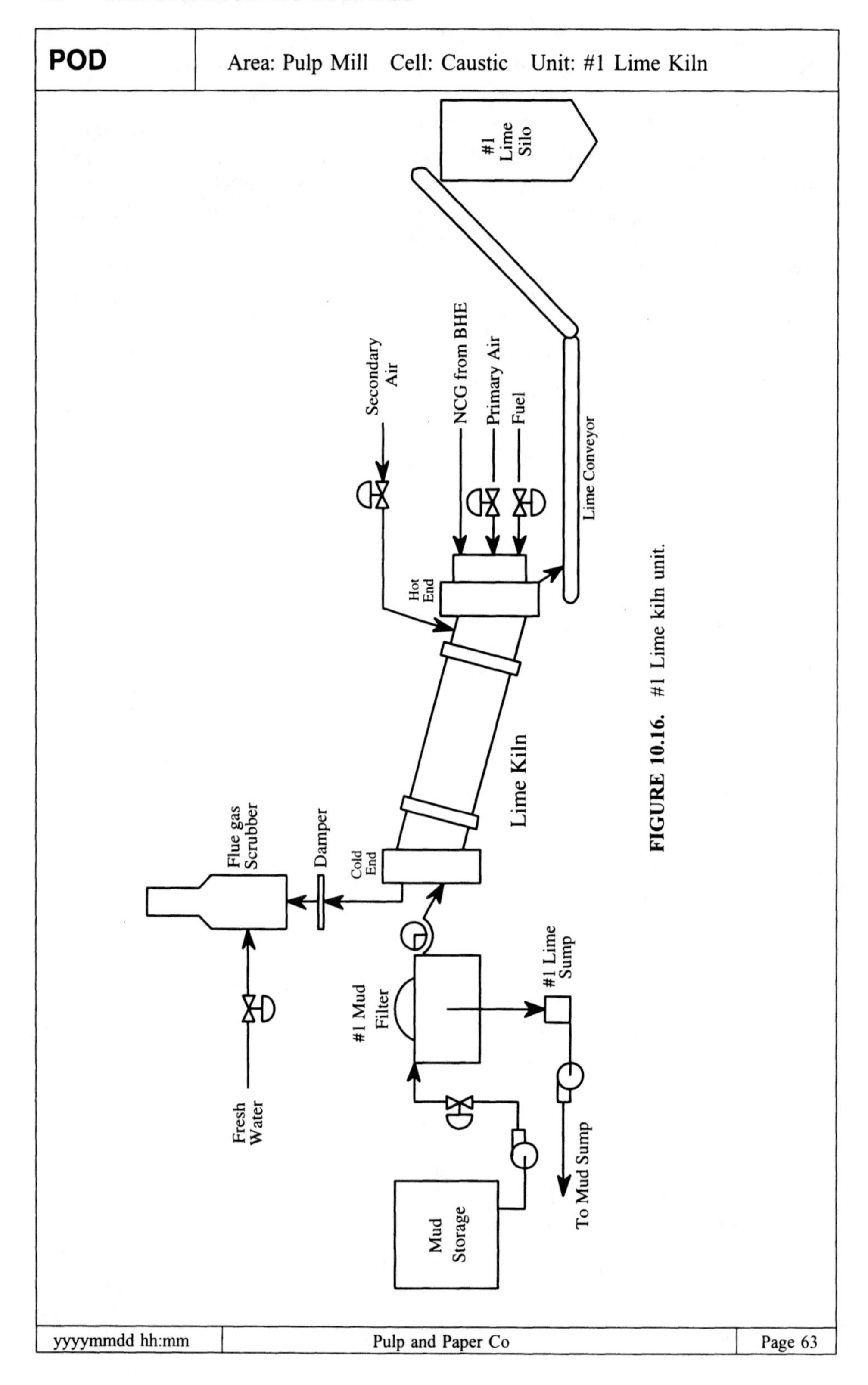


**FIGURE 10.16.** #1 Lime kiln unit.

**POD** | Area: Pulp Mill Cell: Caustic Unit: #1 Lime Kiln

*Primary Control Objectives*

None.

*Performance Information*

None.

*State End Conditions*

All flows started.

**Operating State Name: Running**

*Routine Activities*

Sample stack gas for total reduced sulfur dioxide (TRS).

Sample lime silo.

*Exception Handling*

Shutdown if:

High kiln hot-end temperature for 30 min

High level on lime silo

Low level on mud storage tank

*Primary Control Objectives*

Adequate inventory in lime silo #1

Kiln hot-end temperature

Kiln cold-end temperature

Kiln flue gas oxygen concentration

Total reduced sulfur dioxide in stack gas

Lime activity

*Performance Information*

Tons of lime produced per shift

Total reduced sulfur dioxide in stack gas

Lime activity

*State End Conditions*

Shutdown requested.

**Operating State Name: Shutting Down**

*Routine Activities*

Stop mud flow to kiln.

Stop mud filter.

Wait for lime to trickle through kiln.

yyyymmdd hh:mm | Pulp and Paper Co | Page 64

| POD | Area: Pulp Mill Cell: Caustic Unit: #1 Lime Kiln |
|---|---|

Shut air and gas flow to kiln.

Stop kiln rotation.

*Exception Handling*

None.

*Primary Control Objectives*

None.

*Performance Information*

None.

*State End Conditions*

Kiln rotation stopped.

| POD | Area: Pulp Mill Cell: Caustic Unit: #2 Lime Kiln |
|---|---|

*#2 Lime Kiln Unit.*
Similar to the #1 lime kiln unit.

| yyyymmdd hh:mm | Pulp and Paper Co | Page 66 |
|---|---|---|

| CONTROL CONCEPT | Area: Pulp Mill Cell: Caustic Unit: |
|---|---|

**Caustic Cell Process Management**

*Extent of Automation*

Schedule lime kilns.

Collect performance data from units.

*Flexibility of Automation*

None.

*Control Activity Coordination*

Green liquor unit operation is coordinated with recovery boiler.

*Interaction with Operating Personnel*

Schedule kiln operation once every day. Determined by the expected green liquor production and/or white liquor demand.

Summary information from units displayed for operator.

**Green Liquor Unit Supervision**

*Extent of Automation*

Control all aspects of unit.

Collect performance data.

*Flexibility of Automation*

None.

*Control Activity Coordination*

Coordinate startup and shutdown with recovery boiler.

*Interaction with Operating Personnel*

Handle startup and shutdown commands.

Display primary controlled variables.

Display status information from control modules.

**White Liquor Unit Supervision**

*Extent of Automation*

Control all aspects of unit.

Collect performance data .

*Flexibility of Automation*

None.

*Control Activity Coordination*

Coordinate startup and shutdown with lime mud unit.

| CONTROL CONCEPT | Area: Pulp Mill Cell: Caustic Unit: |
|---|---|

*Interaction with Operating Personnel*

Handle startup and shutdown commands.

Display primary controlled variables.

Display status information from control modules.

**Lime Mud Unit Supervision**

*Extent of Automation*

Control all aspects of unit.

Schedule #1 and #2 mud washer.

Determine #1 mud washer feed flow setpoint.

Determine #1 mud washer sump flow and #2 mud washer sump flow setpoints.

Collect performance data.

*Flexibility of Automation*

None.

*Control Activity Coordination*

Coordinate startup and shutdown with white liquor unit, #1 lime kiln unit, and #2 lime kiln unit.

*Interaction with Operating Personnel*

Handle startup and shutdown commands.

Display primary controlled variables.

Display status information from control modules.

**#1 and #2 Lime Kiln Unit Supervision**

*Extent of Automation*

Control all aspects of unit.

Collect performance data.

*Flexibility of Automation*

None.

*Control Activity Coordination*

Coordinate startup and shutdown with lime mud unit.

*Interaction with Operating Personnel*

Handle startup and shutdown commands.

Display primary controlled variables.

Display status information from control modules.

| CONTROL STRATEGY | Area: Pulp Mill   Cell: Caustic   Unit: Green Liquor |
|---|---|

*Green Liquor Unit*

**Unit Procedure**

The unit procedure phases are the same as the operational states:

Start green liquor clarifier sump pump.

Start green liquor clarifier outlet pump.

Start dregs washer drum.

Start dregs washer sump pump.

**Loops**

Green clarifier level:

Maintain constant level by manipulating flow to green liquor storage.

Clarifier sump flow:

Maintain flow as ratio of flow of green liquor from recovery boiler.

Dregs washer level:

Maintain level by adjusting drum rotation speed.

Dregs washer sump level:

Maintain 50% by adjusting flow to green liquor clarifier.

**Devices**

Green liquor clarifier sump pump:

Start/stop by unit supervision or operator.

Green liquor clarifier outlet pump:

Start/stop by unit supervision or operator.

Dregs washer sump pump:

Start when dregs washer level is above 60%.

Stop when dregs washer level is below 30%.

**Interlocks**

Stop green clarifier outlet pump if green liquor storage level >95%.

| CONTROL STRATEGY | Area: Pulp Mill Cell: Caustic Unit: White Liquor |
|---|---|

*White Liquor Unit.* A possible supervisory control scheme for this unit is described in Section 3.6.1.

**Unit Procedure**

The unit procedure phases are the same as the operational states.

**Loops**

Green liquor flow:
Maintain constant flow as set by caustic cell supervision.
White liquor flow to causticizers:
Maintain constant flow as a ratio of green liquor flow.
White liquor flow to white liquor clarifier:
Maintain constant flow, same as white liquor flow to causticizers.
Clarifier level:
Maintain constant level by manipulating flow to white liquor storage.
White liquor flow to white liquor storage:
Maintain flow, as commanded by clarifier level controller.

**Devices**

Green liquor feed pump:
Start/stop by unit supervision or operator.
Slaker agitator
Start when level is above 30%.
Stop when level is below 10%.
Causticizer feed pump
Start when standpipe level is above 30%.
Stop when standpipe level is below 10%.
Causticizer agitators:
Start/stop by unit supervision or operator.
White liquor clarifier feed pump:
Start/stop by unit supervision or operator.
White liquor storage feed pump
Start when white liquor clarifier level is above 50%.
Stop when white liquor clarifier level is below 30%.

| yyyymmdd hh:mm | Pulp and Paper Co | Page 70 |
|---|---|---|

**CONTROL STRATEGY** Area: Pulp Mill Cell: Caustic Unit: White Liquor

White liquor sump pump

Start when white liquor storage feed pump is started.

Stop when white liquor storage feed pump is stopped.

**Interlocks**

Green liquor feed pump allowed to run if green liquor storage level >40%.

Stop liquor feed pump if green liquor storage level <5%.

**CONTROL STRATEGY** Area: Pulp Mill Cell: Caustic Unit: Lime Mud

*Lime Mud Unit*

### Unit Procedure

The unit procedure phases are the same as the operational states.

### Loops

Mud flow from white liquor clarifier:
Maintain constant flow as set by caustic cell process management.
Weak wash flow to storage:
Maintain constant flow as a ratio of mud flow from white liquor clarifier.
Eighty-foot mud washer sump flow:
Maintain constant pump motor torque as a ratio of the mud density.
#1 and #2 Mud washer feed flow:
Maintain constant flow as commanded by lime mud unit supervisor.
#1 Mud washer sump flow:
Maintain constant flow as commanded by lime mud unit supervisor.
#2 Mud washer sump flow:
Maintain constant flow as commanded by lime mud unit supervisor.

### Devices

Mud mix tank agitator:
Start/stop by unit supervision or operator.
Wash water recirculation pump:
Start/stop by unit supervision or operator.
Weak wash to storage pump:
Start/stop by unit supervision or operator.
Eighty-foot mud washer sump feed pump:
Start/stop by unit supervision or operator.
#1 and #2 Mud washer feed pump:
Start/stop by unit supervision or operator.
#1 Mud washer feed valve
Start when #1 mud washer is started.
Stop when #1 mud washer is stopped.
#1 Mud washer sump pump
Start when #1 mud washer is started.
Stop when #1 mud washer is stopped.

**CONTROL STRATEGY** Area: Pulp Mill Cell: Caustic Unit: Lime Mud

#2 Mud washer feed valve
Start when #2 mud washer is started.
Stop when #2 mud washer is stopped.

#2 Mud washer sump pump
Start when #2 mud washer is started.
Stop when #2 mud washer is stopped.

**Interlocks**

To be determined (TBD)

**CONTROL STRATEGY** Area: Pulp Mill Cell: Caustic Unit: #1 Lime Kiln

*#1 Lime Kiln Unit* (similar to #2 Lime Kiln Unit)

**Unit Procedure**

The unit procedure phases are the same as the operational states.

*Kiln Unit Multivariable Control.* The nature of any strategy to control the lime kiln operation necessarily involves several interacting loops. Loop pairing and decoupling are possible approaches to the construction of a kiln control strategy. The kiln control example summarized in Section 3.6.1 uses dynamic decoupling with feedforward to control the kiln. A model predictive approach to the kiln control problem is described here.

The primary control objectives of the kiln control are:

Hot-end temperature
Cold-end temperature
Flue gas excess oxygen

The available manipulated variables are:

Fuel flow
Primary air flow
Secondary air flow
Damper position

The kiln air supply consists of a primary air flow and a secondary air flow. The primary air flow and fuel flow enter the kiln at the hot end and produce the flame. The primary air flow controls the flame length and consequently the hot-end temperature. The secondary air is injected through a ring of jets about 20 ft from the hot end of the furnace and is primarily used to adjust the flame width. The position of the cold-end damper is a control primarily for the cold-end temperature. The percent of excess oxygen in the flue gas is controlled by adjusting the air–fuel ratio. Because of the interactive nature of this process, a multivariable model predictive controller is constructed with the block diagram shown in Figure 10.17. The hot- and cold-end temperatures are target values selected by the operator after analyzing a sample of the lime from the kiln hot end. The excess oxygen target is usually 4%. The total reduced sulfur dioxide (TRS) concentration is used to fine tune the excess oxygen target. If the TRS is high, the oxygen target is increased. If the TRS is low, the oxygen target is decreased. The multivariable controller outputs are setpoints to single-loop controllers. The fuel flow and air–fuel ratio are used to set the fuel flow and primary air flow (through ratio control). The secondary air flow is a ratio of the primary air flow, with the ratio set by an operator.

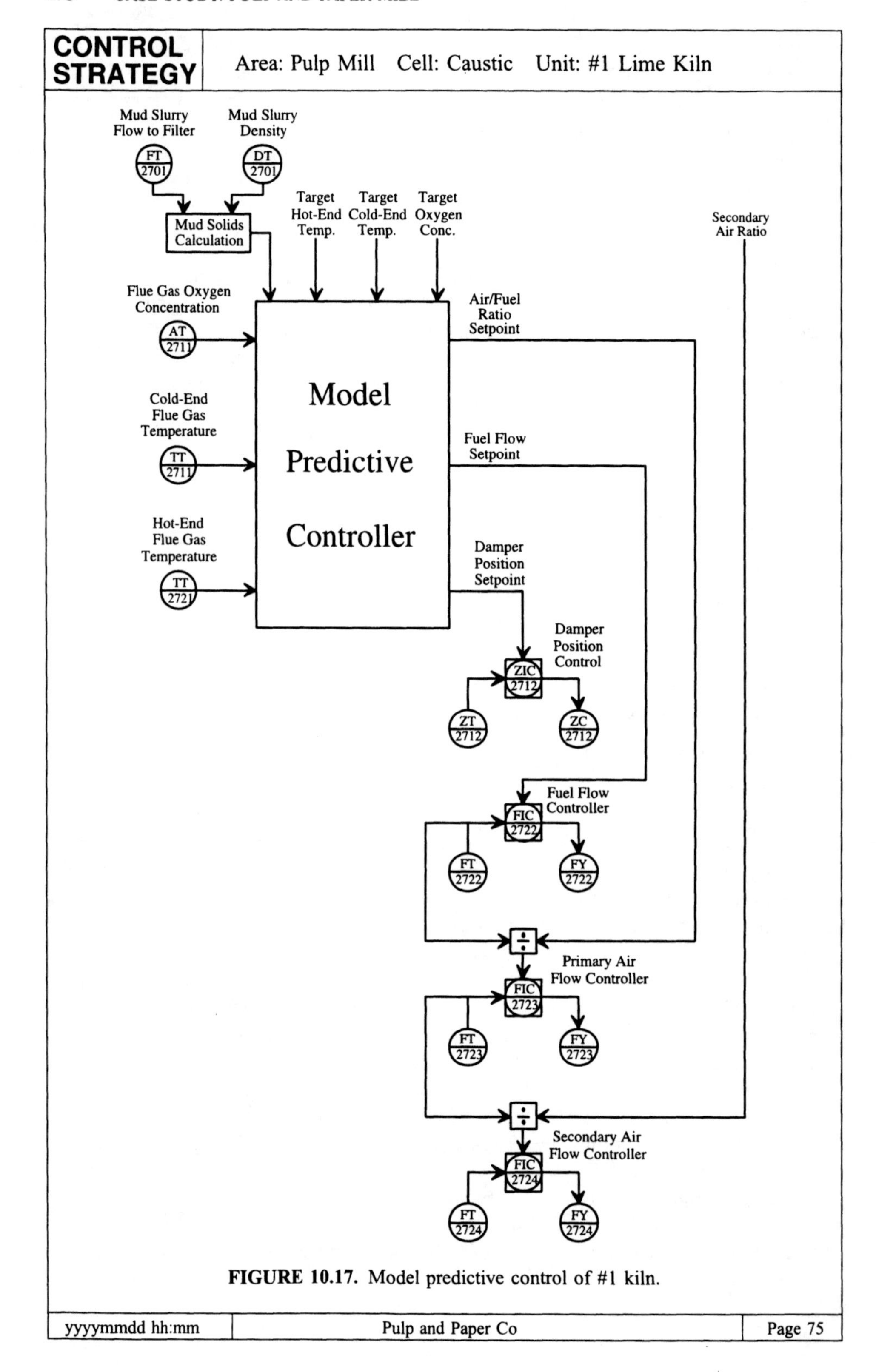


**FIGURE 10.17.** Model predictive control of #1 kiln.

| CONTROL STRATEGY | Area: Pulp Mill Cell: Caustic Unit: #1 Lime Kiln |
|---|---|

**Loops**

Mud slurry flow from mud storage:
Maintain constant flow as set by unit supervision.
Fuel flow:
Maintain constant flow as set by unit supervisory controller.
Primary air flow:
Maintain constant flow as ratio of fuel flow. Ratio set by unit supervisory controller.
Secondary air flow:
Maintain constant flow as ratio of primary air flow. Ratio set to maintain desired flame shape.
Damper position:
Maintain constant position as set by unit supervisory controller.
Mud filter level:
Maintain level by adjusting drum rotation speed.
#1 Kiln rotation speed:
Maintain constant speed as set by cell process management.

**Devices**

#1 lime sump pump:
Start/stop by unit supervision or operator.
#1 mud screw conveyor:
Start/stop by unit supervision or operator.
#1 flue gas scrubber water feed valve:
Start/stop by unit supervision or operator.
#1 NCG block valve:
Start/stop by unit supervision or operator.
#1 Kiln product conveyor:
Start/stop by unit supervision or operator.

**Interlocks**

Shutdown unit if flame not present.
Shutdown unit if negative hood pressure.

| yyyymmdd hh:mm | Pulp and Paper Co | Page 76 |
|---|---|---|

| CONTROL STRATEGY | Area: Pulp Mill Cell: Caustic Equipment Modules |
|---|---|

*White Liquor Storage Equipment Module*

TBD.

*Green Liquor Storage Equipment Module*

TBD.

*#1 Lime Silo Equipment Module*

Holds reburned lime. TBD.

*#2 Lime Silo Equipment Module*

Holds reburned and fresh lime. TBD.

| POD | Area: Pulp Mill Cell: Bleach Unit: |
|---|---|

**Bleach Cell**

A simplified PFD of the bleach cell is depicted in Figure 10.18. The cell consists of a unit that bleaches softwood pulp and a unit that bleaches hardwood pulp. A chlorine dioxide generator supplies chlorine dioxide, an important bleaching chemical that is too hazardous to transport and so is manufactured on-site. Both bleach units operate independently of each other.

The operating states of this cell are shown in Figure 10.5:

Idle
Starting
Running
Shutting Down

Normal operation is continuous in order to maintain adequate inventories in the low- and high-density storage tanks that supply the paper machines. The Idle state is similar to the Idle state for the chip unloading unit and thus is not described. The other states are described as follows:

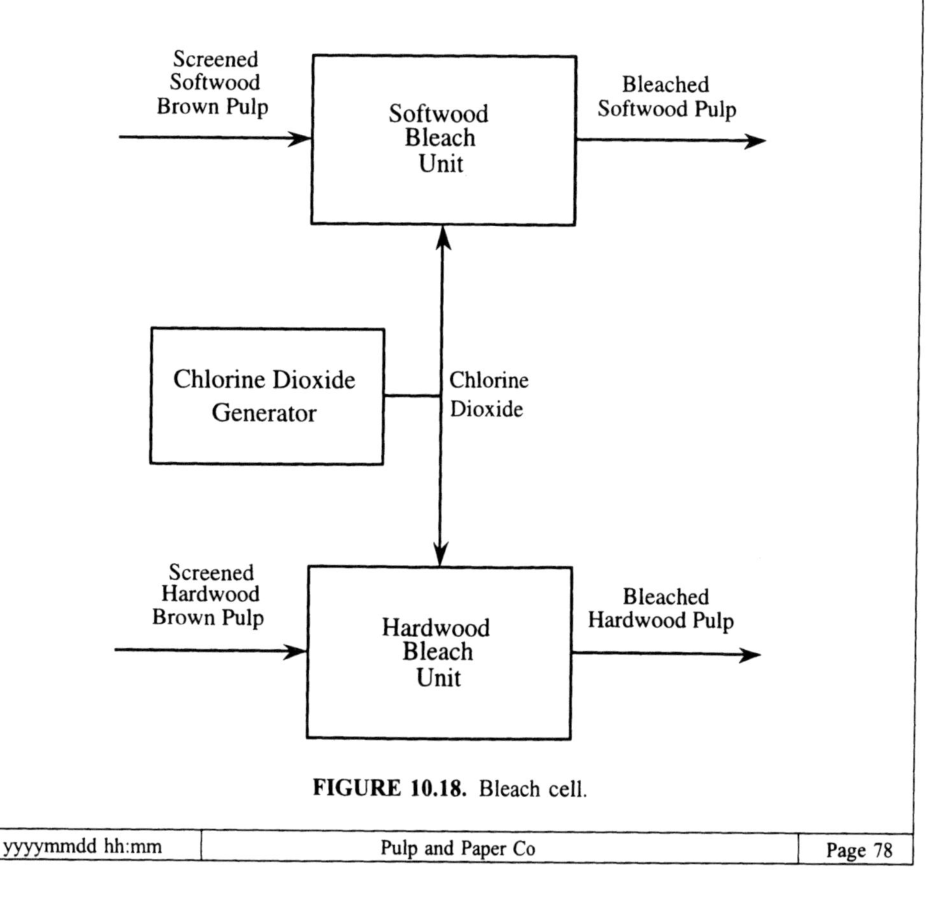


**FIGURE 10.18.** Bleach cell.

| POD | Area: Pulp Mill Cell: Bleach Unit: |
|---|---|

**Operating State Name: Starting**

*Routine Activities*

Prestart checks.

Start chlorine dioxide generator.

Wait for one bleach unit to be running.

*Exception Handling*

Shutdown if:

Low brown pulp inventory

High bleached pulp inventory

*Primary Control Objectives*

None.

*Performance Information*

None.

*State End Conditions*

At least one bleach unit started.

**Operating State Name: Running**

*Routine Activities*

Monitor primary control objectives.

Monitor productivity.

*Exception Handling*

Shutdown if:

Low brown pulp inventory

High bleached pulp inventory

*Primary Control Objectives*

Adequate inventory of low-density bleached softwood pulp

Softwood pulp consistency

Softwood pulp brightness

Softwood pulp pH

Adequate inventory of low-density bleached hardwood pulp

Hardwood pulp consistency

Hardwood pulp brightness

Hardwood pulp pH

| yyyymmdd hh:mm | Pulp and Paper Co | Page 79 |
|---|---|---|

| **POD** | Area: Pulp Mill Cell: Bleach Unit: |
|---|---|

*Performance Information*

Tons of softwood pulp produced per day

Softwood pulp brightness

Softwood pulp pH

Tons of hardwood pulp produced per day

Hardwood pulp brightness

Hardwood pulp pH

*State End Conditions*

Shutdown requested.

**Operating State Name: Shutting Down**

*Routine Activities*

Shutdown chlorine dioxide generator.

*Exception Handling*

None.

*Primary Control Objectives*

None.

*Performance Information*

None.

*State End Conditions*

Chlorine dioxide unit shut down.

| POD | Area: Pulp Mill Cell: Bleach Unit: Softwood Bleach |
|---|---|

*Softwood Bleach Unit.* A simplified P&ID of the softwood bleach unit is shown in Figure 10.19. The device tags are not shown to minimize the diagram clutter. The hardwood bleach unit is similar, except that the last stage is absent. The bleach unit treats the pulp in a series of chemical stages. Each stage typically consists of a mixing unit to mix chemicals with the pulp, a retention tank to provide time for the bleaching chemicals to react with the pulp, and a washer to remove the bleaching chemicals. The four stages are:

Chlorine dioxide
Hydrogen peroxide
Chlorine dioxide
Hydrogen peroxide

The four rotary vacuum drum washers operate in a similar manner to the brown stock washers. The wash water for the last stage comes from the mill water supply. The bleached pulp proceeds to high-density storage. The pulp from the high-density storage is combined with water from the fourth washer to constitute the desired pulp density in the low-density storage tank, from which it is pumped to the paper mill.

The operating states of the unit are shown in Figure 10.5:

Idle
Starting
Running
Shutting Down

Normal operation is continuous in order to maintain adequate inventories in the low- and high-density storage tanks that supply the paper mill. The Idle state is similar to the Idle state for the chip unloading unit and thus is not described. The other states are described as follows:

**Operating State Name: Starting**

*Routine Activities*

Prestart checks.

Start wash water pump.

Start up service pumps, in reverse order of the pulp flow, to initially charge washer vats and service tanks.

Start low-density outlet pump in softwood brown stock washer.

Start washer drum drive motors.

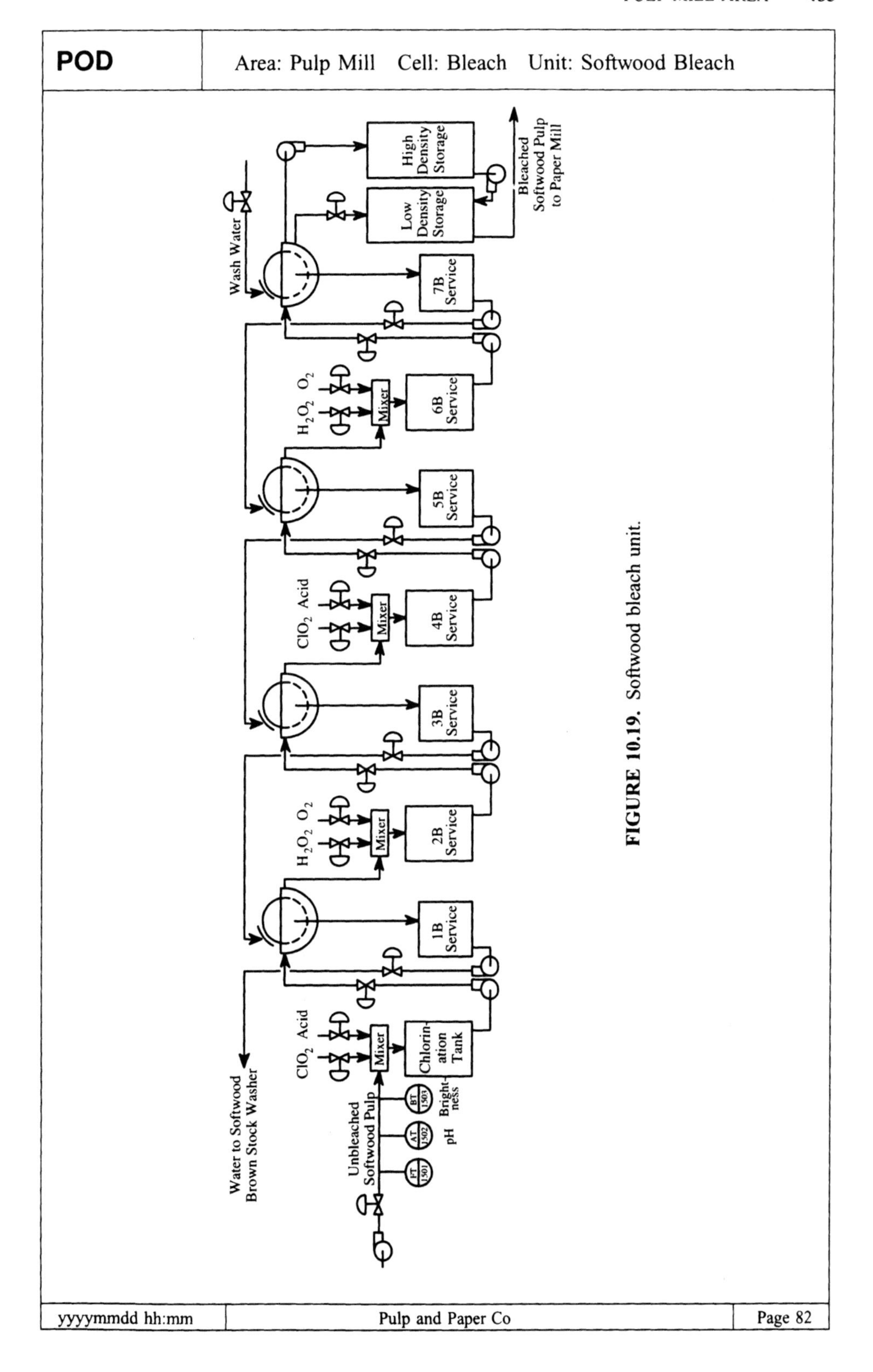


**FIGURE 10.19.** Softwood bleach unit.

| POD | Area: Pulp Mill Cell: Bleach Unit: Softwood Bleach |
|---|---|

*Exception Handling*

Shutdown if:

High level on softwood bleach low- or high-density tank

Low level on softwood brown stock washer low- or high-density tank

*Primary Control Objectives*

Bleached pulp consistency

Bleached pulp pH

Bleached pulp brightness

Bleached pulp Kappa number

Bleached pulp conductivity

*Performance Information*

None.

*State End Conditions*

Washer drums running.

**Operating State Name: Running**

*Routine Activities*

Sample bleached pulp for laboratory brightness test.

*Exception Handling*

Shutdown if:

High level on softwood bleach low- or high-density storage

Low level on softwood brown stock washer low- or high-density tank

Chlorine dioxide generator shut down

*Primary Control Objectives*

Bleached pulp consistency

Bleached pulp pH

Bleached pulp brightness

*Performance Information*

Tons of bleached pulp produced per day

Bleached pulp brightness

Bleached pulp pH

Dilution factor

*State End Conditions*

Shutdown requested.

| **POD** | Area: Pulp Mill Cell: Bleach Unit: Softwood Bleach |
| --- | --- |

**Operating State Name: Shutting Down**

*Routine Activities*

Stop low-density outlet pump in softwood brown stock washer.

Stop wash water pump.

Empty service tanks.

Stop service pumps, in reverse order of the pulp flow.

Stop washer drum drive motors.

*Exception Handling*

None.

*Primary Control Objectives*

None.

*Performance Information*

None.

*State End Conditions*

Washer drum drive motors stopped.

| POD | Area: Pulp Mill   Cell: Bleach   Unit: Hardwood Bleach |
| --- | --- |

*Hardwood Bleach Unit.* The operation of the hardwood bleach unit is similar to the softwood bleach unit, except that only three bleach stages are present. The three stages are:

Chlorine dioxide
Hydrogen peroxide
Chlorine dioxide

| POD | Area: Pulp Mill Cell: Bleach Unit: Chlorine Dioxide |
|---|---|

*Chlorine Dioxide Generator.* TBD.

| yyyymmdd hh:mm | Pulp and Paper Co | Page 86 |
|---|---|---|

| CONTROL CONCEPT | Area: Pulp Mill Cell: Bleach Unit: |
|---|---|

**Bleach Cell Process Management**

*Extent of Automation*

Schedule bleach units.

Collect performance data from units.

*Flexibility of Automation*

None.

*Control Activity Coordination*

Coordinate cell operation with brown pulp cell and paper machines.

*Interaction with Operating Personnel*

Schedule bleach units once every day in order to maximize production. Determined by expected paper machine schedule and brown pulp mill operation.

Summary information from units displayed for operator.

**Softwood Bleach Unit Supervision**

*Extent of Automation*

Control all aspects of unit.

Collect performance data (pulp consistency, pH, brightness).

*Flexibility of Automation*

None.

*Control Activity Coordination*

Coordinate with Softwood Brown Stock Washer.

*Interaction with Operating Personnel*

Accept startup and shutdown commands.

Operator determines brown pulp flow.

Display primary controlled variables.

Display all flows.

Display status information from control modules.

**Hardwood Bleach Unit Supervision**

Similar to softwood bleach unit supervision.

**Chlorine Dioxide Generator Unit Supervision.**

TBD.

| CONTROL STRATEGY | Area: Pulp Mill Cell: Bleach Unit: Softwood Bleach |
|---|---|

*Softwood Bleach Unit.* All flows in the unit are adjusted in accordance with the brown pulp inlet flow rate. In addition, the flows of the bleaching chemicals are adjusted in accordance with the pH and brightness of the brown pulp.

**Unit Procedure**

The unit procedure phases are the same as the operational states.

**Loops**

Brown pulp stock flow:
Maintain flow as set by operator.
Chlorine dioxide flow to first mixer:
Maintain flow as a ratio of the brown pulp stock flow and feedforward of the brown pulp brightness.
Acid flow to first mixer:
Maintain flow as a ratio of the chlorine dioxide flow and feedforward of the brown pulp pH.
Flow to first washer:
Maintain as constant ratio of the brown pulp stock flow.
Flow from 1B service tank to softwood brown stock washer:
Maintain as constant ratio of the brown pulp stock flow.
Hydrogen peroxide flow to second mixer:
Maintain flow as a ratio of the brown pulp stock flow and feedforward of the brown pulp brightness.
Oxygen flow to second mixer:
Maintain flow as a ratio of the hydrogen peroxide flow.
Flow to second washer:
Maintain as constant ratio of the brown pulp stock flow.
Flow from 3B service tank to first washer:
Maintain as constant ratio of the brown pulp stock flow.
Chlorine dioxide flow to third mixer:
Maintain flow as a ratio of the brown pulp stock flow and feedforward of the brown pulp brightness.
Acid flow to third mixer:
Maintain flow as a ratio of the chlorine dioxide flow and feedforward of the brown pulp pH.
Flow to third washer:
Maintain as constant ratio of the brown pulp stock flow.

| CONTROL STRATEGY | Area: Pulp Mill Cell: Bleach Unit: Softwood Bleach |
|---|---|

Flow from 5B service tank to second washer:

Maintain as constant ratio of the brown pulp stock flow.

Hydrogen peroxide flow to fourth mixer:

Maintain flow as a ratio of the brown pulp stock flow and feedforward of the brown pulp brightness.

Oxygen flow to fourth mixer:

Maintain flow as a ratio of the hydrogen peroxide flow.

Flow to fourth washer:

Maintain as constant ratio of the brown pulp stock flow.

Flow from 7B service tank to third washer:

Maintain as constant ratio of the brown pulp stock flow.

Washer level (all washers):

Maintain washer water level by adjusting drum rotation speed.

Bleached pulp consistency:

Maintain consistency by manipulating the ratio of the flow from the high-density storage to the low-density storage to the flow of water from the fourth washer.

Wash water flow:

Maintain flow as a ratio of the brown pulp stock flow.

**Devices**

Chlorination tank outlet pump

Start when commanded by unit supervision.

Stop when commanded by unit supervision.

Service tank (1B, 2B, 3B, 4B, 5B, 6B, 7B) outlet pumps

Start when commanded by unit supervision.

Stop when commanded by unit supervision.

High-density storage inlet pump

Start when commanded by unit supervision.

Stop when commanded by unit supervision.

**Interlocks**

Stop a service tank outlet pump when the service tank level is below 5%.

| CONTROL STRATEGY | Area: Pulp Mill Cell: Bleach Unit: Hardwood Bleach |
|---|---|

*Hardwood Bleach Unit.* Similar to softwood bleach unit control strategy.

| **CONTROL STRATEGY** | Area: Pulp Mill Cell: Bleach Unit: Chlorine Dioxide |
|---|---|

*Chlorine Dioxide Generator Unit.* TBD.

| POD | Area: Pulp Mill Cell: Groundwood Mill Unit: |
|---|---|

**Groundwood Mill Cell**

The POD and Control Concept are documented only at the cell level.

Figure 10.20 shows a simplified PFD of the groundwood mill cell. Debarked logs are received from the woodyard cell and are ground in grinders. Water is used as a cooling medium. The ground pulp is first screened in a bull screen to separate the larger pieces, which are recycled back to the grinder. The coarsely screened pulp is then passed through a low-pressure screen and stored. The rejects from the low-pressure screen are further refined before being conveyed to storage. The screened pulp is cleaned and then passed through deckers that remove the excess water from the pulp before being stored.

The operating states of this cell are shown in Figure 10.5:

Idle
Starting
Running
Shutting Down

Normal operation is continuous in order to maintain an adequate inventory in the groundwood pulp storage tank that supplies the paper machine. The Idle state is similar to the Idle state for the chip unloading unit and thus is not described. The other states are described as follows:

**Operating State Name: Starting**

*Routine Activities*

Prestart checks.
Start white water storage outlet pump.
Start grinder unit.
Start screening unit.
Start cleaning unit.

*Exception Handling*

Shutdown if:

Grinding, screening, and cleaning units shut down
High level on groundwood pulp storage tank
Low level on white water storage tank

*Primary Control Objectives*

None.

*Performance Information*

None.

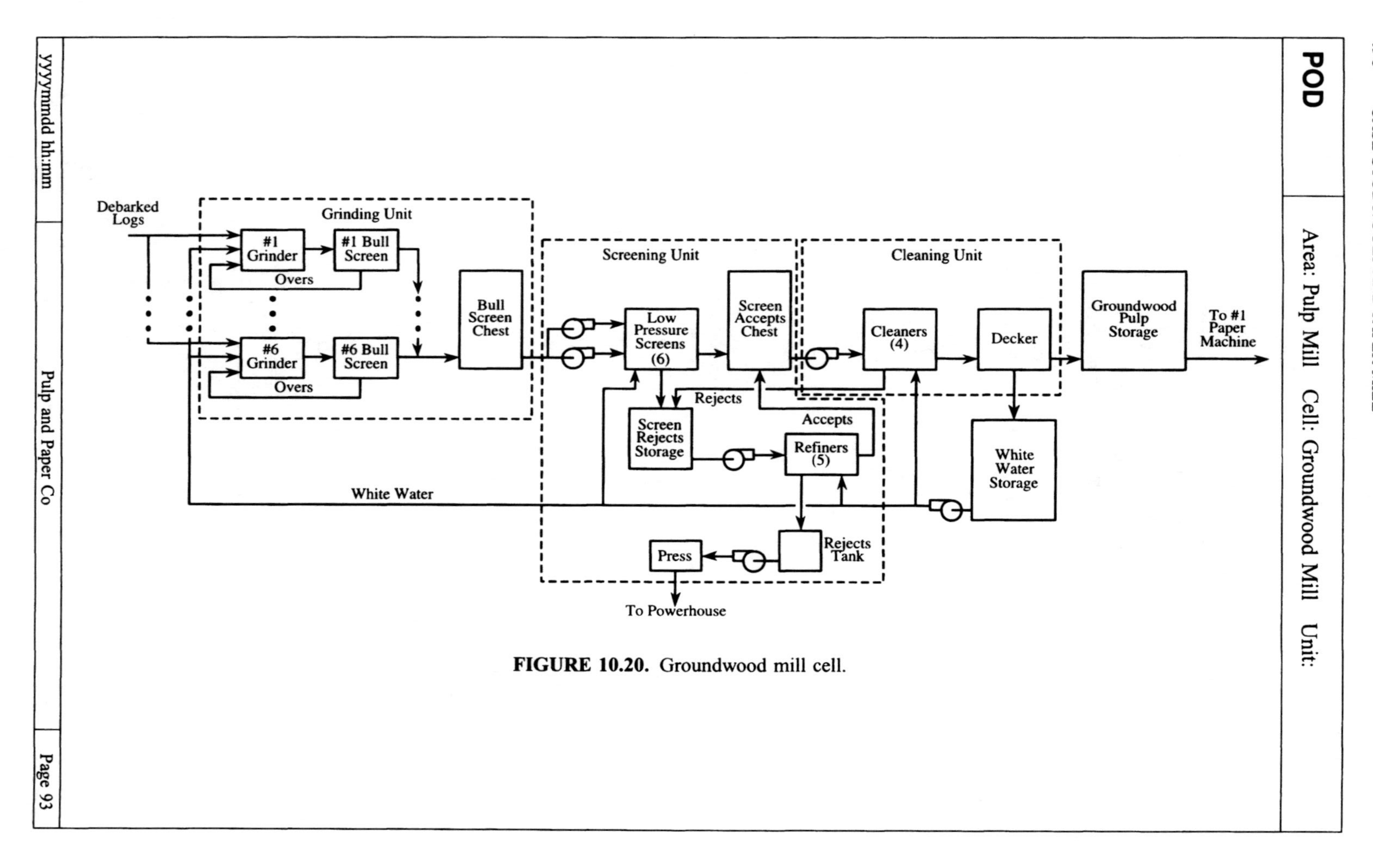


FIGURE 10.20. Groundwood mill cell.

| POD | Area: Pulp Mill Cell: Groundwood Mill Unit: |
|---|---|

*State End Conditions*
All units running

**Operating State Name: Running**

*Routine Activities*
Sample pulp every shift.

*Exception Handling*
Shutdown if:
Grinding, screening, and cleaning units shut down
High level on groundwood pulp storage tank
Low level on white water storage tank

*Primary Control Objectives*
Adequate inventory in groundwood pulp storage
Groundwood pulp consistency

*Performance Information*
Tons of groundwood pulp produced per day
Groundwood pulp consistency

*State End Conditions*
Shutdown requested.

**Operating State Name: Shutting Down**

*Routine Activities*
Shutdown grinding unit.
Shutdown screening unit.
Shutdown cleaning unit.
Stop white water storage outlet pump.

*Exception Handling*
None.

*Primary Control Objectives*
None.

*Performance Information*
None.

*State End Conditions*
All units shutdown.

| CONTROL CONCEPT | Area: Pulp Mill Cell: Groundwood Mill Unit: |
|---|---|

**Groundwood Mill Cell Process Management**

*Extent of Automation*

Collect performance data from units.

Control white water tank.

*Flexibility of Automation*

None.

*Control Activity Coordination*

Coordinate startup and shutdown of grinding unit with woodyard log handling unit.

*Interaction with Operating Personnel*

Summary information from units displayed for operator.

| POD | Area: Paper Mill Cell: Unit: |
|---|---|

## PAPER MILL AREA

The paper mill area is subdivided into the process cells of #1 paper machine, #2 paper machine, and packaging, as shown in Figure 10.1. The process operation and Control Concept are only developed at the cell level.

| POD | Area: Paper Mill Cell: #1 Paper Machine Unit: |
|---|---|

**#1 Paper Machine Cell**

This cell produces catalog-grade glossy paper. An overall view of the units in the paper machine is shown in Figure 10.21. Individual units are shown in Figures 10.22–10.26. The pulp for the paper machine is prepared in the stock preparation unit (Figure 10.22). In this unit, softwood pulp, groundwood pulp, and broke (repulped paper machine rejects) are refined and then combined in the proper ratio into the blend chest. From the blend chest, the pulp is combined with white water from the Fourdrinier to produce pulp of a certain consistency. The machine chest serves as a buffer between the blending and the paper machine. The ticklers perform a final refining of the pulp fibers. After the ticklers, contaminants are removed from the pulp slurry with a series of vortex cleaners. A high percentage of usable fiber occurs in the reject stream from the vortex cleaners; hence the rejects from the primary cleaners are further treated with secondary cleaners, the rejects from the secondary cleaners are treated with a tertiary cleaner, and the rejects from the tertiary cleaners are treated with quaternary cleaners. A chest acts as a small buffer to the wet end of the paper machine.

A simplified diagram of the wet-end unit is shown in Figure 10.23. The wet-end unit consists of the head box, which meters the slurry onto a horizontal, moving, fine-mesh woven wire cloth. The wire cloth runs over the table rolls, suction boxes, and then the couch roll, where the web of fibers leaves the Fourdrinier table. The paper web then proceeds through two presses that mechanically remove water and compress the sheet. The web then proceeds through a set of steam-heated dryer rolls (Figure 10.24) that remove the water by evaporation. The dryer also contains a size press that applies sizing, which improves the water resistance of the paper. Finally, the paper is collected on a reel.

The supercalender (Figure 10.25) and coating (Figure 10.26) units are separate from the paper machine so that different customer requirements can be handled. Depending on the customer, a roll may pass through either unit, both units, or neither unit. The supercalender unit improves the smoothness of the paper. The coater unit is used to apply a coating to the paper. For example, certain papers are coated with clay to impart glossy properties to the paper.

The operating states of this cell are shown in Figure 10.5:

Idle

Starting

Running

Shutting Down

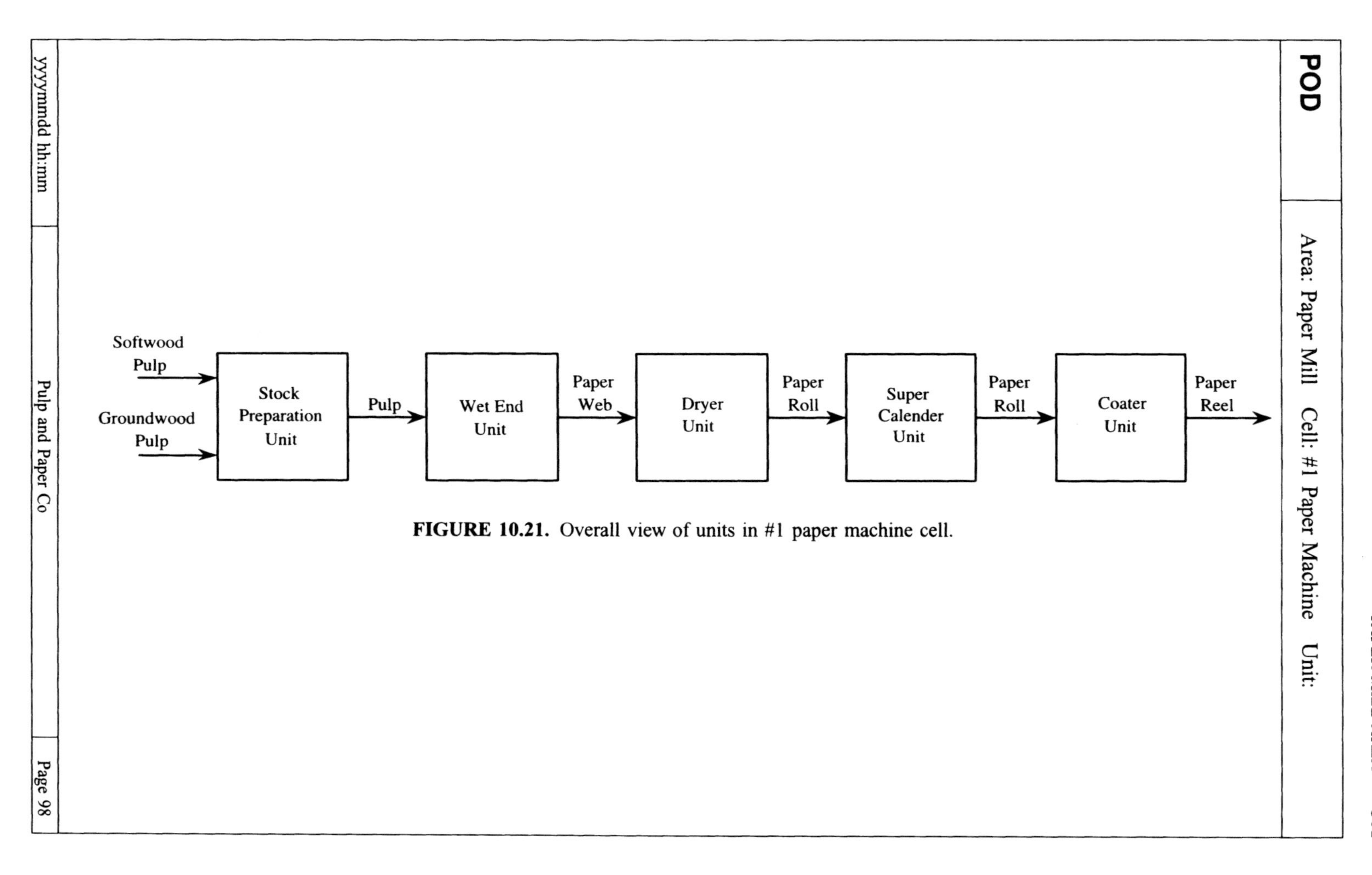


**FIGURE 10.21.** Overall view of units in #1 paper machine cell.

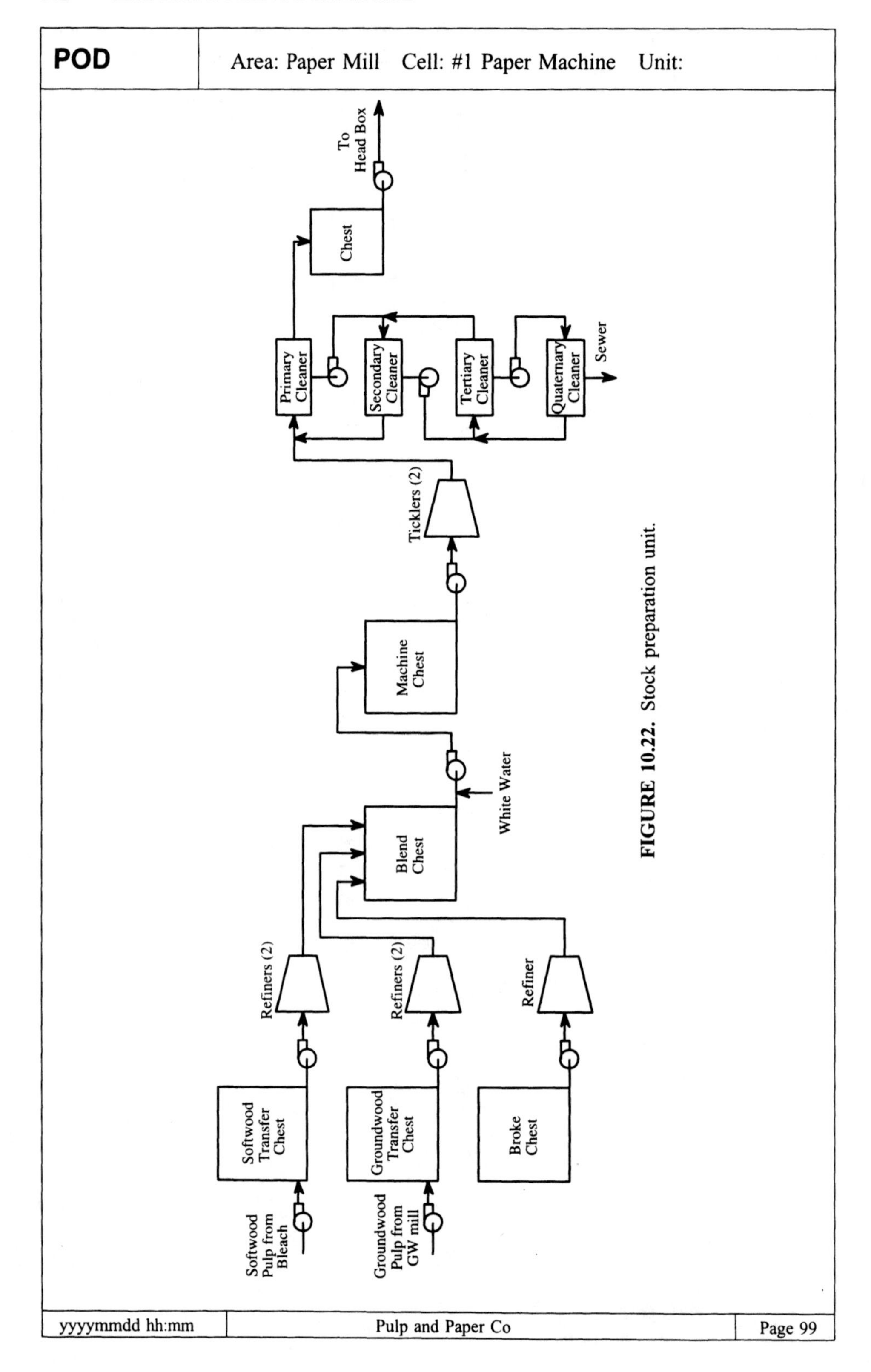


**FIGURE 10.22.** Stock preparation unit.

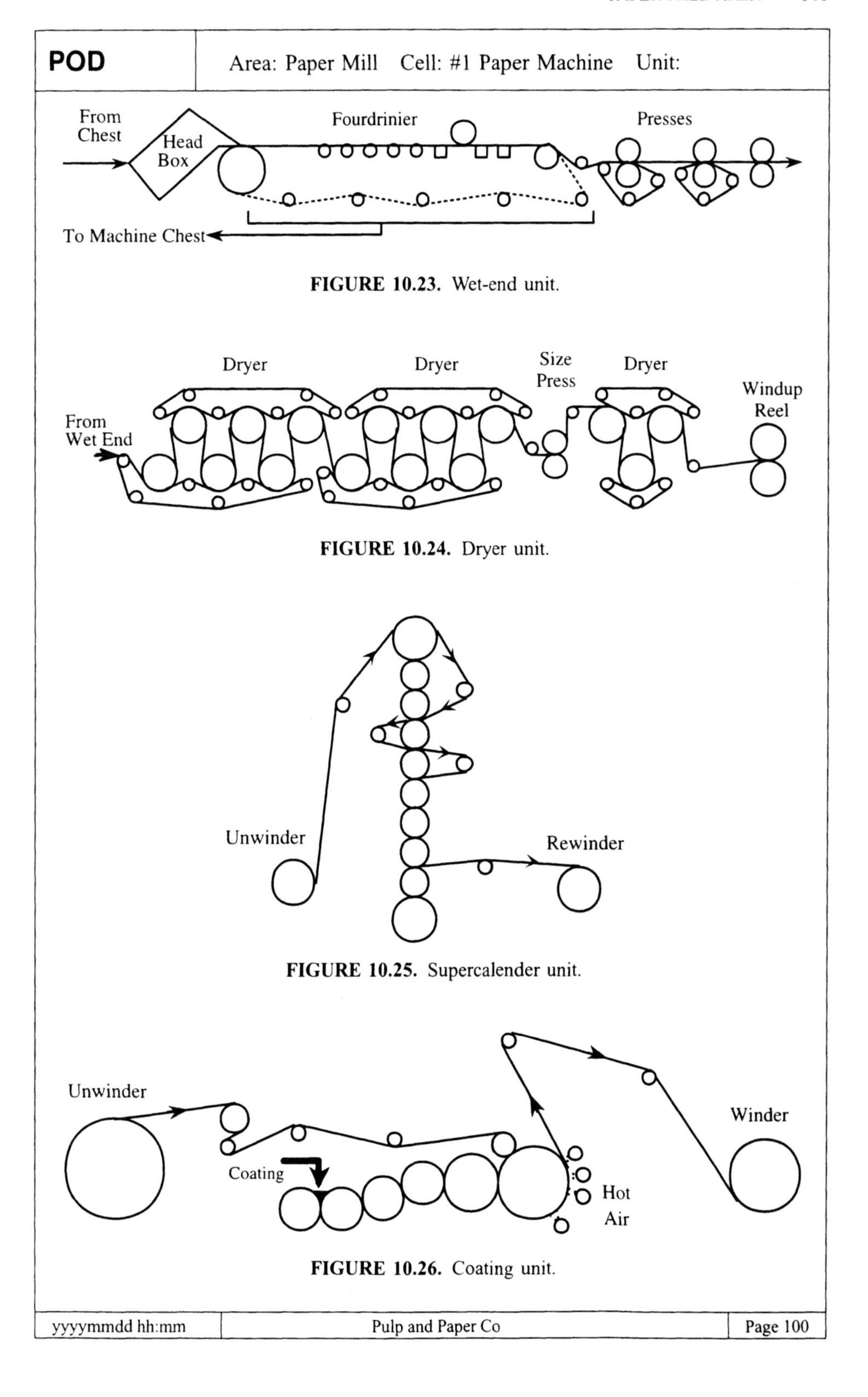


**FIGURE 10.23.** Wet-end unit.

**FIGURE 10.24.** Dryer unit.

**FIGURE 10.25.** Supercalender unit.

**FIGURE 10.26.** Coating unit.

| **POD** | Area: Paper Mill Cell: #1 Paper Machine Unit: |
|---|---|

Normal operation is continuous in order to fulfill orders for reels of catalog-grade paper. The states are described as follows:

**Operating State Name: Starting**

*Routine Activities*

Prestart checks.

Start stock preparation unit.

Start dryer unit at slow speed.

Start wet-end unit to produce a 6-in.-wide web.

Wait for web to be threaded through all dryers and onto winder.

Command wider webs on wet end and increase web speed.

*Exception Handling*

Shutdown if:

Dryer unit shut down

Wet-end unit shut down

Stock preparation unit shut down

*Primary Control Objectives*

Fines count

Brightness

Consistency

*Performance Information*

None.

*State End Conditions*

Paper machine at proper speed and web width.

**Operating State Name: Running**

*Routine Activities*

Collect performance data from units.

*Exception Handling*

Shutdown if:

Dryer unit shut down

Wet-end unit shut down

Stock preparation unit shut down

| yyyymmdd hh:mm | Pulp and Paper Co | Page 101 |
|---|---|---|

| POD | Area: Paper Mill   Cell: #1 Paper Machine   Unit: |
|---|---|

*Primary Control Objectives*

Paper thickness
Paper opacity
Paper brightness
Paper consistency

*Performance Information*

Paper reels produced per shift.

*State End Conditions*

Shutdown requested.

**Operating State Name: Shutting Down**

*Routine Activities*

Shutdown stock preparation unit.
Shutdown wet-end unit.
Shutdown dryer unit.

*Exception Handling*

None.

*Primary Control Objectives*

None.

*Performance Information*

None.

*State End Conditions*

All units shut down.

| **CONTROL CONCEPT** | Area: Paper Mill Cell: #1 Paper Machine Unit: |
|---|---|

### #1 Paper Machine Cell Process Management

*Extent of Automation*

Collect performance data from units.

*Flexibility of Automation*

None.

*Control Activity Coordination*

Coordinate startup and shutdown of stock preparation unit with bleach unit and groundwood cell.

*Interaction with Operating Personnel*

Summary information from units displayed for operator.

| POD | Area: Paper Mill Cell: #2 Paper Machine Unit: |
|---|---|

#### #2 Paper Machine Cell

An overall diagram of the #2 paper machine cell is shown in Figure 10.27 and is similar to the #1 paper machine cell. This cell produces paper to be used in liquid packaging. The #2 paper machine cell blends hardwood and softwood pulp. In addition, this cell does not have a supercalender and a coating unit but has an extruder to layer certain materials onto the paper in order to make it waterproof.

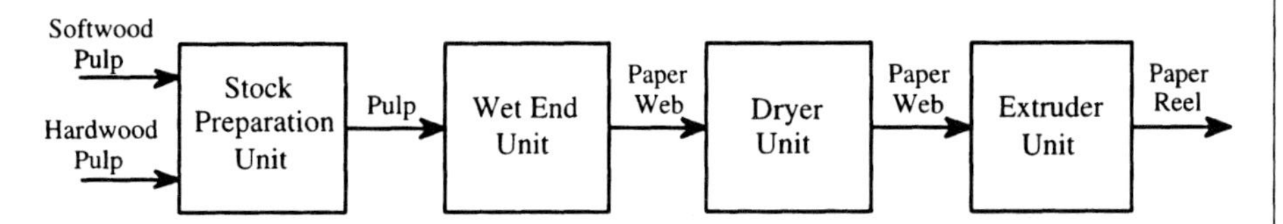


**FIGURE 10.27.** Overall view of units in #2 paper machine cell.

| POD | Area: Paper Mill Cell: Packaging Unit: |
|---|---|

**Packaging Cell**

The packaging cell packages the paper machine reels, which are 20 ft wide, into rolls that are 12–18 in. wide. An overall view of this cell is shown in Figure 10.28. Each paper machine has a slitter unit dedicated to it that slits each reel as it is transferred from the unwind reel to the winding rolls. The wrapping unit receives rolls from the slitter units and packages them. Before it is packaged, each roll is weighed and the diameter is checked. The roll is packaged by wrapping brown paper around the roll, crimping the brown paper, and then heat sealing a paper disk on either side of the roll. The roll is then labeled with its weight, diameter, and lot number and picked up with fork lifts to be taken to the warehouse.

The operating states of the packaging cell are the same as the woodyard cell (Figure 10.4):

Idle
Running

The various packaging cell units operate relatively autonomously, and so there is no special logic to transition between states. If any of the units are not Idle, the cell is running. If all units are Idle, the cell is Idle. The Running state is described as follows:

**Operating State Name: Running**

*Routine Activities*

Collect performance data from units.

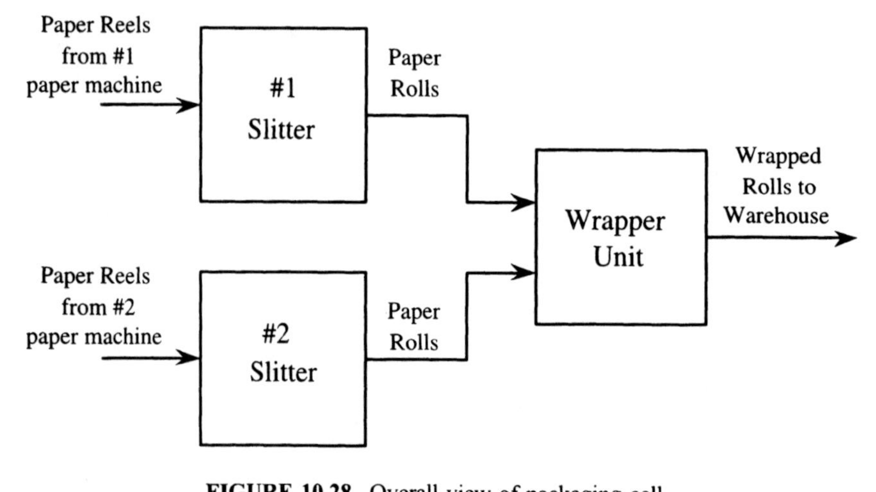


**FIGURE 10.28.** Overall view of packaging cell.

| **POD** | Area: Paper Mill   Cell: Packaging   Unit: |
|---|---|

*Exception Handling*

None.

*Primary Control Objectives*

Maximize production.

*Performance Information*

Paper rolls wrapped per shift.

*State End Conditions*

All units shut down.

| CONTROL CONCEPT | Area: Paper Mill Cell: Packaging Unit: |
|---|---|

**Packaging Cell Process Management**

*Extent of Automation*

Collect performance data from units.

*Flexibility of Automation*

None.

*Control Activity Coordination*

None.

*Interaction with Operating Personnel*

Summary information from units displayed for operator.

## REFERENCES

Biermann, C. J., *Handbook of Pulping and Papermaking*, 2nd ed., Academic, San Diego, CA, 1996.

MacDonald, R. G., and J. N. Franklin, Eds., *Pulp and Paper Manufacture, Vol. 1: The Pulping of Wood*, 2nd ed., McGraw-Hill, New York, 1969.

MacDonald, R. G., and J. N. Franklin, Eds., *Pulp and Paper Manufacture, Vol. 3: Papermaking and Paperboard Making*, 2nd ed., McGraw-Hill, New York, 1970.

| yyyymmdd hh:mm | Pulp and Paper Co | Page 107 |
|---|---|---|

# APPENDIX A
# Symbols Used in Piping and Instrumentation Diagrams

The symbols used in the simplified P&IDs in this text are explained in this appendix. For the most part, they adhere to the ISA standard S5.1 (ISA, 1986).

Each measuring device and control equipment module on a P&ID is identified with a tag that is two or three letters followed by a number. The letter abbreviations are used to identify the various types of measuring devices and control equipment. The first letter of the tag indicates the variable type and the succeeding letter(s) provide more information about the function being performed. The abbreviations used in this text are summarized in Table A.1. The only letter that does not conform to the S5.1 standard is the letter X, which is unclassified in S5.1. Examples of common abbreviations are:

| | |
|---|---|
| FT | Flow transmitter |
| LAH | Level alarm, high level |
| FIC | Flow controller with the measurement indicated |
| XV | Discrete (on–off) valve |
| EX | Electric motor |

**TABLE A.1 Identification Letters**

| | First Letter | Succeeding Letters |
|---|---|---|
| A | Analyzer | Alarm |
| C | Concentration | Control |
| D | Density | |
| E | Voltage | |
| F | Flow | |
| H | | High |
| I | | Indicating |
| L | Level | Low |
| P | Pressure | |
| Q | | Totalize |
| S | Speed, slide gate | Switch |
| T | Temperature | Transmitter |
| V | | Valve |
| X | Discrete (on–off) | Motor |
| Y | | Compute, convert |
| Z | Position | |

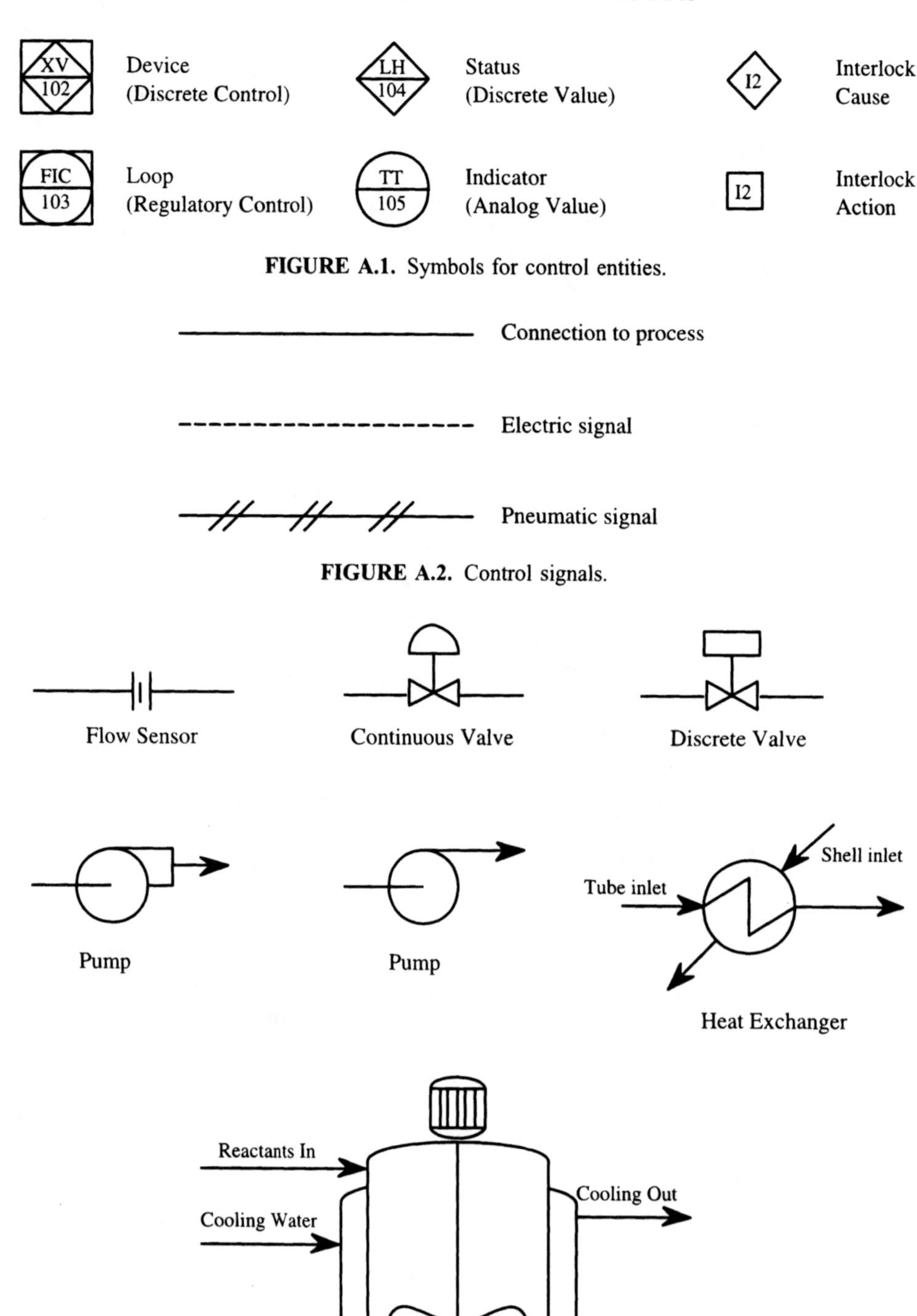


**FIGURE A.1.** Symbols for control entities.

**FIGURE A.2.** Control signals.

**FIGURE A.3.** Process equipment symbols.

Each tag is located inside a circle (regulatory entity) or a diamond (discrete entity). A square is placed around the circle or diamond if the entity performs control functions. These symbols are summarized in Figure A.1. A diamond inside a square represents a device, a discrete-control object. A diamond alone represents a status, a discrete value. A circle inside a square represents a loop, a regulatory control object. A circle alone represents an indicator, an analog value, which is typically an analog transmitter.

In addition, interlock causes and actions are attached to the circles and/or diamonds. An interlock cause is indicated by the letter I followed by a number within a smaller diamond. The associated interlock action uses the same identification as the cause, except within a small square. The specific interlock cause and action information is usually documented as a note in the diagram or in another document.

The signal connections between control entities are indicated as different line styles, as shown in Figure A.2. Symbols for common process elements are shown in Figure A.3.

## REFERENCE

Instrument Society of America, *ISA-S5.1-1984, Instrumentation Symbols and Identification*, Instrument Society of America, Research Triangle Park, NC, 1986.

# APPENDIX B
# Ladder Logic

The IEC 1131-3 international standard (International Electrotechnical Commission, 1993) defines five PLC languages: ladder logic, sequential function charts, function blocks, instruction list, and a text language. Currently, ladder logic is the most prevalent language, though the others are quickly gaining acceptance. The format of sequential function charts is explained in Appendix C.

The ladder logic symbology was developed from the relay ladder logic wiring diagram. In order to explain the symbology, simple switch circuits will be converted to relay logic and then to PLC ladder logic. Consider the simple problem of turning on a lamp when two switches labeled A and B are closed (Figure B.1*a*). All possible combinations of the two switches and the consequent lamp action are shown as a truth table in Figure B.1*b*. To implement this function using relays, the switches A and B are not connected to the light directly but control relay coils whose normally open contacts control the light (Figure B.1*c*). The switches A and B are the inputs to the circuit. The output (lamp in this case) is driven by another relay to provide voltage isolation from the relays implementing the logic. The switches A and B control relay coils to isolate the inputs from the logic. Also, with this arrangement one connection to the switch input can be used multiple times by using the multiple poles (contacts) on the relay for that input. The ladder logic notation (Figure B.1*d*) is shortened from the relay wiring diagram to show only the relay contacts and the coil of the output relay. The ladder logic notation assumes that the inputs (switches in this example) are connected to relay coils and that the actual output is connected to a set of normally open contacts controlled by the rightmost coil. The label shown above the contact symbol is not the contact label but the control for the coil that controls the contact. Also, the output for the rung occurs on the extreme right side of the rung and power is assumed to flow from left to right. The ladder logic rung is interpreted as: "When input (switch) A is on and input (switch) B is on, the lamp is on." If the example is changed to turn on a lamp when either switch A or B is closed, then the two normally open relay contacts are placed in parallel.

As a second example, consider the implementation of a logical NOT function. Suppose a lamp needs to be turned on when switch A is on (closed) and switch B is off (open). Figure B.2 shows the truth table, relay implementation, and ladder logic for this example. The logical NOT for switch B is accomplished with the normally closed contact in the ladder. The ladder logic rung in Figure B.2*c* is interpreted as: "When input (switch) A is on and input (switch) B is off, the lamp is on." This particular example is impossible to implement with only two normally open switches.

A more complicated ladder logic diagram, shown in Figure B.3, demonstrates more obviously why it is called a ladder logic diagram. Each rung has a connection to the left (power) rail and a connection to the right (common) rail. Actually, the ladder logic diagram

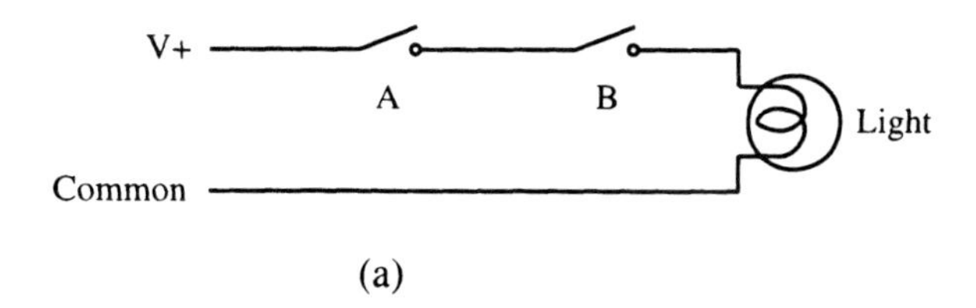


(a)

| A | B | Light |
|---|---|---|
| OFF | OFF | OFF |
| OFF | ON | OFF |
| ON | OFF | OFF |
| ON | ON | ON |

(b)

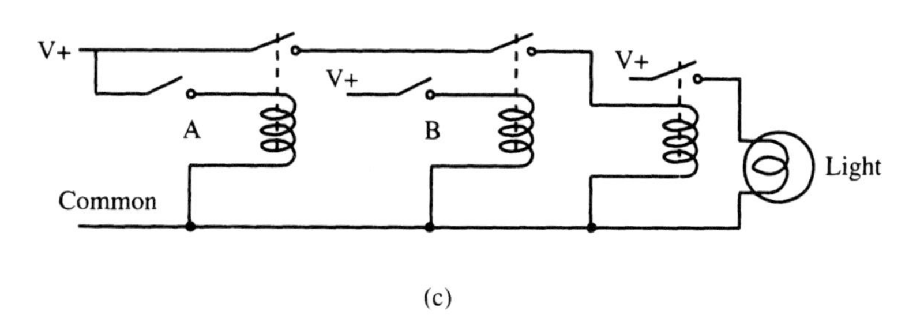


(c)

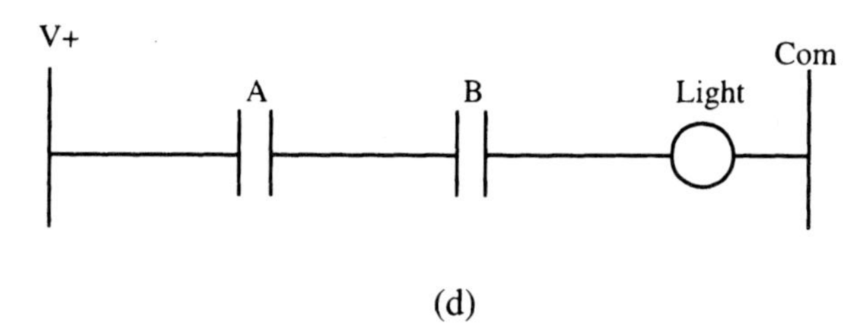


(d)

**FIGURE B.1.** Series circuit ladder logic: (*a*) switch circuit; (*b*) truth table; (*c*) equivalent relay circuit; (*d*) equivalent ladder logic. (Reprinted by permission. Copyright © 1998, Kelvin T. Erickson.)

is only a symbolic representation of the computer program. Power does not really flow through any actual contacts, but the concept of power flowing through contacts is useful when explaining the program operation. The three basic ladder logic symbols are:

Normally open (NO) contact —| |—
Normally closed (NC) contact —|/|—
Coil (output) —○—

The output coil is energized when any left-to-right path of input contacts is closed. For example, in Figure B.3, the output LA_1 is on whenever A and B and C are

| A | B | Light |
|---|---|---|
| OFF | OFF | OFF |
| OFF | ON | OFF |
| ON | OFF | ON |
| ON | ON | OFF |

(a)

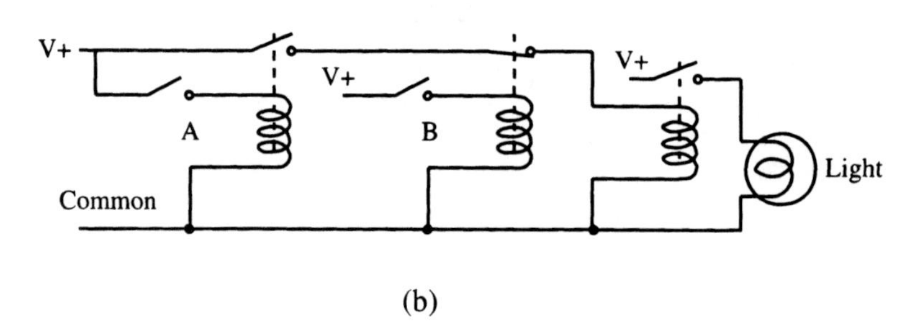


(b)

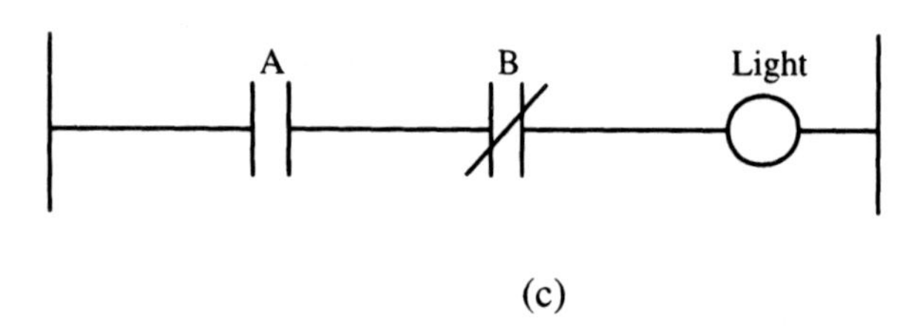


(c)

**FIGURE B.2.** Implementation of logical NOT in ladder logic: (*a*) truth table; (*b*) equivalent relay circuit; (*c*) equivalent ladder logic. (Reprinted by permission. Copyright © 1998, Kelvin T. Erickson.)

simultaneously on or D is off and A and C are on. Symbols are being used for all inputs and outputs in the examples to avoid having to deal with I/O addressing, which is generally different for each PLC manufacturer.

There are three classes of ladder logic instructions: contact instructions, coil instructions, and function blocks. The main contact instructions have already been introduced. There is also a contact instruction to detect positive transitions and one that detects negative transitions. The coil instruction always occurs on the extreme right side of the rung. In the examples used so far, the only coil instruction is –○– . Other coil instructions are:

| | |
|---|---|
| Inverted coil | –⊘– |
| Set (latch) coil | –Ⓢ– |
| Reset (unlatch) coil | –Ⓡ– |

Not all instructions are contacts or coils. All other types of instructions are often called function block or "box instructions" because that is how they appear in the symbology.

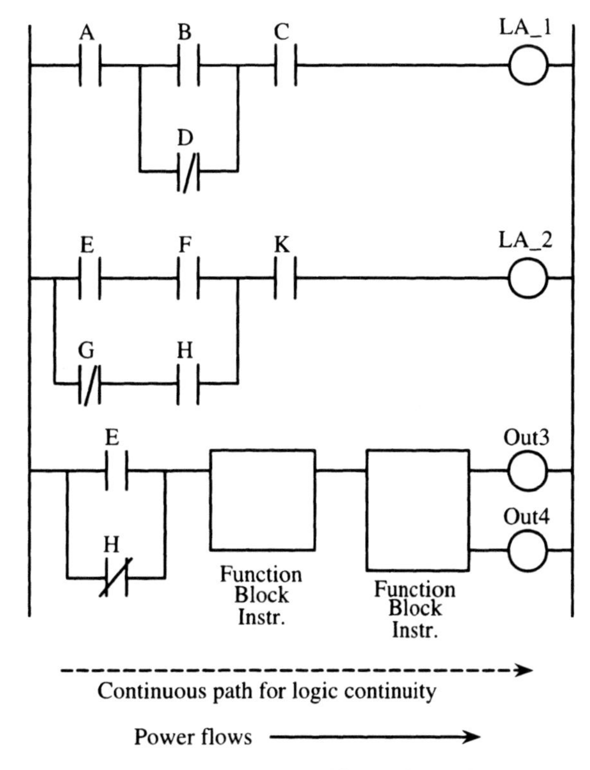


**FIGURE B.3.** General ladder logic diagram.

Timers, counters, comparison, and computation are the most common function blocks, but sequencers, shift registers, and data move instructions are also function blocks. A generic rung with two function block instructions is shown as the last rung in Figure B.3.

During operation, the PLC repeatedly executes a scan, during which the input channels from all of the input modules are copied into the internal memory, the ladder logic is scanned, updating the outputs being held in internal memory, and then the internal outputs are copied to the actual output modules. After the actual outputs have been updated, the scan is repeated. The time to execute a scan, called the scan time, depends on the number of I/O channels and the length of the ladder logic program and is on the order of tens to hundreds of milliseconds.

The previous examples used external (switch) discrete inputs and an external (lamp) discrete output. However, it is not required that all contacts be controlled by external discrete-input devices. The contacts can also refer to an output (like reading the state of an output). In addition, internal one-bit memory locations, often called internal coils, are provided to store information that is not connected to any external output channel.

One aspect of ladder logic that is often confusing to the new user is the use of the normally closed contact. The contact symbol in the ladder does not necessarily correspond to the actual switch type used in the field. The PLC does not know how the switch is wired in the field, only whether the switch is open (off) or closed (on). So, a normally open switch does not require a ⊣ ⊢ in the ladder logic and a normally closed switch does not require a ⊣/⊢ in the ladder logic. Regardless of the type of switch in the field, when one wants "action" (something to be logically true, or on) when the switch is closed (on), use the ⊣ ⊢ symbol. When one wants "action" (something to be logically true, or on) when the switch is open (off), use the –|/|– symbol. A ladder logic diagram must be read as symbols and not as switch contacts.

A common application uses two momentary switches to control a device, for example, a pump. One switch, called START_PB, is a momentary normally open switch that, when pressed, starts the pump. The pump must continue to run after START_PB is released. The second switch STOP_PB is a momentary normally closed switch that, when pressed, stops the pump. The switches are specified in this manner for safety purposes. If START_PB has any faulty wiring, the pump cannot be started. In addition, the pump will automatically stop when STOP_PB has any faulty wiring connections. The ladder logic diagram to accomplish the above function is shown in Figure B.4. Note the contact symbol used for STOP_PB is the NO contact even though STOP_PB is wired normally closed. Remember, the PLC does not know how the switch is wired in the field, only whether the switch is open (off) or closed (on). When START_PB is on and STOP_PB is on (not pressed), the pump is turned on. The contact labeled PUMP in parallel with the START_PB contact ensures that the pump remains on, even after START_PB is released and the PLC reads it as off. When STOP_PB is pressed (turns off), the pump is turned off and remains off until START_PB is pressed again. This type of ladder logic rung is often called a "seal circuit" or "latching circuit." Often, in a real application, there are multiple conditions, called permissives, that must be satisfied before the pump can be turned on, and there will be a multitude of conditions, called interlocks, any one of which will cause the pump to turn off.

For the last example, consider a two-speed pump control. The specifications for the application are:

(a) The pump can only be started in low speed.

(b) The pump is then switched from low speed to high speed after a 10-sec delay.

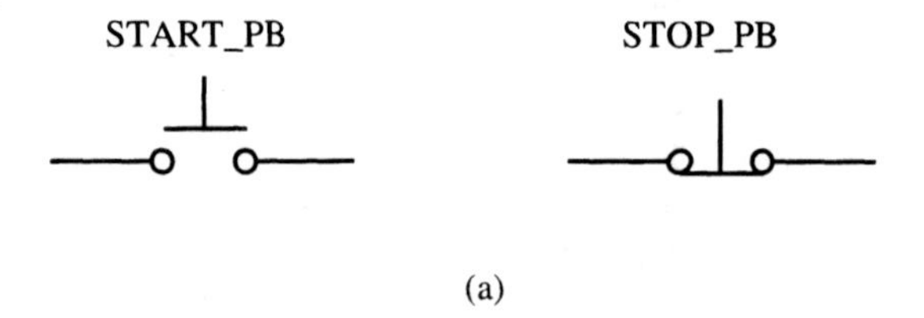


(a)

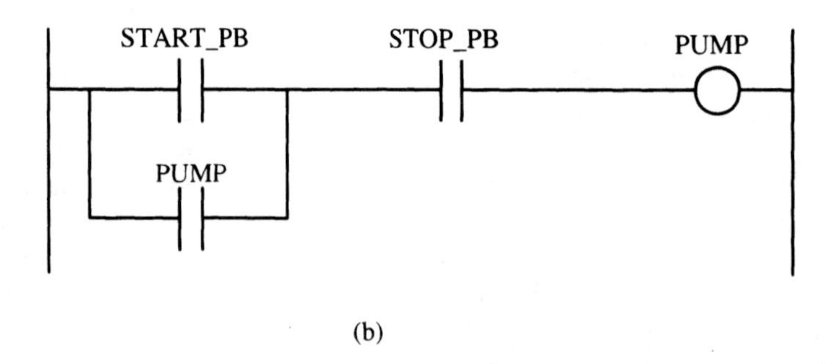


(b)

**FIGURE B.4.** Start–stop application: (*a*) switches; (*b*) ladder logic rung.

(c) The pump cannot be switched from high speed to low speed.
(d) The low- and high-speed controls to the pump cannot be on simultaneously.
(e) If excessive vibration occurs, the pump must stop and cannot be restarted (is locked out) until a reset button is pressed.
(f) If the stop button is pressed when the pump is running in either speed, the pump will stop but will not lock out.

Assume the following input and output assignments (only symbols are used here, to avoid any explanation of an I/O addressing scheme):

INPUTS:

| | |
|---|---|
| START_PB | Start push-button, NO, on when starting |
| STOP_PB | Stop push-button, NC, off when stopping |
| RESET_PB | Reset, NO, on (closed) when resetting |
| VIB_SENS | Vibration sensor, NC, off when vibration occurs |

OUTPUTS:

| | |
|---|---|
| LOW_SPD | Pump low speed |
| HI_SPD | Pump high speed |

The two outputs are assumed to be inputs to a motor controller that directly controls the pump.

Ladder logic that will fulfill the specifications is shown in Figure B.5. RUN and VIB_ON are internal coils (one-bit memory locations) and are not output channels. The TON instruction is an on-delay timer function block instruction. The timing interval is shown as the PT input to the block. When the input conditions to the left of the timer become logically true (continuity through contact), the timer accumulator counts up. When the accumulator equals the PT value, the timer output energizes the TDONE internal coil. If the input condition to the timer becomes logically false at any time, the timer is reset and the accumulator is set to zero. It does not retain the accumulator value. The TDONE bit is used on another rung to turn on the HI_SPD output. The first rung is the normal start–stop rung with an additional condition for stopping. The RUN internal coil is used because there is not a single output that defines the pump operation. When vibration occurs (VIB_ON turns on), RUN turns off. The second rung defines the delay timer operation. The third and fourth rungs drive the outputs that control the pump. As long as RUN is on and the timer has not finished the 10-sec timing interval, LOW_SPD is on. When the 10-sec interval has elapsed, LOW_SPD is turned off and HI_SPD is turned on. The last rung implements a latch–unlatch for the vibration sensor (VIB_SENS). It is different from the start–stop rung for safety reasons. If the RESET_PB normally closed contact is placed in series with VIB_SENS on the upper part of the rung (in the same position as the STOP_PB in the first rung), then holding the reset switch on (pushing the normally open push-button) will override the vibration sensor and allow the pump to run even when vibration continues to occur. Obviously, this is a situation one would want to prevent for safety reasons.

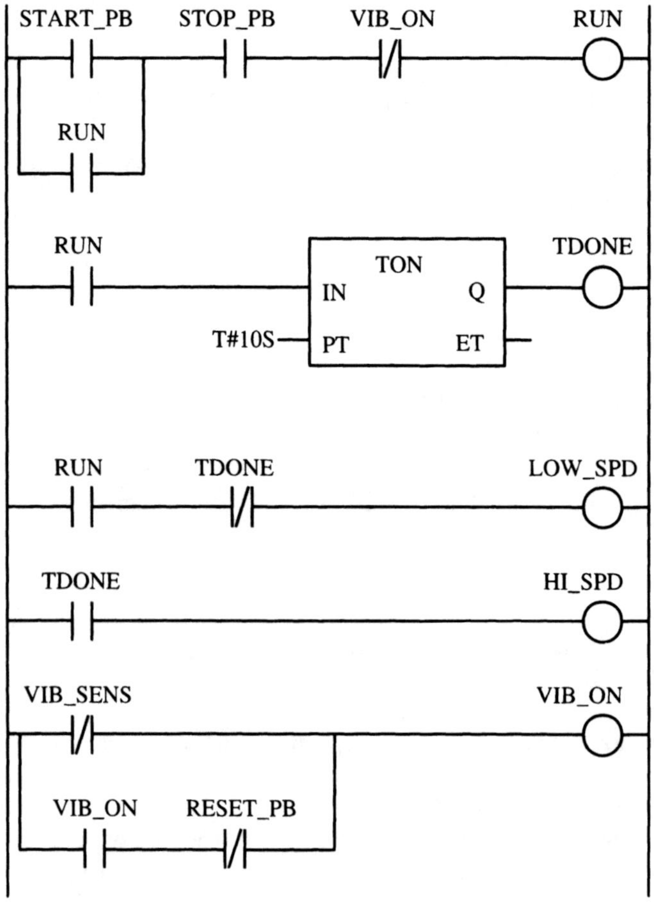

START_PB STOP_PB VIB_ON RUN
RUN
RUN TDONE
TON
IN Q
T#10S PT ET
RUN TDONE LOW_SPD
TDONE HI_SPD
VIB_SENS VIB_ON
VIB_ON RESET_PB

**FIGURE B.5.** Ladder logic for two-speed pump application.

## REFERENCE

International Electrotechnical Commission, *IEC 1131-3: Programmable Logic Controllers—Part 3: Programming Languages*, International Electrotechnical Commission, Geneva, Switzerland, 1993.

# APPENDIX C
# Sequential Function Charts

The IEC 1131-3 international standard (International Electrotechnical Commission, 1993) defines five PLC languages: ladder logic, sequential function charts, function blocks, instruction list, and a text language. Ladder logic, the most prevalent language, is explained in Appendix B. The IEC 1131-3 sequential function chart (SFC) is derived from the IEC 848 function chart standard (International Electrotechnical Commission, 1988) that has been used to define sequence control logic. This format has emerged as a major programming tool in modern control systems, especially batch control systems.

The SFC is a diagram of interconnected steps, actions, and transitions as illustrated in Figure C.1. An SFC begins with an initial step (box enclosed in double line, Figure C.2*a*) followed by an ordered set of numbered (Figure C.2*b*) or labeled (Figure C.2*c*) steps configured to perform the desired sequential control scheme. Associated with each step is a transition condition, shown as a horizontal bar. If the step is active and the transition condition is true, the logic advances to the next step or steps. The stepwise flow continues until the end of the diagram. At this point, the sequencing ends (Figure C.3) or it may recycle back to the initial step waiting to resume the next cycle.

There may be one or more action boxes attached to a step (Figure C.4). The action box is used to perform a process action such as opening a valve, starting a motor, or calculating an endpoint for the transition condition. Generally, each step issues a command, although in cases where a step is only waiting for a transition (e.g., waiting for a limit switch to close) or executing a time delay (step 7 of Figure C.1), no action is attached.

Each step action may have up to three parts (Figure C.5):

a—Action qualifier
b—Symbolic or textual statement describing action
c—Symbol of a feedback variable

The first and last parts of the action are shown only if necessary. The action qualifier is a letter or a combination of letters describing how the step action is processed. Possible action qualifiers are:

N—Nonstored. Action is active only when the step is active.
S—Stored (set). Action becomes active when the step becomes active. The action continues to be executed even after the step is inactive. In order to stop the action, another step must have an R qualified block that references the same action.
R—Overriding reset. Action becomes inactive when the step becomes active.
P—Pulse. Action becomes active for the single scan in which the step becomes active.

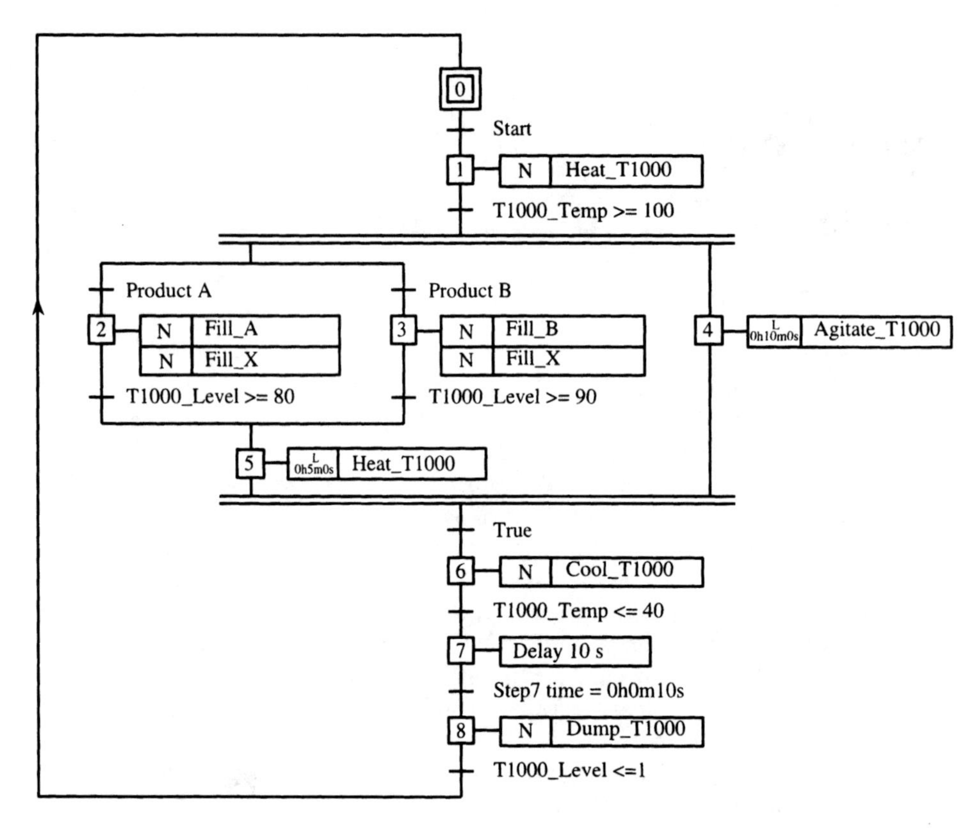


**FIGURE C.1.** Example SFC. (Reprinted by permission. Copyright © 1998, Kelvin T. Erickson.)

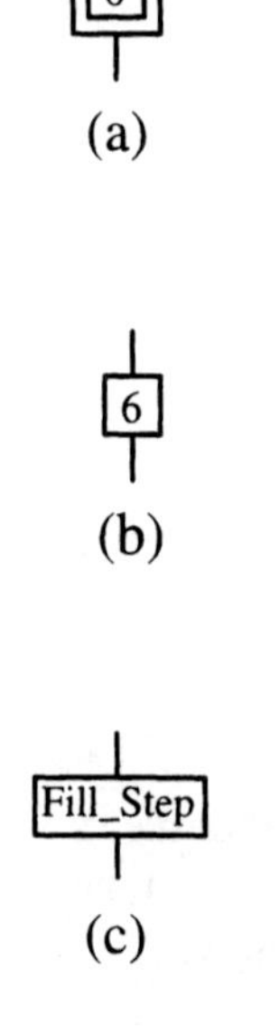


**FIGURE C.2.** SFC step symbols: (*a*) initial step; (*b*) numbered step; (*c*) labeled step.

**FIGURE C.3.** Symbol for end of SFC.

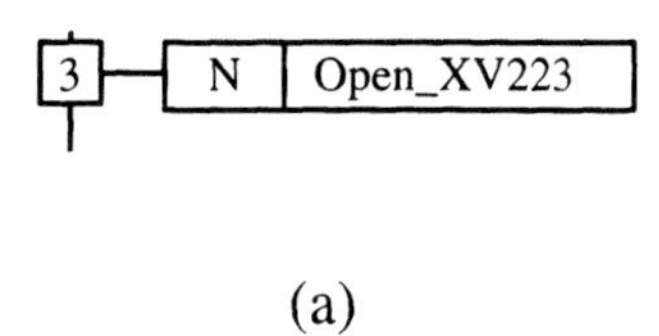


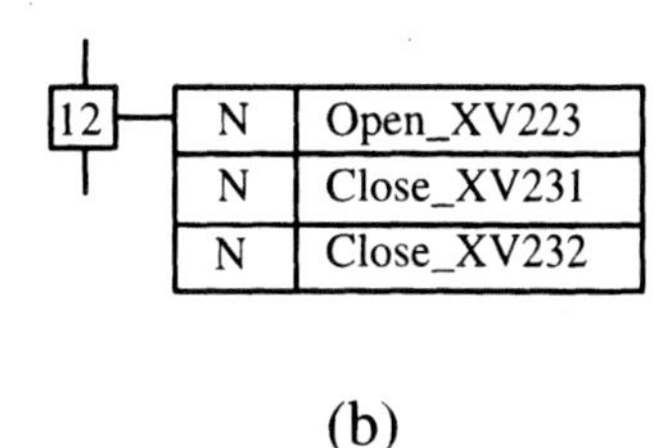


**FIGURE C.4.** SFC step actions: (*a*) single action; (*b*) multiple actions.

| a | b | c |
|---|---|---|

**FIGURE C.5.** General format of action block.

L—Time limited. Action becomes active when the step becomes active. The action becomes inactive when a set length of time elapses or the step becomes inactive.

SL—Stored and time limited. Action becomes active when the step becomes active. After a set length of time elapses or an R qualified action references the same action, the action is disabled. A step with an R qualified block must refer to this action before this action can be used again.

D—Time delayed. Enabling the action is delayed for a set time after the step becomes active.

SD—Stored and time delayed. Enabling the action is delayed for a set time after the step becomes active. The action is stored after the time delay even if the step that has the SD qualifier for this action becomes inactive. After the time delay, the action is enabled and remains enabled until another step with an R qualified block references the same action.

DS—Delayed and stored. Enabling the action is delayed for a set time after the step becomes active only if the step is still active after the time delay has elapsed. After the action is enabled, another step with an R qualified block that references the same action must disable the action.

The condition that causes a transition from one step to the next is indicated by the text to the right of the horizontal bar below the step box. Transition conditions may be represented by textual statements, Boolean expressions, or graphical symbols. In this text, transition conditions are represented by textual statements. Logical operations are indicated by placing a period before and after the word, such as, ".and.", ".or.", and ".not."

The evolution, or sequence, of the steps is indicated by the vertical links between steps. Normally, the sequence proceeds from top to bottom. Arrows are used if this convention is not respected, as in the path from step 8 to step 0 in Figure C.1. As defined, the evolution of an SFC may be a single sequence (Figure C.6), a selection of one sequence (Figure C.7), or simultaneous sequences (Figure C.8). The example in Figure C.1 contains all three possibilities.

Referring to the single sequence of Figure C.6, if step 3 is active and the XV301A Open condition is true, step 3 becomes inactive and step 4 becomes active. Likewise, if step 4 is active and the EX310 Running condition is true, the logic advances to step 5.

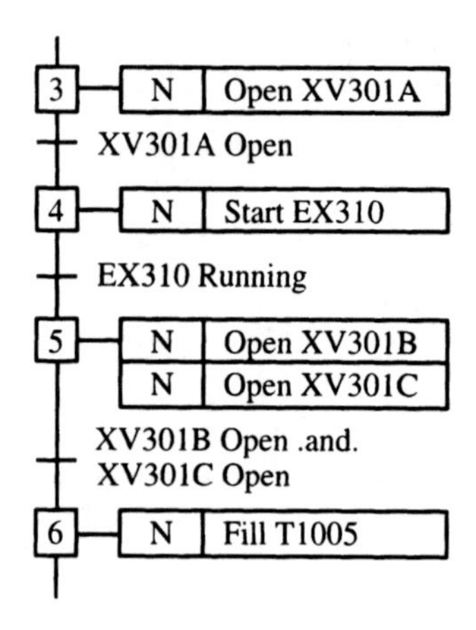


**FIGURE C.6.** Example of single sequence. (Reprinted by permission. Copyright © 1998, Kelvin T. Erickson.)

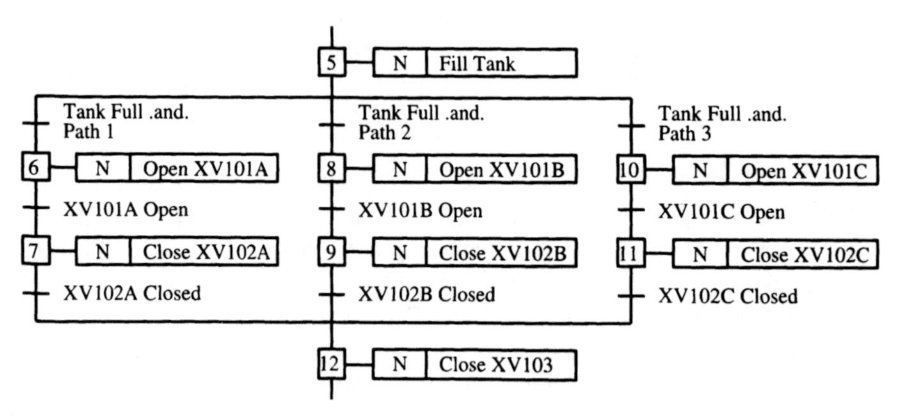


**FIGURE C.7.** Example of sequence selection. (Reprinted by permission. Copyright © 1998, Kelvin T. Erickson.)

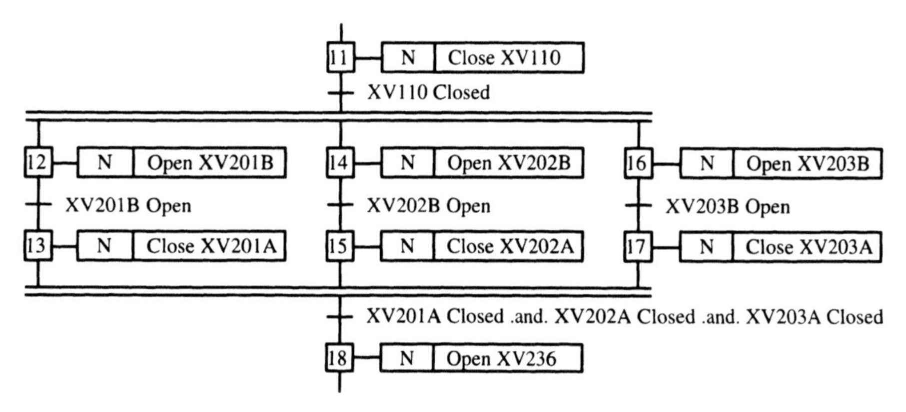


**FIGURE C.8.** Example of simultaneous sequences. (Reprinted by permission. Copyright © 1998, Kelvin T. Erickson.)

A selection of one sequence out of more than one sequence is called an exclusive divergent path and is represented by multiple transitions below the single horizontal line, as in the top part of Figure C.7. Each possible sequence path contains a transition condition. No common transition condition is permitted above the horizontal line. If step 5 is active, then there are three possible transition conditions. If the left path condition is true, then the logic advances to step 6. Otherwise, if the middle path condition is true, then the logic advances to step 8, or if the right path condition is true, the logic advances to step 10. In order to select only one succeeding step, the transition conditions must be mutually exclusive. The several sequences must also converge to a common sequence, as in the bottom part of Figure C.7. There must be as many transitions above the horizontal line as sequences to be regrouped. No common transition condition is permitted below the horizontal line. If step 7 is active and XV102A Closed is true or if step 9 is active and XV102B Closed is true or if step 11 is active and XV102C Closed is true, then step 12 becomes the active step.

If the transition out of a step causes more than one sequence to be activated simultaneously, called simultaneous divergence, these simultaneous sequences are represented as in the top part of Figure C.8. The simultaneous sequences are differentiated from the sequence selection by the double horizontal line. Also, only one common transition condition is permitted above the double horizontal line and no transitions are permitted below the horizontal line. When step 11 is active and the condition XV110 Closed is true, then step 11 becomes inactive and steps 12, 14, and 16 become active simultaneously. The sequences converge with a double horizontal line having a common transition symbol under the double horizontal line. Step 18 will become the active step only when all the steps above the double horizontal line are active and the transition condition XV201A Closed .and. XV202A Closed .and. XV203A Closed is true.

## REFERENCES

International Electrotechnical Commission, *IEC 848: Preparation of Function Charts for Control Systems*, International Electrotechnical Commission, Geneva, Switzerland, 1988.

International Electrotechnical Commission, *IEC 1131-3: Programmable Logic Controllers—Part 3: Programming Languages*, International Electrotechnical Commission, Geneva, Switzerland, 1993.

# Glossary

*Note*: Definitions quoted from other sources are so indicated. References with similar definitions are also noted. Italicized words in a definition are defined in this glossary.

**Air-to-close/air-to-open** Describes control valve fail-safe action. When the air supply fails, an air-to-close valve will fully open so flow through the valve is unobstructed. Upon an air supply failure, an air-to-open valve will fully close, obstructing flow though the valve.

**AP221** Abbreviation for *ISO 10303 Industrial Automation Systems and Integration, Product Data Representation and Exchange—Part 221: Functional Data and Their Schematic Representation for Process Plant*, Working project draft, International Standards Organization, February 23, 1997.

**AP227** Abbreviation for *ISO 10303 Industrial Automation Systems and Integration, Product Data Representation and Exchange—Part 227: Plant Spatial Data*, Working project draft, International Standards Organization, December 9, 1994.

**AP231** Abbreviation for *ISO 10303 Industrial Automation Systems and Integration, Product Data Representation and Exchange—Part 231: Application Protocol: Process Engineering Data: Process Design and Process Specification of Major Equipment*, Committee draft, International Standards Organization, July 17, 1997.

**Area** A geographical grouping smaller than a *site*. An area is a physical, logical grouping of process equipment that performs the desired cell operations (thermodynamic, chemical, biological) to make a product. Plant areas are often set by enterprise policy such as operator jurisdiction, product, or other criteria (ISA, 1995).

**ARMA** Abbreviation for autoregressive moving average, a type of discrete-time empirical model.

**Automatic feedback control** *Feedback control* in which a controller device monitors the controlled variable of interest and commands a manipulated variable in order to maintain a desired value of the controlled variable.

**Basic control** Includes *regulatory control, discrete control, interlock control,* monitoring, and *exception handling* (ISA, 1995).

**Basic process control system** The non–safety system technology that interfaces to the process equipment. Also called *basic control*. See also *safety instrumented system.*

**Batch** "1.) The material that is being produced or that has been produced by a single execution of a batch process. 2.) An entity that represents the production of a material at any point in the process." (ISA, 1995, p. 14).

**Batch control** "Control activities and control functions that provide a means to process finite quantities of input materials by subjecting them to an ordered set of processing activities over a finite period of time using one or more pieces of equipment" (ISA, 1995, p. 14).

**Batch process** "A process that leads to the production of finite quantities of material by subjecting quantities of input materials to an ordered set of processing activities over a finite period of time using one of more pieces of equipment" (ISA, 1995, p. 14). In a batch process, finite quantities (batches) of material are produced by subjecting quantities of input materials to a defined order of processing actions using one or more pieces of equipment. Batch processes are discontinuous processes from a material flow standpoint. Batch processes are neither discrete nor continuous, though they have characteristics of both.

**Batch schedule** "A list of batches to be produced in a specific *process cell*" (ISA, 1995, p. 14; italic added).

**Block diagram** A representation of the process where major pieces of equipment or process stages are shown as rectangular blocks with the material and/or information flow between them.

**BPCS** Abbreviation for *basic process control system.*

**Bumpless transfer** When the controller state is switched from Manual to Automatic, the calculated change in *manipulated variable* is either filtered or ramped to the calculated value. This gradual change in the *manipulated variable* is better tolerated by the equipment and the process.

**Campaign** *Production run* for a *batch process.*

**Capital project** A project in which manufacturing facilities are constructed and/or improved.

**Cascade control** Control scheme where the *manipulated variable* output of one controller becomes the *setpoint* of another controller. Used in the control of any non–flow process variable that uses a valve as its final control element. In this case, the non–flow controller output is connected to the *setpoint* of a flow controller that manipulates the valve. Used in other situations where the process can be described as a cascade combination of a fast process followed by a slow process and the secondary process disturbance effect needs to be mitigated (ISA, 1979).

**Cell** See *process cell.*

**Closed-loop control** Same as *feedback control.*

**CLST** Abbreviation for closed-loop settling time, one of the major tuning parameters for the forward modeling controller, a type of model predictive controller.

**Common resource** "A resource that can provide services to more than one requester. Note—Common resources are identified as either *exclusive-use resources* or *shared-use resources*" (ISA, 1995, p. 14; italic added).

**Condition number** The ratio of the largest *singular value* to the smallest *singular value.* A large condition number indicates that the multivariable system may be difficult to control.

**Continuous control** A control system whose signal inputs and outputs are *continuous-time signals*. The control devices are either analog electronic or pneumatic controllers.

**Continuous model** The model system variables are represented by an infinite number of values. Generally refers to models used by analog controller algorithms.

**Continuous process** A process in which material passes in a continuous stream through the processing equipment. Once the process has established a steady operating state, the nature of the process does not depend on the length of time the process is operating (ISA, 1995).

**Continuous-time signal** A signal whose values are defined for all time. See also *discrete-time signal.*

**Control Concept** A section of the *control requirements definition* that defines the extent of automation for the plant system.

**Controllable** A multivariable process is controllable if the relationships between the *manipulated variables* and *process variables* are mathematically independent. For the purposes of this text, a system is controllable if the determinant of the gain matrix is nonzero (Marlin, 1995).

**Controlled variable** "In a control loop, the variable the value of which is sensed to originate a feedback signal" (ISA, 1979, p. 31).

**Control module** A plant item whose control scheme is generally a discrete or regulating loop. A control module is a collection of sensors, actuators, and associated processing equipment that is operated as a single entity. A control module can also consist of other control modules (ISA, 1995).

**Control recipe** "A type of *recipe* which, through its execution, defines the manufacture of a single batch of a specific product" (ISA, 1995, p. 14; italic added).

**Control requirements definition** Control information package containing the *process operation description, Control Concept, control strategy,* and detailed control design.

**Control state** The current condition of the control entity, e.g., process management, unit supervision, or process control. The control state also defines how the control entity will operate and how it will respond to commands.

**Control strategy** The control strategy is composed of *continuous-control* and *discrete-control* schemes. The control schemes are composed of descriptive information that identifies the *process control* types of objects: *equipment module control, loop, device, indicator,* and/or *status.*

**Control type** Any of the *process control* types of objects: *equipment module control, loop, device, indicator,* or *status.*

**Coordination control** "A type of control that directs, initiates, and/or modifies the execution of *procedural control* and the utilization of equipment entities." (ISA, 1995, p. 14; italic added).

**CRD** Abbreviation for *control requirements definition.*

**CSTR** Abbreviation for continuous-flow stirred-tank reactor.

**Damping ratio** The ratio of the second peak to the first peak in the *process variable* response. Only defined if the process variable response is *underdamped* (ISA, 1979).

**DCS** Abbreviation for *distributed control system.*

**Deadtime** "The interval of time between initiation of an input change or stimulus and the start of the resulting observable response" (ISA, 1979, p. 30).

**Decoupling control** An extension of *multiloop control* where the *interaction* in the multivariable process, as perceived by the controllers, is reduced. The decoupling may be accomplished by (1) detuning one (or more) of the loops, (2) redefining the *manipulated variables* and/or *process variables*, or (3) constructing an *explicit decoupler.*

**Derivative action** Calculation of the controller *manipulated variable* is based on the rate of change of the *process variable.* Also called rate action (ISA, 1979).

**Derivative gain** For PD or PID controllers, the derivative gain is the proportionality constant relating the change in controller output to the *derivative action.*

**Derivative time constant** Same as *derivative gain.*

**Device** A type of basic process control object that has a single output with two possible values (e.g., on–off, open–close). A typical device has one or two inputs. Example devices: discrete valve, motor.

**Digital control** A control system whose control devices are digital controllers or digital computers. Also called discrete-time control; the signal inputs and outputs are *discrete-time signals.*

**Direct acting** A process or controller is called direct acting if an increase in the *input signal* causes an increase in the *output signal.*

**Direct action** See *direct acting.*

**Discrete control** The control type used to prevent process upsets and enforce correct equipment operation in a stepwise manner. It is often referred to as "on–off" control and includes *interlocks* and *sequence control.*

**Discrete model** The model system variables are represented by a finite number of values. For example, the *input signal* to a motor device has two possible values, start or stop, and the *output signal* has two possible values, running or stopped.

**Discrete-parts manufacturing process** A type of process where a specified quantity of material moves as a unit (part or group of parts) between work stations and each unit maintains its unique identity (ISA, 1995).

**Discrete-time control** See *digital control.*

**discrete-time model** Models used by controller algorithms that execute on digital computers. For a discrete-time model, the *input signals* and *output signals* are *discrete-time signals.*

**Discrete-time signal** A signal whose values are defined only at the sampling instants. Any process measurements must be converted to discrete-time signals before being used by a digital controller. See also *continuous-time signal.*

**Distributed control system** In the process control industry, the distributed control system traditionally provided regulatory control functions for the process control system. The functionality of these systems is currently merging with the *programmable logic controller.* See also *programmable electronic system.*

**Distributed model** Model of a system whose properties are dependent on position, usually represented by partial differential equations.

**Disturbance** Process influence that affects the *process variable* but is not manipulated by the controller (ISA, 1979).

**DMC** Abbreviation for dynamic matrix control, a type of *model predictive control.*

**Dynamic decoupler** A type of *explicit decoupler* where the decoupler equations are expressed as ratios of the multivariable process *transfer functions.*

**Dynamic model** *Mathematical model* that attempts to capture the output response of a system as it changes as a function of time.

**Empirical model** *Mathematical model* of a process based on experimental data.

**Enterprise** A commercial or government organization that coordinates the operation of one or more sites to produce products (ISA, 1995). This book focuses on profit-making enterprises that require thermodynamic, chemical, or biological unit operations to make products.

**Enterprise management** The control entity responsible for coordinating the management of all enterprise *sites*.

**Enterprise resource planning system** The information technology of an *enterprise* that supports its business activities of forecasting, costing, production planning, process definition, human resource management, inventory management, purchasing, and distribution.

**Equipment entity** "A collection of physical processing and control equipment and equipment control grouped together to perform a certain control function or set of control functions" (ISA, 1995, p. 14).

**Equipment module** A plant item whose control scheme is a *phase* that coordinates other plant items whose function is *discrete* or *regulatory control*. An equipment module carries out a finite number of specific minor processing activities, e.g., weighing, mixing, and ratioing of feed streams. The equipment module contains all necessary physical processing and control equipment required to perform those activities (ISA, 1995).

**Equipment module control** The batch or sequential control of an *equipment module*.

**Equipment procedural control** *Procedural control* for a plant system that must be operated in a specific manner to produce the desired outcome. Control parameters may change, but the procedure remains fixed for long periods. See also *recipe procedural control*.

**Equipment procedure** "A *procedure* that is part of equipment control" (ISA, 1995, p. 15; italic added).

**Equipment unit procedure** "A *unit procedure* that is part of equipment control" (ISA, 1995, p. 15; italic added).

**ERPS** Abbreviation for *enterprise resource planning system*.

**Exception handling** Those control functions that deal with plant or process contingencies and other events that occur outside the normal or desired behavior of the control system (ISA, 1995).

**Exclusive-use resource** "A *common resource* that only one user can use at any given time" (ISA, 1995, p. 15; italic added).

**Explicit decoupler** Redefines the *manipulated variables* as perceived by the single-loop controllers so that the process has little or no interaction. Used in one approach to decoupling control. Implemented as a *static decoupler* or a *dynamic decoupler*.

**FCCU** Abbreviation for fluidized catalytic cracking unit.

**Feedback control** Control scheme that uses knowledge of the output to take corrective action. See *automatic feedback control* and *manual feedback control* (ISA, 1979).

**Feedforward control** Control scheme that eliminates or reduces the *disturbance* effect on the *process variable* by using a measurement of the *disturbance* to modify the calculation of the *manipulated variable* (ISA, 1979).

**FMC** Abbreviation for forward modeling control, a type of *model predictive control.*

**FODT** Abbreviation for first-order plus deadtime, a type of *empirical model.*

**Formula** "A category of *recipe* information that includes *process inputs*, process parameters, and *process outputs*." (ISA, 1995, p. 15; italic added).

**Full multivariable control** A *multivariable control* scheme that uses all *process variables* to simultaneously calculate the *manipulated variables.*

**Fundamental model** A mathematical process model based on fundamental concepts such as the conservation of material and/or energy.

**Gain** In this text, the gain is more appropriately called the static gain and is calculated as the ratio of the change in the *steady-state* output to a step change in the input, provided the output does not saturate (ISA, 1979).

**General recipe** "A type of *recipe* that expresses equipment and *site*-independent processing requirements" (ISA, 1995, p. 15; italic added).

**GUI** Abbreviation for graphical user interface.

**High-signal selector** A device that automatically compares two or more input signals and allows the signal with the higher value to be the output.

**HOA switch** Type of switch having three positions: (1) hand, (2) off, and (3) auto. It is often used between the automatic control system and a discrete device, such as a motor. In the "hand" position, the operator manually starts the device. In the "off" position, the device is manually stopped. In the "auto" position, the automatic control system starts and stops the device.

**HSS** Abbreviation for *high-signal selector.*

**IAE** Abbreviation for the integral of the absolute error, a control loop performance measure.

**IDEF0 diagram** Integration definition for function modeling diagram. Representation of the activities during the life of a plant.

**IE** Abbreviation for the integral of the error, a control loop performance measure.

**IEC** Abbreviation for International Electrotechnical Commission, an international standards-making organization.

**IEC 1131** Abbreviation for *IEC 1131: Programmable Logic Controllers*, International Electrotechnical Commission, 1993.

**IEC 1131-3** Part 3 of the IEC 1131 standard. Defines the standard PLC languages: *ladder logic, sequential function chart*, function block, structured text, and statement list.

**IEEE** Abbreviation for the Institute of Electrical and Electronics Engineers.

**IEEE 802.3** The IEEE standard that defines the Ethernet data communication protocol, which is a carrier sense multiple access with collision detection type of protocol.

**IEEE 802.4** The IEEE standard that defines a token bus communication protocol.

**IMC** Abbreviation for internal model control, a type of *model predictive control.*

**Increase close/increase open** Controller parameter provided to compensate for valve actuator action (air-to-close or air-to-open). Use increase close for an air-to-close valve and increase open for an air-to-open valve.

**Indicator** A single analog value, typically in engineering units. A type of basic process control object.

**Input signal** "A *signal* applied to a device, element, or system" (ISA, 1979, p. 28).

**Integral action** The controller *manipulated variable* calculation is based on the integrated error, the *setpoint* minus the *process variable* (ISA, 1979).

**Integral gain** For PI or PID controllers, the integral gain is the proportionality constant relating the change in controller output to the *integral action* (ISA, 1979).

**Integral time constant** The reciprocal of the *integral gain* (ISA, 1979).

**Integral windup** When a controller has *integral action*, a persistent error will cause the integral term to increase (or decrease) to a value of large magnitude.

**Interaction** Process interaction is evident when a change in one *manipulated variable* affects more than one *process variable.*

**Interlock** Designed to detect an abnormal process condition and take action to prevent an undesirable or hazardous event.

**Interlock control** The type of *discrete control* that prevents undesirable events. It may also be used in situations that are not related to safety.

**ISA** Abbreviation for Instrument Society of America.

**ISE** Abbreviation for the integral of the squared error, a control loop performance measure.

**ITAE** Abbreviation for the integral of the product of time and the absolute error, a control loop performance measure.

**Ladder logic** A symbolic programming language traditionally used in *programmable logic controllers*. It was developed from the relay ladder logic wiring diagram.

**Linear element** A control system element is linear if the output is proportional to the input (ISA, 1979).

**Linear system** A system in which the time response to several simultaneous inputs is the sum of their independent time responses. A linear system is generally represented by a set of linear differential equations (ISA, 1979).

**Loop** A single-input, single-output controller that monitors a transmitter and manipulates a physical quantity (usually a control valve position) in order to force a *process variable* to the desired *setpoint*. The controller is typically a PID controller, though other single-input, single-output algorithms (e.g., *model predictive control*) may be used (ISA, 1979).

**Loop pairing** The process of pairing a process variable with the appropriate manipulated variable to form an *automatic feedback control* loop. This process is used to design a *multiloop control* scheme.

**Low-signal selector** A device that automatically compares two or more input signals and allows the signal with the lower value to be the output.

**LSS** Abbreviation for *low-signal selector.*

**Lumped model** *Mathematical model* of a system whose properties are not dependent on position, usually represented by algebraic and ordinary differential equations.

**MA** Abbreviation for moving average, a type of discrete-time empirical model.

**MAC** Abbreviation for model algorithmic control, a type of *model predictive control.*

**Manipulated variable** A quantity varied by the controller in order to affect the *controlled variable* (ISA, 1979).

**Manual feedback control** *Feedback control* in which operating personnel monitor the *controlled variable* of interest and take corrective action in order to maintain the desired value.

**Manufacturing Execution System** Handles higher level manufacturing functions such as order processing, inventory management, and purchasing.

**Master recipe** "A type of *recipe* that accounts for equipment capabilities and may include *process cell*-specific information" (ISA, 1995, p. 15; italic added).

**Mathematical model** A system of equations whose solution, given specific input data, is representative of the response of a process to a corresponding set of inputs.

**MES** Abbreviation for *Manufacturing Execution System.*

**MIMO** Abbreviation for multi-input, multi-output.

**Model predictive control** A digital control algorithm that uses a model of the process to predict the trend of the process outputs and to compute the required change in the process inputs to bring the outputs to their desired values. The most popular model predictive control algorithms use an impulse response or step response model.

**MPC** Abbreviation for *model predictive control.*

**Multiloop control** A form of *multivariable control* where single-loop controllers are connected to the process in such a way as to reduce the process *interaction.*

**Multivariable control** A control scheme whose objective is to simultaneously control more than one *process variable* in a process *unit* or *cell.* A multivariable controller is generally part of *unit supervision* or cell supervision.

**MV** Abbreviation for *manipulated variable.*

**Nonlinear** A system that is not linear. See *linear system.*

**Offset** The difference between the *setpoint* and the *process variable* when the system has reached *steady state* (ISA, 1979).

**OLST** Abbreviation for open-loop settling time, one of the major tuning parameters for the forward modeling controller, a type of model predictive controller.

**One-way interaction** A process with two inputs and two outputs where one *manipulated variable* affects only one *process variable* but the other *manipulated variable* affects both *process variables.* This type of *interaction* is evident if only one of the off-diagonal terms in the process gain matrix is nonzero.

**Open-loop control** A control strategy where the *manipulated variables* are set to their design values and held constant (ISA, 1979).

**Operating point** The normal operating value for the system variables.

**Operating state** The current condition of the equipment entity, e.g., *process cell, unit,* or *equipment module.* The operating state also defines how the equipment entity will operate and how it will respond to commands.

**Operating window** The possible range of the system variable values.

**Operation** "A *procedural element* defining an independent processing activity consisting of the algorithm necessary for the initiation, organization, and control of *phases*" (ISA, 1995, p. 15; italic added).

**Output signal** "A *signal* delivered by a device, element or system" (ISA, 1979, p. 28).

**Override control** Scheme used to protect process equipment or personnel. See *high-signal selector* and *low-signal selector.*

**Overshoot** "The maximum excursion beyond the final *steady-state* value of output as the result of an input change" (ISA, 1979, p. 31).

**P&ID** Abbreviation for *piping and instrumentation diagram.*

**pdXi** Abbreviation for Process Data Exchange Institute.

**Peak time** The time from the *setpoint* step change to the time of the first peak of an underdamped *process variable* response.

**Percent overshoot** The amount of the *overshoot*, expressed as a percentage of the *process variable* final value.

**PES** Abbreviation for *programmable electronic system.*

**PFD** Abbreviation for *process flow diagram.*

**Phase** "The lowest level of *procedural element* in the procedural control model" (ISA, 1995, p. 15; italic added). The phase is the sequence control logic that commands the actions of the control modules.

**Physical model** A hierarchical organization of the physical assets of an *enterprise* (ISA, 1995).

**PI control** Abbreviation for proportional-plus-integral control. A controller whose output is the sum of *proportional action* and *integral action.*

**PID control** Abbreviation for proportional-plus-integral-plus-derivative control. A controller whose output is the sum of *proportional action*, *integral action*, and *derivative action.*

**Piping and instrumentation diagram** A symbolic drawing of a *process*, or portion of a process, showing the piping and instrumentation.

**Plant area management** The control entity that coordinates the *process cell* operations.

**PLC** Abbreviation for *programmable logic controller.*

**POD** Abbreviation for *process operation description.*

**PRBS** Abbreviation for pseudorandom binary sequence, a common random input signal used in least-squares identification of *empirical models.*

**Procedural control** "Control that directs equipment-oriented actions to take place in an ordered sequence in order to carry out some process-oriented task."(ISA, 1995, p. 15). See also *equipment procedural control* and *recipe procedural control.*

**Procedural element** "A building block for *procedural control* that is defined by the procedural control model" (ISA, 1995, p. 15; italic added). The four procedural elements are (1) *procedure*, (2) *unit procedure*, (3) *operation*, and (4) *phase.*

**Procedure** "The strategy for carrying out a process" (ISA, 1995, p. 16). It is the *sequence control* logic to make a complete *batch* of product.

**Procedure control model** Hierarchically organized model of the control strategy of a plant (ISA, 1995).

**Process** "Physical or chemical change of matter or conversion of energy; e.g., change in pressure, temperature, speed, electrical potential, etc." (ISA, 1979, p. 24). For this text, the process is the system to be controlled.

**Process action** "Minor processing activities that are combined to make up a *process operation*" (ISA, 1995, p. 16; italic added).

**Process cell** A logical grouping of equipment required to process one stream or manufacture one product or group of products. A process cell is a set of cooperating *units* (ISA, 1995).

**Process control** The control entity that encompasses the basic discrete, regulatory, and equipment module procedural control elements. Process control encompasses the types of objects that can be supervised at the unit level: *equipment module control*, *loop*, *device*, *indicator*, and *status* (ISA, 1995).

**Process flow diagram** An abstract drawing of a process or plant showing equipment groupings (*areas*, *cells*, *units*, etc.) and the material and/or information flows between them.

**Process input** When considering the *process*, an input is generally material moving into the process. From the standpoint of control, the process input is a *manipulated variable*.

**Process management** The control entity responsible for *cell*-wide control and coordination of the *unit* operations (ISA, 1995).

**Process model** (1) An overall model of the *process* that describes the processing actions required to convert the raw materials into finished product (ISA, 1995). (2) A more detailed *mathematical model* of the process used to design *regulatory control* systems.

**Process operation** Process operations represent major processing activities, which usually result in a chemical or physical change in the material being processed. A *process stage* consists of one or more process operations (ISA, 1995).

**Process operation description** Design information package that defines the operating strategy and control objectives of the plant systems.

**Process output** When considering the *process*, an output is generally material moving out of the process. From the standpoint of control, the process output is a *controlled variable* or a *process variable*.

**Process stage** "A part of a process that operates independently from other process stages and that usually results in a planned sequence of chemical or physical changes in the material being processed" (ISA, 1995, p. 16).

**Process unit** See *unit*.

**Process variable** The measured value of the *controlled variable* that, along with the *setpoint*, is used by the controller to calculate a value of the *manipulated variable*.

**Production run** The operation of a plant system to manufacture a quantity of production to fulfill a production order.

**Programmable electronic system** Any system that implements the control functions in the *process management* layer or lower control layers. A programmable electronic system encompasses the functions of both a traditional *distributed control system* (DCS) and a traditional *programmable logic controller* (PLC).

**Programmable logic controller** In the process control industry, the programmable logic controller traditionally provided *discrete control* functions for the process control system. The functionality of these systems is currently merging with the *distributed control system*. See also *programmable electronic system.*

**Proportional action** The controller *manipulated variable* is calculated as a constant (called the *proportional gain*) multiplied by the error, the *setpoint* minus the *process variable*. (ISA, 1979).

**Proportional band** The change in error required to produce a full range, 100%, change in the *manipulated variable*, due to *proportional action*. It is equal to 100 divided by the *proportional gain* (ISA, 1979).

**Proportional gain** The ratio of the change in the *manipulated variable* due to *proportional action* to the change in the error. See also *proportional band* (ISA, 1979).

**PV** Abbreviation for *process variable*.

**RACI chart** Defines the roles and responsibilities of the *capital project* team members. RACI is an abbreviation for "responsible, accountable, consult, inform."

**Rate action** See *derivative action*.

**Ratio control** Control scheme used in blending or other applications where two streams need to be held in constant ratio to each other.

**RDG** Abbreviation for *relative disturbance gain*.

**Recipe** "The necessary set of information that uniquely defines the production requirements for a specific product" (ISA, 1995, p. 16).

**Recipe management** "The control activity that includes the control functions needed to create, store, and maintain general, site, and master recipes" (ISA, 1995, p. 16).

**Recipe operation** "An *operation* that is part of a *recipe procedure* in a master or *control recipe*" (ISA, 1995, p. 16; italic added).

**Recipe phase** "A *phase* that is part of a *recipe procedure* in a master or *control recipe*" (ISA, 1995, p. 16; italic added).

**Recipe procedural control** *Procedural control* for a plant system that provides flexibility in the operation of the plant system to produce differing products. See also *equipment procedural control.*

**Recipe procedure** "The part of a *recipe* that defines the strategy for producing a batch" (ISA, 1995, p. 16; italic added).

**Recipe unit procedure** "A *unit procedure* that is part of a *recipe procedure* in a master or *control recipe*" (ISA, 1995, p. 17; italic added).

**Regulatory control** The control type used to maintain *process variables* at desired values. It is often referred to as *continuous control* or *discrete-time control.*

**Relative disturbance gain** An indication whether the *interaction* resulting from a particular *disturbance* is favorable or unfavorable.

**Relative gain array** A measurement of the *interaction* in a multivariable process, providing a tool for *multiloop control* design.

**Reset action** See *integral action*.

**Reset windup** See *integral windup*.

**Reverse acting** A process or controller is called reverse acting if an increase in the *input signal* causes an decrease in the *output signal*.

**Reverse action** See *reverse acting*.

**RGA** Abbreviation for *relative gain array.*

**Rise time** The time required for the process variable to go from 10 to 90% of the *steady-state* change (ISA, 1979).

**S88.01** Abbreviation for *ISA-S88.01, Batch Control, Part 1: Models and Terminology*, Instrument Society of America, 1995.

**Safety instrumented system** The control technology for automatic safety protection.

**SAMA** Abbreviation for the Scientific Apparatus Makers' Association.

**Sequence control** Enforces the correct operation of process equipment in a stepwise fashion.

**Sequential function chart** A diagram of interconnected steps, actions, and transitions used to define *sequence control* logic.

**Setpoint** "An input variable which sets the desired value of the *controlled variable*" (ISA, 1979, p. 28; italic added).

**Settling time** The time from the *setpoint* change to the time that the process variable response has settled within a certain percentage band of the final value, usually 2 or 5% (ISA, 1979).

**SFC** Abbreviation for *sequential function chart.*

**Shared-use resource** "A *common resource* that can be used by more than one user at a time" (ISA, 1995, p. 17; italic added).

**Signal** "In process instrumentation, physical variable, one or more parameters of which carry information about another variable (which the signal represents)" (ISA, 1979, p. 28).

**Singular value** A diagonal element of the middle matrix of the *singular-value decomposition*.

**Singular-value decomposition** Determines the condition and rank of a matrix and other information that can help avoid numerical computation errors. Can be used as a tool to support *multivariable control* design.

**SIS** Abbreviation for *safety instrumented system.*

**SISO** Abbreviation for single-input, single-output.

**Site** A geographical grouping of equipment and buildings determined by the *enterprise*. All physical equipment in a site share common geographic and meteorological data (ISA, 1995).

**Site management** The control entity responsible for coordinating the management of the plant *areas*.

**Site recipe** "A type of *recipe* that is site specific" (ISA, 1995, p. 17; italic added).

**SP** Abbreviation for *setpoint.*

**Split-range control** Control scheme when process has one *process variable* and more than one *manipulated variable*.

**Stable system** A system is considered to be stable if a bounded *input signal* always results in an *output signal* that is also bounded.

**State** The current condition of the physical or control entity. The state also defines how the entity will operate and how it will respond to commands (ISA, 1995).

**Static decoupler** A type of *explicit decoupler* where only the gains of the *dynamic decoupler* are implemented.

**Status** A single discrete value. A type of basic *process control* object.

**Steady state** The long-term output response of a system after it has been disturbed. Generally refers to the long-term output due to a *manipulated variable* or *disturbance* change (ISA, 1979).

**Steady-state model** *Mathematical model* where only the long-term output response is captured.

**STEP** Abbreviation for the STandard for the Exchange of Product Data, ISO 10303.

**SVD** Abbreviation for *singular-value decomposition*.

**TBD** Abbreviation for 'to be determined'.

**Time constant** In an expression for linear system time response, the time constant is the value $\tau$ in the response term $Ae^{-\tau t}$. In a transfer function, the time constant is the value $\tau$ in the denominator term $1 + s\tau$. For the output of a first-order system whose input is a step signal, the time constant is the time required to complete 63.2% of the total output change (ISA, 1979).

**Transfer function** For a continuous-time system, the transfer function is the ratio of the Laplace transform of the output variable to the Laplace transform of the input variable, with all initial conditions assumed to be zero. For a discrete-time system, the transfer function is the ratio of the *Z*-transform of the output variable to the *Z*-transform of the input variable, with all initial conditions assumed to be zero (ISA, 1979).

**Underdamped** The time response of the system to a step signal input has *overshoot* (ISA, 1979).

**Unit** A set of *equipment modules* and *control modules* usually centered on a major piece of equipment, e.g., a reactor. A unit combines all necessary physical processing and control equipment required to perform one or more major processing activities, e.g., react or separate (ISA, 1995).

**Unit procedure** "A strategy for carrying out a contiguous process within a *unit*. It consists of contiguous *operations* and the algorithm necessary for the initiation, organization, and control of those *operations*" (ISA, 1995, p. 17; italic added).

**Unit recipe** "The part of a *control recipe* that uniquely defines the contiguous production requirements for a *unit*." (ISA, 1995, p. 17; italic added). The *unit procedure* coupled with the recipe parameters and equipment information of a *unit* is sometimes called the unit recipe.

**Unit supervision** The control entity that supervises the *unit* and the unit resources. For *batch control* systems, the unit procedure executes the *unit recipe* and is concerned with such issues as resource coordination and allocation. For *continuous processes*, the various single-loop controllers are integrated into a functional unit. Multivariable and advanced control algorithms are generally executed at this level (ISA, 1995).

**Zero-order hold** A device used in *digital control* that changes the *discrete-time signal* at the output of a digital controller into the *continuous-time signal* needed by the device (e.g., valve) that manipulates a *process input*.

**ZOH** Abbreviation for *zero-order hold*.

## REFERENCES

Instrument Society of America, *ISA-S51.1, Process Instrumentation Terminology*, Instrument Society of America, Research Triangle Park, NC, 1979.

Instrument Society of America, *ISA-S88.01, Batch Control, Part 1: Models and Terminology*, Instrument Society of America, Research Triangle Park, NC, 1995.

Marlin, T. E., *Process Control: Designing Processes and Control Systems for Dynamic Performance*, McGraw-Hill, New York, 1995.

# INDEX


Air-to-close, 185, 245–246, 249–250, 526
Air-to-open, 185, 245–246, 248, 526
Alarms, 12–13, 74–75
Antireset windup, 179, 226, 265
AP221 Standard, 10, 23–28, 526
AP227 Standard, 10, 23, 526
AP231 Standard, 10–11, 23–28, 526
Architecture, system, 14–16, 30–32, 34, 336–337
Area, 6, 48, 54, 72, 404, 526
Autoregressive moving average (ARMA), 156–162, 526

Balance
  energy, 102, 105–106
  mass, 102, 104–105, 273–275
Basic control, 11, 45, 58–60, 330, 526
Basic process control system, 336–337, 526
Batch, 12, 40, 372, 526
Batch control, 56–59, 70–71, 75, 264–266, 371–401, 526
Batch operation, 56, 373–375, 395–400
Batch process, 40, 46, 371–373, 384, 527
Batch reactor, 51, 224–226, 244–247, 265, 269, 375
Batch report, 375, 384–385
Batch schedule, 384, 527
Bias, of PID controller, 181–183, 245
Bleach, 80, 82, 481–494
Blending, 242–244, 265, 273–279, 293–294
Blow heat evaporator, 80–83, 438–441, 443, 451
Brown pulp mill, 404–406, 424–451
Brown stock washer, 80–81, 433–437, 443, 448–450
Bumpless transfer, 184–186, 527

Campaign, 383–385, 527
Capital project, 4, 11, 21–35, 337–338, 527
  detailed engineering, 27, 32–33
  first production startup, 27, 35
  implementation, 27, 34
  installation, 27, 34
  preliminary engineering, 28–32
  training, 35
Cascade control, 68, 124, 214–228, 270, 527
Caustic, 404–406, 452–480
Cell, *see* Process cell
Chip screening, 408, 413–415, 418–419, 422
Chip unloading, 408, 410–412, 418, 420–421
Cohen–Coon tuning, 198–199, 210–211
Common resource, 55–56, 380, 442, 444, 527
Communication protocol
  Ethernet, 77, 531
  token bus, 77, 531
Condition number, 300–305, 527
Constraints, 174, 297, 321. *See also* Override control
Continuous-flow stirred tank reactor (CSTR), 103–107, 112, 115, 136–137
Control
  automatic feedback, 168–169, 170–172, 526
  basic, 11, 45, 58–60, 330, 526
  batch, 56–59, 70–71, 75, 264–266, 371–401, 526
  cascade, 68, 124, 214–228, 270, 527
  continuous, 70, 99, 527
  coordination, 12, 45, 375, 377, 380, 528
  decoupling, 86–89, 269, 283, 292–299, 528

digital, 99, 128–129, 177–180, 183–184, 252, 311
discrete, 12, 16, 169, 329–369, 513, 529
emergency, 169–170. *See also* Control, interlock; Control, override
equipment module, 58–59, 376–377, 393–395, 530
equipment procedural, 375–377, 530
feedback, 168–178, 530
feedback–feedforward, 236–239
feedforward, 167–168, 214, 227–242, 294, 531
interlock, 12, 19–20, 330, 335–339, 532
manual feedback, 168, 533
model predictive, 214, 252–266, 311–327, 533
multiloop, 269–292, 533
multivariable, 56, 100, 268–327, 477–478, 533
on–off, 52, 59, 330–332, 511
open-loop, 167, 533
override, 69–70, 86–90, 214, 248–251, 534. *See also* Control, interlock
PID, 59, 67, 178–187, 211–212, 264–265, 534. *See also* PID controller tuning
procedural, 12, 56–59, 375–380, 534
ratio, 214, 242–244, 293–294, 536
recipe procedural, 375–376, 381, 536
regulatory, 12, 45, 166–169, 372, 536
sequence, 33, 330–335, 350–369, 375–380, 521–525, 537
split-range, 214, 244–248, 265, 537
supervisory, 12, 56. *See also* Unit supervision
Control Concept, 25, 28–32, 363, 390, 403, 528
Control model (S88.01), 54–62, 72–74, 100. *See also* Procedural control model (S88.01)
Control module, 51–54, 86–90, 93, 330, 391, 528
Control recipe, 375, 381, 384, 528
Control requirements definition (CRD), 28–32, 45, 402–510, 528
Control state, 63, 67–71, 340, 528. *See also* Operating state
Control strategy, 30–33, 59, 167–169, 402–403, 528
Control technology, 13–14, 16, 76–77, 336–337
Control type, 12–13
device, 52, 59, 73–75, 99, 339–350, 529
equipment module, 58–59, 376–377, 393–395, 530
indicator, 59, 269, 403, 513, 532
loop, 52, 59, 74–75, 513, 532
status, 59, 74, 269, 339, 513, 537
Controllable, 276, 528
Controlled variable, 98, 168–173, 528
Coordination control, 12, 45, 375, 377, 380, 528
Crude tower, 286–288, 307–308

Damping ratio, 146–147, 176, 192–194, 210–211, 528
Deadtime, 107–108, 139, 142, 231, 528
Decoupler
dynamic, 86–89, 283, 294–296, 529
explicit, 294–296, 530
singular-value decomposition (SVD), 308–311
static, 295–299, 308–311, 537
Decoupling control, 86–89, 269, 283, 292–299, 528
Degrees of freedom, 101
Derivative action, 179–184, 190, 529
Derivative gain, 180, 190, 529
Derivative time constant, 180, 529
Device, 52, 59, 73–75, 99, 339–350, 529
Digester, 79–81, 428–432, 442–446
Digital control, 99, 128–129, 177–180, 183–184, 252, 311
Direct-acting, 184–185, 245, 248–249, 529
Discrete control, 12, 16, 169, 329–369, 513, 529
Discrete model, 99, 529
Discrete-parts manufacturing process, 40–41, 50
Discrete-time control, *see* Digital control
Discrete-time model, 99, 125–132, 155–165, 529
Distillation column
cascade control, 224–225

Distillation column (*Continued*)
  decoupling control, 292, 295–297, 310–311
  Dynamic Matrix Control (DMC), 320–322
  feedforward control, 240–242
  Forward Modeling Control (FMC), 316–317, 321–322
  model, 284, 286–287, 289
  multiloop control, 286–292, 307–308
  multivariable control, 316–317, 320–322
  singular-value decomposition (SVD) decoupler, 310
  singular-value decomposition (SVD) loop pairings, 307–308
Distributed control system (DCS), 16, 77, 329, 529
Distributed model, 99, 529
Disturbance, 167–173, 177–178, 211, 214, 529
Dynamic decoupler, 86–89, 283, 294–296, 529
Dynamic Matrix Control (DMC)
  MIMO, 317–322, 324–326
  SISO, 260–264
  tuning parameters, 262, 319
Dynamic model, 99–100, 120, 530

Emergency control, 169–170. *See also* Interlock control; Override control
Empirical model, 99–100, 120–125, 137–165, 198, 530
  first order, 122–123, 139–142, 187, 198, 531
  integrating, 125, 151–155, 187, 206
  second order, 123–125, 142–151, 187
Endpoint, 265, 332, 378, 393–395
Energy balance, 102, 105–106
Enterprise, 2, 21, 47–48, 54, 375, 530
Enterprise management, 54, 530
Enterprise resource planning system (ERPS), 72–73, 530
Equal-percentage valve, 166
Equipment entity, 63, 379, 385, 530
Equipment module, 50–54, 80, 95, 391, 530
Equipment module control, 58–59, 376–377, 393–395, 530
Equipment procedural control, 375–377, 530
Equipment unit procedure, 351–354, 377–380, 395–400, 530
Exception handling, 12–13, 45, 530. *See also* Interlock control; Override control
Exclusive-use resource, 52, 444, 530. *See also* Common resource
Explicit decoupler, 294–296, 530

Feedback control, 168–178, 530
Feedback–feedforward control, 236–239
Feedforward control, 167–168, 214, 227–242, 294, 531
Fertik tuning, 198, 202–207, 210–211
Final value theorem
  Laplace transform, 115
  Z-transform, 128
First-order plus deadtime (FODT), 123, 139–142, 198, 531
First principles, for modeling, *see* Fundamental model
Fluidized catalytic cracking unit (FCCU), 38–39, 530
Formula, 375, 380–382, 386, 400, 531
Forward Modeling Control (FMC)
  MIMO, 311–317, 321–322
  SISO, 252–260, 264
  tuning parameters, 257, 316
Fundamental model, 99, 100–120, 531
Fundamental principles, for modeling, *see* Fundamental model
Furnace control, 224–225

Gain, 117, 123, 138–139, 163, 531
General recipe, 381, 531
Graphical user interface (GUI), *see* User interface
Green liquor, 81–83, 86–87, 455–457, 470, 472
Groundwood mill, 404–406, 495–498

Heat exchanger
  cascade control, 216–218, 221–224, 226
  Dynamic Matrix Control (DMC), 262–263

feedforward control, 227–229, 232–235, 237–239
Forward Modeling Control (FMC), 258–260
modeling, 221
Heat exchanger system, 285–286, 307
High-signal selector (HSS), 69–70, 248–251, 531. *See also* Low-signal selector; Override control
HOA switch, 73, 340–342, 531

IDEF0 diagram, 23–28, 373–375, 531
Identification, empirical models
autoregressive moving average (ARMA), 156–162, 526
first order, 139–142, 531
integrating, 151–155
moving average (MA), 156–161, 164–165, 533
second order, 142–151
IEC 1131 standard, 16, 77, 333, 531. *See also* Ladder logic; Sequential function chart
IEEE
802.3 standard, 77, 531
802.4 standard, 77, 531
Increase close, 185, 532
Increase open, 185, 532
Indicator, 59, 269, 403, 513, 532
Information management, 11, 21, 55, 72, 375
Input signal, 117, 133, 158–159, 532. *See also* Manipulated variable
Integral absolute error (IAE), 176–177, 210–211, 531
Integral action, 179, 180, 187, 283, 532
Integral error (IE), 177
Integral gain, 179, 190, 532
Integral squared error (ISE), 176–177, 532
Integral time absolute error (ITAE), 177, 198, 532
Integral time constant, 179, 191–192, 532
Integral windup, 179, 226, 532. *See also* Antireset windup
Integrating process model, 125, 151–155, 187, 206
Interaction, 70
multivariable control, 269–270, 272–283, 532
one-way, 282, 533
PID equation, 180–184
Interlock, 332, 338–339, 358, 362, 513, 532
Interlock control, 12, 19–20, 330, 335–339, 532
Inverse Laplace transform, 113–114, 116
Inverse Z-transform, 127

Ladder logic, 16, 77, 331–333, 514–520, 532
Laplace transform, 102, 113–120
final value theorem, 115
inverse, 113–114, 116
table of transforms, 114
time delay property, 114
Least-squares estimation, 155–165
autoregressive moving average (ARMA) model, 156–165, 526
moving average (MA) model, 156–161, 164–165, 252, 533
Lifecycle, 20–28
Lime kiln, 85–90, 465–469, 471, 477–479
Lime mud, 86–90, 462–464, 475–476
Linearization, 108–113
multivariable, 111–113
one variable, 108–111
Linear system, 101, 115, 133, 532. *See also* Linearization
Log handling, 416–417, 419, 423
Loop, 52, 59, 74–75, 513, 532
Loop pairing, 269, 283–288, 305–308, 532
Low-signal selector (LSS), 250–251, 532. *See also* High-signal selector; Override control
Lumped model, 99, 533

Management
enterprise, 54, 530
plant area, 54–55, 74, 77, 534
process, 55–56, 74, 375, 390, 535
recipe, 55, 375, 383, 536
site, 54, 77, 537
Manipulated variable, 98, 168–171, 177, 178–184, 533

Manipulated variable overshoot, 177, 210–211
Manual feedback control, 168, 533
Manufacturing Execution System (MES), 33, 55, 72, 533
Mass balance, 102, 104–105, 273–275
Master recipe, 56, 375, 381–383, 533
Mathematical model, 99–100, 533
Maximum deviation, 177–178, 211
Model, *see also* also Empirical model
  control (S88.01), 54–62, 72–74, 100. *See also* Model, procedural control (S88.01)
  discrete, 99, 529
  discrete-time, 99, 125–132, 155–165, 529
  distributed, 99, 529
  dynamic, 99–100, 120, 530
  fundamental, 99, 100–120, 531
  lumped, 99, 533
  mathematical, 99–100, 533
  physical (S88.01), 44, 47–54, 534
  procedural control (S88.01), 45, 56–59, 351–355, 375–380, 534. *See also* Model, control (S88.01)
  process (S88.01), 46–47, 535
  steady-state, 99–100, 122, 138, 538
Model Algorithmic Control (MAC), 252, 311, 533
Model predictive control, 533
  MIMO, 311–327, 477–478
  SISO, 214, 252–266
Moving average (MA), 156–161, 164–165, 252–253, 312
Multiloop control, 269–292, 533
Multivariable control, 56, 100, 268–327, 477–478, 533
  decoupling, 86–89, 269, 283, 292–299, 300, 528
  Dynamic Matrix Control (DMC), 260–264, 317–322, 324–326
  Forward Modeling Control (FMC), 252–260, 311–317, 321–322
  interaction, 70, 269–270, 272–283, 533
  loop pairing, 269, 283–288, 305–308, 532
  model predictive control, 252–266, 311–327, 477–478, 533
  singular value decomposition (SVD) analysis, 299–311, 537

Niederlinski stability theorem, 283, 285, 286, 288, 308
Non-self-regulating, *see* Integrating process model

Offset, 174, 179, 226, 233–234, 533
One-way interaction, 282, 533
On–off control, 52, 59, 330–332, 511
Open-loop control, 167, 533
Operating module, 6–9, 25, 30–31, 372, 375, 402–403
Operating point, 101–102, 108–113, 159, 321, 533
Operating state, 9, 30, 63–67, 388–389, 402–403, 533. *See also* Control state
Operating window, 277, 533
Operation, 58, 70, 90–95, 534
  batch, 56, 373–375, 395–400
  process, 46–47, 535
  recipe, 380–383, 536
Operator interface, *see* User interface
Orifice plate, 109–111
Output signal, 129–130, 172, 534. *See also* Controlled variable
Override control, 69–70, 86–90, 214, 248–251, 534. *See also* Interlock control
Overshoot, 174–175, 177, 210–211, 265–266, 534

Packaging, 41, 50, 332, 404, 508–510
Pairing, loop, 269, 283–288, 305–308, 532
Paper machine, 81, 84–85, 404–405, 500–507
Peak time, 175, 534
Percent overshoot, 174–175, 177, 210–211, 534
Performance measure
  damping ratio, 146–147, 176, 192–194, 210–211, 528
  integral absolute error (IAE), 176–177, 210–211, 531
  integral error (IE), 177

integral squared error (ISE), 176–177, 532
integral time absolute error (ITAE), 177, 198, 532
maximum deviation, 177–178, 211
manipulated variable overshoot, 177, 210–211
offset, 174, 179, 226, 233–234, 533
overshoot, 174–175, 177, 210–211, 265–266, 534
peak time, 175, 534
percent overshoot, 174–175, 177, 210–211, 534
rise time, 174, 210, 537
settling time, 176, 210–211, 537
Pharmaceutical, 40, 90–96
Phase, 51, 58–59, 70, 351–355, 534
examples, 95, 358–369, 391–400, 445–446
recipe, 376–377, 536
Physical model (S88.01), 44, 47–54, 534
PID controller, 59, 67–69, 178–187, 211–212, 264–265, 534
PID controller tuning
Cohen–Coon, 198–199, 210–211
damped oscillation, 196–198
Fertik, 198, 202–207, 210–211
initial settings, 188–189
integrating process, 206–210
process response, 191–193
quarter wave damping, 192, 196
Ziegler–Nichols closed loop, 192–196, 210–211
Ziegler–Nichols open loop, 198–202, 206–211
PID equation, 180–184
Piping and instrumentation diagram (P&ID), 6, 27, 33, 511–513, 534
Plant area management, 54–55, 74, 77, 534
Prediction, 160, 253–254, 260, 264, 325–326
Primary loop, cascade control, 214–215, 219–220
Procedural control, 12, 56–59, 375–380, 534
equipment, 375–377, 530
examples, 94–95, 356–369, 391–400
Procedural control model (S88.01), 45, 56–59, 351–355, 375–380, 534. *See also* Control model (S88.01)
Procedural element, 58, 70–71, 376–380, 383, 534
operation, 58, 70, 90–95, 373–375, 534
phase, 51, 58–59, 70, 351–355, 534. *See also* Phase, examples
procedure, 56–58, 70–71, 377, 380, 534
unit procedure, 56–58, 351–355, 377–380, 402, 538
Procedure, 56–58, 70–71, 377, 380, 534
equipment, 375–377, 530
recipe, 380–383, 536
unit, 56–58, 351–355, 377–380, 402, 538
Process action, 47, 535
Process area, *see* Area
Process cell, 48–50, 54, 372, 386–390, 535
Process Data Exchange Institute (pdXi), 23, 373, 534
Process flow diagram (PFD), 4–5, 10, 28, 406, 535
Process input, *see* Manipulated variable
Process management, 55–56, 74, 375, 390, 535
Process model (S88.01), 46–47, 535
Process operation, 46–47, 535
Process operation description (POD), 28–31, 402–403, 535
Process output, *see* Controlled variable
Process stage, 46–47, 535
Process variable, 137–138, 170–171, 178–185, 339, 535
Production run, 12–13, 383–384, 535
Programmable electronic system (PES), 16, 77, 337, 535
Programmable logic controller (PLC), 16, 77, 332, 514–525, 535
Project team, 21, 31, 35–36
Proportional action, 178–180, 191, 536
Proportional band, 179, 536
Proportional gain, 178–179, 187–191, 536
Pseudo-inverse, 299, 313–314
Pseudo-random binary sequence (PRBS), 158–159, 161

Pulp and paper mill
  bleach cell, 481–494
  bleach unit, 80, 82, 484–488, 490, 491–493
  blow heat evaporator unit, 438–441, 443, 451
  brown pulp mill cell, 404–406, 424–451
  brown stock washer unit, 433–437, 443, 448–450
  caustic cell, 404–406, 452–480
  chip screening unit, 408, 413–415, 418–419, 422
  chip unloading unit, 408, 410–412, 418, 420–421
  digester unit, 79–81, 428–432, 442–446
  green liquor unit, 81–83, 86–87, 455–457, 470, 472
  groundwood mill cell, 404–406, 495–498
  lime kiln unit, 85–90, 465–469, 471, 477–479
  lime mud unit, 86–90, 462–464, 471, 475–476
  log handling, 416–417, 419, 423
  packaging cell, 404, 508–510
  paper machine cell, 81, 84–85, 404–405, 500–507
  white liquor unit, 458–461, 470–471, 473–474
  woodyard cell, 79–80, 404–406, 407–423

RACI chart, 35–36, 536
Rate action, *see* Derivative action
Ratio control, 214, 242–244, 293–294, 536
Reaction curve, 137–138
Recipe, 50, 372–373, 375–376, 380–383, 536
  control, 375, 381, 384, 528
  general, 381, 531
  master, 56, 375, 381–383, 533
  site, 54, 375, 381, 383, 537
  unit, 56, 377, 538
Recipe management, 55, 375, 383, 536
Recipe operation, 380–383, 536
Recipe phase, 376–377, 536
Recipe procedural control, 375–376, 381, 536
Recipe procedure, 380–383, 536
Regulatory control, 12, 45, 166–169, 372, 536
Relative disturbance gain (RDG), 284, 285, 536
Relative gain array (RGA), 279–283, 536
  applications, 284–288
  loop pairing, 283–288
  measure of interaction, 281
  properties of, 282
Reset, *see* Integral action
Reset action, *see* Integral action
Reset windup, *see* Integral windup
Resource
  common, 55–56, 380, 442, 444, 527
  exclusive-use, 52, 444, 530
  shared-use, 52, 380, 386–387, 537
Reverse-acting, 184–185, 245, 248–249, 536
Rise time, 174, 210, 537

Safety instrumented system (SIS), 21, 32, 335–339, 537
Sampled signal, 126–127
Schedule, batch, 384, 527
Secondary loop, cascade control, 214–215, 219–220
Second-order process model, 123–125, 142–151, 187
Sequence control, 33, 330–335, 350–369, 375–380, 521–525, 537
Sequential function chart (SFC), 333–335, 521–525, 537
Setpoint, 168–171, 173, 178–185, 537
Settling time, 176, 210–211, 537
Shared-use resource, 52, 380, 386–387, 537. *See also* Common resource
Signal
  continuous-time, 99, 528
  discrete-time, 99, 125–132, 529
  input, 117, 133, 158–159, 532. *See also* Manipulated variable
  output, 129–130, 172, 534. *See also* Controlled variable
Simulation, 34, 133–137, 338
Singular value, 300, 537

Singular value decomposition (SVD), 537
decoupler design, 308–311
definition, 299–301
loop pairing, 305–308
sensor/manipulated variable selection, 301–305
Site, 3, 47–49, 54, 72, 537
Site management, 54, 77, 537
Site recipe, 54, 375, 381, 383, 537
Split-range control, 214, 244–248, 265, 537
Stage, process, 46–47, 535
State, 63–71
control, 63, 67–71, 340, 358, 528
operating, 9, 30, 63–67, 388–389, 402–403, 533
Static decoupler, 295–299, 308–311, 537
Status, 59, 74, 269, 339, 513, 537
Steady-state, 117, 125, 138, 265, 538
Steady-state model, 99–100, 122, 138, 538
Step response, 115–116, 122–125, 157, 252
STEP standards
AP221, 10, 23–28, 526
AP227, 10, 23, 526
AP231, 10–11, 23–28, 526
Strategy, control, 30–33, 59, 167–169, 402–403, 528
Supervisory control, 12, 56. *See also* Unit supervision
System architecture, 14–16, 30–32, 34, 336–337

Tablet manufacturing, 90–96
Technology, 13–14, 16, 76–77, 336–337
Time constant, 123, 139–143, 538
Transfer function, 115–117, 128–132, 538
Tunnel dryer, 302–303

Ultimate gain, 194–195
Ultimate period, 194–195
Underdamped, 124–125, 146, 538
Unit, 9, 50–54, 268–269, 402–403, 538
Unit procedure, 56–58, 351–355, 377–380, 402, 538
Unit recipe, 56, 377, 538
Unit supervision, 56–58, 269, 330, 375–376, 403, 538
User interface
batch control, 385–386
cascade control, 225–228
discrete control, 331, 344, 347
model predictive control, 264, 322–327
multivariable control, 322–327
PID controller, 211–212, 225–228
safety instrumented system, 336–337, 339

Valve
air-to-close, 185, 245–246, 249–250, 526
air-to-open, 185, 245–246, 248, 526
Valve position control, 224, 250
Variable
controlled, 98, 168–173, 528
manipulated, 98, 168–171, 177, 178–184, 533
process, 137–138, 170–171, 178–185, 339, 535

White liquor, 85–87, 458–461, 470–471, 473–474
Wild stream, 242–243. *See also* Ratio control
Windup,
antireset, 179, 226, 265
integral, 179, 226, 532
Woodyard, 79–80, 404–406, 407–423

Zero-order hold, 129–130, 538
Ziegler–Nichols tuning
closed loop, 192–196, 210–211
open loop, 198–201, 206–211
Z-transform
definition, 126–127
final value theorem, 128
inverse, 127
table of transforms, 128
time delay property, 127–128

CPSIA information can be obtained at www.ICGtesting.com
Printed in the USA
LVOW09*0935280114

371288LV00002B/11/A

9 780471 178354